ROUTLEDGE LIBRARY EDITIONS:
THE ECONOMICS AND POLITICS OF OIL AND GAS

Volume 8

THE ECONOMICS AND POLITICS OF THE UNITED STATES OIL INDUSTRY, 1920–1990

THE ECONOMICS AND POLITICS OF THE UNITED STATES OIL INDUSTRY, 1920–1990

Profits, Populism, and Petroleum

STEVE ISSER

Routledge
Taylor & Francis Group
LONDON AND NEW YORK

First published in 1996 by Garland Publishing, Inc.

This edition first published in 2016
by Routledge
2 Park Square, Milton Park, Abingdon, Oxon OX14 4RN

and by Routledge
605 Third Avenue, New York, NY 10017

Routledge is an imprint of the Taylor & Francis Group, an informa business

British Library Cataloguing in Publication Data
A catalogue record for this book is available from the British Library

ISBN: 978-1-138-64127-3 (Set)
ISBN: 978-1-315-62232-3 (Set) (ebk)
ISBN: 978-1-138-65485-3 (Volume 8) (hbk)
ISBN: 978-1-315-62293-4 (Volume 8) (ebk)

THE ECONOMICS AND POLITICS OF THE UNITED STATES OIL INDUSTRY, 1920–1990

PROFITS, POPULISM, AND PETROLEUM

STEVE ISSER

GARLAND PUBLISHING, INC.
NEW YORK & LONDON / 1996

Library of Congress Cataloging-in-Publication Data

Isser, Steve.
The economics and politics of the United States oil industry, 1920–1990 : profits, populism, and petroleum / Steve Isser.
p. cm. — (Garland studies on industrial productivity)
Includes bibliographical references and index.
ISBN 0-8153-2307-7 (alk. paper)
1. Petroleum industry and trade—Government policy—United States—History—20th century. I. Title. II. Series.
HD9566.I85 1996
338.2'7282'0973—dc20 95-46812

Printed on acid-free, 250-year-life paper
Manufactured in the United States of America

To the two women in my life, my wife Rhonda who has suffered through this ordeal with the patience of Job's wife, and my mother Natalie, who never stopped believing I would become a mensch someday.

CONTENTS

Tables *xi*

Preface *xiii*

1 Development of the System of Control *3*
The US Oil Industry After WWI *3*
Control of International Oil *6*
Chaos in the Domestic Oil Market *10*
Development of the Prorationing System *12*
Oklahoma City and East Texas *21*
The New Deal and Proration *34*
East Texas, the US Oil Market and the NRA *40*
Putting the Lid on East Texas *44*
Aftermath *55*

2 The Battle Over the Oil Import Quota *73*
Trade Policy Under Truman *73*
International Oil Market 1945-1955 *76*
The Economic Impact of Prorationing *79*
Congress and Oil Tax Policy *82*
Development of Pressure for Oil Import Quotas *86*
Voluntary Controls *97*
Mandatory Oil Import Quotas *101*
The Oil Import Quota: 1960-1970 *105*
The Rising Tide of Criticism *115*

3 The Oil Crises *127*
Marching Slowly to Disaster *127*
The Oil Embargo of 1973 *135*
President Ford: Picking Up the Pieces *139*
The Carter Energy Plan *145*
Decontrol and the Windfall Profits Tax *151*
The Reagan Revolution *160*
Strategic Petroleum Reserve *164*
The Oil Price Collapse *167*
The Oil Crisis That Never Happened *172*

4 Economists, Energy and Public Policy 189
Exhaustible Resource Theory 190
Conservation 193
Oil Policy Pre-WWII 195
Post WWII to the Mandatory Import Program 198
The Great Depletion Allowance Debate 203
Oil Policy and Oil Import Quotas 206
The World Oil Market 211
Oil Policy and National Security 212
Supply and Demand Estimates 214
Modeling Post Embargo Oil Market Behavior 218
Post 1980 Price Shock Literature 221
Energy Models 224
Economists and Policy 225

5 Public Opinion and Oil Politics 237
Public Opinion and Politics 237
Public Opinion and Voting Behavior 240
Public Opinion Toward Oil Issues 245
The Media and Public Opinion on Energy 250
Public Opinion and Energy Policy 260

6 Congressional Voting On Oil Issues 271
Theories of Congressional Voting 271
Voting on Oil Issues 273
The Data 274
Methodology 275
Vote #1: Trade Agreement Extentions Act of 1949 278
Vote #2: Amend Natural Gas Act of 1938 281
Vote #3: Trade Agreement Act of 1953 282
Vote #4: Natural Gas Bills 284
Vote #5: Votes on the Oil Depletion Allowance: 1951-1969 285
Vote #6: Votes on Fuel Allocation–1973 290
Vote #7: Votes on the National Energy Emergency Act –1973 and 1974 294
Vote #8: Oil Depletion Allowance–1975 300
Vote #9: House Votes on Oil Imports and Energy Taxes –1975 304
Vote #10: Oil Price Controls–1975 307
Vote #11: Natural Gas Deregulation–1976 316
Vote #12: Miscellaneous Senate Votes–1976 321
Vote #13: Carter Energy Plan Votes–1977 323
Vote #14: House Energy Department Authorization Votes –1979 331
Vote #15: Windfall Profit Taxes 334

Vote #16: Miscellaneous Reagan Era Votes: 1981-1982 *341*
Vote #17: Miscellaneous Reagan Era Votes: 1983-1987 *344*
Vote #18: Natural Gas Votes During the Reagan Era *346*
Vote #19: Oil Import Fee/Quota Votes *349*
Conclusions *352*

7 Interest Groups, Rent Seeking, and Oil Politics *359*
Interest Group Theory *359*
Public Choice Theory *364*
The Political Economy of Rent Seeking *369*
Constituents, Interest Groups and Lobbys *374*
Congressional Politics *381*
Congressional Reform *387*
Political Entrepreneurs and Agenda Setting *391*
Oil Politics and Economics: 1945-1968 *400*
Oil Politics and Economics: 1969-1976 *406*
The Economics of Oil Politics: 1977-1980 *410*
Politics of the Post-Crises Years *413*

Conclusion *423*

Bibliography *439*

Index *463*

TABLES

1.1	Discovery of Major US Oil Fields	*6*
1.2	Foreign Oil Production by Region	*7*
1.3	Prices and Petroleum Stocks: 1920-1935	*11*
2.1	Distribution of US Production and Proved Reserves	*80*
2.2	Stripper Well Production 1945-1960	*81*
2.3	US Oil Production and Imports: 1946-1960	*88*
2.4	Impact of Oil Import Controls	*99*
3.1	Domestic Oil Prices 1974-1980	*152*
3.2	Filling Up the Reserve	*165*
3.3	Production and Prices	*169*
4.1	Projections of Petroleum Demand for 1975	*217*
4.2	Projected vs. Actual OPEC Behavior	*219*
4.3	Forecasts of Energy Consumption	*225*
5.1	Is the Energy Shortage For Real?	*247*
5.2	Who Deserves Blame for Energy Crisis	*248*
5.3	Rate of Occurrence of Articles on Energy Issues	*251*
6.1	Probit Runs for HR 1211	*279*
6.2	Probit Runs for HR 1758	*282*
6.3	Probit Runs for HR 5894	*283*
6.4	Probit Runs for S 1853 and HR 7125	*285*
6.5	1958-1962 Depletion Allowance Votes	*287*
6.6	Probit Runs for the 1969 Depletion Allowance Votes	*289*
6.7	Probit Runs for S 1570–Senate Fuel Allocation Votes	*292*
6.8	Probit Runs for House Oil Allocation Votes–1973	*293*
6.9	Probit Runs for the House NEEA Votes	*296*
6.10	Probit Runs for the Senate NEEA Votes–1974	*297*
6.11	Probit Runs for House Oil Price Votes–1974	*298*
6.12	Logit Runs for Oil Allocation and Oil Price Votes	*299*
6.13	Probit Runs for House Votes on Depletion Allowance	*302*
6.14	Probit Runs for Senate Votes on Oil Depletion	*303*
6.15	Probit Runs for the House Energy Tax Votes	*306*
6.16	Probit Runs for S 622–Standby Energy Power Votes	*309-10*
6.17	Probit Runs for S 621–Decontrol of Oil Prices	*311*
6.18	Probit Runs for House Oil Price Control Votes	*312*

6.19 Probit Runs for HR 7014–Energy Policy and Conservation Act *313*
6.20 Logit Runs for Oil Price Decontrol Votes *314*
6.21 Probit Runs for S 2310–Senate Natural Gas Votes *318-9*
6.22 Probit Runs for HR 9464–House Natural Gas Votes *320*
6.23 Logit Runs for Natural Gas Price Decontrol Votes *321*
6.24 Probit Runs For Miscellaneous Senate Votes–1976 *322*
6.25 Probit Runs for 1977 House Enrgy Votes *326*
6.26 Probit Runs for S 2104–Senate Natural Gas Pricing *328*
6.27 Probit Runs for HR 5263–Senate Energy Tax Votes *329*
6.28 Probit Runs for HR 3000–Energy Department Votes *333*
6.29 Probit Runs for HR 3919–House Windfall Profits Tax *337*
6.30 Probit Runs for HR 3919–Senate Windfall Profits Tax *338-40*
6.31 Logit Analysis of Senate Windfall Tax Votes *341*
6.32 Probit Runs for Miscellaneous House Votes *343*
6.33 Probit Runs for Miscellaneous Senate Votes *344*
6.34 Probit Runs for Miscellaneous Senate Votes–1983-1987 *345*
6.35 Probit Runs for Senate Natural Gas Votes *348*
6.36 Probit Runs for Oil Import Votes *351*

Preface

This investigation into the economics and politics of the oil industry began with a simple question: why were oil import restrictions imposed during the 1950s in response to pressure from the domestic oil industry but not during the 1980s when similar pressure was applied to the political system? A superficial glance at these two time periods would seem to present more similarities than differences: Congressmen from oil producing states were highly influential during both periods (Rayburn, Kerr and Johnson during the 1950s, Wright, Bentsen, Dole and Long during the 1980s). National security, used as a rationalization for import quotas in the earlier period, would seem to be a more valid motivation for import controls in the latter period in light of the oil embargo of 1974, the Iran-Iraq war and its impact on shipping in the Persian Gulf, and the Gulf War. The economic power of the industry and the number of states in which oil production made a significant contribution to their economies were both greater during the 1980s than the 1950s. Thus, on the surface, the inability of the domestic oil industry to obtain political relief for its economic troubles during the 1980s presents an interesting puzzle.

It quickly became apparent that solving this problem would require examining the relationship between the political system and the economic system. National economic policy is a product of our divided system of government, requiring action by Congress and acquiescence, if not active cooperation, by the Executive branch The oil industry is closely involved with natural gas production, and is a competitor of the coal industry. The major oil companies were traditionally the largest source of foreign investment by American multinationals, while oil has been for a number of decades the most important component of international trade. World War II was won on a sea of oil while a minor war was fought over control of Persian Gulf oil. Thus the issues of trade policy, regulatory policy, energy policy, tax policy and national security are all relevant to the politics of oil.

There are four major venues for interaction between the political system and the market system: (1) Macroeconomic policy, relatively unimportant for understanding oil policy (though oil market disruptions can have significant macroeconomic consequences); (2) Public Finance, taxes and government expenditures; (3) Economic Regulation; and (4) Trade and National Security. It is the areas of regulation, taxation and trade policy where the most important interactions between the petroleum industry and the government have taken place. Regulatory impacts on the industry include the development and operation of the system of production control, petroleum price controls and allocation, and price control of natural gas. Tax issues include the oil depletion allowance, expensing provisions, taxation of foreign profits of multinational oil firms, and the imposition of a windfall profits tax. Trade policy and national security issues include government intervention in opening foreign oil fields to American companies, voluntary and mandatory oil import restrictions, and the development of a strategic petroleum stockpile.

Chapter One traces the development of government policy toward the oil industry during the 1920s and 1930s when the domestic system of production control was established. The legal environment under which the producing sector of the industry operated was defined by the "rule of capture," established by court rulings based on the treatment of wild game under English common law. By granting ownership to the producer of oil inefficient production practices resulted in physical and economic waste. At the same time, the geology of oil in the United States provided numerous small shallow pools which encouraged the entry of small producers into the industry. The boom and bust cycles which resulted created pressure for government intervention, which became overwhelming as the East Texas oil field threatened to bankrupt large portions of the industry. The final shape of the prorationing system was the result of the interplay of conservationist, laissez-faire and Populist ideologies, filtered through Texas politics. A small group of politically connected oil producers in Texas and California were able to block a coalition of major oil companies and the majority of independent oil producers in determining the final shape of the prorationing system.

Chapter 2 describes the development and collapse of oil import controls, and the wild scramble for economic rents generated by Government regulation. Oil import controls were a product of the need to protect the domestic system of control from competition with less expensive foreign oil. Internally, the industry had to balance the interest of three groups, international oil companies with existing imported supplies, newcomers to the international oil market, and domestic producers. The need to maintain a semblance of unity to protect tax advantages it had acquired over the years kept the issue from dividing the industry, while a coalition of coal and oil producing states were able to pressure the Eisenhower administration into making an exception to the general trend toward free trade. The dynamics of the market forced changes in company strategy in adapting to existing policies, as the multinational firms which had been opposed to import controls in the 1950s changed the focus of their political activity in the 1960s toward restricting pursuit of the resultant economic rents by actors outside of the industry.

Chapter 3 covers the two oil crises and the "phantom" Gulf War crisis. Oil policy had become a salient part of the policy agenda even before the oil embargo of 1973, as rising prices and energy "shortages" caused by price regulations placed energy under the spotlight of public scrutiny. Oil policy in the 1970s was dominated by Neopopulist attitudes and utopian consumerism, based on the belief that if only the monopoly of the oil companies could be eliminated, cheap oil would return. The attacks on the oil industry during the 1970s were grounded in the Populist tradition of hostility toward big business, despite the lack of solid evidence showing that the energy crisis or oil shocks were the product of collusion by "Big Oil." The new oil politics put the short-run interests of consumers above efficiency considerations and even national security, as Congress overwhelmingly voted down taxes or other measures which would reduce consumption or imports at the cost of higher prices. President Carter, at great personal political cost, managed to end interventionist regulatory policy by obtaining decontrol of oil prices from Congress using a carrot called the windfall profit tax. Despite the acceptance of deregulation in the 1980s, Congress has made it clear that no policy explicitly designed to raise oil or gasoline prices would be acceptable.

The pervasive influence of underlying Populist attitudes toward business can be seen in the support for small refiners. While independent oilmen had sizable resources to expend on gaining political favors, and dominated the politics of a number of states, small refiners had limited financial resources and no political base. Yet both of these groups consistently received special attention, often at the expense of the large oil companies, supposedly the dominant actors in oil politics. Small refiners were the beneficiaries of the small refiner bias of the oil import program during the 1960s and the entitlements program of the 1970s. Independent producers were granted a phase out of the depletion allowance (when it was eliminated for larger companies) and exemptions to the windfall profits tax.

Chapter 4 reviews the resource, energy and oil economics and policy literature from the 1930s through the 1980s. Policy is influenced by the scribblings of academics, as those in power seek rationales which legitimize their actions. Conservationist theories, although soon out of favor in the academic community, provided an intellectual fig leaf for the prorationing system and oil import controls. As the majority of economists reached a consensus critical of the oil control system, their views gradually leaked into the political debate. However, the limit of academic influence can be seen in the resistance to market solutions in the 1970s, though the eventual triumph of deregulation can be partially attributed to the persistent preachings of the profession. Economics as a guide to policy demonstrated a serious empirical weakness; it became obvious that while economic principles might have merit in shaping policy, economic predictions of prices, quantities and market behavior were grossly unreliable.

Chapter 5 examines the importance of public opinion in shaping the policy agenda, and reviews the attitudes of the American people toward business in general and the oil industry in particular during the 1970s. A primary source of these attitudes can be found in the Populist movement of the 1890s, persisting as an underlying theme of American political discourse. The Populists inherited the traditions of Jacksonian democracy, seeing a natural coalition between the farmer, the worker and the small businessman producing the wealth of the nation, allied against the monied interests, the trusts, and the corporations. This antagonism toward large business is rooted in the underlying American ideology of individualism, which, while fiercely protective of property rights, is uncomfortable with unfettered competition and the centralizing effects of industrialism. A secondary source was Progressivism, which sought to alleviate the consequences of competition through scientific management, and both beliefs shaped the perception by many of the energy crises of the 1970s. The Populist strain of belief explains the hostility to the oil companies, while the Progressive tradition explains reliance on governmental control of prices and quantities and elaborate plans to encourage energy conservation and stimulate the production of new energy supplies.

Chapter 6 provides an in-depth study of Congressional oil votes from the 1950s through the 1980s, including a description of the politics behind the vote, the parlimentary maneuvers which led to the sequence of votes, and an econometric analysis of the factors behind Congressmen's voting decisions. The dominant influences on voting were Party, the Congressman's ideology as expressed by their ADA rating, and oil production variables. Constituent ideology and economic characteristics were generally of minor importance.

Chapter 7 discusses the formation of oil policy, beginning with theories of economic regulation, the role of interest groups in developing the policy agenda, and the role of money in politics. Two factors which had a significant influence on oil politics were congressional reform and the public interest movement. Congressional reform weakened the power of committee chairmen allowing oil company opponents to bring bills to the floor of both houses and force Congressmen to put their votes on record. The breakdown of the discipline of the House also made it difficult to develop a cohesive energy policy, while increasing access points for interest groups. This increased access was exploited by both industry groups and the rising number of public interest groups, polarizing the energy debate, and raising the cost of supporting the industry for representatives from non-oil producing states. The consumer movement provided a voice for a group long absent from economic politics, though it led many to confuse low prices with the "public interest." The result was that the pluralist heaven became the pluralist hell, as every group had its say and no consensus could be reached.

This book should be accessible to the general reader, as I have attempted to avoid excessive rigor. It is primarily an economic and political history, but it is impossible to talk about economic policy without some discussion of economics. My focus is on policy formation, as opposed to the mechanics and implications of specific policies. However, economic policy can never be divorced from economics, and, the politics of economic policy cannot be understood without reference to the economic implications of policy. For this reason, economic history and policy are areas of scholarly interest which lend themselves to interdisciplinary studies. I hope the reader will find this book to be a worthy addition to such a literature.

The Economics and Politics of the United States Oil Industry, 1920–1990

I

Development of the System of Control

THE US OIL INDUSTRY AFTER WORLD WAR I

The roots of the struggle over oil imports in the 1950s can be found in the emergence of the modern US oil industry in the 1920s. Independent producers had been at the heart of the industry from its beginnings in Pennsylvania in the nineteenth century, but the economics of petroleum led to the growth of large integrated firms, especially after the break-up of the Standard Oil Trust. After WWI the major oil companies became international, integrated giants (Exxon, Mobil, Texaco, Chevron, and Gulf), while a second tier of major companies emerged during this period. It would be these second tier companies who would undermine attempts at accommodation during the 1950s between the Majors and the independents. Finally, the cycle of recurring gluts on the market during the 1920s, by driving prices down and stocks up to the point where even the integrated firms wanted regulation, would result in the formation of the prorationing system. It was the need to protect this system of production control, and those interest groups whose fortunes were tied to it, that would prompt the Eisenhower Administration to establish restrictions on the importation of oil.

From the beginning the industry was characterized by a cyclic pattern of boom and bust, with most of the risks incurred by producers, especially independents. The producers' weakness was due to a combination of physical and legal factors. Oil in a reservoir usually flows under pressure partially supplied by associated gas. When a well is drilled oil flows to it from regions of higher pressure in the reservoir. At the same time the land above most oilfields is owned by numerous landowners. Given the ease of obtaining leases and the low cost of drilling in the shallow oilfields which were the first to be discovered there were few barriers to entry for oil producers. Once the producer had purchased a lease or the landowner realized that he had oil under his property he was obliged to drill and produce oil as quickly as possible since otherwise it would flow to competitors' wells. The courts made this situation a permanent feature of the industry with a number of rulings in the 1880s and 1890s establishing the "rule of capture." This doctrine of property rights made petroleum analogous to wild game, granting ownership to the landowner or lessee who produces the oil regardless of its point of origin.[1]

The rule of capture would have a strong influence on the structure of the producing sector of the petroleum industry. The basic pattern was a spasm of overproduction in newly discovered fields, followed by a rapid decline in output as reservoir pressure was depleted. The reason for this production cycle was twofold: the common-property nature of oil reservoirs under the rule of capture and the lack of available credit for the small producer. Under the rule of capture the firm is affected three ways: (1) the value of a unit of reserves declines since if it is not produced today it may not be available tomorrow; (2) there are rapidly increasing marginal costs of production as other firms deplete the reservoir, independent of the individual firm's production decision; (3) given the surge in production, unless the reservoir is amply served by suitable transportation (i.e., pipelines) the high marginal cost of transporting surge production to markets will cause the price at the wellhead to decline even when this additional production is insufficient to effect market price.

The first two factors encouraged each firm to maximize its short-run level of production, usually at the expense of long-term production from the reservoir and certainly with regard to the optimal path of production. The third factor left the small producer at the mercy of the larger companies. Most small producers had limited capital and no collateral and were unable to invest in storage facilities. The larger companies, aware of the inability of the small producer to transport his production to markets, would invest in storage facilities near the newly discovered fields, buy oil at depressed prices, and wait until the production surge ended with the drop in reservoir pressure. The cost of storage was usually far less than the eventual rebound in prices, providing a handsome profit without incurring the risks of exploration. These companies would also buy out producers during the period of lowest prices, obtaining producing properties at bargain prices.

The low cost of drilling wells to the shallow depths of the early fields allowed the small wildcatter to survive during this period. In 1922 wells were commonly drilled at a cost of $40,000 to $60,000. Lifting costs were around $0.16 per barrel in 1919, though they were considerable less for wells in flush production. The first wildcatters in a field could make a considerable amount of money with limited capital if they received the market price (which varied between $1-2 per barrel during the 1920s). The trick was to finance additional wells with the revenue from the initial well, produce oil as fast as possible, sell out once the price dropped and use the proceeds to finance the search for another field. The large companies took advantage of the independents' lack of risk aversion through two leasing strategies. In checkerboard leasing companies would purchase 10-year leases scattered in a checkerboard pattern over wide areas of prospective oil country. Spread plays were used by independents or promoters to finance test wells by selling leases surrounding the hole, allowing the large companies the opportunity to purchase protection leases in areas of special interest.[2] The major companies let the independents take the risks and then moved in once oil was found.

The small producer was still prominent during this period in the US, though his role would steadily decline. In 1919 there were 9,814 producers, with an average capitalization of $246,737, and annual production of 35,674 barrels. In the same year however, the 32 largest firms produced almost 60 percent of total output, of which 35 percent was produced by 15 integrated firms. Only a thousand firms in that year had revenues exceeding $100,000, but they accounted for 96 percent of the value of the oil produced.[3]

The genesis of the companies which would dominate the industry was in the discovery of oil in Texas, which resulted in the emergence of Sun Oil, Texaco and Gulf Oil, and the breakup of the Standard Oil Trust. The Standard Oil Trust had been under constant attack in state courts since 1904 when the federal government filed a bill in equity in the US Circuit Court for the Eastern district of Missouri against the holding company, Standard Oil Company of New Jersey. The defendants were charged with violations of the Sherman Anti-Trust Act of 1890. In November 1909, Standard was found guilty as charged and in 1911 the decision was upheld by the Supreme Court. The Court ordered Jersey Standard to transfer to stockholders a prorata share of stock in each subsidiary. Jersey Standard and California Standard were the only two subsidiary companies which were fully integrated from pipeline to marketing. During the 1920s and 1930s the remnants of the Standard Oil empire as well as the newly emergent oil companies scrambled to integrate either forward and backward. Integration was accomplished by investing in facilities and purchases of existing companies. Most of the companies purchased were in the second rank of firms in their sector of the industry.[4]

The emergence of new technologies for finding and refining oil was partially responsible for the growth of these newly integrated companies. The shallow, easily noticeable fields had been found and new discoveries would require systematic and deeper exploration. Drilling capability increased rapidly with wells drilled to previously inaccessible depths of 5,000 to 10,000 feet. The rotary drill provided means for controlling the higher pressures encountered with deep wells, increased drilling times, and improvements in sampling quality, furnishing reliable data for subsurface analysis. Introduction of geophysical techniques, gravimetric, seismic and magnetic, along with the acceptance of geological methods, reduced risk in exploration.[5] The new techniques improved the competitive position of the major integrated firms since only a large company could afford the staff and equipment required to make use of these techniques. Coordinated use of aerial photography, surface data, core drilling, micropaleontology, and geophysical techniques required a large staff of professionals and competent managers. Exploration by small operators concentrated on stratigraphic traps difficult to detect with the new techniques or small fields and extensions. Table 1.1 illustrates this trend.

TABLE 1.1 Discovery of Major US Oil Fields

Period	Independent Producers		Major Companies	
	# of fields	million bbls	# of fields	million bbls
1871-1900	9	4,086		
1901-1910	17	4,898		
1911-1920	17	4,082	16	6,276
1921-1925	9	3,258	9	2,348
1926-1930	21	8,980	30	6,694
1931-1935	10	2,791	17	3,765
1936-1940	10	1,820	33	9,137
1941-1945	3	516	19	3,279

SOURCE: Owen, *Trek of the Oil Finders*, p. 472

CONTROL OF INTERNATIONAL OIL

In the Middle East Anglo-Persian (British Petroleum) was becoming a major power in world oil due to its Persian oilfields while the Iraq Petroleum Company had commenced development by the end of the 1920s. In the Far East the Dutch East Indies were responsible for 75 percent of the region's production. Control of world supplies was firmly in the hands of the International Majors, sometimes referred to as the "Seven Sisters" (the five American Majors plus BP and Shell) by the end of this period. Although Mexico had been open to all comers because of private ownership of mineral rights, its production peaked early in the 1920s (see Table 1.2). Mineral rights were obtainable in the rest of South America, but they were relatively expensive and the Majors were able to purchase the rights directly or indirectly from local actors. In the Middle East the mineral rights were owned by governments, allowing the Majors to obtain total control of exploration over vast regions. In the Far East and Africa colonial powers controlled access to potential oil lands, which gave a significant advantage to BP and Shell.

Competition for control of world oil became intertwined with national policy in the aftermath of World War I. The importance of petroleum for supplying fuel oil for the fleet, aviation fuel for the air force, and gasoline for motorized vehicles made access to crude oil a military necessity. While the British and French governments were well aware of their dependency on external supplies, the existence of large domestic supplies had caused the US government to be complacent with regard to supporting American oil companies in their attempts to gain access to foreign oilfields. The perception of shortages after the war resulted in a temporary change in the government's

TABLE 1.2 Foreign Oil Production by Region (thousand barrels)

Country	1920	1925	1929	1932	1940
Mexico	157,069	115,515	39,530	32,805	44,036
Venezuela	457	19,687	137,472	116,541	185,570
Other S.A.	6,611	21,119	53,295	51,219	83,194
Iran	12,230	34,038	42,145	49,471	66,317
Other M.E.	0	0	798	837	31,929
Far East	29,145	35,868	55,339	53,244	81,740
United States	442,929	763,743	1,053,128	785,159	1,353,214

SOURCE: E. DeGoyler and L.W. McNaughton, *Twentieth Century Petroleum Statistics* (Dallas, 1950).

attitude, persisting long enough to apply sufficient pressure to enable the American companies to enter the Middle East fields. The publication of the Edgar statement in May 1920, gloating over British control of world oil, combined with the Polk Report to the United States Senate, documenting the exclusion of American companies from various countries, motivated the Congress and the State Department into action. The State Department protested the exclusion of Standard Oil from the former Ottoman Empire and Dutch territories. Congress moved to bar foreign citizens from acquiring oil leases on public lands. Between the application of diplomatic pressure and the realization by Shell that it was risking large American investments for wildcat territory of uncertain value, entry of American companies into British and Dutch territory was accomplished.[6]

Seven American companies expressed interest in Mesopotamian oil, but due to the extended negotiations and financial pressures only three companies joined the Iraq Petroleum Company, Jersey and Socony each with 9.9 percent and Gulf with 3.96 percent of ownership. The "self-denying clause" prohibited any involvement in crude oil production in the Ottoman Empire by the owners of the IPC outside the auspices of the company. The final agreement of July 31, 1928, became known as the "Red Line" agreement because of a red line on a map delimiting the area of the self-denying clause.[7]

The Middle East was dominated by the Majors from the beginning. Major Frank Holmes, operating through the Eastern and General Syndicate, acquired an option on oil rights in Saudi Arabia in 1923, the Saudi-Kuwait Neutral Zone in 1924 and on Bahrein Island in 1925. The Syndicate sold their rights to Gulf in 1927, which, bound by the Red Line agreement which included Bahrein and Saudi Arabia, allowed Socal to take over the Bahrein commitments in 1928, and oil was found in May 1932. Bahrein Island had similar geology to Saudi Arabia, and the discovery of oil renewed interest in the mainland. Socal, lacking markets near the newly discovered oil, sold a half interest to Texas in return for the assignment of eastern hemisphere marketing concerns to the newly formed California-Texas Oil Company (Caltex).[8] The California Arabian Standard Oil Co., which later became the Arabian American Oil Co. (ARAMCO), was incorporated in 1933, and the company struck a major find in March 1938. Before the finding of commercial quantities of oil Texas had purchased a half interest for $3 million cash and $18 million out of future production, maybe the greatest bargain of all time.[9]

When Gulf relinquished its rights to Bahrein to Socal it retained its interests in Kuwait. Competition between Gulf and Anglo-Persian stalemated negotiations with Kuwait so the two companies decided to work together, forming the Kuwait Oil Company Ltd. which obtained a concession from the Sheikh of Kuwait in December 1934.[10] The Burgan field was discovered in 1938, a supergiant field with ultimate reserves of 72 billion barrels, second largest in the world. While development was delayed until after the second World War it was obvious that the two companies were in possession of vast reserves.

The Red Line agreement fit in with the attempts of the Majors to bring order to the world oil market. A desire to avoid price wars became the motivation behind a series of meetings during the early months of 1928, culminating in a meeting of Teagle of Jersey, Sir John Cadman of Anglo-Persian, and Deterding of Shell at Achnacarry Castle in Scotland in August. The document adopted there was titled the Pool Association of September 17, 1928, more commonly referred to as the "As Is" or Achnacarry agreement. The agreement was an attempt to control the market by the pooling of industry resources. Each member was to maintain the status quo with reference to market shares, make existing facilities available to competitors on a favorable basis, and add new facilities only to cover increased requirements of existing customers. Markets should be supplied from the nearest producing area and surplus production from any area should not be allowed to drive prices down in other regions. Crude short companies agreed to purchase additional supplies from crude long firms, with no discounts permitted on sales to outside marketers.[11]

The agreement established an internal pricing system for cartel members which operated parallel with the world pricing system. The cartel pricing method designated each port of shipment as a basing point; US Gulf port published prices would prevail at each shipping point; and freight rates were assigned for each shipping path. In contrast, the world pricing method up to World War II used the United States Gulf ports as the single governing basing point. Standard tanker charges from the United States Gulf to ports of destination were added to the basing point price to determine prices at these destinations. The common acceptance of this pricing method by all parties in world trade resulted in a uniform delivered price in each market, allowing favorably located producers to collect a freight premium. As long as the United States and Caribbean areas were the world's principal sources of crude, there was no reason to undercut the Gulf-plus price in any foreign market since this would trigger a price war which the dominant American companies, because of their size and location, would probably win. Since US domestic production tended to be high-cost compared with the flush production of Venezuela and Iran, and Gulf port prices reflected the higher cost of US output, foreign producers were provided with additional profits.[12]

The Achnacarry agreement did not end competition in the world oil market. There was a fringe of producers and marketers who were not bound by the agreement, and even the Majors could not always agree among themselves. Attempts were made to patch over differences, first with the Memorandum for European Markets in 1930 and then the Heads of Agreement for Distribution, adopted in December 1932. The principle of "as is" was to continue to apply to supply and distribution of crude oil and refined products outside the United States. Supply problems would be handled by a committee sitting in New York while distribution would be determined by a London committee, with a central committee, also in London, coordinating the decisions made concerning supply and distribution.[13]

The key to making the "as is" system work was for the Majors to control all foreign supplies of oil. South America was the region where the entrance of independent producers was most likely. While oil production peaked in Mexico during the early 1920s relations between the foreign oilmen and the Mexican government gradually deteriorated. With the success of the Mexican Revolution in 1917, Article 27 was inserted into the Constitution, resurrecting the traditional Spanish position on mineral rights. Under the administration of President Cardenas, who took office in December 1934, the petroleum industry of Mexico began its progressive march to nationalization, culminating in expropriation in 1938.

Venezuela had developed into a far more important petroleum province. The Venezuelan dictator, General Juan Vincente Gomez, encouraged foreign investment in oil, offering liberal conditions and even using the army to quash oilworkers' strikes for higher wages and better working conditions. Shell had already made substantial investments in obtaining concessions before Jersey Standard entered the country in response to its competitor's moves. Jersey began by buying concessions in 1919, and purchased a majority interest in the Creole syndicate in 1928 in order to obtain producing oil properties in and around Lake Maracaibo. When Shell's No. 2 Los Barrosos Well in the La Rosa field blew out, gushing at over 100,000 barrels per day (bpd) for nine days in 1922, oil companies flocked to Venezuela. US company investment rose from $75 million in 1924 to $162 million in 1929. Despite the involvement of numerous companies only Gulf was able to come up with a substantial share of Venezuelan production. The capital short independent British firms began selling their concessions in 1922, and further consolidation occurred when surplus production and falling prices put the squeeze on the independent producers in the 1930s. Most of these newly discovered oil fields were in Eastern Venezuela, held by subsidiaries of Jersey Standard and Gulf. Gulf was included in an agreement to control production by transforming Mene Grande, the Gulf operating subsidiary, into a jointly owned enterprise of Jersey, Shell and Gulf in 1937.[14]

CHAOS IN THE DOMESTIC OIL MARKET

The fear of a crude shortage in the US, which had enabled the US Majors to enlist the US government in supporting their entry into the Middle East, was due partially to projections by the US Geological Survey (USGS) and the Bureau of Mines. The USGS had projected a total of little more than nine billion barrels of recoverable oil in 1920, and rising prices and increased imports from Mexico reinforced this perception of impending exhaustion. Ironically, at the time when talk of a shortage was at its height, new oil fields were being discovered which would lead to a persistent glut on the market. Drilling activity had taken a sharp upswing after the war due to high prices and increases in production soon followed. The source of the first oil glut of the 1920s was in California where three giant fields were discovered in the period 1919-1921, though the impact of this oil was not felt until 1923, when tank steamers began using the Panama canal, cutting shipping costs by 50 percent.[15] The arrival of this California oil coincided with peak production by five mid-western oil fields.

In 1924, the output of these eight fields of 1923 dropped 81 million barrels, but the total decline in the United States was only 18.5 million. The increase in production would have been even higher if there had been some place to store the oil. The large companies which counted on storing oil until the glut ended found themselves with 300 million barrels of deteriorating oil in storage.[16] The only respite came in 1926 when prices recovered briefly before diving in the wake of a new flood of oil. Table 1.3 presents the cycle of stock build-up and decline which caused prices to fluctuate during the decade. Note that it would require the establishment of production controls before stocks were reduced to the level of the early 1920s.

In January 1926 flush wells were struck in the Texas panhandle. At the same time the Permian basin in West Texas was also beginning development. But it was the Great Seminole field in Oklahoma which ended the brief respite from overproduction. Discovered in July, 1926, output peaked in July 1927 at 525,847 bpd from less than 700 wells. One-fourth of the ultimate production from Seminole came in 1927, a total of about 130 million barrels.[17] During the three-year period from 1928 to 1930, 27 oil fields with recoverable reserves of 11 billion barrels were discovered, but exploitation of most of the new fields was restricted by limited pipeline outlets, unit ownership, voluntary agreements or development lags, delaying peak production until they came under some

TABLE 1.3 Prices and Petroleum Stocks: 1920-1935

Year	*Stocks 12/31 (Million bbls)*	*Average Price*	*High Price (\$/barrel)*	*Low Price*	*Gasoline Price (¢/gallon)*
1920	149.45	3.40	3.50	2.75	29.74
1921	217.32	1.79	3.50	1.00	26.11
1922	295.71	1.72	2.00	1.25	24.82
1923	340.64	1.58	2.00	1.00	21.06
1924	360.48	1.60	2.00	1.00	19.46
1925	345.86	1.85	2.04	1.25	20.09
1926	315.03	2.13	2.29	1.79	20.97
1927	379.66	1.42	1.90	1.28	18.28
1928	392.63	1.31	1.36	1.28	17.90
1929	428.45	1.36	1.45	1.20	17.92
1930	411.88	1.25	1.45	0.95	16.16
1931	370.19	0.64	0.95	0.18	12.98
1932	339.88	0.87	0.92	0.69	13.30
1933	354.22	0.61	1.00	0.25	12.41
1934	337.25	1.00	1.00	1.00	13.64
1935	314.86	1.00	1.00	1.00	13.55

SOURCE: API, *Petroleum Facts and Figures*, 1950

form of production control. The serious threats to an orderly oil markets were the fields with a single producing reservoir which lead to the kind of flush production which upset the market in 1923 and 1927. Even then, few fields were large enough to disrupt markets by themselves.

However, two fields with that kind of market power were discovered at the end of the decade, driving prices below a dollar a barrel and providing the impetus for reaching an agreement on prorationing. The Oklahoma City field, discovered in 1928, began flush production in 1929. The Oklahoma State government, which had already tried to control the Seminole field, fought with the oil producers to restrain production to the point of sending in troops. East Texas was opened in late 1930 and production climbed to over 100 million barrels annually starting in 1931 and peaked at over 170 million barrels in both 1933 and 1935 (compared to a national total of 905 million and 966 million barrels in these years, respectively) despite numerous attempts by the Texas Railroad Commission to control output.

DEVELOPMENT OF THE PRORATIONING SYSTEM

The development of the prorationing system does not easily fit the body of theory concerning regulation and rent seeking.[18] While a desire to control production and thus prices was a primary motivation of oil industry actors, the bitter conflict over the shape of a system of control was impelled by ideology as much as self-interest. There were three major issues which had to be resolved; how would production be controlled at the field level, who would decide what the proper level of production should be, and what combination of private and government institutions would enforce these decisions. In addition, it was also necessary that any domestic agreement to control production be integrated into the system of control that had been established over the world oil market.

The attempt to formulate a national policy for the oil industry had its roots in the experience of WWI, both with regard to specific elements of government-oil industry cooperation and the general experience of wartime corporatism. In March 1917, Bernard Baruch, then a member of President Wilson's Advisory Committee to the Council for National Defense, approached A.C. Bedford, chairman of the board of directors of Jersey Standard, and asked him to form an advisory committee to represent the oil industry. When Wilson dissolved the Council for National Defense to create the War Industries Board in July, 1917, the Petroleum Industry Advisory committee continued its existence under the new agency as the National Petroleum War Service Committee. Membership included Walter Teagle, president of Jersey Standard, William Farish of Humble Oil, G.S. Davidson of Gulf, Harry Sinclair, and C.F. Lufkin, president of the Texas Company.[19]

President Wilson also established a new United States Fuel Administration, with an Oil Division with authority to impose controls on the production, marketing and pricing of petroleum products. Mark Requa, a petroleum engineer and a protégé of Herbert Hoover, was chosen as director. Requa was an advocate of cooperation among businessmen and between businessmen and government, and he subscribed to Teddy Roosevelt's doctrine of regulated competition. Requa encouraged cooperation among oilmen in production, insisting on the practice of efficient, sound conservation methods and the prevention of waste. Bedford and Requa agreed on the need for cooperation in the industry to increase efficiency, as an alternative to the extremes of cut-throat competition supported by some industry members and nationalization, threatened by a Federal Trade Commission report accusing oil companies of profiteering.[20]

With the end of the war there was no enthusiasm in the industry for the immediate abolition of the system of cooperation. The National Petroleum War Service Committee, led by Henry Doherty of Cities Services and Lufkin, called a meeting in New York City in March 1919 to draft a constitution and by-laws for a new petroleum trade association. On March 26, in Chicago, the American

Petroleum Institute (API) was established, with the directors composed of the members of the War Service Committee with the addition of Requa. Its aims were to facilitate cooperation within the industry and between the industry and government. Requa called for the new organization to support conservation through increased technical efficiency.[21]

In 1923, Henry Doherty, head of Empire Gas and Fuel, became a pioneer for conservation after opposing government intervention during the postwar oil "shortage." The motivating factor behind his conversion was the arrival of California oil and the subsequent decline in oil prices, but he was also concerned with the physical waste associated with existing practices in the industry, caused by the rule of capture. Doherty entered into correspondence with oil executives and federal officials to push for government action. Doherty wanted to replace the rule of capture by requiring mandatory unitization (where operators are required to operate a field as a single economic unit) as the normal operating procedure for the industry. Patterned after the irrigation and drainage districts in the west, Doherty's plan would require production to be supervised by oil districts, staffed by the various land owners in the area. No drilling would be allowed in a newly discovered field until an oil district was formed by the majority acreage.[22]

Doherty's ideas received some support in the API from Ernest W. Marland, the future governor of Oklahoma. However, Walter Teagle, while publicly expressing interest in the plan, selected Amos Beaty of the Texas Company, who proved to be a major opponent of unitization, as the head of a committee to investigate Doherty's proposal. Beaty's report, instead of examining unitization, proposed to give state commissions the power to approve voluntary agreements between oil companies in the same field. The objection of Teagle and Beaty to compulsory unitization was that it violated the due process and the obligation of contract clauses of the Constitution. Beaty stated that the oil district scheme would deprive the landowner of his royalty without due process.[23]

The API was split between different factions which had various interpretations of the term cooperation. Teagle focused on barriers to intra-industry cooperation such as the antitrust laws. Requa saw cooperation as the application of guidelines by the industry to conserve oil and stabilize the industry, to be jointly enforced by the industry and the oil-producing states. The younger Pew brothers, J. Howard and J.N. Jr., of Sun Oil, along with Davidson of Gulf, disliked working with the government and welcomed a return to unfettered competition. Their older cousin, J. Edgar Pew, working with oil producers, developed a position closer to the rationalization and scientific management ideas represented by Requa. He supported voluntary agreements but drew the line at governmental involvement.[24]

Doherty was unable to gain the support of the API over Beaty's opposition and turned to the federal government to press his ideas. He went first to the head of the Bureau of Mines and then to President Coolidge. Coolidge, prompted by both Doherty and Secretary of the Interior Hubert Work, established the Federal Oil Conservation Board (FOCB) in December 1924, a cabinet level committee of inquiry, to investigate and report on conditions in the oil industry. His motives, stated in a letter to Work, were based on considerations of national security, though a secondary motive was to handle the political fallout from the Teapot Dome Scandal. President Harding's Secretary of the Interior, Albert Fall, had persuaded the President to transfer jurisdiction over three naval oil reserves from the Navy to Interior in May 1921. Within a year Fall issued leases for the Elk Hills reservoir to E.L. Doheny and the Teapot Dome reservoir in Wyoming to Harry Sinclair. The leases were justified by the need to drill offset wells to protect the government's interests, and the two contracts received little publicity. Senator La Follette, acting on a tip, began to investigate the disposal of the reserves. It turned out that numerous politicians were on the payroll of Doheny, and Doheny and Sinclair had loaned Fall $404,000. While Fall's concerns about drainage and the need to make arrangements with private companies to ensure supplies and storage for Naval Reserves were later justified by events, the stench of political corruption precluded any discussion of policy.[25]

The Congressional debate over the $50,000 appropriation for the FOCB centered around accusations by Democrats attempting to tie the Coolidge Administration to Teapot Dome. This disturbed the oilmen, who were wary of the new board and its attempts to investigate conditions in the industry. They were reassured by the Coolidge administration that the FOCB was not interested in muckraking, and that it would work with the industry to determine the industry's needs. A questionnaire was sent in January 1925 to the major oil companies who responded that there was no possibility of shortages, no waste in current oil operations and no need to alter the existing structure of property rights. Teagle actually claimed that the drilling of a large number of wells in a field resulted in a larger recovery of oil from formations than with a smaller number, and that unitization would not result in conservation. Beaty joined with Teagle in the suggestion that the federal government legalize agreements between oil companies to restrict output in periods of overproduction.[26]

In January 1925, J. Edgar Pew, president of the API, appointed a committee of eleven to gather data and report to the FOCB. In May they issued a report, *American Petroleum: Supply and Demand.* Among its conclusions was that there was no danger of exhaustion of petroleum reserves, that price will correct any imbalance in the market, and that "Waste in the production, transportation, refining and distribution of petroleum and its products is negligible." The report did mention that there were beneficial effects of edge water, and that maintenance of back pressure on wells increased recovery, but ignored any connection between these effects and field production. With regard to storage: ". . . but, as indicated, there is no waste of oil itself, for methods of handling and storage are now such that the natural losses through leakage and evaporation are nominal."[27]

Dissenting voices included those of Doherty and Requa, and George Smith, head of the Geological Survey. Numerous scientists expressed doubts, and the Rocky Mountain Association of Petroleum Geologists unanimously questioned the API report. When the FOCB conducted a hearing in February 1926, future Supreme Court Justice Charles Hughes, as Counsel for the API, testified that Congress had no power to control production of petroleum within the states and that it was doubtful that states had the power to control drilling on private property. The subsequent report to the President attempted to straddle the line between Doherty and his critics, admitting the importance of gas in oil production and the "urgent need of concerted action in development and operation of an oil pool," but also agreeing that the Federal government had no power to regulate oil production unless the national defense was impaired. The report supported voluntary cooperative agreements and called for legislation which removed barriers to such action without supporting compulsory unitization. The report also mentioned the need for cooperation between State regulatory agencies and Federal bureaus, including the possibility of interstate compacts, while cooperation in the industry to better balance supply and demand was recommended.[28]

The resistance of the major actors in the industry to federal involvement in controlling production was based on their confidence in informal methods of control. The major companies owned extensive producing properties and were usually able to obtain a majority of ownership in new fields due to their superior capital resources relative to independent producers. While interstate pipeline companies were subject to control by the Interstate Commerce Commission, in practice they were usually free to set their own rates. The major companies also set prices for sales by independents, providing further control over supply. A temporary respite from falling prices in 1926 convinced oilmen that this system of informal controls was working.

Late in 1926, the petroleum industry was shaken by the opening of the Seminole field in Oklahoma, causing oilmen who had seen no need for controls in the past to become converts to the cause of conservation. The API created a committee of seven to examine the waste of gas during the production of oil, headed by Oklahoma oilman E.W. Marland. Marland had been an opponent of Doherty, but now became an ardent supporter of conservation. The committee hastily produced a tentative draft for a state law to prevent the waste of gas, aimed at Oklahoma. Despite Marland's insistence that the bill would aid the small oil operator by eliminating the boom and bust cycle, other Seminole well owners as well as a committee of the Mid-Continent Oil and Gas Association condemned the proposal. Marland's lawyers, in 1927, were unable to secure passage of the bill in the Oklahoma legislature. Walter Teagle responded to Marland's failure by calling a conference of oil executives at which agreement was reached for a temporary restriction of drilling operations in the Seminole area. Three months passed before voluntary restriction was accomplished due to the large number of different interests operating in the field. Some major oil firms, including, ironically, Doherty's company, favored rapid drilling because they had millions of dollars tied up in oil leases that might be forfeited if they did not drill, while other operators believed that it was the rest of the nation that was overproducing and that other fields should be curtailed first.[29]

The API's board of directors split over the government recommendations at the annual meeting in December of 1926, although the group went on record endorsing the FOCB report. The Institute accepted the need for more effective conservation practices to be adopted by members of the industry, motivated by the emerging oil glut. R.L. Welsh, general counsel of the API, pointed out that this would allow producers to restrict their output, and an increasing number of oilmen were coming to favor conservation because it promised to curb overproduction. However, most of these operators envisioned some system of voluntary cooperation and were opposed to federal intervention. The advocates of voluntary cooperation were concerned with possible Justice Department enforcement of the antitrust laws. Thus, a number of executives from the integrated oil companies as well as some members of the Coolidge Administration favored amendment of antitrust legislation as the best means of allowing the industry to solve its own problems.[30]

API President E.W. Clark declared early in 1927 that the industry desired legal approval of cooperative agreements to restrict production and improve conservation, and he sent Walter Teagle to confer with Secretary of the Interior Work. Work referred Teagle, Clark, and the other oil leaders to Assistant Attorney General William J. Donovan, head of the Department of Justice's antitrust division and an advocate of antitrust revision and business cooperation. While Donovan expressed support for the industry's efforts, his superiors were not as enthusiastic. Coolidge told Requa in February 1927 if large, integrated oil companies secured legal authority to make voluntary agreements restricting production, they could easily manipulate prices. This could lead to accusations that the federal government was furthering monopoly. Coolidge dashed the hopes of the API and other groups for amendment of the antitrust laws.[31]

While industry leaders were now convinced of the need for production controls there was still disagreement over what kind of controls. William Farish, though in favor of unitization in theory, opposed federal unitization for two reasons; as a supporter of states rights he opposed federal involvement and he felt the independents were deeply opposed to compulsory unitization. Since the vote and influence of the independents were critical in the industry's battle to maintain its preferential tax position there was good reason to avoid alienating them. Doherty persisted in his support of unitization, including federal laws to provide legal sanctions for the execution of voluntary cooperative efforts and mandatory return of all excess gas to the reservoirs as an additional conservation method.[32]

The industry was now split between those members who felt that only federal control could bring order to the market; those who felt the solution was elimination of antitrust restrictions on cooperation; and those who wanted state regulation of the oil fields. A group led by Marland, including many independent operators, favored state regulation as the best means to meet existing problems. Marland was afraid that federal regulation would lead to the domination of public policy by the large integrated oil companies. State regulation, however, would ensure that the influence of local groups would still be strong. The only thing that most oilmen could agree upon was their opposition to Secretary of the Treasury Mellon's statement that natural laws of supply and demand would solve the problem of surplus oil. At the API meeting in December 1927, the convention urged the enactment of federal and state laws to authorize producers to enter into unit agreements free of the threat of antitrust prosecution.[33]

A factor influencing attitudes toward conservation was the steadily increasing technical knowledge concerning reservoir behavior. Publications concerned with production problems and addresses and articles by engineers emphasized the need for conservation, the extent of waste, and the role of gas in the recovery of oil. Such articles also emphasized the potential contribution of unit operations to the conservation of gas and oil and to economical production. An API technical report in December 1927 summarized these findings: "The data show conclusively that the conservation and efficient utilization of gas are of paramount importance in the conservation and economical production of oil . . . . The data show finally that many, probably most, of the best methods for gas utilization can be practiced advantageously only when they can be applied systematically to a whole field . . ." Two years later the publication of a study by H.C. Miller, commissioned by the API, *The Function of Natural Gas in the Production of Oil*, put the official stamp of approval on the new knowledge, emphasizing the importance of gas in the production of oil and the costs and waste associated with standard industry practices.[34]

The API began discussions for developing a plan for voluntary restriction of crude production in August 1928, and a preliminary plan was presented in December, calling for limiting the daily rate of production in 1929 to that of the last nine months of 1928. In February 1929, the API established a Committee on World Production and Consumption of Petroleum, chaired by R.C. Holmes of the Texas Company. At a meeting in Houston Texas, on March 16, the committee devised a national program for oil production, dividing the country into four producing areas and assigning a regional committee of major oil executives to each area. These regional committees were to divide production among the various oil companies, as well as to promote adoption of state regulations and the development of voluntary unitization agreements.[35]

Although George Smith of the Geological Survey supported this proposal, E.S. Rochester of the FOCB felt it was politically dangerous for the new administration, as federally approved stabilization might raise prices, possibly resulting in charges of supporting monopoly. Rochester preferred the more indirect method of federal support for the development and coordination of state conservation programs. The new Secretary of the Interior, Ray Lyman Wilbur, a physician by trade, was left with a difficult decision, and turned to the Attorney General, William Mitchell. Mitchell made it quite clear that the FOCB had no authority to approve action contrary to federal and state antitrust laws. Mitchell's position was strengthened by President Hoover's statements in

April that he could not condone the API agreement because it violated the Sherman Act, and that any action by Congress to deal with the oil surplus would result in price manipulation.[36]

The new president, Herbert Hoover, opposed amending the antitrust laws to allow oilmen to make agreements restricting production, proposing voluntary cooperation for promoting conservation by individuals and among the states. Hoover's Secretary of the Interior advised the industry to seek production control through uniform legislation by individual oil-producing states, and Interior officials even drafted an interstate compact and set up a governors' oil conference. In April, 1929, the API, realizing there was no hope of obtaining Justice Department approval for a voluntary industry agreement to curtail production, also came out for the enactment of state regulatory laws and an interstate compact.[37] The Interior Department's interstate oil compact would have the oil-producing states set up an interstate conservation committee with wide-ranging authority to control output to prevent waste, specifically including "economic waste." The emphasis had shifted from conservation to price maintenance.

Rochester and Wilbur agreed that the states should control output within their boundaries, and Wilbur proposed a conference of state governors. When the FOCB meet with the API on April 2, Wilbur suggested that the oil problem should be handled by coordinated state-sponsored drilling agreements. Although a number of executives were skeptical, convinced that state and regional rivalries would prevent cooperation, the administration forged ahead with Mark Requa as its point man. Requa drafted an interstate oil agreement, met with governors and oilmen, working with Attorney James Veasey of Jersey Standard, who preferred to stay in the background. The compact drafted in May 1929 by Interior Department officials called for Oklahoma, Texas and California to set up an interstate conservation committee. The committee would have the power to control output to prevent waste, including "economic waste." President Hoover invited the governors of the western states and interested oil producers to a conference on June 10, 1929, at Colorado Springs. Approximately one hundred delegates attended, including leading members of the API and spokesmen for the oil state governors. The meeting was a fiasco, for a number of reasons. The decision in March 1929 to stop the issuance of leases on public lands alienated the Governors of Colorado, Utah, and Wyoming, who thought they were being sacrificed to appease Texas and Oklahoma. The smaller operators were opposed to any restrictions on production and opposed a compact, but supported a tariff on foreign petroleum. Requa, when it became clear that the administration would oppose both Congressional action and a tariff on imports, resigned as chairman of the conference in late July.[38]

The issue of imports was a major stumbling block in the way of an agreement. The independent oil producers felt that state and federal programs were being designed primarily to make room for foreign imports. The Independent Petroleum Association of America (IPAA), led by Wirt Franklin and Russell Brown, was the primary lobbying group for the tariff. In hearings before the Ways and Means Committee in February, 1930, Franklin presented the arguments which would be repeated almost without modification over the next decade. He noted that almost all the materials used by oilmen were protected by tariffs, and it was only fair that oilmen also receive protection. He argued that the benefits of imports were limited to only "the manufacturing industries who want cheap fuel . . . and those industries, gentlemen, are the industries to whom you have given such wonderful protection from foreign competition." The cost of imports were the loss of the production of the "stripper" wells forced to close because of low crude prices as well as the loss of work in the coal industry due to the substitution of fuel oil for coal. The oversupply of crude oil which was driving crude prices down was due to imported oil, not overproduction.[39]

Franklin went to Washington in February, 1930, on a special train with a group of Mid-Continent oil producers, financed by a $100,000 fund raised by the IPAA. Only seven Senators were reported supporting an oil duty, excluding both Texas Senators, Morris Sheppard and Tom Connally. Secretary Wilbur opposed the tariff on the grounds that the country was still a net exporter of oil products, and a tariff might cost foreign markets. Among domestic integrated firms only the smaller firms such as Barnsdall, Phillips, and Skelly supported the independents' campaign. Franklin became involved in a heated controversy triggered by the independent lobby's free wheeling use of funds. While Franklin concentrated on his contention that the "so-called conservation policy" was a plot to shut in American oil to make room for more imports, the Senate lobby committee grilled him on his Washington activities. He struck back at accusations that he was strong arming Senators by blaming the oil crisis on Royal Dutch-Shell, the Mellon group (Gulf Oil) and the Standard Oil group. However, Franklin's admission that his net wealth was around $50 million damaged his credibility as the spokesman for the "poor beleaguered oilman."[40]

Congressional supporters of the tariff echoed Franklin's arguments, constantly referring to the big "three" or the big "four" group of oil companies, suggesting that the motive behind imports was to drive the independent firms out of business. There were numerous references to the 300,000 stripper wells and the 500,000 bpd which would be lost if imports were allowed to continue. The chief supporters were the Senators from Oklahoma, Elmer Thomas and William Pine. One of their targets was a Wall Street Journal editorial which stated that a tariff would operate as an export bounty, that higher domestic prices would merely bring forth more production, and that the "only hope for oil stabilization is further centralization of ownership. Oil lands must pass more and more into the hands of corporations able and willing to defer recovery." This was seen as proof that the real motive behind conservation was monopoly, and that the independent oilman was the only thing between the consumer and monopoly pricing. Tom Connally said he based his decision to support a tariff upon the apparent effort of the great oil companies to crush the independent oil producers.[41]

Opponents of the tariff, led by Senator Millard Tydings (D-MD) focused on the increased cost to consumers, scoffing at the contention that a dollar a barrel tariff would have no impact on product prices, asserting that it would cost up to a billion dollars per annum. The threat of foreign oil was credited with keeping the price of domestic oil low, and thus the price of gasoline for the farmer and working man. Ironically, the opponents of the tariff also accused its supporters of being in league with "monopoly." Despite a number of attempts to pass a tariff bill in the Senate, all that tariff supporters could obtain was a order directing the Tariff Commission to investigate the comparative costs of production in the US and Venezuela. The major oil companies had accepted the need to placate the independents as early as 1929 when James Veasey privately suggested restricting oil imports in a company memorandum which was forwarded to President Hoover. An informal pact was negotiated in which the domestic producers, in exchange for voluntary import restrictions, would cooperate in strengthening the state conservation programs. The voluntary curtailment program did not end lobbying for an oil tariff, and in 1932 a 21¢ a barrel tariff on crude and a 2.5¢ a gallon tariff on gasoline was established.[42]

The Hoover administration was less accommodating to renewed efforts to stabilize the industry by circumventing the antitrust laws. In March, 1930, representatives of the API, led by C.B. Ames, general counsel of the Texas Company, came to Washington to attempt to secure FOCB endorsement for a production-restriction agreement. When the oil men appeared before the FOCB on April 4, they were met with a blunt rejection, on the grounds that they were seeking an "immunity bath."[43]

Despite the unwillingness of the Administration to accept cartelization by the API, it encouraged cooperation between the states to control production. Wilbur moved to facilitate coordinated production control by announcing on March 1930 the creation of a voluntary Committee on Petroleum Economics which would report to the FOCB on the status of petroleum demand. This followed a suggestion by Farish and Requa for an authority to determine the market demand for petroleum. The Committee was composed of oil men who volunteered their services, but in 1931 Wilbur appointed a Bureau of Mines economist to chair the committee, granting it increased legitimacy. Between 1930 and 1933 the Committee on Petroleum Economics issued six surveys of national petroleum requirements. The API created its own committee in April 1930, and the two committees arranged to release their statistical surveys on the same day at the Petroleum Institute. The Committee assumed that the demand for gasoline was independent of price and depended upon three factors, the number of cars and trucks and their average consumption of gasoline, the general level of industrial activity, and the weather. The problem of calculating demand was simplified by asking each firm for its individual estimates of consumption, while the proper gasoline inventory was determined by the API. Implied in these calculations were assumptions regarding the proportion of crude which would be "cracked," which in effect required an estimate of the proportion refined by integrated companies vs. smaller refiners.[44]

The first report was released on April 14, after its transmission to Secretary Wilbur on March 25. Estimates were made of gasoline consumption for 1930, and from that estimate the amount of required crude oil was determined. The Committee also assumed that 30 million barrels of oil would be removed from storage. Once the total supply requirement was calculated, the Committee allocated it among the oil producing states, adjusted for actual production for the first three months of the year, providing a daily production figure. Texas was assigned 750,000 bpd, California 610,000 bpd, and Oklahoma 655,000 bpd, as well as 78,000 bpd to be removed from storage.[45]

OKLAHOMA CITY AND EAST TEXAS

Oklahoma had first attempted to control the state's oil output in response to the Seminole field. Production reached 527,000 bpd in July 1927 when the field umpire, Ray Collins, appealed to the Oklahoma Corporation Commission for an order curtailing production. A hearing led to an order halting drilling activity for two weeks, and in August Collins was able to file an agreement between operators limiting production to 450,000 bpd, followed by a statewide order in September, 1928, which asserted that the total demand for oil in Oklahoma was 700,000 bpd, settled pools were taking 275,000 bpd, and Seminole was awarded the remaining 425,000 bpd. The order of April 30, 1929, reflected the Oklahoma Commission's recognition of national efforts to organize production control: "that a general nation-wide scheme of stabilization has been undertaken by the crude oil and natural gas producers of the United States," and stressed the importance of cooperation with this effort.[46]

The Oklahoma field was discovered in December 1928 and production grew steadily during 1929. An oilmen's committee obtained an order shutting down the field for thirty days in September as it became apparent that the new field had the potential to flood the market. Wirt Franklin, a major acreage holder in the field, became head of the Oklahoma City Producers Association and with the cooperation of the major producers in the field helped arrange a proration plan. On October 14, 1929, the Commission issued a new statewide order which included the Oklahoma City field, limiting Greater Seminole to 20% of potential production, Oklahoma City to 40%, and other pools to 50%. In order to avoid the need for measuring well potentials the order specified the percentage of time that wells would be allowed to run wide open.[47]

The Oklahoma Corporate Commission reduced allowables to 25 percent in January, 1930, as perceived statewide demand declined to 600,000 bpd. Given the cost of the relatively deep wells, from $100,000 to $150,000, and the high production potentials, there was tremendous pressure on small producers in the Oklahoma City field to ignore production controls. Aware of the growing pressure against proration, the Commission's June 30 order included a full justification for the severe restriction on the Oklahoma field, based on the excess of potential production over demand which could lead to large quantities of oil in earthen storage, resulting in substantial waste. The Commission assumed the duty of determining production potentials, except for the 60,000 stripper wells on the grounds that it was impracticable to regulate these wells, and curtailment would cause many to be abandoned.[48]

The participation of Texas was the key to a system of production control. The Texas Railroad Commission (TRC) was authorized to supervise the oil and gas industry under a 1919 conservation law. The first test for the Commission came in 1928 when Farish requested proration on a ratable basis in the west Texas fields of Yates and Hendricks, supported by Marland, Indiana and Shell, but opposed by independent oilmen led by Tom Cranfill. Humble argued that uncontrolled production would lower prices, shutting down marginal wells and reducing royalty payments to landowners and taxes to the state. Cranfill, supported by Magnolia and Gulf, argued that the state should act to prevent physical waste only and the market should determine price. The TRC issued prorationing orders for the fields, maintaining prices. Despite this controversy the legislature faced little opposition when it specifically excluded regulating economic waste in the State Conservation Act of 1929.[49]

A Humble price cut (Humble was the largest purchaser of oil in Texas) in early 1930 generated action on the legislative front. Led by Tom Cranfill, independent oil producers enlisted the support of farmers, land owners, chambers of commerce, and university officials in launching a legislative program against the "stranglehold of pipeline companies" on the crude oil market. On March 18, 1930, the Common Purchaser Bill was passed, requiring any purchaser of oil in Texas affiliated with a common-carrier pipeline to purchase oil ratably without discrimination between fields or producers, including the purchasing company's own production, and the Railroad Commission was given jurisdiction over rates and regulations. With Texas output exceeding the Economic Committee's allocation by 13 percent, Humble announced that beginning July 1 it would discontinue purchasing oil in seven North Texas counties. After representatives of various oil and gas associations conferred with the Railroad Commissioners Humble rescinded its decision. It seems that Farish wanted to push the Commission into applying the Common Purchaser Act, which would require some sort of state prorationing.[50]

A meeting of oilmen was held in late July under the auspices of the Railroad Commission. Led by Robert Penn, a committee of Texas producers presented a plan to reduce production by 125,000 bpd for ninety days beginning August 1. While most oilmen went along with the proposals there were a few exceptions, as some producers argued that the Commission had no authority to stop economic waste, only physical waste. Chairman Pat Neff said that he wanted to have the Commission's authority tested under the pipeline law and wished to proceed with proration. On August 14, the Railroad Commission ordered a 90 day period of curtailed oil production in Texas, effective August 27, to prevent waste of natural resources. The Commission based its order on the premise that storage of crude oil above ground led to excessive waste, and that to decrease supply to market demand would eliminate that waste. After hearings the Commission modified its order, increasing permitted output to 764,000 bpd, defined marginal wells and exempted them from any curtailment order.[51]

The decision to justify prorationing on the grounds of avoiding waste from storage was an invitation for a lawsuit. If this was truly the Commission's motivation, it could have banned earthen storage, required adequate steel tankage to prevent evaporation, and insisted on precautions against fire hazards. Oil companies were aware of these storage problems and were improving their facilities in order to reduce evaporation losses. As far as the danger to stripper wells from discontinuing production there was little data on the effects of shutting down or abandoning stripper wells. Halting the pumping operations on an oil well did not necessarily mean that the remaining oil was lost forever, as this depends on the nature of the oil sands, which varied from well to well. The only plausible motive for statewide prorationing was price maintenance.

While Texas was implementing statewide proration Oklahoma was struggling to maintain its system. The Corporation Commission reduced allowables by 25 percent in its July 1930 order. A number of small firms had erected refineries at the edge of the Oklahoma City field, including Champlin Refining and C.C. Julian. Julian brought suit in the Supreme Court of Oklahoma in late August, alleging that the 1915 act violated the Fourteenth Amendment by depriving the plaintiff of its property without due process. The court, in a seven to two majority, sustained the statute and the authority and orders of the Commission on October 14. The next attack came in Federal court at Oklahoma City in September, 1930, by Champlin Refining in an injunction suit against the Commission. The federal district judge refused to grant a temporary restraining order, instead assembling a three-judge court which denied the application for an interlocutory order in a two to one decision on October 1.[52]

The Oklahoma proration hearing of October 10, 1930, was far more heated than previous meetings, and opponents of proration were becoming more vociferous. The Commission contended that price-fixing was not involved in decisions to limit production and that its orders were based solely upon the purpose of preventing waste of crude oil. The true extent of the Oklahoma City field was becoming known as its potential was estimated at 2,000,000 bpd. The field was presenting problems which would recur in the East Texas field; with production control and uneven access to outlets, some leases could not meet their allowances, requiring complicated rules to allow underproducing leases to "catch up." Proration was changed from a time basis to a percentage basis in December, requiring estimates of well potentials. The increasing weakness in the oil market made itself felt through a new round of price cuts as prices for most crudes dropped below the $1 per barrel level in late October.[53]

One of the myths that arose concerning the East Texas field is that "Dad" Joiner found the field after the professional oil men had written off the area as nonproductive. In 1928 a Humble geologist, G.M. Knebel, suggested the possibility of a shoreline field in Gregg county and advocated leasing in the general area. When Pure Oil discovered the Van field in 1929, additional East Texas drilling was planned, and Humble, Gulf and Herbert Oil entered an agreement for a joint test well on acreage which proved to be in the heart of the field, but title difficulties delayed drilling.[54]

Columbus Marion "Dad" Joiner had teamed up with a self-taught "geologist," A.D. Lloyd, who despite his dubious methodology seemed to have a touch for finding oil. "Doc" Lloyd and Joiner just missed discovering the Cement and Seminole fields in Oklahoma. Joiner had bought leases on 320 acres in Rusk County, and armed with Lloyd's imaginative description of East Texas geology, raised funds to drill a wildcat in August 1927 on Daisy Bradford's farm. Twice holes were abandoned and in 1929 Joiner was down to his last chance with the Daisy Bradford No. 3. When the Van field came in, forty miles away, Joiner was able to raise enough money to continue drilling, reaching oil sand on September 5. The rumor mill set off a spurt of activity in the area, which subsided when the Joiner well flowed a disappointing 300 barrels in twenty four hours. Leasing activity slowed in November as operators awaited the results of the fifteen wells initiated in October, and when the first few came in dry interest in the area declined. H.L. Hunt had concluded that the main field lay to the west of the Joiner well, and hustled Joiner into selling his leases using information from a spy at the Deep Rock well, a mile west of the Joiner well, which confirmed his theory, flowing at 3,000 bpd on November 28. A new cycle of drilling exploded across Rusk County with nearly 100 wells under the drill by late December.[55]

Ten miles north of the Joiner well near Kilgore, another oilman-promoter, Ed Bateman, drilled a "po-boy" well, and the Bateman well reached oil sand on December 14. Two weeks later the well came in at an estimated 15,000 bpd. Nineteen miles north of the Bateman well two local businessmen gathered 9,300 acres of leases and brought in John Farrell and oilman W.A. Moncrief to run the operation. On January 23 a core sample was taken showing oil saturated sand and when the well was tested it came in with a potential flow of 17,000 bpd. The Bateman and Farrell wells raised the question of whether there were three separate fields or one large field. Either way it was obvious by the end of January that a huge field had been discovered. Humble geologists correctly concluded that the three wells defined a single, large field, and Humble purchased more lease acreage, with the result that ultimately it was the largest owner in the East Texas field, with an estimated 16 per cent of the acreage of the entire field.[56]

By early February 1931 a full scale oil boom was underway, with camps, loading racks, pipelines and small refineries under construction in all parts of the field. Contract prices declined to 33-40¢/barrel by mid-February. The relatively low cost of drilling, and the speed of drilling, 15 to 30 days, allowed entry by relatively small operators. Farmers stopped leasing their land in blocks and began to subdivide it into small parcels once they became aware how much the small operators would pay for a few acres. Financing wells became an art in and of itself; as banks were too cautious to make oil loans, operators would borrow from friends and pay back double or triple in a few weeks when a well came in. Oil companies soon initiated a variation on these short term loans by lending money against production, often at an effective rate of 10¢/barrel.[57]

While the new field meant prosperity in East Texas, it would add to the misery of the rest of the state. Even before the East Texas field began producing oil the Railroad Commission was decreasing the allowables for other Texas oil fields. In November 1930 the Commission fixed the state allowable at 680,000 bpd, reduced to 644,000 bpd in its January 25 order, after Humble announced price cuts averaging 17-22 cents per barrel. Robert Penn, Chairman of the Texas Central Proration Committee, called for East Texas to cooperate with prorationing and work out a rational development program for the new field. East Texans, led by Carl Estes, editor of the *Tyler Courier-Times*, opposed prorationing for Rusk and Gregg counties. After some wrangling, a temporary truce was called, and Humble agreed to provide pipeline facilities to the new field if operators would agree on a proration plan.[58]

The two major groups opposing prorationing were the Independent Oil Men's Association of Texas (Claude Wild, vice-president), and Tom Cranfill's Independent Producer's Association of Texas, which merged into the Independent Petroleum Association of Texas (IPAT) in July, 1931. In addition, there were numerous local producer and royalty owners groups which fought prorationing. The general consensus was that imports were the cause of lower oil prices, Governor Sterling was an ally of the major companies (as a founder of Humble he was distrusted by the independents), and Harry Sinclair was a hero for purchasing their oil. Ironically, one of the main complaints of the small operators was that the lack of storage and pipeline outlets forced them to sell at distress prices to major companies, which would have been prevented by prorationing.[59]

The deluge of oil pouring out of East Texas drove down prices throughout the Mid-Continent area. "Alfalfa" Bill Murray in late February 1931 called a meeting of nine governors from oil-producing states in Fort Worth, Texas. The Governors and their representatives organized an Oil States Advisory Committee (OSAC) to coordinate the conservation and production curtailment activities of the states represented. They agreed to make voluntary forecasts of production allowables for each state and entertained the idea of an interstate oil compact. Telegrams were then sent to President Hoover as well as the oil importing companies, urging an agreement to restrict imports. Secretary of Commerce Robert Lamont, an opponent of the oil tariff, helped arrange voluntary import restraints. In April Teagle announced that Jersey Standard would cut oil imports by 25 percent, Gulf and Indiana Standard agreed to similar reductions, and Shell offered to reduce imports by 50 percent. With

Lamont taking the lead, the major oil companies could proceed without fear of antitrust prosecution.[60]

The OSAC recommended in April that production be held constant in the member states for the next six months. In September when new recommended allowables were issued for the various states, the only state given an increase was Texas, from 714,000 to 800,000 bpd in recognition of the tremendous potential of East Texas. Further negotiations resulting in setting Texas oil output at 902,000 bpd for the rest of the year on September 21.[61] In effect the other oil states permitted Texas output to grow at their expense in return for the Railroad Commission's cooperation in maintaining oil prices.

National pressure, the continued growth of East Texas production and drastic price discounting in the field pushed both the Commission and the Governor into action. Governor Sterling announced that he would recommend to the legislature the creation of a new conservation commission to supervise oil and gas production and conservation. President Hoover noted that the oil industry could not be stabilized until the East Texas district was prorated. The Railroad Commission announced plans for a hearing on proration in the new field to be held in Austin on March 24. Commissioners C.V. Terrell and Lon Smith issued a statement answering the criticisms of Sterling and Hoover, attacking proration as monopolistic.[62]

The hearing on East Texas prorationing was a raucous affair, threatening to end in fisticuffs. Testimony at the hearing made it clear that most Texas oilmen were in favor of proration. Low prices threatened to shutdown high cost producers in North and West Texas. Oilmen in flush fields defended proration on the grounds that it delayed the use of pumps, which added 30¢ per barrel to their lifting costs. Even the East Texans were divided over the issue. On April 2 the Commission issued its proration order, limiting the East Texas field to 90,000 bpd starting April 10, with graduated increases to 130,000 bpd by June 15. While squabbling continued over proration in East Texas, field developments were moving many operators toward a qualified acceptance of control. Production rose to 190,000 bpd by the second week of April and prices steadily plummeted, falling as low as 15 cents a barrel with some wells accepting 10 cents. By April 21 an agreement had been reached among the East Texas oil men. The Railroad Commission issued a new special order on April 22, increasing output to 130,000 bpd, rising to 160,000 bpd by July.[63]

While proration in East Texas gradually collapsed, the field was continually growing. By the end of April production had reached 340,000 bpd, declining temporarily before reaching 360,000 bpd by late June. In addition to oil shipped out of the field, by the middle of June six independent "skimmer" refineries were completed. These plants produced a low grade gasoline, shipped to major markets by railroad car, sold by the railroad tracks, driving down the price of the better grade gasolines. The combination of increased oil output and lower gasoline prices forced oil prices down, first in East Texas and then nationwide. On May 26, 1931, four major companies lowered their price for East Texas oil to 35¢ per barrel. A week later Humble ceased to post a price for

East Texas crude, but continued to pay 15¢ a barrel although some crude was selling for as little as 6 cents and spot sales were reported as low as 2 cents. On June 1 Stanolind (Indiana Standard) cut its price for 36° oil to 33 cents in Kansas, Oklahoma and Texas and price cuts soon spread across the country.[64]

Although many East Texas operators had opposed legislative attempts to stiffen regulation, there was a growing consensus that real proration was both inevitable and desirable. Pressure from the rest of the state was building, and it became obvious that control would be imposed, with or without the cooperation of the East Texas oil men. One sign of this conversion was the change in attitude by Carl Estes, one of the most vociferous opponents of prorationing in East Texas. He appealed to Sterling to appoint a committee of East Texas citizens to write a valid conservation order and pledged the support of the East Texas Lease, Royalty and Producers Association. A plan proposed by Tom Cranfill was adopted on June 15, dividing the field into 20 acre units, with each unit permitted to produce 300 barrels per day, initially setting production at 210,000 bpd. The plan was to take effect on June 20, but the Railroad Commission refused to issue an immediate order giving the plan its official sanction, guaranteeing its failure. As it became obvious that the Cranfill plan had collapsed the major oil companies moved to erect storage and buy distress oil at prices as low as 10-12¢/barrel with rumors of 8¢ oil.[65]

The Commission finally issued a new East Texas proration order on July 2, based on the Cranfill plan but two weeks late. Exasperated with the struggle between the East Texas mavericks and the Railroad Commission, Governor Sterling called for the consolidation of state departments and bureaus into a conservation commission. As pressure for a special session grew East Texas producers organized for the political battle. The decision by Humble to reduce prices statewide by an average of 15¢ on July 7 finally pushed Sterling into calling a special session on July 8.[66]

While the legislature conducted hearings on the oil control issue, the Federal District Court in Austin handed down its ruling in the MacMillan case on July 24, holding the Railroad Commission's April proration order to be invalid. The opinion was written by Circuit Judge Joseph Hutcheson of New Orleans, and concurred with by Judge Duval West of San Antonio and Judge Randolph Bryant of Paris. The Court found that the purpose of proration was to reduce output to market demand. The Court conceded that there was evidence that a plan of ratable and moderate withdrawals from each well could prolong the life of the field, and thus enable more oil to be ultimately obtained. However, "in the light of present knowledge this was theory and speculation," requiring careful testing before application. The Railroad Commission, by setting an allowable on an arbitrary basis without test or experimentation, made it obvious that its purpose was to deal with economic waste, in the form of lower prices. Since Texas statutes exempt economic waste from the authority of the Commission, the Commission's order was contrary to that legislative policy.[67]

The special session rejected the Governor's proposed resources commission, and passed a bill excluding authority to control economic waste or implement market demand regulation, but including as waste inefficient use of gas or water drives, and inequitable withdrawal from a common pool.[68] With the MacMillan decision operators had concluded that proration no longer applied to the field and a free-for-all ensued. The two week delay in passing a conservation measure provided the time for chaos to rule in East Texas. Production steadily increased while prices plunged to new lows.

Governor Murray, responding to the drop in oil prices, moved to shut down the flush Oklahoma fields. Murray declared martial law and sent the troops into the oil fields to enforce his shutdown order, with the objective of forcing the price of oil up to a dollar a barrel. Murray blamed Harry Sinclair for the crisis, stating that the oilman had intimidated and possibly bribed state legislators into repealing the oil laws. Martial law would continue in the Oklahoma City field until October 10, when the Governor released the field for operation under the Corporation Commission's orders. Murray asked for a similar response from Texas but Sterling replied that Texas was not in a position to follow his lead in shutting down flush oil wells.[69]

Murray's decision may have influenced the prompt decision by a three judge federal panel, delivered the next day, upholding the Oklahoma prorationing statutes. Judge Cotteral cited past decisions that gave the state the right to protect producers from a common pool and conserve the resources of the state as a valid police power. The finding that there was waste, and that prices continued to decline, seemed to refute the allegation of price fixing by the Commission. Judge F.E. Kennamer dissented, rejecting section 2 of the Corporate Commission Act because it was a price fixing statute. He claimed that "economic waste" and "reasonable market demand" had no definite meaning, even to economists. While conceding the police power to prevent waste of a resource, Kennamer suggested that this should be consistent with production at minimum costs, on a competitive basis, to provide the consuming public with the advantage of reduced costs.[70] Until the United States Supreme Court delineated the limits of state authority over oil production, producers had an incentive to shop for state and federal judges willing to issue favorable rulings.

Pressure began to mount on Sterling to act as production increased in East Texas, reaching 648,000 bpd in the first week of August, driving the average price in East Texas down to 14 cents a barrel. On Friday, August 14, production in the field may have exceeded a million barrels a day. That afternoon, 1,500 oil men met at Tyler to protest the "stealing" of oil "by certain violators of the shutdown agreement" and a resolution was adopted asking Sterling to impose martial law. Sterling was influenced by the resolution, a telegram from a sheriff, and a report from a special observer, journalist Ray Dudley, a personal friend, suggesting that violence was a real possibility in the field. National Guard troops were mobilized the next day and sent into the field under the command of General Jacob Wolters as Sterling declared martial law on August 17. The order was written by Wolters who also doubled as a counsel to the Texas Company. The troops shut down production from the 1,600 wells in the East Texas field, and by Friday, August 21, posted prices rose to 70 cents per barrel across the Mid-Continent region.[71]

The Commission moved with unusual speed and had a ruling on East Texas before it, ready to sign and promulgate, on Monday afternoon, August 31. The new ruling provided for proration of oil production in East Texas on a straight per well basis, allowing every well to produce 225 bpd, for a field total of 405,000 bpd, given the 1,800 wells now completed. Well spacing would be 10 acres, but exceptions would be granted upon application. The field was reopened on September 5 under the Commission's order, though martial law continued in effect. As the number of wells grew it was necessary to constantly decrease the allowable per well to keep total production constant. By December, 1931, the number of wells exceeded 3,000 and the allowable per well was down to 100 bpd. By mid-December there were new warnings of weakness in the oil market.[72]

A suit by Constantin and Wrathers challenging martial law was heard in Federal Court in Houston in front of a three judge panel (including Judges Bryant and Hutcheson from the MacMillan case, and Judge William Grubb) on January 4, 1932. The outcome of the suit was a foregone conclusion as two of the three judges described martial law in East Texas as a "subterfuge and pretext" to legally cover restriction of production. Judge Hutcheson left the status quo in effect pending a final decision, but all parties expected martial law to be ended. The Court, ruling on February 18, found that at no time was there any actual uprising in the territory, and that without martial law there would have been no state of insurrection, but merely breaches of the peace to be suppressed by the militia as a civil force.[73]

On February 25, the Railroad Commission resumed control of the East Texas oil field, setting the allowable at 75 barrels per well per day. The Railroad Commission's authority was strengthened by the ruling in the Danciger case by the Third Court of Civil Appeals in Austin on March 23. Associate Justice J.H. Baugh delivered the opinion, explicitly differing with the verdict of the Federal Court in the MacMillan case in deciding that underground waste was not mere "theory and speculation." He concluded that the Commission was authorized to curtail production below the potential of the field and to prorate production to make withdrawals uniform throughout the field. The Justice reasoned that the Commission could take into account economic conditions if they result in physical waste, and unequal withdrawals causing underground waste could justify limiting production to market demand.[74] It seemed that East Texas, and therefore the national oil market, would finally be stabilized.

Additional confidence in the ability of the state to enforce prorationing came with the Supreme Court's ruling in the Champlin case on May 16. The Court found that there was a serious potential overproduction throughout the United States and that if no curtailment were applied, crude oil would inevitably go into earthen storage and be wasted. It also ruled that none of the Corporate Commission's orders were made for the purpose of fixing the price of crude oil, primarily because the price continued to decline. The Court noted that if the plaintiff were to engage in uncontrolled production there would inevitably result great physical waste due to the decrease in ultimate recovery as well as surface waste. The Court affirmed the power of the State to prevent unnecessary loss, destruction or waste.[75]

Meanwhile the number of wells in East Texas grew to 6,500 by July 1, encouraged by the Railroad Commission's willingness to grant exemptions for small lots, resulting in additional wells drilled and also additional offset wells drilled on neighboring tracts, as operators had a legal obligation to drill offsets. As output per well declined, the payback periods for new wells grew longer, squeezing the small producers and increasing tension between these producers and the larger companies who wanted orderly development. While many companies attempted to challenge the Railroad Commission's authority to curtail production in the courts, they also resorted to more subtle techniques of circumventing state control. The term "hot oil" was used to describe oil produced in excess of allowables. According to Clark and Halbouty, when Sergeant W.D. Johnson, working undercover, complained that a sudden rain had made the night chilly, an operator, suspected of violations, told him to lean against a tank, saying "It's hot enough to keep you warm." Sergeant Johnson used the term in his investigative report and it became part of the common oilfield vocabulary.[76]

Initially, hot oil was usually run with the knowledge and acquiescence of the landowner, who received his royalty on the excess production. As the Commission's field men became more familiar with the methods used by hot oil runners, the amateurs were replaced by specialists. As the cost of hiring specialists increased, operators began neglecting to inform landowners of their activities, and landowners became increasingly hostile to hot oil runners. A few hot oil runners began independent operations, tapping producers' lines, draining off the oil and selling to the sources that had purchased their clients' oil. The major companies were the primary targets of these activities. Some of the small refiners became hot oil runners to obtain a source of cheap crude. Running hot oil was a generally acceptable practice in East Texas, to quote a judge in 1955: "so-called hot oil runners were not thieves, stealing oil, but they were just plain ordinary, hard working honest Americans, who believed that they had the right to produce oil from their own property, which they did."[77] However, stealing oil was frowned up, even from major companies, since it cost the landowner his royalty payments.

Overt violations of proration orders was initially limited to a few operators. Of the 691 operators in the field, 33 violated proration orders between September 1931 and February 1932. Despite the accusations of many independents, the major companies were found to produce above allowables only when faced with an offset well producing excess oil. In order to counteract rumors that the major company pipelines were carrying hot oil, the API established a committee to audit the records of the major purchasing companies, including J.R. Parten, a noted opponent of prorationing in East Texas. The Railroad Commission gradually tightened the pressure on hot oil, filing penalty suits to enforce its regulations, but by September 1932 the progress made on constraining hot oil production seemed to be reversed. Lease and royalty owners held meetings to demand immediate action. Humble Oil responded to increased output and declining gasoline prices by announcing a 50 percent reduction in its purchases of oil. Colonel Ernest O. Thompson, who had replaced Pat Neff in June, announced plans for a vigorous campaign to prosecute hot oil runners. Hearings in the State legislature on oil theft resulted in new laws to provide the Commission with additional powers to enforce its orders. At a conference in Tyler in September, the new Commissioner pledged vigorous cooperation of the commission in preventing waste.[78] Despite some price weakness in the oil market it seemed in the fall of 1932 that the crisis in East Texas was over.

In September, cases of seven different independent oil companies and operators were consolidated for the hearing before a three judge federal panel, composed of Judges Hutcheson, Bryant and Grubb, in order to clear the slate of East Texas oil field litigation. The Court handed down its decision in the Peoples case on October 24, ruling that the Commission had acted to control the supply of oil and maintain prices by maintaining production within the amount fixed by agreement of the Oil States Advisory Committee. The Court upheld the validity of the conservation statutes, but rejected the Commission's orders. It found that the Commission had never made experiments, tests, or inquiries to determine the maximum amount which each producer could withdraw from his wells without damage to the field.[79]

The Court left fifteen days before the proration of the Railroad Commission would be voided. Governor Sterling resisted pressure for a special session but soon changed his mind and called a special session, to start on Thursday, November 3. Sterling urged the legislature to enact laws to permit the Railroad Commission to consider market demand in his message to the assembly. On November 10 the oil conservation bill passed both Houses, but mustered only 84 votes in the House, short of the necessary margin for immediate implementation. In conference the Senate bill was adopted, striking out the "economic waste" provision of the House bill, and limiting duration to September 1, 1935. The final bill passed the House by a vote of 108-20, which allowed it to go into effect immediately. Following passage of the bill the Railroad Commission called for a hearing on East Texas for November 25.[80]

At the hearing the Commission decided to adopt a plan basing two-third of the allowable on a well basis and one-third on a formula taking into account acreage and bottom hole pressure. The Railroad Commission's new convoluted formula did not win universal approval. The IPAT went on record in opposition to proration methods involving an acreage basis, arguing that this would lead to compulsory unitization. Ernest Thompson, in a paper to the IPAT annual meeting, condemned the proposed interstate compact and Federal Regulation, making it clear that while he might disagree with some IPAT positions he was in general agreement with their position on unitization, claiming that "unitization means the extermination of the little man."[81] Thompson was carefully building alliances with both independent camps as he stumbled toward a solution which would be acceptable politically as well as by the federal court.

On December 17 Thompson ordered the field shut down to take measurements, and the Commission's engineers concluded that the field could produce 450,000 bpd without endangering the reservoir mechanism. However, the Commission decided to limit output to 290,000 bpd from January through March, 1933, to keep the state allowable under 760,000 bpd, and wells in East Texas were restricted to between 28 and 35 bpd. At the meeting in Tyler on January 12, 98 operators and companies were cited as violators, out of a total of 750 operating in the field. Detailed examination of the Atlantic Pipe Line records revealed a discrepancy of more than 1 million barrels between September 1931 and January 1933. Thompson announced that meters would be required to be placed at every direct outlet for oil to market. In response to the announcement of a price cut to 50 cents a barrel on January 18, Attorney General James Allred blamed withdrawals of oil from storage by major companies. Proration supporters claimed price cuts were due to "distress" oil and "cut price" gasoline, and that enforcement of proration in East Texas was lax.[82]

A legislative committee appointed to investigate proration enforcement also decided to hold hearings in Tyler. Luther Swift, in charge of proration enforcement from October 1931 to October 1932, estimated that hot oil was produced at a rate of 60,000 to 100,000 bpd during this period. Most of the refineries and loading racks lacked meters and some pipelines had refused to comply with Commission orders requiring installation of meters. Commission employees said they needed more men in the field, as well as tougher laws to enforce proration. The growing number of operators producing under injunctions was increasing unrest in the field, as hot oil was selling for 15-20 cents per barrel. A bill was introduced in the House to establish a separate natural resource commission to administer oil and gas regulation in Texas. Thompson responded to the threat to his authority by returning to East Texas and tightening enforcement.[83]

By March 1933 the situation was beginning to get out of hand. The number of wells in the field was approaching 10,000, with 25 local refineries. The railroad companies had obtained a temporary injunction to prevent the Commission from interfering with shipping oil by tank cars. There were 7½ million barrels of steel storage waiting to be filled in the event proration collapsed, and the potential to ship 1½ million barrels a day through pipelines and railcars. Hot oil was estimated at 125,000 to 150,000 bpd. Locals stepped up their dynamite campaign, hitting violators of proration orders and major company pipelines. On March 16, Humble dropped its price to 25¢ per barrel. The final blow came with the March 20 ruling of a three-judge Federal Court in

Tyler striking down the Commission's orders, invalidated because the method for allocating production between wells was considered improper. It was reported that oil was selling in the field for as low as 13¢ a barrel. The Commission announced a hearing on April 3 to develop an order which would definitely be legal.[84]

The Commission issued an order to close the field for an indefinite period, and Judge Bryant permitted the RTC to shut down the ET field to April 24 to give the Commission time to conduct tests of potentials, analyze its findings and issue a new order. Starting with 40 bpd as the minimum per well output, due to the legislative requirements of the Marginal Well Act, the output per well was set at 15% of the potential of test wells in that region of the field. The total output of the field was expected to be in the range of 750- 775,000 bpd. Although numerous petroleum engineers had suggested an optimum production rate of 325,000 to 450,000 bpd, the Commission decided on the formula to avoid conflict with Judge Bryant. It was expected that pressure would soon drop in the field, causing many wells to produce under the pump instead of by natural flow.[85]

When the field reopened on Monday, April 24, prices soon tumbled to 10¢ a barrel. The Commission charged that the drop in price was an attempt to extort the legislature into establishing a new Oil and Gas Commission. Production on May 1 reached 1.1 million barrels, and even with some wells shut in, average output remained at around 900,000 bpd in early May, overwhelming the capacity of oil companies to absorb this volume. Humble moved to cut prices throughout Texas and New Mexico and the Dallas Morning News and Fort Worth Star-Telegram blamed the Commission for 10 cent oil.[86]

There had been mutterings about reviving Sterling's proposal for an appointive natural resource commission, but dissatisfaction with the performance of the Railroad Commission engendered concrete action. Governor Ma Ferguson in her address to the House of Representatives on Monday, April 3, evoked the specter of interference from Washington. The address was careful not to accuse the Commission of malfeasance, but hinted at a situation that was out of control. A bill to establish an appointive commission whose sole duty would be to regulate the production of oil and gas moved through the House. The bill was headed for passage in the Senate when an ardent supporter of the IPAT was beaten up in the lobby of an Austin hotel by proponents of the new commission, and the resulting outrage killed the bill. The marginal well bill, which lowered the marginal well allowance for East Texas wells to 20 bpd, passed the bill by a two-thirds majority on April 22. By the time the bill was signed by the Governor on the twenty-seventh the Railroad Commission had already issued its new order, ignoring the new lower limit for marginal wells.[87]

With the legislative threat ended, the Railroad Commission called for a new hearing on the East Texas field for May 15, to determine whether the current order was conducive to physical waste. The same day saw an increase in the East Texas price to 25¢, while price cuts spread throughout the mid continent region, reflecting the displacement of other oil by that from East Texas. Purchasing companies began filling all available storage to take advantage of the low priced crude, and numerous storage tanks were under construction. Pressure to lower the allowable increased as field pressure decreased. There was growing awareness that many wells would soon go on the pump at an average cost of $5,000.[88]

The Commission hearing on May 15 revolved around contentions of damage to the field from the new proration orders. The Commission decided to continue its wide-open policy, bolstered by a decision of the Fifth Court of Civil Appeals in Dallas which held that the proration orders were constitutional. By early June there was an average decline in potential flow of 11% since the field was opened in April, and the allowable had risen to 854,000 bpd. At a new hearing on June 12, the Commission began consideration of a cut in the allowable. Gordon Griffin, a Commission engineer, testified that a majority of wells would go on the pump within sixty days. The Commission announced a reduction in the allowable to 573,000 bpd on June 13, triggering an increase in the East Texas oil price to 50 cents per barrel.[89]

THE NEW DEAL AND PRORATION

Chaos in East Texas encouraged a search toward a general solution for the perceived problem of overproduction. The Legal Sub-Committee of the OSAC drafted a proposed federal bill, which was approved by the API and actively pressed in Congress in March, 1932 as the Thomas-McKeon Bill. It proposed creation of an interstate oil compact for the conservation of petroleum and also provided for establishment of an Interstate Oil Conservation Board to cooperate with the FOCB to recommend production quotas, approve private unitization agreements, and to promote uniform state conservation laws. After hearings on the measure, the House Judiciary Committee reported it favorably. But since the bill threatened to interfere with the impending passage of an oil tariff, the Oil States Advisory Committee reluctantly agreed to have it withdrawn. In December 1932, the committee made arrangements for informal discussions with the new President-elect to acquaint him with the seriousness of the oil problem and to solicit his support for the Thomas-McKeon bill. Roosevelt promised to give careful consideration to their requests, and to take some form of action. Soon thereafter, the committe went out of existence and eagerly awaited the action of the new Administration.[90]

The national market had been gradually weakening during the tail end of 1932, even before the floodgates were opened in East Texas. A twelve million barrel decline in oil imports was negated by a twenty-one million barrel decline in oil exports. The continued decline in oil held in storage, from the high of 431 million barrels in March 1930 to 340 million barrels by the end of 1932, crowded out new production. This was a bone of contention with the independents, who felt that their production was being restricted to enable the majors to dump their excess oil on the market. The large companies in turn, felt no need to apologize for liquidating unneeded stocks at a loss. The trend toward lower stocks was influenced by two other factors; the increased output of gasoline from a barrel of crude and the growing integration of the petroleum transportation network. Stocks would decline through WWII and stabilize around 230 million barrels in the late 1940s.[91]

Standolind began the new year by announcing a 25 cent price cut in the Mid-Continent region, blaming sales at reduced prices, the growing quantity of bootleg crude, and the production of cheap gasoline. Frustration with the current state of affairs had some oilmen suggesting that proration be abandoned and flush fields allowed to produce wide open in the hope that they would soon decline, ridding the industry of the unruly elements which persisted in violating proration statutes. One reason for this disgust was the revelation of widespread corruption in the Oklahoma City Oil Fields, with allegations that Colonel Cicero Murray, head of military operations in the field, was soliciting bribes. Supposedly, protection for "hot oil" production could be purchased for 10 cents a barrel.[92]

While the federal government was in a state of paralysis awaiting the new administration, oilmen had high hopes of influence with the new regime. Two oilmen prominent among the contributors to the Democratic Party's campaign of 1932, James Moffett and Michael Benedum, aspired to enjoy the same kind of advisory capacity in the Roosevelt administration that Mark Requa exercised under Coolidge and Hoover. "Jimmy" Moffett had been Walter Teagle's fair-haired boy at Jersey, joining the Board in 1919, becoming vice-president in 1924, but his career accession stalled as another contender to the throne, Bill Farish, emerged. Benedum was an oil millionaire and a back room supporter of Roosevelt at the 1932 Democratic convention, as well as a friend of the Mellons and Teagle. Both oilmen believed in the use of federal authority to aid the depressed petroleum industry.[93]

On March 7, 1933, the Executive Committee of the API appointed James Moffet to represent their views to Roosevelt. Moffett arranged a meeting with the President, who was urged to listen carefully to the API's envoy: "You may recall that Mr. Moffett is a Democrat and was very active in the campaign helping us both financially and in other ways." Gaining an audience on March 13, Moffett told the President that the oil industry with nearly fourteen billion dollars invested was in disorder and only prompt federal action would save it from collapse. The states were unable to control the minority of oil men who were refusing to cooperate with production control plans, and hot oil was driving down the price of crude petroleum and gasoline. Moffett concluded by requesting consideration of a federal program to reduce output. At the next day's Cabinet meeting, Roosevelt instructed Interior Secretary Harold Ickes to invite representatives of the governors of Oklahoma, Texas and California to meet in ten days to confer on curtailment of oil production. Ickes also asked Moffett to nominate industry spokesmen for the meeting.[94]

Wirt Franklin, as head of the IPAA, called a parallel conference of independents, which met in the Mayflower Hotel on Sunday, March 26. The API plan for production control was presented, inciting a row culminated by the walkout of the Independent Petroleum Association of California, led by John Elliott, protesting any agreement to curtail production. At the formal meeting the next day the resolution carried by a vote of 13-1, with the IPAT under J.R. Parten the only association to oppose the plan. The IPAA resolution called for the Federal Government to prevent the transportation in interstate commerce of oil or products produced in violation of state laws, continue the FOCB as an advisory body, impose a competitive tariff on oil imports, and to cooperate with the states in allocating production. The states were requested to enact adequate laws, issue valid conservation orders, equitably allocate production between pools and limit production to market demand. A goal of 2 million bpd was set as the national limit, with Texas assigned 800,000 bpd.[95]

The three groups, independents, major companies, and governors' representatives, meet separately and appointed committees of five to confer with Ickes. After conferring with Secretary Ickes, Governor Landon presented their unanimous recommendations including: that the President call upon the governors of California, Kansas, Texas, Oklahoma, and New Mexico to close all flush pools until April 15; recommend to Congress passage of a statute prohibiting the interstate transportation of hot oil; and submit to Congress a request for emergency legislation. Other provisions restricted imports to the level of the second half of 1932, asked states to limit production to market demand, and called for a 2 million bpd production limit.[96]

The Parten-Elliott group opposed these recommendations and formed the Independent Petroleum Association Opposed to Monopoly (IPA-OM). Elliott met with Ickes on March 28, asked that nothing be done to curtail oil production, and proposed divestment of the major firms. Elliott stressed his contacts with the administration due to his management of Roosevelt's California campaign. Elliott managed to influence Ickes to the extent of raising doubts concerning the independence of the IPAA. The IPA-OM group used Elliott's connections to meet with President Roosevelt and present a resolution from the Texas House condemning federal controls. Roosevelt, in his telegrams to the governors of oil states sent April 3, disavowed a moratorium on production, supported a bill to curtail interstate hot oil shipments, encouraged the Governors to consult and take action, and approved the IPA-OM call for pipeline divorcement.[97]

The California independents had managed to stalemate the vast majority of the industry. The IPA-OM represented the interest of small refiners in the Southwest and Southern California who depended on cheap crude oil to compete with the large refineries, along with some independent well owners whose oil properties were advantageously located in the high-production fields, East Texas, Southern California and Oklahoma City. These groups wanted a program of antitrust reform, contending that the ills of the industry stemmed from "monopolistic control, agreements in restraint of trade, price-fixing agreements, burdensome taxation, false propaganda and excessive government regulation." While the arguments against monopoly had a few supporters in the initial phase of the New Deal, the primary reason for the kid glove treatment of a self-interest minority was Elliott's California connections. Elliott was a close friend of Senator McAdoo and on friendly terms with the state's other Senator, Hiram Johnson, and the new administration could ill afford to alienate potential supporters of New Deal.[98]

Roosevelt was also indebted to James Garner and Sam Rayburn for orchestrating his nomination in 1932. Rayburn was chosen by Texan partisans to head the Garner for President Campaign. While Rayburn felt that Roosevelt was the best candidate for the party, he saw Garner as the viable alternative if a stop Roosevelt movement were to succeed. After the convention had stalled on the third ballot, with Roosevelt a hundred votes short of nomination, Rayburn and Garner agreed to release Garner's delegates. Rayburn ran into McAdoo, told him of their decision, and McAdoo arranged to release the California delegates. Rayburn overcame the opposition of Texans who wanted to deadlock the convention to nominate Garner and managed to sway the Texas caucus, 54-51, and Roosevelt's nomination was assured.[99]

In early May the Interior Department drafted a measure to vest the Secretary of the Interior with the power to assign production quotas to states, set prices to protect stripper wells and prohibit hot oil shipments. The administration bill was introduced by Marland and Capper on May 19, along with a letter from the President calling for federal legislation, suggesting that action relating to the oil industry be incorporated with other industrial legislation, an implied request to submit the oil bill as an amendment to the National Industry Recovery bill. Opposition arose in the House Ways and Means Committee, and the President informed Chairman Doughton that the Committee could proceed upon the Industrial Recovery bill without action on the oil proposal. The bill also ran into a roadblock in the Senate Finance Committee, where the Marland-Capper bill was defeated 17-1 due to McAdoo's opposition. Senator Connally of Texas, another bitter opponent of the Marland-Capper bill, introduced an amendment to prohibit the shipment of hot oil which was accepted. The two Senators pressured Ickes to accept only the hot oil provision and to desist from new attempts to add production control authority to the bill. The final bill included three provisions in Section 9, (a) authorizing the president to initiate proceedings to fix reasonable rates by pipelines before the ICC; (b) authority to institute divorcement proceedings; and (c) the Connally Hot Oil provisions.[100]

Roosevelt, preoccupied with banking, agriculture, and relief, was jarred into action by the unexpected passage in the Senate of the Black thirty-hour-week bill on April 6, 1933. Combined with a Senate drive for public works, the administration faced the prospect of independent Congressional action if it continued its procrastination. Roosevelt now told Raymond Moley to get in touch with the people working on business-government cooperation plans around town. It seems evident that two groups were especially important. One centered in Senator Wagner's office on Capitol Hill, the other in the office of John Dickinson, the Undersecretary of Commerce. Business leaders rallied behind Dickinson, hoping to defeat federal regulation of wages and hours and to use the issue to secure "industrial self-regulation." The National Association of Manufacturers (NAM), appears to have taken the lead in defeating the Black bill, fearful that it would become the basis for regulation of all aspects of business operations.[101]

In the meantime Roosevelt had asked Senator Wagner to prepare a bill, and the Senator assembled a group of bankers, industrialists, economists, and lawyers. Dickinson had drafted a measure that would make use of trade associations as instruments of national planning. From the beginning he established contact with the Wagner group, and the two groups worked on a common draft, combining a program of public works and government loans with a plan for industrial self-government through trade associations. Moley, running into Hugh Johnson late in April, begged him to get into the picture: "Nobody can do it better than you. You're familiar with the only comparable thing that's ever been done—the work of the War Industries Board." By early May there were two main drafts—a Johnson draft, based on a tough government licensing system, and a Dickinson draft, looking to industrial self-government through trade associations. At a White House meeting Johnson and Dickinson wrangled over the respective drafts, and Roosevelt told the two groups to work out an agreement. Wagner, Dickinson, Perkins, Johnson and Jerome Frank fought it out before the drafting job ended in Johnson's hands. Wagner yielded to Johnson's insistence that Congress delegate legislative powers relating to resource allocation, production, wage and hour scales, and relations between labor and management. Johnson insisted that codes of fair competition should apply to all industries involved in interstate commerce, not just to a limited number like steel and automobiles.[102]

In his second fireside chat, on May 7, Roosevelt talked about "a partnership in planning" between government and business, with government having the right to prevent unfair practices. The House passed the bill with few changes in just over a week. In the Senate, however, the bill in general and the collective bargaining provisions in particular had a stormier time. The NAM and the Chamber of Commerce assailed the labor provisions, but the more important part of the Senate discussion turned on the significance of the suspension of the antitrust laws. William Borah of Idaho led the attack on the bill from the viewpoint of an old-fashioned trust buster, contending that suspension of the antitrust acts would promote the concentration of wealth and power. Wagner replied that the urgent need was to outlaw sweatshops, long hours, and low wages, and that this could only be done by allowing business cooperation, as a direct federal attack on wages and hours was presumed unconstitutional. The question of price-fixing rapidly assumed critical importance. Wagner stated that it would not be a price fixing measure, but exceptions including banning sales below cost, and permitting cooperation might be necessary to prevent harmful price fluctuations.[103]

Brand suggests that the major oil companies supported the NIRA as an alternative to the Marland-Capper Bill. This was more a matter of choosing the lessor of two evils. The NIRA contained elements with which the large companies were uncomfortable, such as the Section 9 provisions fixing reasonable pipeline rates and providing the authority to institute divorcement proceedings. Since the President could impose a code of fair competition if the industry could not come to an agreement, NIRA placed the majors in the uncomfortable situation of having to work with the IPAA to prevent the IPA-OM from employing their influence with the chief executive. A code conference was called for June 14-17 in Chicago, and delegates from 39 associations attended, with only the IPA-OM absent. An Emergency National Committee was established, and a subcommittee assigned to publish forecasts of market demand, split equally between majors and independents. J. Edgar Pew presided over the general meeting, which after much discussion developed a proposed code to be submitted to the National Recovery Administration (NRA). The primary conflict concerned price fixing, strongly supported by Franklin, Moffet, Sinclair and Kingsbury, but opposed by Teagle and Farish. When the API board of directors approved the Chicago code on June 26 a number of companies attached reservations to the price fixing section of the code, including Gulf, Shell, Indiana, Jersey, Texas, Sun, Atlantic, Mid-Continent, Ohio Oil, and Magnolia.[104]

Oil code hearings were held in Washington from July 24 through August 2. The IPA-OM attempted to bring up its set of issues, but found a far less sympathetic audience than in Congress. They wanted pipeline divorcement, elimination of patent restrictions, free access to service stations, guaranteed access to crude supplies, a ban on oil imports and withdrawals from storage, and production quotas based on a state's production potential. Unitization was opposed on the grounds it would result in the drilling of fewer wells, decreasing oil field employment. Hugh Johnson, who was the NRA administrator, realized there was no reconciling these "radicals" with the remainder of the industry and on July 31 presented a code of fair practices drafted by his department. He replaced the Emergency National Committee with a new federal agency for the calculation of market demand. Johnson made it clear that the Federal government would not interfere with state control of oil, limiting intervention to interstate shipments. However, if a state produced more oil than its quota the excess would not go into interstate commerce. The General refused to ensure supplies for small refineries, commenting that in limiting production someone was bound to get squeezed.[105]

Although they had no desire to emasculate the code, Texas officials, lead by Ernest Thompson, objected to the provision that a government agency set the state production quotas. Thompson wrote Roosevelt on August 17 that the production control clause violated Texas sovereignty, and requested that the code designate a federal agency to 'recommend' production quotas. East Texas oilmen, along with Allred, joined Thompson in his demand that Texas retain ultimate control of its oil. They were opposed by Wirt Franklin, the major companies and the Interior Department, but the power of the Railroad Commission to thrust the market into chaos prevailed. The change was more symbolic than real as the President still had the authority to regulate shipments of oil from states. Thompson said that the Commission would consider the Texas recommendation, but accept it only if convinced that the allocation was equitable.[106]

Roosevelt had decided in early August that Ickes, not Johnson, would administer the oil code, though the formal appointment did not come until August 29. Ickes headed the Planning and Coordinating (P&C) Committee in administering the code and also supervised price fixing. Ickes and Roosevelt removed Sinclair from the committee for political reasons, and added three government representatives, Moffett, Donald Richberg and Benedum. Although the committee favored price fixing, with a goal of one dollar oil, Ickes was reluctant to immediately implement price controls. The Secretary wanted to see if supply reduction could raise prices without resorting to politically controversial measures.[107]

The Petroleum Administrative Board (PAB) was created to aid Secretary Ickes administer the Petroleum Code. Nathan Margold, one of Frankfurter's "happy hotdogs," chosen by Ickes to serve as Solicitor of the Interior Department, was made the PAB chairman. Its most important function was to cooperate with the Bureau of Mines in compiling and publishing the monthly market demand projections, continuing the work of the now defunct FOCB. The first set of allocations was set by Secretary Ickes in early September, limiting national production to about 2.4 million bpd and Texas to 975,000 bpd. This was followed by new allocations reducing output to 2.2 million bpd nationwide and 888,000 bpd for Texas in December. Oil prices rose to close to the one dollar a barrel level by early October, but Ickes continued to feel pressure to fix prices. Ickes finally conceded to the pressure, but delayed hearings until late November, before announcing that price control had been indefinitely postponed.[108]

EAST TEXAS, THE US OIL MARKET AND THE NRA

The spectre of federal control of oil production increased pressure on the Railroad Commission to demonstrate that it could control the East Texas field. While the Railroad Commission gradually reduced the allowable, hot oil production, estimated at around 90,000 bpd (compared to official Commission estimates of 5-10,000 bpd) threatened to end price increases in the Mid-Continent.[109]

Thompson realized that enforcement of prorationing in East Texas would require federal assistance. On July 10 the Railroad Commission sent a telegram to President Roosevelt requesting that he prohibit the shipment of illegally produced oil in interstate and foreign commerce. Roosevelt responded on July 12 by issuing an order banning the transport of hot oil and Ickes proceeded to promulgate new regulations three days later. Secretary Ickes engaged in a touch of hyperbole with his statement that hot oil had reached 500,000 bpd, but it was clear that the Railroad Commission had been less than candid concerning the extent of the problem. The initial impact of the new regulations was to virtually eliminate hot oil shipments.[110]

Help also came from the Texas oil industry, as a Texas Petroleum Council (TPC) was formed at a meeting of a hundred oilmen in Houston on July 10. Their efforts were given formal approval by the attendance of Thompson and Terrell. The Council was a nonfactional group, including both major oil companies and members of the IPAT. The Council offered assistance to both the Commission and the federal authorities in their efforts to control hot oil. In April 1934 the Council hired an attorney to represent them in the field, W.F. Weeks of Tyler, to work with the Attorney General's office. Since the local district and country attorneys were responsible for bringing criminal actions for violations of Commission regulations, local pressure was necessary to ensure action. The Council also sought to bring pressure on violators through boycotts and encouraging banks to withhold credit.[111]

By early August, 1933, estimates of hot oil production were down to 20,000 bpd and Secretary Ickes restricted the 'Hot Oil' ban to the Oklahoma City and East Texas fields. Ickes sent fifty federal agents into the field in September, but the initial success of the federal regulatory effort soon evaporated as the wily natives learned how to evade another foreign invasion force. The federal agents had little experience with the oil business, and a number were as easy to bribe as Commission employees. By the end of October ominous signs were beginning to appear on the horizon, and in November hot oil production was on the upswing. The decline in the East Texas allowable from 466,000 bpd in October to 400,000 bpd in November put increased pressure on enforcement efforts. January 1934 saw renewed efforts to crack down on hot oil shipments, as Ickes and Attorney General Cummings moved to expedite proceedings against violators and added twenty additional investigators. Allred established a daily court of inquiry at Longview to obtain facts with which to prosecute offenders. Fines were charged to purchasers and producers, but most were too small to have much impact, and hot oil producers saw them as a cost of doing business.[112]

There was increasing evidence that control over East Texas was beginning to decay. By December 1933, 45 refineries were operating in the field, with five additional plants added in January. Most of these refineries were operating on hot oil, as it made no sense to refine legal oil and sell the gasoline for half the legal value of the oil. Some reports claimed that hot oil production reached 90,000 bpd before the Amazon Petroleum case was decided, as many operators felt the court would invalidate the Commission's orders. The PAB, represented by Marshall, won an early victory when Judge Bryant denied the oil companies' motions for a preliminary injunction. After Judge Bryant issued the injunction sought by the companies in January 1934 the government appealed to a three-judge district court. The cases were then bifurcated.[113]

In the majority opinion by the three judge panel on February 12, Judges Hutcheson and T.M. Kennerly refused the plaintiffs' request for an injunction. The two judges accepted the Commission argument that use of the work and figures of the Bureau of Mines was not an abdication of the Commission's delegated authority, but merely the use of the best available information. They placed the burden of proof upon the plaintiffs to show that the orders were arbitrary and confiscatory in order to obtain an injunction. Judge Bryant dissented, saying that the Commission exercised its powers to keep Texas production down to an arbitrary amount, "fixed as the quota of Texas oil under agreements and arrangements." The court held itself without jurisdiction with regard to the related Panama Refining case which challenged the validity of the NRA and the oil code. Judge Bryant in a separate opinion held that the regulations of the oil code and the oil administrator under the code were invalid with regard to producing and refining operations conducted entirely within the state. He noted that the regulations to be enforced applied to only the East Texas and Oklahoma City fields, and therefore were not regulations of commerce but merely an attempt to limit production, a state function.[114]

The locus of action then shifted to the Texas legislature, as it was obvious that closer monitoring of refineries was the new key to containing hot oil. Governor Ferguson submitted a bill to enable the Railroad Commission to supervise refinery runs. The small East Texas refiners organized opposition, aware that it would put a crimp in their activities. The bill was rejected by the House, and then rewritten with a strengthened secrecy provision, and it passed both Houses. The Governor signed the refinery bill and also a bill making officers of companies liable for violations of Commission orders. A bill was introduced in the House to increase and assign the barrelage tax to the Commission. The oil operators were more than willing to pay additional taxes to adequately fund the Railroad Commission, and they requested that the tax rate be increased to 1/8 cent. The bill easily passed both Houses and the Governor signed it on March 9.[115]

Following the passage of these laws there was a new wave of optimism. There was good reason for oilmen to hope for a more successful effort to curtail hot oil; the majority of hot oil was being run through the newly regulated refineries. A Dallas hot oil meeting resulted in an ultimatum to the Commission to stop hot oil production. Carl Estes asserted that "if you don't do the job in the next fifteen days, we're going to ask the big, bad federal wolf to come in and do the job you fell down on." Colonel Thompson defended the Commission's record, and said that the organization had been handicapped for lack of finances with which to hire sufficient men. He promised to relieve "every untrained man" in the Commission's employ. Responding to the oilmen's demands, the Commissioners appointed R.D. Parker as administrative chief with control of the East Texas field and the authority to hire and fire employees.[116]

Allred conferred with the district court judges of Rusk and Gregg counties, and obtained their promised assistance of opening their courts to suits against proration violators. Parker added 45 new men, all with at least four years experience, dismissed several employees and expanded the tender department. Oil run to refineries in East Texas declined by 75,000 bpd during the week ending April 15, indicating the effectiveness of the enhanced regulatory effort. The Commission attempted to control drilling in the field by requiring a public hearing before granting an exception to the spacing rule, Rule 37, as only 18 hearings could be held per week.[117]

The decision by the Fifth Circuit Court of Appeals, in late May 1934, upholding the Oil Code and NRA legislation, allowed federal agents to resume their activities in East Texas. The Court threw out an injunction granted to Amazon Petroleum and other oil companies in the Texas District Court. Early in its fall term the Supreme Court granted the oil companies their petition for certiorari, with argument set for December. The Circuit Court made it clear that control of interstate shipments of oil required that producers provide information on production and transactions, and that this did not constitute self-incrimination. Ickes dispatched Louis R. Glavis, his chief investigator, to East Texas to initiate a new campaign against hot oil production.[118]

The first crack in the Commission's new image came in May, 1934, when R.D. Parker's jurisdiction was limited to East Texas, followed by his dismissal on June 19. Parker's discharge was due to his firing of incompetent political appointees of the Commissioners, and the arrest of Lon Smith's son Frank as part of an oil theft ring with the aid of Commission investigators. E.N. Stanley, a protégé of Thompson, was put in charge East Texas. Stanley had recently returned to the Commission after a five-month leave of absence, used to partner with Jerry Sadler, a young lawyer, in a firm called Stanley-Sadler Engineering, the primary business of which was peddling influence with the Commission. The discharging order said that Parker had been given a free hand and had shown himself unable to prevent hot oil production from rising to 100,000 bpd. The TPC absolved Parker of any blame for the increase in hot oil.[119]

The continued ineptitude of the Railroad Commission and other state authorities in controlling East Texas oil production created new pressure for national legislation in 1934, but the political power of Texans in Washington prevent any attempts to circumvent state authority from reaching fruition.

Ickes attempted to use the worsening situation to expand his delegated powers. The legal staff of the PAB prepared a new bill, including provisions establishing quotas for motor fuels, ensuring crude supplies for small refiners, pipeline divorcement, and crude oil production control. Representatives of the major companies met with Ickes on April 2, supporting provisions to control oil production and interstate shipments but objecting to proposals to regulate refinery runs of gasoline. Walter Teagle suggested solving the production problem by making East Texas a federal oil reserve. The actual bill, introduced to Congress on April 30 by Senator Thomas, was bereft of the more controversial provisions. The Secretary of the Interior was authorized to collect data and determine the demand for petroleum and products in the United States. Imports were to be limited, and the Secretary could prescribe quotas for movement in commerce, and if necessary, production quotas, for States, pools, leases and storage, equal to the reasonable market demand. The bill established the PAB as the designated body to administer the act.[120]

The oil control bill was introduced in the House on May 17 by Representative Disney of Oklahoma, after a White House conference where Disney sought FDR's support for the measure. The bill was originally routed to the Committee on Mines and Minerals before Rayburn interceded with Speaker Rainey to route it to his committee. Roosevelt said he wanted the legislation but did not give it "must" status for the current session. Senator Connally, speaking in the executive committee hearing of the Mining Committee, said he didn't want federal intervention in internal oil affairs in Texas so long as they were properly handled by the state officials. Efforts to facilitate passage of the bill in the House ran into an immovable roadblock in the Committee on Interstate and Foreign Commerce. Rayburn first employed delaying tactics, and then killed the bill in his committee. Instead of the bill, the Committee voted a resolution authorizing an investigation of the oil situation to report back to the next Congress. When Speaker Rainey announced he would not allow the bill to be called up under suspension of the rules unless the Commerce Committee gave its approval the bill was dead.[121]

PUTTING THE LID ON EAST TEXAS

With the defeat of the Thomas-Disney bill attention shifted back to the Texas Railroad Commission and the East Texas oil field. Thompson met with a group of bankers in Dallas and arranged for them to finance a three man board, consisting of East Texas bankers, to oversee the issuance of tenders. The TPC convinced the Commission to make the tender hearings open to the public to increase scrutiny of potential violators. Hot oil runners flaunted the authority of the Tender Board by signing tender forms as Franklin D. Roosevelt, Will Rogers, Julius Caeser, etc. The primary problem was the unwillingness of local officials to vigorously prosecute violators, especially during an election year. Attempts to call courts of inquiry were consistently blocked by officials, uninterested in "cracking down" on violators who were friends and political contributors.[122]

The Commission reversed a trend with its announcement of the new allowables, effective July 1, 1934. The Commission had attempted to relieve pressure on enforcement by gradually allowing the East Texas allowable to creep upwards, above the Bureau of Mines allocations, and by the end of June the allowable had climbed back to 510,000 bpd. The July allowable reduced the field to 460,000 bpd in response to reports that pressure was dropping in the field.[123]

In August the Commission issued a new order requiring legal tenders on shipments of refined products from East Texas, hoping to stem the flow of hot oil leaving the field as topped crude, fuel oil and other products. The federal government also moved to increase the pressure on the hot oil runners. The Department of Justice and the PAB reached an agreement for joint enforcement of the oil code, with Justice handling prosecution of violators. Federal officials in East Texas complained to Washington that hot oil merchants were bribing telephone and Western Union employees and tapping telephones used by the Interior Department. The head of the Oil Enforcement Unit concluded that there was sufficient evidence of corruption to warrant a Congressional investigation of the Railroad Commission.[124]

On September 8, Judge E.M. Dodson in District Court in Austin held that the Commission had no power to restrict movement of crude oil products. This left the small refiners free to run oil, divide it into crude tops and topped crude, and ship them to be refined into gasoline at another refinery without informing the Commission. It was estimated that 600 wells and 40 refineries refused to pass their transactions before the Tender Committee. The Railroad Commission requested a refined products bill in response to the court decisions granting injunctions. The bill wrote into law the recent order of the Railroad Commission, prohibiting the shipment of refined products without a tender attesting to the legality of the crude from which it was produced. A proposed 50 cent tax on hot oil received a cool reception from East Texas representatives. The bill was reported out without the tax by the House Committee on September 18, passed the Senate, but languished for a couple of days in the House before passing on September 25, but only by 90 votes, delaying implementation by 90 days after the Governor signed it on September 28.[125]

A third gasoline buying program was established in the field, attempting to absorb overproduction (Four programs were instituted in 1934, on June 23, August 13, September 8, and November 2. They were all short-lived, due to objections from the Justice Department and the continual flow of hot oil). Despite the new tools available to state and federal authorities conditions worsened in the field in early October, with overproduction estimated between 120-130,000 bpd.[126]

While the need for cooperation between State and Federal authorities became more urgent, the opening of the Cole Committee hearings emphasized their differences. Ickes, testifying before the Committee on September 18, made the maximum recovery of oil the top priority of oil regulation and said conservation justified higher prices and high profits for flush producers. He made it clear that he required the power to force Texas to toe the line. Ickes also asserted that it would be far easier to regulate production than control interstate commerce of oil produced in excess of a state's quota. While a few large companies supported his position, most of the major firms were in favor of some sort of interstate compact and hot oil legislation.[127]

Allred and Thompson missed no opportunity to blast the idea of federal control of Texas production. The two men were encouraged by the decision by Judge Vaught, US District Judge, Oklahoma City, in the Eason Oil case. Judge Vaught ruled that the Secretary of the Interior overstepped his constitutional authority when he attempted to regulate drilling in an Oklahoma field. He said that if the motive was to protect oil resources for the common defense then Congress has ample power under the Constitution to purchase oil lands and hold them for the benefit of the government. Otherwise the power to regulate and control the production of oil and gas within a state exceeds its constitutionally delegated powers. Thompson and Allred hailed the decision as a victory for state's rights.[128]

Gasoline price wars spread out of the Mid-Continent to the East Coast, with prices declining a penny per gallon in New Jersey and Pennsylvania. Governor-designate E.W. Marland of Oklahoma engaged in a long telephone conversation with the President, warning that "business in general" would suffer from a crude price slash. The oil commissions in Kansas, Oklahoma and Texas discussed the possibility of shutting down oil production if prices began to collapse. On Tuesday, October 23, oilmen were near panic as Atlas Pipeline Company lowered its price to 60 cents per barrel, following a week of rumors that Oklahoma prices would be cut 50 percent. The Railroad Commission responded by preparing an order for a complete shutdown of every oil field in Texas "for use in the event of an emergency." In Washington, the Oil Administration obtained pledges from the major companies that they would not lower prices until success or failure of the new Federal drive was proved. In Kilgore, Norman Meyers, chairman of the new Federal Tender Committee and Ernest Thompson had a meeting that was a virtual "lovefest," as the Commission offered its full support. Meyers agreed to Thompson's suggestion that state and federal tender committees sit jointly, passing cases from state to federal jurisdiction if interstate sale was contemplated.[129]

The Federal Tender Board went into action on Thursday, rejecting most of the initial applications because information deemed necessary had not been supplied. Working together with Commission employees and the State Tender Board, tenders were consistently denied to shippers of hot oil, even when a penalty had been paid to the courts, by using dubious technicalities. The objective was to buy time and slow the rate at which stored crude and gasoline left East Texas, as well as piling up storage fees for those handling hot oil. With railroads and pipelines cut off refineries switched to trucks to circumvent the Federal orders, but the Railroad Commission issued an order requiring an approved tender with each truck load of gasoline.[130]

The key to the success of Federal control was ending the epidemic of corruption which pervaded both state and federal agents in the East Texas field. The Department of the Interior hired Tom Kelliher from the FBI to supervise the enforcement effort. After spending three months in the field, working undercover, Kelliher was ready to act. He lured a dozen FBI agents from the agency and told the fifty Interior agents in the field that they would be given a clean slate, but he expected strict enforcement of the tenders. By arresting two agents and obtaining indictments against the men on the charge of accepting bribes, Kelliher made it clear that the tender program would be taken seriously. A hot oil man attempted to bribe Howard Marshall of the Tender Board, offering ten cents a barrel for a tender for a million barrels. Kelliher was offered $50,000 by a lawyer representing one of the larger refineries in the field. The numerous arrests overloaded the docket of Judge Bryant, who accepted guilty pleas and $1,000 fines to clear the caseload.[131]

Judge Bryant had shifted from an attitude of skepticism to one of cooperation with federal authorities, refusing to issue injunctions against the Federal Tender Board. By mid-November estimates of hot oil had declined to an estimated 35,000 bpd. Dockets were so crowded in East Texas that the state began filing receivership suits in Travis County against East Texas violators. The tender program was so successful that the gasoline buying program was delayed, since few small refiners wanted to sell 5 cent gasoline made from one dollar oil. This was fortunate as the Justice Department had concluded that the program violated the antitrust acts.[132]

When the tender program cut down on shipments out of Texas, hot oil men shifted their attention to dumping their stored products before the expected order of the Commission to check shipments. By December 20 a significant slump in product shipments was noted, as tank car shipments fell to close to zero by the end of the grace period. The Commission shifted its attention to tank trucks, the last resort of the small refiner attempting to evade Commission regulation. The number of operating refineries fell to 26 by the end of the month and overproduction declined to an estimated 12-13,000 bpd by the last week in December and as low as 3-5,000 bpd by the new year.[133]

The fall crisis had prompted new calls for Federal control of the industry, but this was no longer a majority view among oilmen. The large companies preferred limited regulation of their activities, while the more aggressive and/or fortunate independents preferred a system without the constraints which federal control was sure to impose. Awareness that the oil code primarily benefited labor did not increase its popularity among oilmen, not generally known for being socially progressive. Ickes, speaking before the API, attacked the concept of an interstate compact. He defended his administration of the code, noting that profitability had returned to the industry and boasted of the increase in labor expenditures. He castigated certain parties in Texas who "merely pretended to co-operate with us." Ickes hinted that if the industry did not get its act together it would eventually be made a public utility. Ickes failed to convince a majority of the API, as a resolution favoring an interstate compact, Federal determination of market demand, legislation to ban hot oil shipments across state lines, and control of oil imports, passed the board of directors by 42 to 17. The minority report, presented by Beaty and Franklin, endorsed a Thomas-Disney type bill.[134]

On November 17 Governor E.W. Marland of Oklahoma invited the governors of California, New Mexico, Louisiana, Wyoming and Michigan to gather in Ponca City for a review of the situation. He had already visited with James Allred and conferred with Alf Landon of Kansas. Marland said he intended to ask Congress to pass an act to permit states to enter into oil compacts. At the Governors' conference on December 3, strong opposition to any form of Federal control of the oil industry was expressed, except for Landon. The conference approved enactment into permanent legislation of Section 9(c) of the NIRA, and granting the President power to limit imports of crude oil or refined products. Allred, while in favor of a compact, made it clear that Texas would not become a part of any agreement which would permit any outside body to set its production allowables.[135]

The test of the new system of control came with the Panama Refining decision in early 1935. On January 7, 1935 the Supreme Court handed down its decision, with Chief Justice Charles Hughes, the former API counsel, writing the majority opinion. The Court stated that Congress had declared no policy concerning circumstances or conditions under which transportation of petroleum or petroleum products should be prohibited, giving the President an unlimited authority to determine policy. Therefore the Executive Orders and Regulations issued by the Secretary of the Interior lacked constitutional authority. Cardozo's dissent questioned the Court's logic, noting that the President had been given a specific task under 9(c) "to eliminate unfair competitive practices," and to "promote the fullest possible utilization of the present productive capacity of industries." The unfair advantage provided hot oil producers and the resultant demoralization of prices were contrary to these stated goals.[136]

Despite the removal of federal authority regulation in East Texas did not immediately collapse. The order requiring tenders on refined products was sufficient, combined with controls on crude production, to deter a massive increase in interstate shipments. Railroads continued to refuse to accept oil or products for shipment without a State Tender Board order. The highway patrol was covering all roads leading from the field and checking up on tank trucks. Even F.W. Fischer, the lawyer in the Panama case, was willing to admit that the laws of Texas covering shipments were valid and constitutional, and that these laws could prevent shipment of hot oil in interstate as well as intrastate commerce. The situation worsened on January 19 when the District Court denied the Commission the power to prohibit the shipment of 2.2 million barrels of oil products in storage.[137]

Meanwhile in Congress Representative Disney introduced a bill on January 10 to regulate commerce in petroleum, submitted to the Committee on Interstate and Foreign Commerce, while Senator Connally introduced a similar bill in the Senate, sent to the Committee on Mines. The Mines Committee gave a favorable ruling on January 16, and the bill passed the Senate on January 22 by an overwhelming majority on a voice vote. The bill's passage was delayed in the House, as Rayburn permitted the Cole Subcommittee on oil legislation to deliberate before reporting the bill to the Interstate Commerce Committee. Cole wanted to see if the states made progress in developing an Interstate Compact before recommending legislation. The report of his investigatory committee had recommended a Compact as the solution which raised the least constitutional questions. Representatives of eight states voted for an Interstate Compact on January 5, but the details remained to be negotiated. The Oklahoma legislature authorized Marland to negotiate a compact on January 29, and the Texas legislature followed on February 1, but included a resolution specifying that proposed compacts should be confined to prevention of physical waste of natural resources, excluding price fixing. Negotiations for a compact were conducted with a sense of urgency, prompted by Ickes's attacks on the oil industry, including a request for a federal control bill. In early February Disney presented a proposed version of a federal oil control bill to Roosevelt at the White House.[138]

A conference of the potential members of the proposed Compact was held in Dallas starting February 15. Allred's proposals included allocating votes based on the producing ability of states, prevention of physical waste, and a declaration against price fixing and monopoly. Waste was defined as the operation of any oil well with an inefficient gas-oil ratio, waste of natural gas, the spacing of wells so as to bring about loss in the ultimate recovery, and the inefficient use of reservoir energy. Control of oil production was authorized only to prevent waste and not to stabilize or fix price or to match supply to market demand. Realizing that a Compact could not be obtained without Texas, and that the alternative would be federal control, Marland accepted Allred's proposals. The Texas Governor in turn accepted a few changes, including granting the Compact Commission the power to make recommendations to promote the maximum ultimate recovery of oil, an euphemism for protecting stripper wells. The Texas legislature passed bill ratifying the Oil States Compact on April 11.[139]

The Connally bill finally reached the House on February 18, after an agreement at the Interstate Compact Conference had been announced. Four stormy sessions in the Interstate Commerce Committee resulted in a bill that was basically a rewrite of section 9(c) amended to Connally's statement of policy. Interstate commerce in hot oil was prohibited, but this limitation could be lifted by the President if a shortage of oil existed, a clause added to assuage the fears of consumer states. In the debate in the House, Representatives skeptical about the resolve of the Texas government to restrain production were assured by Representative Blanton that he and the other West Texas Representatives were interested in protecting the stripper wells in their districts. The bill passed the House on the eighteenth and returned from the Conference Committee on February 22, 1935, with the House amendments intact, and was unanimously passed by both Houses and signed by the President that evening.[140]

The Federal Tender Board was re-established at Kilgore, resuming operations on March 4, while the state effort to control oil continued unabated, reflected in the gradual decline in shipments and refinery activity, interrupted by a temporary jump in anticipation of the return of federal tenders. Realizing that state control would never be firmly established unless the Courts could be restrained from issuing injunctions, a bill to block injunctions was introduced in the Texas legislature. The bill flew through the House in mid-February, passing the Senate on February 26 and Governor Allred signed it into law on March 6.[141]

A new market demand bill was introduced in the Texas House on March 11, signed by 79 members. The bill made tender orders discretionary of the Commission, authorized receiverships for violators of the tender law, and specified that the recovery of a penalty did not authorize violations. This was added to counter the tendency of sympathetic district judges to impose a small fine, making hot oil marketable. The House passed the new market demand bill on March 25, with the Senate concurring on April 8, both by overwhelming majorities. Allred signed the bill April 13, despite his opposition to market demand. A bill providing for the confiscation of hot oil moving by trucks was passed by the House on April 15, but was stalled in the Senate until amended to include other forms of transportation. The conference report finally cleared both Houses on May 10. The bill also provided that applicants who had tenders rejected could appear in Travis County district courts to appeal, thus assuring they would be less likely to find a sympathetic judge.[142]

Despite these efforts overproduction gradually inched upward again by April as new refineries opened and intrastate shipments increased, especially tank trucks carrying gasoline. The Railroad Commission responded by moving additional agents to East Texas to intercept trucks leaving the field. The Commission's efforts were given an assist on May 2 when a three-judge Federal Court in Houston decided the state had power to prohibit movement of oil without tenders. Nine refineries immediately shut down responding to the decision by a District Court Judge to put a refinery in receivership under the new conservation law. The Commission, working with the Attorney General's office, moved to seize illegal oil under the new act and offer it for sale at public auction. By early June a million barrels of oil had been confiscated, and 430,000 barrels were sold for 40 cents per barrel. To prevent additional crude being pumped into pits, once the oil was seized the pits were to be plowed in. By eliminating excess storage, hot oil producers would be limited to the capacity of the defiant East Texas refineries.[143]

The primary factor motivating action by both the Texas legislature and the Railroad Commission was a renewed call for federal control of the oil industry. As Russell Brown noted "The hot oil situation in East Texas had greatly improved since the hearings on the Thomas bill closed. This seems to be the standard strategy of the 'hot oilers' as part of their program to persuade Congress." Hearings on the Thomas bill before a subcommittee of the Committee on Mines provided another pulpit for supporters of federal regulation. Although the Committee reported the bill to the Senate, it had little chance of passage in the House unless Texas allowed the oil market to be thrown into chaos. The Commissioners were well aware of this danger, as Jack Blalock had informed them as early as January that there was little chance of legislation passing as long as East Texas was kept under control. Roosevelt told Allred that he wanted him to cooperate with Ickes so as to head off oil legislation.[144]

The May 27th decision of the Supreme Court invalidating the National Industrial Recovery Act eliminated the threat to the Commission's authority from the Thomas oil control bill. The Court clearly stated that an industry whose activity only indirectly affected intrastate commerce was subject only to state control. The only area of immediate concern was the elimination of the PAB, since this meant that some other source of allocation estimates would be required to coordinate state activity. Robert Hardwicke suggested that Thompson ask Ickes for the Bureau of Mines to send to the Commission its findings about market demand. Since these numbers were generated by an "impartial agency," they could be used as evidence of market demand by the Commission.[145]

Still, the perceived need for new regulation opened the door for a final attempt to establish federal control. Roosevelt endorsed a plan for voluntary observance of agreements and regulations under the invalidated oil code, but avoided declaring for petroleum legislation, sending oilmen to Rayburn. A few days later the President endorsed the Interstate Compact, backstopped by a federal petroleum board to bring non-signers into line. Ickes attempted to push the revised Thomas bill, claiming that the industry could not regulate itself by voluntary agreements. Rayburn made it clear he would not accept the Thomas bill or the legislation proposed by Disney, which would include a Federal fact-finding agency to recommend allocations and limitations on imports.[146]

At a White House conference in early August, attended by Ickes, Chairman Fahy of the PAB, Disney, Thomas, Rayburn and Cole, a compromise was hammered out on an oil bill. The proposed measure would include ratification of the Interstate Compact, permanent enactment of the Connally act, limitation of imports, a new federal fact finding board to recommend allowables, and approval of trade practice agreements. Roosevelt sent a message to Congress on August 9 recommending legislation giving consent to the Interstate Compact, but with no other guidance. Cole and Disney introduced identical bills in the House on August 6, while Thomas introduced a similar bill in the Senate the same day. The Cole measure was reported with amendments by the Interstate Commerce Committee on August 13. Senator Connally introduced a substitute bill on August 14, in the form of an amendment to the Thomas bill, eliminating the federal board and the clause exempting voluntary agreements from the antitrust laws, which unanimously passed the Senate. The House refused to consider the Connally bill and Congress adjourned. With the two Houses deadlocked over the different versions of the oil bill, a resolution ratifying the Interstate Compact passed Congress on August 24 and was signed into law by Roosevelt five days later.[147]

While Texas adroitly managed to stymie outside interference with its oil industry, there was still a lively debate within the state concerning whether the Railroad Commission should continue as the ruling body. A lack of confidence in the Commission was expressed by the Texas House on April 23 when it voted 85 to 49 for an investigation of hot oil running in Texas. An investigative committee was named by Speaker Stevenson, but an attempt to increase its funding from $1,000 to $5,000 was soundly defeated, suggesting that enthusiasm for the probe was less than overwhelming.[148]

The House Oil Committee meet in Longview on June 4, and announced in a press conference that it had heard testimony strongly indicating "the sale of influence rather than service in several phases of oil regulation. The securing of permits to drill exemptions to Rule 37 is one of the fields in which this character of sale of influence has been apparently most prevalent." More than half of the drilling permits granted in recent months were exceptions to this rule. The Texas Supreme Court on June 12, in *Brown v. Humble Oil*, interpreted Rule 37, as Justice John Sharp found that allowable production of a well drilled as an exception should be commensurate with the size of the tract and the rule could not be evaded by voluntary subdivision of the land. The case made it obvious that exceptions to Rule 37 were granted arbitrarily, not to protect vested rights.[149]

Pressure on the Commission was renewed in late July as the "Hot Oil" Investigating Committee held hearings through September. Owen White, in an article in *Colliers*, claimed that one hot oilman purchased 16 state senators at five hundred dollars apiece to kill House Bill 99, establishing a criminal district court in the East Texas field. The same oilman claimed that a state official had made at least a million dollars in graft. A more balanced article in the *New York Times* provided realistic descriptions of hot oil activities, but also claimed that legislators could be bought for $500 a head.[150]

The Investigation revealed that the suspicions of the IPAA that hot oil seemed to decline when Congress was considering federal control were justified. Commission enforcement personnel in the East Texas field asked hot oil refineries to cease illegal activities when Thompson was in Washington, testifying to the Senate that the Commission had East Texas under control. According to H.A. Miles, the chief of enforcement in the field: "I told them they were doing everything they could to hasten Federal control and asked them to help us. They wanted to know how, and I told them that I wanted them to shut down their refineries and keep them down until Monday of the following week, which they did." One reason they didn't want federal control was that corruption and influence peddling had become a fine art with the Commission and the Courts. There was a general pattern, involving high Commission personnel, former members of Allred's Attorney General department, various Judges and former Commission employees, of interference with enforcement efforts.[151]

The activities of Allred's former assistant Attorneys General were at the very least, unethical, if not outright illegal. Former assistant Attorneys General used their influence to get obtrusive investigators dismissed, interfere with prosecution of violators, as well as knowingly file fraudulent suits to obtain tenders to move oil. Exceptions to Rule 37, the well spacing rule, were peddled by Frank Smith with the help of Commission employees. A small group of well connected men, including Stanley and Sadler, were able to obtain drilling permits on a consistent basis, charging $1,000 for a permit. Thompson refused to testify at one permit hearing because he would make his Deputy Supervisor, a brother-in-law of Assistant Attorney General Upchurch, out to be a liar. There were also some accusations of outright corruption and bribery. A former hot oilman who went undercover for the Commission reported that he was told that the former Chief Investigator, Renshaw, would provide protection for "$20 a thousand." A member of the state Tender Committee said he knew of Commission employees paid $50 a week to allow the running of hot oil. Witnesses in cases were bought off by oil operators and given jobs in West Texas, and when cases were ready for trial, they could not be located.[152]

The activities of Commission employees investigating violations was not only hindered by attempts to circumvent their authority, but also through personnel policies dictated by the Commissioners. Phantom workers were common, as it was typical practice to force unqualified but politically connected personnel upon managers who lacked the authority to hire and fire. The previous experience of employees included: drug clerk, car salesman, baseball player, school teachers, stock raiser, postal clerk, embalmer, farmer, traveling salesman, livestock dealer, candy factory operator, cotton dealer, and ice cream maker. The Commission's Chief Engineer made it clear that he dared not dismiss employees for fear of his job, no matter how incompetent. The man who calculated refinery yields used to determine tenders for movement of refined products was a former dancing instructor. Thompson said that many jobs did not require special qualifications, and that good results had been obtained from men who came without experience.[153]

A turnaround in enforcement seemed to have begun in January 1935, with William McCraw as the new Attorney General. At that point there were 300 unresolved lawsuits, some of them dating as far back as 1933. Commission records were in terrible shape, due to incompetence, understaffing and corruption. John Johnston, the new Commission Auditor in East Texas, replaced D.W. Power in June 1935, after Power resigned to work for a hot oil producer. Johnston admitted that the Commission had failed to check reports submitted by refineries against the records of the Comptroller's office until he joined the Commission. Generally, they accepted sworn reports and rarely audited a refinery's books. Since the Comptroller's office collected both the gross production tax on crude oil and the gasoline tax, its figures provided a check on Commission estimates of production. A significant discrepancy existed statewide during the all of 1935, extending into early 1936.[154]

A source of hot oil, both in terms of past illegal production and a means of disguising current activities, was the large volume of oil stored in open pits in the field. The Commission conducted a survey of pits in the field, dated April 1,

1935, just before the legislature finally passed a statute banning open pit storage. A list of the refineries with the largest open pit capacity reads like a list of the most serious proration violators in the field. Eight refineries had a total pit storage capacity of 4 million barrels of oil, while other refineries and reclamation plants in the field had an additional 2.5 million barrels of storage capacity. One impact of the pit building campaign was to overwhelm the ability of the Commission's field men to keep track of stored oil. By claiming large volumes of stored oil, and paying a fine, tenders could be obtained which would allow movement of products refined from hot oil yet to be produced. In August 1935, to combat fraud, the Attorney General's office began requesting a gauge of the pit (hot oilmen then dug pits that were deep along the edge, where the gauger would measure, and shallow in the middle). A new provision in court orders required that no more oil be moved than the contents of the pit when actually emptied.[155]

By September the Attorney General's office had filed 27 confiscation suits, resulting in the sales of almost 2 million barrels of crude oil and products. When the oil was sold the buyer paid the price set by the sheriff, supposedly determined by open auction, but the attorney for the defendant would threaten the title if any one but the owner bought the oil. Despite the fact that illegal production of oil was subject to a $1,000 dollar fine for each instance, the Attorney General's office usually failed to charge owners. The pits were supposed to be cleared and destroyed in sixty days, but four months after these suits commenced only eight or nine pits had been destroyed. One reason for the large volume of fuel oil reported in storage was that it usually sold for about 20 cents per barrel, as opposed to 40-50 cents for fresh crude.[156]

Reacting to the hearings the Commission moved once again to tighten its regulation of East Texas. On August 5 it issued an order requiring a certificate from the State Comptroller showing gasoline tax payments as a prerequisite to all products tenders. Attorney General McCraw obtained judgments to destroy 58 open storage pits as their 60 to 90 day grace periods expired. The Commission added an order requiring applicants for tenders to display certificates showing payments of the gross oil production tax. An employee was discharged for falsifying records to increase per well potentials. Thompson issued a preemptive attack against the major oil companies, accusing them of threatening a cut in the price of gasoline and crude oil.[157]

With the legislature convening in mid-September, hot oil production declined below 40,000 bpd by the end of the month. Announcement of the Interstate Compact, agreed upon by six oil producing states on September 12, removed the impetus for oil legislation in the new session. Thompson asked for a law allowing confiscation of trucks carrying hot oil, but otherwise maintained his usual stance that everything was under control. He advocated a graduated oil tax in front of the House Appropriations Committee, from five cents a barrel scaled up to 20 cents for large producers. When he announced for re-election to the Commission a week later, Thompson judiciously omitted the oil tax from his statement.[158]

The final report of the "Hot Oil" Committee was far more charitable to the Commission. It found that hot oil runners, aware that the Commission was checking the Comptroller's records, paid their production taxes under fictitious names to escape detection. The Comptroller's understanding with operators of fictitious companies that no information with reference to excess oil transactions would be revealed to the Commission was discontinued. The Committee praised the Commission for its efforts to accurately gauge oil stored in earthen pits, eliminating abuses associated with confiscation sales. Stung by accusations that it had not vigorously pursued the major companies, the Committee proclaimed its diligent effort to track all leads that might show the production and handling of excess oil by major companies. The Committee arranged to have access to the files of the Federal Petroleum Agency, the investigative arm of the Federal Tender Board. They found that an extensive Federal investigation of pipelines had only revealed an excess of 5,000 barrels, far less than the 1% allowable pipeline overage.[159]

The investigations by the Texas legislature revealed the extent to which state oil regulation became a means of surreptitiously transferring wealth from the large landowners and major oil companies to independent oilmen. The Commission, and the Attorney General's office under James Allred, were co-conspirators. While many of the actors were motivated by personal gain, the populist animus toward large corporations encouraged Judges, lawyers, assistant Attorney Generals and Railroad Commissioners to wink at violations of the law. Most of the recipients of this largesse were not poor men, as entry into this game required thousands of dollars, whether to drill wells, build refineries or make payoffs. Only after a new Attorney General, William McGraw, was elected, and the veil of secrecy was removed, did this system begin to collapse.

AFTERMATH

Thompson continued his opposition to the use of acreage in determining allowables during a hearing in December, 1935. By the next hearing in January the number of wells in the field had reached 20,000. A Commission geologist testified that the average field pressure was declining excessively when producing at 436,000 bpd. Support grew for acreage allowables due to the realization that it was necessary to find some means of curtailing drilling in the field, as there was an engineering consensus that the field be limited to 425,000 bpd. Attempts to protect small producers had resulted in longer payback periods and increased confiscation of oil from larger tracts. The obvious solution was to deny Rule 37 exception requests to owners who refused to pool their tracts. Other states started to enact compulsory pooling legislation at this time, Oklahoma and New Mexico in 1935, and Louisiana in 1936. By 1952, sixteen states had enacted compulsory pooling laws. The absence of compulsory pooling, a policy which the Texas legislature adamantly rejected, resulted in thousands of excess wells in the field.[160]

During the next few years Thompson began to face the internal contradictions in his previous statements. In an address to the API in November, 1935, he admitted that the real reason for fighting federal control was the fear that Congress, dominated by consuming states, would not support higher prices for oil, which was the goal of Texas. He also made it clear that the Commission would continue to protect small oil producers at the expense of the larger firms by opposing acreage proration. However, the problem of marginal wells caused a partial change of heart. By January 1, 1938, the average production of the 23,951 wells in the East Texas field was only 20.5 bpd, with little additional output available for the most prolific producers once the minimum production was assigned to marginal wells. Thompson began to shift on the issue of permitting close drilling. In 1939 he admitted that the field had been drilled excessively but absolved himself and the Commission of responsibility, blaming small fee holdings and past patterns of spacing.[161]

An important factor in the shift in attitudes was the impact of the opening of the Louden fields in Illinois in 1938. Since these fields were shallow, and the state lacked legislation to control producers, output quickly rose to levels threatening national markets. By August 1938 the impact of the spread of low grade gasoline from the field reached East Texas, where the price for crude dropped ten cents to $1.25. In October prices declined by 20 cents across the Mid-Continent region. Finally, in August, 1939, a new wave of price cuts hit the region led by Sinclair and Humble.[162]

Thompson proposed a thirty day shutdown of Texas oil fields, but settled for a fifteen day moratorium on production beginning August 15. The Colonel called a meeting of the Interstate Oil Compact Commission in Oklahoma City for the same day, and Oklahoma and New Mexico issued their own shutdown, followed by Arkansas and Kansas. Thompson in a radio address on August 14 denied there was overproduction and accused Humble of wanting to fill its storage tanks with cheap oil. Jerry Sadler, who defeated Terrell in 1938, accused the "Standard Oil Company" of attempting, through its "Texas Puppet company" to deplete the state's oil reserves for the benefit of "the money barons of Wall Street." On August 28 Humble rescinded its price cut retroactively, and the Commission immediately terminated the shutdown. The company learned its lesson, refusing to be the first to cut prices when the market weakened in 1940.[163]

Fortunately for the new system of oil production control the Illinois oil field exhibited the usual behavior of flush fields; a quick peak in production followed by a rapid decline in output. The new fields were drilled up by the summer of 1940, and crude oil production fell off sharply after that time. As the network of gathering systems and connecting lines was completed, producers were given the opportunity to sell their output at posted prices, decreasing the supply of oil to small refiners, and pressure on gasoline prices.[164]

The Commission also survived a judicial challenge to its authority in the Rowan & Nichols case. On June 12, 1939, Judge McMillan of the Western District of Texas ruled that the proration of the field on a per well basis discriminated against the owners of the best property, as it took no account of the difference in wells, thickness of sand, location upon the structure, or acreage. The Circuit Court of Appeals affirmed the decree, and the case went to the Supreme Court. The Court's decision, presented by Judge Frankfurter, rejected application of the court's conception of fairness and reasonableness to the challenged order, on the grounds that it was a question of administrative and not judicial judgment. He also gave the seal of approval to setting allowables in order to make small wells profitable as "their effect on the state's economy is an appropriate factor to be taken into account when plans are devised to keep the wells open." Justice Roberts in his dissent pointed out that the issue was not only whether the order was unfair, but that it resulted in confiscation of property.[165]

Price wars at the retail level continued to plague the industry, as access to inexpensive crude tempted refiners to penetrate new markets, and even when crude was controlled, conflicts between company outlets, independent stations and trackside dealers often triggered discounting as well as indirect competition through premiums and additional services. Tax evasion was another contributing factor, as it was estimated that five percent of gasoline was sold without paying the two to seven cent tariff, putting additional pressure on the market. Most observers tended to blame the overabundance of outlets for the instability of the gasoline market.[166]

The most effective method of stabilization was the attempt to remove distress gasoline from the market through agreements between the majors and the independent refiners. A number of buying programs had been initiated during 1933 and 1934, but the failure to control hot oil doomed these attempts, as small refineries gleefully sold their legal allotment at market prices to the majors, while shipping hot gasoline from the refinery backdoor. The eventual success of oil regulation, and the increase in gasoline demand from its nadir in 1933, provided the opportunity to design a successful program. In July of 1934 the PAB asked Charles Arnott of Socony-Vacuum to head a voluntary movement to deal with price wars. A Mid-Continent buying program was organized by Arnott in March, 1935, soon followed by an East Texas buying program. Each of the major companies was to select one or more independent refiners as a "dancing partner," assuming responsibility for purchasing its distress supply. The programs increased the wholesale price of Mid-Continent third-grade gasoline by almost 2 cents a gallon.[167]

The legal umbrella for the buying programs was removed by the demise of the NIRA after the Supreme Court decision in the Schechter Poultry case, but the companies continued the programs with the blessing of Secretary Ickes. This did not help when a group of disgruntled oil jobbers went to Washington in June 1935 and complained of collusion among large companies in fixing jobber margins. Attorney General Robert Jackson announced the initiation of an antitrust investigation of the oil companies, and in July, 1936, a grand jury in Madison, Wisconsin, returned an indictment against 23 oil companies and 58 individuals who allegedly conspired to fix prices. Most frustrating to the oil companies' case was their inability to secure testimony from Ickes concerning his sanctioning of their arrangement. Ickes did not wish to appear as a voluntary witness lest he embarrass the Roosevelt Administration. The court adjudged the defendants guilty of a misdemeanor under the Sherman Act in September 1938, upheld by the Supreme Court two years later.[168]

The difficulty of enforcing proration diminished during the decade as changes in the gasoline market limited the viability of "skimmer" refiners. The gradual upgrading in octane numbers brought about by the widespread adoption of thermal cracking increased the quality differential between third-grade gasoline and the regular grades of gasoline. Third-grade gasoline did not perform well in the high compression engines which were introduced during this period, but its marked price advantage made it attractive to price conscious customers, who were more likely to be operating older vehicles. As engines improved, the value of third grade gasoline continued to decline. Only plants with cracking facilities could achieve sufficient realizations on their crude and produce a high enough quality product to survive. Invest or die seemed to be the dominant theme in refining after 1935, as 140 firms without cracking plants went out of business between 1940 and 1946, whereas only 30 companies with cracking facilities disappeared. The most serious decline was suffered by firms in the Mid-Continent area, as the Gulf Coast refining region emerged as the most important in the nation. Gulf Coast refineries grew steadily larger, to an average capacity of 65,000 bpd in 1950, reflecting economies of scale in refining and transportation.[169]

Attacks on the major oil companies as monopolists would be a constant theme throughout the rest of the decade, to be revived once again in the early 1950s, and repeated during the oil crisis years of the 1970s. Part of this animosity simply reflected the vestiges of Midwestern progressivism, evidenced by the chainstore taxes implemented in various states. At the marketing level conflict centered around the desire of large companies to control outlets to protect their investments in brand recognition, and the fear of independent refiners that they would be denied customers for their products. At the refinery level the conflict involved access to and the price of crude oil supplies. Distributors wanted independent refiners who would undercut the prices offered by the majors, but gasoline marketers were wary of price cutting, as it triggered price wars which lowered their margins.[170]

In April 1938, Senator Guy Gillette introduced a bill to compel separation of marketing from other lines of the petroleum business. This attempt had little success as it became obvious that marketers and independent distributors had conflicting interests, while the likely loser was the consumer. In every Congress from 1938 to 1941, Congressmen, lead by Senator Borah, introduced bills to divorce pipelines from oil producing companies. While these efforts also bore little fruit, since pipeline divorcement was opposed by independent oilmen who feared losing assured customers for their crude, they prompted action by the ICC. Ironically, the ICC had initiated an investigation of pipeline rates after a small refiner complained about *reductions* in rates in 1934. A preliminary report in 1936 concluded that pipeline rates were unreasonable and minimum tenders were unjustifiably high. The Commission held additional hearings before issuing its final order in 1941, requiring carriers to adjust their tariffs to yield a maximum of eight per cent return on the ICC valuation of investment and to reduce minimum tenders to 10,000 barrels.[171]

June, 1938, saw the creation of the Temporary National Economic Committee, headed by Senator Joseph C. O'Mahoney. Relying on testimony presented at the TNEC hearings, Roy Cook, a member of Thurman Arnold's staff, prepared a monograph about the oil industry. Cook emphasized the dominant role of large corporations and concluded that the industry bore all the earmarks of a public utility and thus should be subject to strict federal and state controls. The climax to the New Deal's antitrust policy regarding the oil industry came with the "Mother Hubbard" Case, an all-embracing antitrust suit against 22 major oil companies and some independents, charging them with 69 violations of the Sherman Act. Since the action was so broad in scope, it was dubbed as a Mother Hubbard suit. A related case accused three pipelines of violating the Elkins Act by giving rebates through the payment of dividends to parent companies which shipped oil through these lines. The original Justice Department petitions in both cases called for divestiture, but the National Defense Advisory Commission questioned whether divestiture proceedings might hinder the defense effort. Agreement on the Elkins Act cases was obtained in July, 1941, with a consent decree limiting payments to shipper-owners to seven percent of the ICC valuation of carrier property. The Mother Hubbard suit was suspended by the President in April, 1942, for the duration of the war, and dismissed in June, 1951.[172]

There is little hard evidence that the oil companies had been successful oligopolists in the domestic market. If this was the case we would expect that their profits would exceed that of other lines of business in the long run (monopolistic competition would lead to overcapacity and average profits, but this is unlikely in an industry where product differentiation is at best weakly successful). National City Bank estimated that petroleum companies' earnings were below the average of manufacturing companies between 1930 and 1950, only approaching the latter group in the postwar period. There was little evidence that integrated oil firms achieved higher profits than other companies. The primary advantage of integration was stability of earnings, as periods when one sector of the industry was in turmoil was usually a time of higher earnings for another sector.[173]

With the entry of the United States into World War II production controls were temporarily suspended, and the increase in output made the extent of control self-evident. In 1940, national production was about 3.7 million bpd, about 22 percent below the estimated Maximum Efficient Rate (MER). Production was limited by transportation constraints at first, but by 1944 a massive investment in pipelines and tankers allowed output to reach and then exceed the MERs of fields. The expanded web of crude and product pipelines across the eastern two-thirds of the country would tie various oil markets closer together after the war. The extent to which the Commission had restricted Texas production to maintain prices was revealed by the war, as production rose from 1.3 million barrels per day to slightly over 2 million bpd by December 1943. WWII saw the formation of the Petroleum Administration for War (PAW), which adopted policies to increase efficiency such as requiring 40-acre well spacing for oil wells, a materials allocation policy which encouraged unitization, and the development of data on MERs for fields. The three years of operation under PAW regulations encouraged adoption of wider well spacing and use of MERs as the basis of production restraint.[174]

The system of production required cooperation by the Texas Railroad Commission, purchased at a substantial cost in inefficient allocation of resources, higher prices for consumers, and significant wealth transfers to well connected Texans. The Commission's policies, by making Texas a high-cost producing state despite abundant low-cost oil reserves, shifted investment to other states, and in the 1940s and 1950s, to other countries. Despite attempts to protect the small producers, most of these oilmen would be driven out of the industry over the next decade, due to higher costs of operation. To maintain the prorationing system, a costly system of oil import restrictions would be imposed on the nation. This whole chain of events can be traced to the ability of a small group of oilmen to block any attempt to rationalize the Texas oil industry, both in Austin and in Washington.

NOTES

[1] Stephen L. McDonald, *Petroleum Conservation in the United States: An Economic Analysis* (Baltimore: Johns Hopkins University Press, 1971), p. 31.

[2] Harold F. Williamson, R.L. Andreano, A.R. Daum, and G.C. Klose, *The American Petroleum Industry: The Age of Energy 1899-1959* (Evanston, Illinois: Northwestern University Press, 1963), pp. 34-36; Edgar Wesley Owen, *Trek of the Oil Finders: A History of Exploration For Oil* (Tulsa: American Association of Petroleum Geologists), 1975, pp. 312-14.

[3] Williamson et al., *Age of Energy*, pp. 60-63.

[4] Melvin De Chazeau and Alfred Kahn, *Integration and Competition in the Petroleum Industry* (New Haven: Yale University Press, 1959); and John McLean and Robert Haigh, *Growth of the Integrated Oil Companies* (Boston: Harvard University Press, 1954). Among the companies created by the breakup were Standard Oil of New Jersey (Exxon), Standard New York (Socony, now Mobil), Standard Indiana (Amoco), Standard California (Socal, now Chevron), Standard Ohio (Sohio), Atlantic Refining (now part of ARCO), Continental Oil (Conoco), and Ohio Oil (Marathon).

[5] Owen, *Trek*, p. 461.

[6] Owen, *Trek*, pp. 1291-92.

[7] Owen, *Trek*, p. 1301; and U.S., Federal Trade Commission, Staff Report, *The International Petroleum Cartel*, 82nd Cong., 2nd sess., (Wash., D.C.: GPO, 1952), pp. 40-41, 58-63, henceforth referred to as FTC-1952.

[8] Owen, *Trek*, pp. 1304-9.

[9] Owen, *Trek*, pp. 1319-25, 1330-31.

[10] Owen, *Trek*, pp. 1336-38.

[11] FTC-1952, pp. 200-5.

[12] FTC-1952, pp. 206-7.

[13] FTC-1952, pp. 270-72.

[14] Franklin Tugwell, *The Politics of Oil in Venezuela* (Stanford: Stanford University Press, 1975), p. 38; Jonathan C. Brown, "Why Foreign Oil Companies Shifted Their Production from Mexico to Venezuela during the 1920s," *American Historical Review* 90 (Apr. 1985): 378-380; and FTC-1952, pp. 163, 190-91.

[15] Norman E. Nordhauser, *The Quest For Stability: Domestic Oil Regulation 1917-1935* (New York: Garland Publishing, 1979), p. 8.

[16] Owen, *Trek*, pp. 449-452. Oil in storage loses the most volatile fractions, the most valuable part, to evaporation if not tightly sealed. Earthen storage was often used, with losses in volume up to 5% a year, and more in terms of value, while steel storage, though more expensive, if properly constructed limited losses to about 1%. Additional costs are the interest paid on the capital to build the tanks and purchase oil above normal inventory requirements.

[17] Williamson et al., *Age of Energy*, pp. 322-28.

[18] George Stigler, "The Theory of Economic Regulation," *Bell Journal of Economics* 2 (1971): 3-19; and Richard A. Posner, "Theories of Economic Regulation," *Bell Journal of Economics* 5 (Aug. 1974): 335-57.

[19] Gerald Nash, *United States Oil Policy 1890-1964* (Pittsburgh: University of Pittsburgh Press, 1968), pp. 24-28.

[20] Nash, *United States Oil* Policy, pp. 30-37.

[21] Nash, *United States Oil* Policy, pp. 39-46.

[22] Nordhauser, *Quest For Stability*, pp. 10-12.

[23] Nordhauser, *Quest For Stability*, pp. 12-14.

[24] John G. Clark, *Energy and the Federal Government: Fossil Fuel Policies, 1900-1946* (Urbana: University of Illinois Press, 1987), pp. 156-57; and August W. Giebelhaus, *Business and Government in the Oil Industry: A Case Study of Sun Oil, 1876-1945* (Greenwich, Conn.: JAI Press, 1980), pp. 118-25.

[25] Nordhauser, *Quest For Stability*, pp. 18-20; Nash, *United States Oil* Policy, pp. 86-87; and Burl Noggle, *Teapot Dome: Oil and Politics in the 1920's* (New York: W.W. Norton, 1965).

[26] Nordhauser, *Quest For Stability*, pp. 20-22.

[27] American Petroleum Institute, *American Petroleum: Supply and Demand* (New York: McGraw-Hill, 1925), pp. 3-5, 24-25, 34.

[28] Nordhauser, *Quest For Stability*, p. 24; Nash, *United States Oil Policy*, p. 88; *Federal Oil Conservation Board, Public Hearing, May 27, 1926* (Washington D.C.: GPO, 1926); and *Report of the Federal Oil Conservation Board to the President, September, 1926* (Wash., D.C.: GPO, 1926).

[29] Nordhauser, *Quest For Stability*, pp. 27-30. Marland was motivated by the overexpansion of his operations and the resulting threat to his company's solvency. He was forced to borrow $30 million in June 1927 from the Morgan bank. The Morgan people then forced Marland up to the position of chairman of the board in October 1928, influenced by the company's loss of $7,690,000 in 1927. In 1929 the company would be merged with Continental Oil and the name changed to the latter. Samuel D. Myers, *The Permian Basin: Petroleum Empire of the Southwest* (El Paso: Permian Press, 1973), pp. 585-88.

[30] Nash, *United States Oil* Policy, pp. 91-92.

[31] Nash, *United States Oil* Policy, pp. 93-94.

[32] Nash, *United States Oil Policy*, pp. 89-91; and Owen, *Trek*, pp. 466-67.

[33] Nash, *United States Oil Policy*, pp. 94-96; and Nordhauser, *Quest For Stability*, p. 33.

[34] "The Conservation of Gas," Report of the Technical Sub-Committee of the American Petroleum Institute Gas Conservation Committee, Chicago, Dec. 2, 1927 in *Federal Oil Conservation Board, Report No. 3, Appendix A*, February 25, 1929 (Washington D.C.: GPO, 1929), pp. 19-20; Williamson et al., *The Age of Energy*, pp. 328-29; Robert Hardwicke, *Antitrust Laws, et al. v.*

Unit Operation of Oil or Gas Pools, Revised Ed., (Dallas: Society of Petroleum Engineers of AIME, 1961), pp. 29, 41-42; and Henrietta M. Larson and Kenneth W. Porter, *History of Humble Oil and Refining Company* (New York: Harper & Brothers, 1959), pp. 257-58.

[35] Nordhauser, *Quest For Stability*, pp. 37-38; Nash, *United States Oil Policy*, p. 102.

[36] Nordhauser, *Quest For Stability*, pp. 38-39; Nash, *United States Oil Policy*, pp. 102-3; Robert F. Himmelberg, "Chapter Three," in *Herbert Hoover and the Crisis of American Capitalism*, ed. J. Joseph Huthmacher and Warren Susman (Cambridge, Mass: Schenkman Publishing Co., 1973), p. 71.

[37] Nash, *United States Oil Policy*, pp. 99-103; and Nordhauser, *Quest For Stability*, pp. 34-35.

[38] Nash, *United States Oil Policy*, pp. 104-5; and Nordhauser, *Quest For Stability*, pp. 41-46.

[39] U.S. Congress, House, *Tariff on Petroleum*, *Hearings before the Committee on Ways and Means*, 71st Cong., 2nd sess., Feb. 8, 1930, pp. 11-13, 35, 48, 51.

[40] *Congressional Record*, 71st Cong., 2nd Sess., 1930, pp. 4503-4506; *Oil and Gas Journal (O&GJ)*, Feb. 6, pp. 29-30, Mar. 6, 1930, p. 55, 206; *Houston Post-Dispatch*, Mar. 1, 1930; Nordhauser, *Quest for Stability*, pp. 51-52.

[41] *Congressional Record*, 71st Cong., 2nd Sess., 1930, pp. 4312-15, 4472-81, 4491-97, 4502, 5593; *Houston Post-Dispatch*, Feb. 28, 1930.

[42] *Congressional Record*, 71st Cong., 2nd sess., 1930, pp. 4484, 4489, 4497-98, 4517-18, 5600, 5603, 5610, 5785, 5791; *O&GJ*, Mar. 27, 1930, p. 149; Nash, *United States Oil Policy*, pp. 106-110; and Nordhauser, *Quest For Stability*, pp. 50-56, 75-79.

[43] Robert F. Himmelberg, *The Origins of the National Recovery Administration* (New York: Fordham University Press, 1976), p. 100.

[44] Nordhauser, *Quest For Stability*, pp. 58-61; Himmelberg, *Origins*, pp. 101-102; *OG&J*, Apr. 17, 1930, pp. 37, 82.

[45] *O&GJ*, Apr. 17, 1930, pp. 37, 82.

[46] W.P.Z. German, "Legal History of Conservation of Oil and Gas in Oklahoma," in *Legal History of Conservation of Oil and Gas* (American Bar Association, December 1938), pp. 151-154; and Leonard Logan, *Stabilization of the Petroleum Industry*, (Norman: University of Oklahoma Press, 1930), pp. 160-62.

[47] Carl Rister, *Oil! Titan of the Southwest*, (Norman: University of Oklahoma Press, 1949), pp. 250-58; and German, "Legal History in Oklahoma," pp. 156-59.

[48] German, "Legal History in Oklahoma, " pp. 160-65.

[49] Barbara Day, "The Oil and Gas Industry and Texas Politics, 1930-1935" (Ph.D. Dissertation, Rice University, 1973), pp. 63-70.

[50] *Houston-Post Dispatch*, Feb. 24, 1930, State News Sec., p. 1; Day, "Oil and Texas Politics, " pp. 69-71; Larson and Porter, *History of Humble Oil*, pp. 323-324; *OG&J*, June 19, 1930, p. 38.

[51] *Houston Post-Dispatch*, Aug. 15, Aug. 24, Aug. 26, Aug. 27, 1930; *O&GJ*, Aug. 7, pp. 29, 110, Aug. 14, pp. 37, 149-50, Aug. 21, 1930, pp. 37, 73.

[52] German, "Legal History in Oklahoma, " pp. 166-69; and *O&GJ*, Aug. 21, 1930, p. 151.

[53] German, "Legal History in Oklahoma, " pp. 170-76; *O&GJ*, Oct. 9, p. 51, Oct. 23, 1930, p. 29, 135; *Houston Post-Dispatch*, Oct. 17, Oct. 26, State/Auto p. 4, Oct. 29, 1930.

[54] Larson and Porter, *History of Humble Oil*, pp. 395-97.

[55] Presley, *Saga of Wealth*, pp. 111-14; James Clark and Michel Halbouty, *The Last Boom* (New York: Random House, 1972), pp. 6-9, 18-19, 34-39, 56-61, 83-95; *Houston Post-Dispatch*, Sept. 14, 1930, Auto/Radio p. 14; *O&GJ*, Sept. 11, p. 70, Sept. 25, p. 150, Dec.4, 1930, p. 74.

[56] Clark and Halbouty, *Last Boom*, pp. 85-86, 97-99, 106-107; Presley, *Saga of Wealth*, pp. 122-23; *Houston Post-Dispatch*, Dec. 28, 1930, Local p. 2; *O&GJ*, Jan. 1, 1931, p. 22-23; Larson and Porter, *History of Humble Oil*, pp. 398-99.

[57] Clark and Halbouty, *Last Boom*, pp. 125-29; Presley, *Saga of Wealth*, pp. 127-31; *O&GJ*, Feb. 26, 1931, p. 21.

[58] *Houston Post-Dispatch*, Nov. 21, 1930; *OG&J*, Jan. 1, pp. 23, 107, Jan. 22, p. 111, Jan. 31, 1931, pp. 169-70; Ruel McDaniel, *Some Ran Hot* (Dallas: Regional Press, 1939), pp. 76-81; Clark and Halbouty, *Last Boom*, pp. 157-58.

[59] McDaniel, *Some Ran Hot*, pp. 82-86, Clark and Halbouty, *Last Boom*, p. 160; and Day, "Oil and Texas Politics, " p. 89.

[60] *O&GJ*, Feb. 5, 1931, p. 116, Mar. 5, p. 39, Mar. 12, pp. 29, 143, Mar. 19, 1931, p. 21; *New York Times*, Mar. 9, Mar. 10, 1931; *Houston Post-Dispatch*, Feb. 15, 1931.

[61] *O&GJ*, Apr. 23, p. 23, Sept. 17, p. 13, Sept. 24, 1931, p. 19.

[62] *Houston Post-Dispatch*, Mar. 15, Mar. 16, Mar. 17, 1931.

[63] Clark and Halbouty, *Last Boom*, pp. 161-62; McDaniel, *Some Ran Hot*, pp. 90-96; *Houston Post-Dispatch*, Mar. 21, Mar. 26, Mar. 27, Apr. 22, 1931; *O&GJ*, Apr. 2, pp. 21, 122, 124, 135, Apr. 16, 1931, pp. 21, 53, 128.

[64] *O&GJ*, May 14, p. 50, May 28, p. 41, June 4, p. 19, 21, June 11, p. 17, 24, July 9, 1931, p. 13; *Houston Post-Dispatch*, May 27, June 3, 1931; Larson and Porter, *History of Humble Oil*, p. 453; Day, "Oil and Texas Politics," p. 95; and Clark and Halbouty, *Last Boom*, p. 163.

[65] *Houston Post-Dispatch*, June 8, June 13, June 16, June 18, June 24, 1931; *O&GJ*, June 18, pp. 13, 105, June 25, pp. 13, 126, July 2, 1931, p. 13; and Day, "Oil and Texas Politics," p. 100.

[66] *Houston Post-Dispatch*, July 3, July 4, p. 2, July 8, July 9, 1931; *OG&J*, July 9, p. 14, 24, July 16, 1931, p. 28.

[67] *Austin-American*, July 25, 1931; *MacMillan v. Railroad Commission of Texas*, 51 F.2d 400.

[68] *Houston Post-Dispatch*, Aug. 13, 1931.

[69] *Austin American*, Aug. 4, 1931; German, "Legal History in Oklahoma," pp. 180; *Houston Post-Dispatch*, Aug. 6, 1931.

[70] *Champlin Refining Co. vs. Corporation Commission, et al*, No. 1156, (Second Opinion) from the Texas State Archives, Attorney General Subject Files, 1989/41-13.

[71] *Austin American*, Aug. 7, 1931; *O&GJ*, Aug. 13, p. 42, Aug. 20, pp. 15, 20, 27, 103, Aug. 27, 1931, p. 25; Warner Mills, *Martial Law in East Texas* (Alabama: University of Alabama Press, 1953), p. 24-31; *Houston Post-Dispatch*, Aug. 11, Aug. 16, Aug. 17, Aug. 18., 1931.

[72] Mills, *Martial Law*, pp. 32-37; *Houston Post-Dispatch*, Sept. 2; *O&GJ* Sept. 3, p. 21, Oct. 29, p. 13, Dec. 17, 1931, p. 22.

[73] Mills, *Martial Law*, pp. 37-39; *Houston Post-Dispatch*, Dec. 16, 1931, Jan. 7, 1932; *O&GJ*, Jan. 14, 1932, p. 12; *Constantin et al. v. Smith et al.*, 57 F.2d 227; *Sterling, Governor of Texas, et al. v. Constantin et al.*, 287 U.S. 378.

[74] *Houston Post-Dispatch*, Feb. 26, 1932; *O&GJ*, Mar. 3, 1932, p. 13; *Danciger Oil & Refining v. Railroad Commission of Texas*, 49 SW.2d 837.

[75] *Champlin Refining Co. v. Corporation Commission of Oklahoma*, 286 U.S. 210.

[76] *O&GJ*, July 28, p. 69; *The Texas Independent*, July 1932, p. 16; Clark and Halbouty, *Last Boom*, p. 173.

[77] McDaniel, *Some Ran Hot*, pp. 112-17; John Griffin and Robert Cotner, "Kilgore--East Texas Boom Town," in *Texas Cities and the Great Depression* (Texas Memorial Museum, 1973), p. 113.

[78] McDaniel, *Some Ran Hot*, pp. 198-99; *O&GJ*, June 23, p. 52, Sept. 15, p. 82, Sept. 22, 1932; Larson and Porter, *History of Humble Oil*, pp. 462-63; *Houston Post -Dispatch*, Sept. 12, Oct. 13, 1932; "Threatened Price Cut Meets Opposition," *IPA of A Monthly*, Sept.-Oct., 1932, p. 13.

[79] *Houston Post-Dispatch*, Sept. 20, 1932; *O&GJ*, Sept. 29, 1932, p. 14; *Peoples Petroleum Producers v. Smith et al.*, 1 F. Supp. 361.

[80] *Houston Post-Dispatch*, Oct. 26, Oct. 27, Nov. 4, Nov. 5, 1932; *Austin American*, Nov. 10, Nov. 11, Nov. 13, 1932; *O&GJ*, Oct.27, p. 8, Nov. 3, pp. 11, 31-32, Nov. 17, 1932, p. 26.

[81] *O&GJ*, Dec. 1, p. 9, Dec. 8, p. 8, Dec. 15, 1932, p. 12.

[82] Clark and Halbouty, *The Last Boom*, p. 195; McDaniel, *Some Ran Hot*, pp. 212-14; *O&GJ*, Dec. 22, p. 33, Dec. 29, 1932, p. 47, Jan. 19, pp. 7, 23, 38, Jan. 26, p. 13, Feb. 2, 1933, p. 30; *Houston Post-Dispatch*, Jan. 21, 1933.

[83] *Houston Post-Dispatch*, Jan. 20, Jan. 21, Jan. 27, 1933; *O&GJ*, Feb. 2, p. 30, Feb. 9, 1933, p. 9; McDaniel, *Some Ran Hot*, pp. 215-18.

[84] *O&GJ*, Mar. 9, p. 37, Mar. 16, pp. 9, 11, 25, Mar. 23, pp. 8, 28, 32, Apr. 6, 1933, p. 44; Larson and Porter, *History of Humble Oil*, p. 475; *Austin American*, Mar. 23, 1933.

[85] *Austin American*, Mar. 27, Apr. 3, Apr. 6, Apr. 24, 1933; *Houston Post-Dispatch*, Mar. 27, Apr. 3, Apr. 6, 1933; *O&GJ*, Apr. 13, p. 45, Apr. 20, p. 43, Apr. 27, 1933, p. 11; *Texas State House Reporter*, Apr. 20, 1933.

[86] *Houston Post-Dispatch*, Apr. 25, Apr. 26, 1933; *O&GJ*, Apr. 27, p. 42, May 4, 1933, pp. 9-10, 14, 16.

[87] *Houston Post-Dispatch*, Apr. 7, 1933; Texas House of Representatives, *House Journal*, 1933, pp. 1140-45; *Texas State House Reporter*, Apr. 5, Apr. 24, Apr. 26, Apr. 27, May 1, 1933; *Austin American*, Apr. 6, Apr. 7, Apr. 26, 1933.

[88] *Texas State House Reporter*, May 3, 1933; *Houston Post-Dispatch*, May 4, May 5, May 7, May 14, May 16, 1933; *O&GJ*, May 11, p. 37, May 25, 1933, p. 42.

[89] *Texas State House Reporter*, May 15, June 12, 1933; *Houston Post-Dispatch*, June 2, June 4, June 13, June 14, June 17, 1933; *O&GJ*, June 1, p. 45, June 8, 1933, p. 44.

[90] Nash, *United States Oil Policy*, pp. 126-27; *O&GJ*, Mar. 31, p. 28, May 12, p. 15, Oct. 27, 1932, p. 9

[91] *Petroleum Facts and Figures, Ninth Edition* (New York: American Petroleum Institute, 1951), pp. 1, 160; *Report V of the Federal Oil Conservation Board* (Wash., DC: GPO, 1932), p. 33.

[92] *O&GJ*, Jan. 19, p. 32, Jan. 26, pp. 12, 16, Feb. 2, p. 7, Feb. 16, 1933, p. 11.

[93] Nordhauser, *Quest for Stability*, p. 97; and Bennett Wall and George Gibb, *Teagle of Jersey Standard* (New Orleans: Tulane University, 1974), p. 282.

[94] Nordhauser, *Quest for Stability*, pp. 100-101; and Harold Ickes, *The Secret Diary of Harold Ickes: The First Thousand Days, 1933-1936* (New York: Simon and Schuster, 1953), p. 6.

[95] Nordhauser, *Quest for Stability*, p. 102; Day, "Oil and Texas Politics," p. 218; *O&GJ*, March 30, 1933, p. 30; *IPA of A Monthly*, April 1933, pp. 10-11.

[96] *IPA of A Monthly*, April 1933, p. 12-14; *O&GJ*, March 30, 1933, p. 30.

[97] Ickes, *Diary*, p. 10-16; Day, "Oil and Texas Politics," pp. 221-22; *IPA of A Monthly*, April 1933, p. 14.

[98] Nordhauser, *Quest for Stability*, p. 104; *O&GJ*, April 6, 1933, p. 13.

[99] D.B. Hardeman and Donald Bacon, *Rayburn* (Austin: Texas Monthly Press, 1987), pp. 139-49.

[100] U.S. Congress, *Congressional Record*, 72 Cong. 1st Session, May 1933, pp. 3776, 3805, 3808, June 1933, p. 5294; *O&GJ*, May 4, p. 23, May 11, p. 8, May 18, p. 13, May 25, pp. 7, 44, June 1, p. 1, June 8, p. 10, June 15,

1933, p. 9; Ickes, *Diary*, pp. 47, 49; Nordhauser, *Quest for Stability*, pp. 109-11.

[101] Ellis Hawley, *The New Deal and the Problem of Monopoly* (Princeton: Princeton University Press, 1966), pp. 22-25; Robert F. Himmelberg, *The Origins of the National Recovery Administration* (New York: Fordham University Press, 1976), pp. 193-95, 200-3; and Arthur M. Schlesinger, *The Coming of the New Deal*, (Boston: Houghton Mifflin, 1955), pp. 95-96.

[102] Schlesinger, *New Deal*, pp. 96-97; Hawley, *Monopoly*, pp. 25-29; Joseph Lash, *Dealers and Dreamers* (New York: Doubleday, 1988), pp. 119-22; Bernard Bellush, *The Failure of the NRA* (New York: W.W. Norton, 1975), pp. 13-14.

[103] Schlesinger, *New Deal*, pp. 98-101.

[104] Donald Brand, *Corporatism and the Rule of Law* (Ithaca: Cornell University Press, 1988), pp. 183-84; Nordhauser, *Quest for Stability*, p. 119; *O&GJ*, June 22, 1933 p. 6-7, 26.

[105] Nordhauser, *Quest for Stability*, pp. 120-125; *Texas State House Reporter*, July 19, 1933; *Houston Post-Dispatch*, July 27, July 28, 1933; *O&GJ*, Aug. 10, 1933, p. 9-10.

[106] Nordhauser, *Quest for Stability*, pp. 126-28; *Texas State House Reporter*, Aug. 23, 1933; *Houston Post-Dispatch*, Aug. 2, 1933; *O&GJ*, Aug. 24, 1933, p. 9.

[107] Ickes, *Diary*, pp. 73, 84-86, Nordhauser, *Quest for Stability*, p. 133; *O&GJ*, Aug. 31, p. 9, Sept. 7, p. 8, Sept. 14, 1933, p. 9.

[108] Ickes, *Diary*, pp. 95-99, 104-8; Nash, *United States Oil Policy*, p. 141; *O&GJ*, Sept. 7, p. 30, Sept. 14, p. 10, Nov. 2, p. 9, Nov. 23, 1933, p. 9.

[109] *O&GJ*, June 22, p. 29, June 29, 1933, p. 39.

[110] *Texas State House Reporter*, July 19, 1933; *Houston Post-Dispatch*, July 13, 1933; *O&GJ*, July 13, p. 45, July 20, p. 7, Aug. 3, p. 33, Aug. 10, 1933, p. 10.

[111] *Texas State House Reporter*, July 10, Aug. 31, 1933; Day, "Oil and Texas Politics," pp. 245-47.

[112] *Texas State House Reporter*, Aug. 31, Sept. 1; *O&GJ*, Sept. 7, p. 41, Dec. 21, 1933, p. 8, Feb. 1, p. 32, Feb. 15, 1934, p. 13; James Clark, *Three Stars for the Colonel* (New York: Random House, 1954), p. 102; *Austin American*, Oct. 26, 1934; *Texas State House Reporter*, November 1, Dec. 22, 1933, Feb. 2, 1934.

[113] *Texas State House Reporter*, Jan. 11, 1934; Peter Irons, *The New Deal Lawyers* (Princeton: Princeton University Press, 1982), p. 66.

[114] *Amazon Petroleum v. Railroad Commission of Texas*, 5 F.Supp 633; *Panama Refining v. Ryan*, 5 F.Supp 639.

[115] *Texas State House Reporter*, Feb. 14, Feb. 19, Feb. 26, Mar. 6, Mar. 9, 1934; *Austin American*, Mar. 7, Mar. 8, Mar. 9, Mar. 10, 1934.

[116] *Texas State House Reporter*, Mar. 14, 1934, Mar. 30, Apr. 2, 1934; *O&GJ*, Apr. 5, 1934, pp. 8, 11.

[117] *Texas State House Reporter*, Apr. 4, Apr. 17, Apr. 23, May 10, May 29, May 30, 1934; *Houston Post-Dispatch*, Apr. 4, Apr. 10, 1934; *O&GJ*, Apr. 12, 1934, p. 46.

[118] Irons, *New Deal Lawyers*, pp. 68-69; *Houston Post-Dispatch*, May 23, 26, 1934; *O&GJ*, May 31, 1934, p. 90.

[119] *Texas State House Reporter*, May 17, May 21, June 11, June 20, 1934; Clark and Halbouty, *The Last Boom*, p. 217-18; *Austin American*, June 20, June 21, June 22, 1934; *Houston Post-Dispatch*, June 2, June 3, June 20, 1934.

[120] Ickes, *Diary*, p. 158-59; *O&GJ*, Mar. 1, p. 12, Mar. 15, pp. 9, 11, Mar. 29, p. 9, Apr. 12, p. 24, May 3, 1934, p. 9; *Congressional Record*, 73 Cong. 1st sess., April 1934, pp. 7633-35.

[121] Day, "Oil and Texas Politics," p. 268-70; *Congressional Record*, 73 Cong., 1st sess., June 7, 1934, p. 10740; *Houston Post-Dispatch*, May 18, May 29, May 30, 1934; *IPA of A Monthly*, July 1934, p. 5; *O&GJ*, May 31, p. 9, June 7, p. 9, June 14, 1934, p. 9; *Texas State House Reporter*, June 14, June 18, 1934.

[122] Clark, *Three Stars for the Colonel*, p. 111-12; Day, "The Oil Industry and Texas Politics," p. 251; *Houston Post Dispatch*, June 27, 1934; *O&GJ*, June 28, 1934, p. 12.

[123] Nordhauser, *The Quest for Stability*, pp. 141-42; *Texas State House Reporter*, May 31, June 29, 1934.

[124] *O&GJ*, July 19, p. 38, Aug. 13, 1934, p. 5, 30.

[125] *O&GJ*, Sept. 13, 1934, pp. 12-13; *Texas State House Reporter*, Sept. 17, Sept. 18, Sept. 21, Sept. 24, Sept. 25, Sept. 28, 1934.

[126] *Houston Post-Dispatch*, Sept. 12, 1934; *O&GJ*, Sept. 13, p. 9, Sept. 20, p. 32, Oct. 11, 1934, p. 8; *U.S. v. Socony-Vacuum Oil* 310 U.S. 150.

[127] U.S. Congress, House of Representatives, Committee on Interstate and Foreign Commerce, *Petroleum Investigation, Hearings before a Subcommittee of the Committee on Interstate and Foreign Commerce*, 73rd Cong., Sept. 17-22, 1934, Part 1, pp. 169-86, 233, 245, 346-48, 377-78, 483-91, 507, 515-16, 674-75.

[128] *Austin American*, Sept. 28, 1934; *O&GJ*, Sept. 27, p. 11, 49-51, Oct. 4, 1934, p. 8.

[129] *Texas State House Reporter*, Oct. 22, Oct. 23, Oct. 25, 1934; *Austin American*, Oct. 22, Oct. 23, Oct. 24, Oct. 25, 1934; *O&GJ*, Oct. 25, 1934, p. 8.

[130] *Austin American*, Oct. 27, Oct. 28, 1934; *O&GJ*, Nov. 1, 1934, p. 8.

[131] Clark and Halbouty, *The Last Boom*, pp. 221-23; Presley, *A Saga of Wealth*, pp. 168-71.

[132] Nordhauser, *The Quest for Stability*, pp. 150-51; *Texas State House Reporter*, Nov. 19, 1934; *O&GJ*, Nov. 15, p. 36, Nov. 22, 1934, p. 14.

[133] *O&GJ*, Dec. 6, p. 11, Dec. 13, p. 14, Dec. 20, p. 15, Dec. 27, 1934, p. 64, Jan. 3, 1935, p. 32.

[134] *O&GJ*, Nov. 15, pp. 38, 40-43, 64, Nov. 22, 1934, pp. 8, 10.

[135] *O&GJ*, Nov. 22, p. 6, Dec. 6, 1934, p. 8, 10.

[136] *Panama Refining Company v. Ryan*, 293 U.S. 388.

[137] *O&GJ*, Jan. 17, p. 8, Jan. 24, 1935, p. 32.

[138] *Congressional Record*, 74 Cong., 1st sess., 1935, pp. 366, 407, 653-54, 761-63; *O&GJ*, Jan. 10, pp. 8, 14-15, Jan. 24, pp. 8, 11, 20, Feb. 7, p. 12, Feb. 14, 1935, pp. 12, 19; *Houston Post-Dispatch*, Jan. 30, Feb. 2, 1935.

[139] *Texas State House Reporter*, Feb. 14, Feb. 15, Apr. 11, 1935; *O&GJ*, Feb. 21, 1935, p. 8-9, 32.

[140] *Congressional Record*, 74th Cong. 1st Sess., 1935, pp. 2128-29, 2133-35, 2144, 2150, 2502; *Houston Post-Dispatch*, Feb. 10, 1935; *O&GJ*, Feb. 21, 1935, p. 10.

[141] *Texas State House Reporter*, Feb. 26, Mar. 6, 1935; *O&GJ*, Mar. 7, p. 10, 18, Mar. 14, 1935, p. 32.

[142] *Texas State House Reporter*, Mar. 11, Mar. 12, Mar. 19, Mar. 25, Apr. 8, Apr. 13, Apr. 15, May 10, 1935.

[143] *Texas State House Reporter*, Apr. 8, Apr. 13, June 1, June 3, June 4, 1935; *O&GJ*, Apr. 11, p. 35, Apr. 25, p. 16, May 9, p. 32, May 16, p. 35, May 23, p. 14, May 30, p. 12, June 6, p. 13, June 27, p. 12, July 4, 1935, p. 16.

[144] U.S. Congress, Senate, Committee on Mines and Mining, *Federal Petroleum Act, Hearings before a Subcommittee of the Committee on Mines and Mining*, 74th Cong., 1st sess., 1935; *Houston Post-Dispatch*, May 3, 1935; *O&GJ*, April 18, 1935, p. 12; Telegram from Jack Blalock to Ernest Thompson, Jan. 31, 1935, University of Texas, Barker History Center, E.O. Thompson Papers, 4Ze 136; letter from Russell Brown to Wirt Franklin, April 23, 1935, Barker History Center, Morris Shepard Papers, 2G 194; copy of resolution mailed out by Marland for the Governor's conference with handwritten notes by unknown party, Barker History Center, E.O. Thompson Papers, 4Ze 139.

[145] *O&GJ*, May 30, 1935, p. 13; letter from R. H. Hardwicke to Railroad Commission, June 10, 1935, University of Texas, Barker History Center, E.O. Thompson Papers, 4Ze 136.

[146] *Congressional Record*, 74th Cong., 1st sess., 1935, pp. 9258-60; *Austin American*, May 30, June 5, June 7, 1935; *O&GJ*, June 20, p. 10, July 11, 1935, p. 10.

[147] *Congressional Record*, 74th Cong., 1st sess., 1935, pp. 12971, 14584-93; *Houston Post-Dispatch*, Aug. 10, Aug. 29, 1935; *O&GJ*, Aug. 8, p. 17, Aug. 15, p. 10, Aug. 25, 1935, p. 25; *IPA of A Monthly*, Sept. 1935, p. 8, 19.

[148] *Texas State House Reporter*, Apr. 23, May 11, 1935.

[149] *Texas State House Reporter*, June 3, June 4, June 12; *Houston Post-Dispatch*, June 5, 1935.

[150] Owen White, "Piping Hot," *Collier's*, Jan. 1935, p. 10-11, 30-31; and William Dupuy, "Once More 'Hot Oil" Gushes Ruinously," *New York Times Magazine*, Jan. 27, 1935, p. 2, 15.

[151] *Hearing Before the Legislative "Hot Oil" Investigating Committee*, July 15, 1935, pp. 24-29, 101, Aug. 30, 1935, pp. 1-6, Texas State Archives, Box 2-10/844.

[152] *"Hot Oil" Hearings*, July 15, pp. 27-29, 45, 100-1, July 16, pp. 192-213, July 17, pp. 321-25, 335-37, Aug. 30, pp. 7-14, Sept. 10, pp. 2-5, 25-26, Sept. 25, 1935, pp. 2-9.

[153] *Texas State House Reporter*, July 31, 1935; *"Hot Oil" Hearings*, July 15, pp. 74-91, Aug. 30, 1935, pp. 2-3.

[154] *"Hot Oil" Hearings*, July 16, 1935, pp. 207-12, July 18, pp. 110-91.

[155] *Survey of Earthen Storage in East Texas Oil Field*, April 1, 1935, Texas State Archives, Box 2-10/845; *"Hot Oil" Hearings*, Aug. 30, 1935, pp. 16, 34-35.

[156] *"Hot Oil" Hearings*, Sept. 10, p. 9-24, 26-31, 36-45.

[157] *Texas State House Reporter*, Aug. 5, 1935; *Austin American*, Sept. 1, Sept. 11, 1935; *Houston Post Dispatch*, Aug., 25, Sept. 1, Sept. 9, Sept. 10, 1935; *O&GJ*, Sept. 19, 1935, p. 14.

[158] *Austin American*, Sept. 13, Sept. 15, Sept. 16, 1935; *O&GJ*, Sept. 19, p. 14, Sept. 26, 1935, p. 26.

[159] *Final Report of the "Hot Oil" Committee*, Oct. 22, 1936, Texas State Archives, Box 2-10/845.

[160] Jacqueline L. Weaver, *Unitization of Oil and Gas Fields in Texas* (Wash., D.C.: Resources for the Future, 1986), p. 104, n427; *Texas State House Reporter*, Dec. 17, 1935, Jan. 16, 1936.

[161] Ernest Thompson, "Proration Laws and Their Application in Texas," address to the API, Nov. 13, 1935, Texas State Archives, Railroad Commission Oil and Gas Division, Box 4-31/338; ibid., *An Administrator's Views on Proration*, 1939; Northcut Ely, "The Conservation of Oil," *Harvard Law Review* 51, 1938, pp. 1226-32.

[162] McLean and Haigh, *Integrated Oil Companies*, pp. 598-99, John McGee, "Price Discrimination and Competitive Effects: The Standard Oil of Indiana Case," *University of Chicago Law Review* 23 (1956): n407-n408; Larson and Porter, *History of Humble Oil*, pp. 530-31.

[163] *Texas State House Reporter*, July 14, 1939; Clark, *Three Stars*, pp. 150-51, 159-69; Larson and Porter, *History of Humble Oil*, pp. 530-34.

[164] McLean and Haigh, *Integrated Oil Companies*, pp. 598-99.

[165] *Texas Law Review*, 18, 1939, pp. 49-50; *Railroad Commission of Texas v. Rowan & Nichols Oil Co.*, 310 U.S. 563 (1939); W.L. Summers, "Does the Regulation of Oil Production Require the Denial of Due Process and the Equal Protection of the Laws?" *Texas Law Review* 19 (1940): 2-3.

[166] W. Earl Masincup, and Donald Sherbondy, "Price and Marketing Practices Under the Petroleum Code," *George Washington Law Review* 3 (March 1935): 319, 323-28.

[167] Bruce Johnson, "Property Rights to Cartel Rents: The Socony-Vacuum Story," *Journal of Law and Economics* (April 1991): 185-88; Williamson et al., *The American Petroleum Industry*, p. 696; *Socony-Vacuum, 310 U.S. 150.*

[168] Nash, *United States Oil* Policy, pp. 152-53; Johnson, "Property Rights to Cartel Rents," pp. 187-88; and Williamson et al., *The American Petroleum Industry*, pp. 697-99.

[169] McLean and Haigh, *Integrated Oil Companies*, pp. 520-35, 579-80, 596-98, 695-96.

[170] Williamson et al., *American Petroleum Industry*, pp. 700-9.

[171] Nash, *United States Oil* Policy, p. 154; and Williamson et al., *American Petroleum Industry*, pp. 596-97.

[172] Nash, *United States Oil* Policy, pp. 154-55, Williamson et al., *American Petroleum Industry*, pp. 597-600.

[173] Joel Dirlam and Alfred Kahn, "Leadership and Conflict in the Pricing of Gasoline," *Yale Law Journal* 61 (1951): 830; McLean and Haigh, *Integrated Oil Companies*, pp. 147-48.

[174] McDonald, *Petroleum Conservation*, pp. 40-41; Williamson et al., *American Petroleum Industry*, pp. 764-66, 778-82; Kenneth Davis and York Wilbern, "Administrative Control of Oil Production in Texas," *Texas Law Review* 22 (1944): 156-61.

II

The Battle Over The Oil Import Quota

TRADE POLICY UNDER TRUMAN

The politics of the oil import quota were intertwined with the politics of general trade policy during the period when pressures were developing for restraints on oil imports. Trade policy determined the parameters of the debate, while extension of the Trade Expansion Act was the vehicle by which Congress both granted the President authority to impose an oil import quota and pressured him into using that authority. Every administration since the second World War has been in favor of free trade, and in general, the executive branch had succeeded in lowering tariffs and other trade barriers for most goods until the 1970s. Until the reversal of the move toward freer trade the oil industry was the only industry able to obtain significant protection from foreign competition.

Schattschneider's classic work on the politics of the tariff described the difficulties Congress faced in developing trade policy before WWII. The Republicans, who controlled Congress in the 1920s, were blatantly in favor of protectionism while the Democratic platform declared in favor of duties using a costs formula. Since trade policy was made in Congress, not by the Executive, the 1930 Smoot-Hawley tariff act attempted to set rates for each product in the Ways and Means Committee. Schattschneider observed that this encouraged every interested industry to petition for relief from competition, overwhelming the ability of legislators to attempt to make intelligent decisions concerning the need for protection, even had they been so inclined. Instead they merely developed a decision rule which granted industries a politically acceptable level of protection.[1]

Big business, particularly the automobile companies, vehemently opposed the bill, and 1,028 economists sent an urgent letter to Hoover asking him to veto the bill. Hoover favored protection of US industries, which was the final reason for Smoot-Hawley. Even in his memoirs, Hoover denied that Smoot-Hawley had anything to do with the deepening of the depression, despite the obvious evidence. Smoot-Hawley resulted in average ad valorem rates on imports of 52.8 percent, the highest American tariffs in the twentieth century. The bill resulted in rapid retaliation from trade partners and a precipitous decline in world trade, from $35 billion in 1929 to $12 billion in 1933, with US exports falling from $488 million to $120 million during the same period. In 1932 the new, predominantly Democratic Congress attempted to lower tariffs but Hoover vetoed the bill.[2]

The appointment of Cordell Hull, who was closely identified with free trade, as Secretary of State, signaled FDR's support for tariff reduction. In March 1934, Roosevelt sent the Reciprocal Trade Agreements Bill, an amendment to the Tariff Act of 1930, to the House. The bill delegated to the President the power to negotiate foreign trade agreements which could reduce US tariffs by as much as 50 percent. The Democrats on the floor of the House adopted a series of compromise amendments, limiting the grant of authority to the President to three years and allowing Congress to terminate agreements after three years. The purpose of setting the time limit was to ensure that the President would have to return to the Congress for new authority, allowing the Congress to exercise oversight. Democratic leadership kept the debate firmly under control, defeating numerous other amendments offered by the Republicans. The amount of attention in 1934 to trade issues was a small fraction of what it was in 1930, with interest groups conspicuous by their absence. Business groups split over the bill, but their influence was not discernable either in hearings or in the final bill. The act irrevocably linked US domestic rule-making to international negotiations, ushering the United States into an era of bilateralism based on reciprocity.[3]

Manley suggests, following Schattschneider's lead, that the reason for the delegation of authority over tariffs to the executive branch was the realization by Congress that the issue of tariffs is so complex that it cannot be handled by that institution. In effect, by transferring the power over tariffs from Ways and Means to the executive branch Congress was acknowledging that it had changed from a revenue issue to a foreign policy question. The effect of this transfer of power is to shift interest group demands to the executive branch.[4] Congress then has the more politically rewarding task of oversight, in which it can take credit for advancing interest group concerns without the responsibility for the consequences of actually meeting these demands. The pressure is on the President to balance the political, economic and foreign policy implications of trade policy.

The trade act was routinely extended until 1948, when the Republicans regained control of Congress. In 1947, the US initiated the first round of multilateral trade negotiations in Geneva, resulting in the signing of a General Agreement on Tariffs and Trade (GATT) that fall, affecting about 54 percent of US imports, with an average reduction in tariff levels of 35 percent. With the Republicans in control of the Congress, there was some trepidation about the future of the trade program. A bill was introduced in the House limiting the Executive's negotiating authority during the Geneva negotiations. Truman worked out a compromise with the Republican leadership, in the form of an executive order establishing a formal mechanism to act on applications of domestic industry for escape clause relief.[5]

Betrand Gearhart (R-CA), chairman of the Ways and Means trade subcommittee, submitted a bill in 1948 extending the Trade Agreements Act one year in defiance of Truman's request for the usual three year extension. The Gearhart bill upgraded the Tariff Commission, giving it the responsibility to determine the point at which a reduction or an increase in the tariff on goods would injure the domestic economy or impair the national defense, referred to as a "peril point" provision. The next year the Democrats regained control of Congress and eliminated the Republican changes, extending the Trade Act three years retroactive to 1948. With the Democratic majority reduced after the 1950 election the peril point and escape clause provisions were restored as amendments to the 1951 Trade Agreements Extension Act. Two factors motivated the defection of Democrats; the increase in protectionism in representatives from newly industrialized border states and failure of Truman to use the escape clause to address perceived grievances.[6]

When Eisenhower took office in 1953 with a Republican Congress he faced a Ways and Means Committee opposed to the reciprocal trade policy, along with about two-thirds of the House Republicans. The chairman of Ways and Means, Dan Reed (R-NY) and the chairman of the Senate Finance committee, Eugene Millikin (R-CO), were both protectionist. The third ranking Republican on Ways and Means, Richard Simpson (PA), was a crusader for protectionism who would lead the fight for restrictions on oil imports for the benefit of the Pennsylvania coal industry. The new administration showed its free trade bias by requesting an extension of the Trade Agreements Act without change. The House Republicans gave Eisenhower a one year extension, added a seventh member to the Tariff Commission and established a Commission on Foreign Economic Policy to study the trade program. A more ominous development was the introduction in Ways and Means of a protectionist bill by Simpson, initiating the battle over oil import controls.[7]

Despite the protectionist sentiments of the Republicans in this period there was no attempt to shift control of trade back to the Congress. At most they attempted to put pressure on the executive branch by establishing mechanisms to identify industries with claims of damage due to free trade. The change in Presidents from Truman to Eisenhower had no effect on the commitment to free trade. The reasons for this support can be traced to the disastrous consequence of Smoot-Hawley. Three reports to the president in the early fifties, the Gray Report (Report to the President on Foreign Economic Policies, November 1950), the Paley Report (Resources for Freedom, June 1952), and the Bell Report (A Trade and Tariff Policy in the National Interest, 1953) all supported the policy of lowered trade barriers and acceptance of increased imports.[8] In this period of emerging American economic hegemony, the executive branch was committed to lowering world trade barriers.

INTERNATIONAL OIL MARKET 1945-1955

The rising flood of imports during the fifties was due to both the needs of the Majors to market their Middle East production and the addition of the second-tier international companies in the world market. The functioning of the world oil market during this period was determined by the attempts of an international cartel composed of the Seven Sisters to control prices and supplies. To operate this system of price control they required a system for determining prices. Before World War Two, basing the price of world oil on the US Gulf price was acceptable simply because it was the main source of supply (along with Venezuela) for the two main markets, the United States and Europe. During this period the Platt's price quotations for Texas well-head oil, adjusted for transportation costs to the Gulf, was used to determine the Gulf f.o.b. price. The Venezuelan crude f.o.b. price was then calculated as the Gulf price for similar quality, minus the US import duty of 10.5 cents a barrel.[9]

During the war the British government, purchasing large quantities of Iranian oil, protested the price charged, since they felt it to be ridiculous to be charged transportation from the Gulf of Mexico for Middle East oil. The outcome of this action was the establishment of a second basing point in the Persian Gulf, using the US Gulf price. In Aramco's negotiations with the US Navy it followed the precedent set by British Petroleum for use of Gulf prices for its oil.[10] The two basing point system required a determination of an equalization port, the destination at which oil from both sources would be priced equally, so to set the actual Persian Gulf price (determined by subtracting freight charges from the Middle East to the equalization point from the price at the equalization point). Without such a price schedule Middle East oil could undersell Gulf of Mexico oil in all European markets because of its lower cost of production, reducing potential profits. Exxon, Gulf and Shell had a vested interest in maintaining a market share for Venezuelan oil in Europe, while a price war had little appeal for anyone.

With the end of the war and price controls the US Gulf crude price underwent a series of increases. As production increased in the Middle East the point of equalization was established at England by June 1948, to provide markets for Middle Eastern oil. Pressure from the Economic Cooperation Administration, which was financing large purchases of crude oil and products under the Marshall plan, combined with the threat of intervention by the American Congress, encouraged the companies to shift the equalization point to New York in September 1949, following a suggestion made by Paul Frankel. By shifting the equalization point westward the European price could be lowered without affecting the Gulf price. Along with the shift of the equalization point, standard freight charges used to calculate equalization prices were lowered.[11]

During the next few years while freight charges rose the Persian Gulf netback remained constant, pricing Middle East oil out of the US East Coast market. Yet producers continued to export to the United States instead of diverting shipments to Europe and competing for market shares, sacrificing immediate gains for continued stability in world markets. This was only partially in agreement with the "parity" theory of pricing expounded by Walter Levy. Under this system the f.o.b. price of oil from each major source will be determined by competition at the destination where they are both competitive, i.e., New York.[12] As freight rates change f.o.b. prices should also adjust or the equalization point should shift. The fact that prices remained steady and export patterns did not change are evidence of collusion. By refusing to compete for markets the Majors were able to base world oil prices on US Gulf prices, with producers of low cost oil obtaining higher profit margins at the expense of greater market share.

M.A. Adelman of MIT considered the parity theory to be incomplete since it ignores comparative production costs. He also rejected a simple monopoly theory on the grounds that increased exports to the United States would have resulted in greater total profits despite the required drop in Persian Gulf netbacks, the expected behavior of a profit maximizing monopolist. The third explanation which he examined was that half of the Persian Gulf output was produced by companies with large interests in US producing and refining (Exxon, Shell, Socal, Texaco, Mobil, Gulf), who had to trade off between increased revenue from Middle East oil versus loss of revenue from domestic sources. However, Adelman felt that they had far more to gain from increasing imports than maintaining their domestic production.[13]

The explanation provided by Adelman for constraint by the Majors is political. As early as 1949, the National Petroleum Council stated that imports in excess of economic needs would hold back domestic exploration and development. Early the following year the Venezuelan government agreed to curtail exports to the United States, and a number of major companies, including Jersey and Shell, made substantial cuts in their shipments. A series of Congressional studies of the oil industry took place in 1949-50, focusing on injuries suffered by independent producers from oil imports. In 1953 the TRC began demanding from all producers a highly detailed statement of import forecasts, which gave each company detailed information on the others' activities. Since the TRC controlled the allowable rate of production from the Majors' existing investments in Texas, and thus their profitability, this was a strong incentive not to rapidly expand imports. The availability of information on each company's activities also made it easier to maintain an oligopolistic structure since it made it more difficult for a company to cheat.[14]

The rise in prices following the removal of price controls in the spring of 1953 illustrates the operation of the world market at that time. Phillips raised prices by 25¢ despite the existence of excess capacity in the United States, due to the willingness of the TRC to support the increase. Within a month producers in Venezuela and Kuwait followed suit, followed by the Aramco members soon afterwards. Two reasons are postulated for the willingness of Middle East producers to follow US prices; import constraints made market expansion impossible, and the change in concession agreements to include profit sharing based on posted prices gave the host governments a vested interest in high posted prices. In addition to restrictions on imports to the US there were also constraints to the growth of European demand. Coal producing countries restrained sales of fuel oil, heavy taxes on gasoline were in effect in all European countries, and gasoline consumption in Europe was inhibited by poor road conditions and high taxes on automobiles.[15]

Thus in the early 1950s the Majors had control of the world market in conjunction with the TRC. Prorationing in the US resulted in an artificially high Gulf price on which the world price was based. While Adelman may be right in attributing import constraint by the Majors to political pressure (along with the need for restraint to maintain the industry's united front and to protect their domestic investments) I suspect that they were also aware that the collapse of the prorationing system, due to imports, would result in a lower US price, and thus a lower world price. Steady, predictable growth in world supply and demand, while not allowing maximization of short-run revenue, allowed minimization of costs due to elimination of uncertainty. Given the historical tendency of the oil market to experience violent fluctuations of supply, profit maximization at this time did not necessary imply maximization of cartel revenues.

A number of American independents had been involved in the foreign oil industry in the period between the World Wars. Cities Service and Continental were exploring and producing in Mexico in the 1920s, Atlantic had integrated operations in Mexico and Venezuela and Sinclair was integrated throughout the Western Hemisphere. Indiana Standard was a major foreign producer through its Pan American company and Union Oil of California was active along the Pacific Rim.[16] However, the combination of Mexican nationalization and the Great Depression caused them to curtail their foreign activities. The best Western Hemisphere prospects had been cornered by Shell, Gulf and Jersey while searching for oil in the Eastern Hemisphere was an expensive and risky proposition for capital short companies without foreign markets.

Toward the end of WWII the independents began looking abroad once again. Prorationing, by restricting production from domestic operations, encouraged larger independent producers to look abroad for uncontrolled oil in the late 1940s and early 1950s. Other factors driving them abroad were the threat of increased imports from the Majors, the profits to be made abroad, and simply because to paraphrase Willie Sutton, "that's where the oil is." A vice president of Chase Manhattan Bank estimated in 1946 that future Middle East discoveries would reach 150 billion barrels, far more than the 27 billion barrels of known reserves or the estimate of 50 billion barrels of future US discoveries.[17]

THE ECONOMIC IMPACT OF PRORATIONING

The flood of imports which arrived during the latter part of the decade would not have been as threatening to the domestic industry had not the prorationing system resulted in the development of a high-cost domestic industry. A number of state regulatory commissions required wider spacing in new reservoirs and encouraged unitization for pressure maintenance and secondary recovery. World War II saw the formation of the Petroleum Administration for War (PAW), which adopted policies to increase efficiency such as requiring 40-acre well spacing for oil wells, a materials allocation policy which encouraged unitization, and the development of data on maximum rates of production (MER) for fields. The three years of operation under PAW regulations encouraged adoption of wider well spacing and use of MERs as the basis of production restraint.[18] Still, despite the measures taken to prevent waste the basic rationale for market-demand restrictions was protection of the small producer. By limiting the sum of field and well allowables to the market demand in a period, buyers were motivated to purchase the full allowable, preventing discrimination by pipelines and purchasers between fields and producers. By limiting the sum of state production the market price was maintained at a level which allowed smaller, more inefficient producers to continue operating.

The operation of the prorationing system required the active cooperation of Texas due to its disproportionate share of production and reserves, as can be seen in Table 2.1. Small producers soon found that they had more influence at the state level than the Major companies. The income of drillers, supply companies and private leaseholders all depended on Railroad Commission policies, giving them a greater influence than Major oil companies financed with out-of-state capital. One of the Commissioners, Olin Culberson, in 1941 stated that approximately one-sixth of the total population of Texas depended on employment in oil production, refining or distribution. The value of production to Texas included wages, royalties, taxes and equipment purchases, paid to citizens or the state government.[19]

Given the political importance of the local oil industry Texas politicians had a strong motivation to maintain prices at a level which allowed small producers to make a profit. Stability of the system was ensured by the participation of all the important producing states in controlling production. Of the seven largest producing states, five were market-demand states, with only California (which had a voluntary system of control) and Wyoming abstaining.

Since California production had only a limited impact on the national market and Wyoming's production was relatively small this meant that the five market demand states could effectively control price by determining supply.

TABLE 2.1 Distribution of US Production and Proved Reserves

	% of Production			% of Reserves	
State	1935	1948	1958	1948	1958
Texas	39.4	44.7	38.4	58.6	48.2
Louisiana	5.1	9.0	12.8	14.8	14.3
California	20.8	16.8	12.7	8.7	11.3
Oklahoma	18.6	7.6	5.6	5.7	6.1
Wyoming	1.4	2.7	4.7	3.1	4.0
New Mexico	2.1	2.4	4.0	2.3	3.6
Kansas	5.5	5.5	4.9	2.9	3.1
Total	92.9	88.7	83.1	96.1	90.6

SOURCE: API, *Petroleum Facts and Figures*, 1950, 1959

While these states apparently didn't actually collude to set price, there was substantial interdependency and degree of parallel behavior.[20] Direct cooperation wasn't necessary as Interior's predictions of market demand, combined with the reports of the majors of their intentions to produce and import in the TRC monthly hearings, provided both the information needed to determine production rates and the mechanism to detect cheating and ensure cooperation. Given an industry in which prices were maintained at a constant level for a period of time despite the continued existence of significant overcapacity one would not require a "smoking gun" to conclude that some kind of collusion existed.

Regulatory agencies attempted to match up purchasers and suppliers. The major integrated companies controlled most of the pipelines and produced a substantial fraction of the crude oil. If left to themselves, the large companies would naturally purchase from either their own fields or sources that were the most convenient and least expensive. The regulatory agencies prevented this by limiting the total allocated production to the expected purchases for the next allocation period. The larger companies responded by shifting exploratory, development and purchasing activities to areas in which they owned a larger part of the producing acreage, disposing of pipeline facilities in areas in which they own little production. They also shifted operations to states which were more lenient in their regulatory stance.

The allocation of allowables among wells and the well spacing requirements were for the most part arbitrary or politically motivated. The formulas used for allocating a state quota among pools and wells were quite diverse but shared three elements: restraining production below the maximum efficient capacity of some pools and fields; taking in account depth of production in determining allowables; and adjusting allowable production for differences in well spacing. Close well spacing was encouraged by allowables which granted a "bonus" for a well regardless of spacing. In Texas, every leaseholder in a field was allowed to drill at least one well until the 1960s. Exceptions from production restrictions were often granted for discovery wells, marginal wells and fields, and waterflood wells.[21]

One result of overdrilling is that demand prorationing states were constantly producing below capacity, partially because of production control but also because short-run capacity is a function of the quantity of producing wells. The bulk of capacity restraint was felt by producers in Texas and Louisiana. What made this system of control even more inefficient was that the actual production restriction was on the most productive wells. The result of these policies was to expand the supply of economically marginal wells. As an oil field ages eventually production from wells will drop to the point where they would be classified as strippers. If a field is overdrilled production per well will be lower and their economic life will be shorter. Therefore prorationing not only kept uneconomic wells in production, but it also encouraged the creation of far more marginal wells than would have occurred under a more efficient system. As can be seen in Table 2.2, the growth of stripper wells coincided with a decline in productivity of "normal" wells during this period.

TABLE 2.2 STRIPPER WELL PRODUCTION 1945-60

Year	# Strippers	Production (1000 bbls)	Per Stripper (barrels per year)	Per Nonstripper (barrels per year)
1945	299,146	231,690	775	12,709
1947	294,672	279,659	949	11,985
1950	321,287	375,551	1,169	11,053
1953	332,552	383,947	1,155	11,859
1957	372,519	474,919	1,275	10,887
1960	403,323	577,151	1,431	10,636

SOURCE: API, *Petroleum Facts and Figures*, 1950, 1959, 1967

Proration also discouraged attempts to apply unitization. Since the opposition of even one leaseholder could prevent pooling of a field some sort of compulsory law was required at the state level. Among the seven large producing states only Oklahoma (1945) and Louisiana (1960) had statutes requiring unitization; with Oklahoma setting 63 percent and Louisiana 75 percent agreement of both operators and royalty owners as the limit needed for unitization. Opposition was due to independent operators who viewed compulsory unitization as "socialistic" and interference with "free enterprise."

Small oilmen were skeptical that a unit operation which would most likely be dominated by one or more major producers would be fairly administered.[22]

The economic waste due to the prorationing system is hard to determine, though it is measured in billions of dollars. In the East Texas field alone, it was estimated that 1,500 wells could have drained the whole field, instead of the more than 17,000 which were drilled and operated. Mckie and McDonald suggest that if excess well drilling nationwide was 50 percent of all wells this would imply excess development costs approximating one billion dollars per annum during the late 1950s. Kahn quotes an industry estimates of $500 million per year as the cost of excess drilling. Adelman, in his 1964 article, estimated that jettisoning stripper wells and unitizing fields would result in a saving of $2 to 2.5 billion. He breaks this numbers down by assigning $1.3 billion to over-drilling costs and $800 million to excess operating costs.[23]

CONGRESS AND OIL TAX POLICY

The battle over oil import quotas was influenced by the politics of oil taxation for a number of reasons. First, the collective interest in favorable tax policies united all domestic oil companies, from the small independent producer to the international major, limiting the intensity of conflict over oil imports. Second, both tax and trade policy were determined in the House Ways and Means Committee and the Senate Finance Committee, requiring careful balancing of issues within the committees to deal with these overlapping interests. Finally, the justification for oil industry tax breaks, the importance of a strong domestic industry, would be used to rationalize oil import quotas.

In the Revenue Act of 1913, oil producers along with other mining interests were able to obtain an allowance for capital consumption set at 5 percent of the gross value of annual production. The Revenue Act of 1916 removed the percentage deduction, replacing it with an allowance for depletion, not to exceed the actual cost of discovery. During WWI, the Senate Finance Committee, advised by industry representatives that deduction of the cost of discovery was insufficient to finance replacement of reserves, recommended reduction of taxes to encourage investment in strategic resources. The War Revenue Act of 1918 allowed producers to base their deduction either on discovery cost or the fair market value at the date of discovery of their properties. In many cases the market value of mineral properties greatly exceeded direct discovery costs. In the Revenue Act of 1921 discovery value depletion was limited to 100 percent of a property's net income while the Revenue Act of 1924 tightened the limitation to 50 percent of a property's net income.[24]

During the early 1920s numerous disputes arose between producers and the Bureau of Internal Revenue over determining the discovery value of new oil wells. A Senate select committee to investigate the Bureau of Internal Revenue held extensive hearings during 1925, and issued a report recommending a revision of existing laws governing the appraisal of oil and gas properties, urging adoption of a depletion allowance based on a percentage of gross sales. In 1926 the Senate Finance Committee included in its report a suggestion by the President of the Mid-Continent Oil Producers Association to Senator David Reed of Pennsylvania (whose father had been an original member of the Mellon syndicate in Gulf) that a deduction of 25 percent of gross income should constitute a minimum. Senator James Couzens of Michigan argued on the Senate floor that only cost depletion should be allowed, but he was overridden. The Senate amended the bill to provide for a 30 percent deduction, as suggested by Senator Matthew Neely (WV). The Conference Committee worked out a compromise that provided for a 27.5 percent depletion allowance, applicable to income from all oil and gas production.[25]

Roosevelt's Treasury Secretary, Henry Morgenthau, first suggested elimination of percentage depletion in 1933, and repeated the message throughout the Roosevelt administration. Congress responded by passing a tax bill in 1944 over Roosevelt's veto which extended the depletion allowance to ten new minerals. The same year Interior Secretary Ickes, while refraining from an endorsement of the depletion allowance, embarked on a heated campaign against a Treasury ruling prohibiting deduction of drilling costs as intangible expenses. The option to expense intangible development costs and the costs of unsuccessful exploration developed through administrative decisions, not legislation. In 1917 incidental expenses of drilling wells were made deductible from gross income as operating expenses and in 1918 expensing of dry holes was allowed. Gradually, incidental expenses of drilling came to include all items of an intangible nature.[26]

President Truman in his annual tax message of January, 1950, pleaded for the reduction of the depletion allowance from 27.5 percent to 15 percent. Secretary of the Treasury John W. Snyder, before the House Ways and Means Committee, argued for the President's position as well as for the abolition of the intangible drilling allowance for producers. Following the Truman tax message Ways and Means held lengthy hearings at which 98 witnesses testified before the Committee either in favor of keeping oil and gas at 27.5 percent or on raising rates for other substances. The only two witnesses favoring lower depletion allowances were from the CIO and the AFL, and in neither case was depletion a main concern of the witness. The 1950 attempt to extend the depletion allowance to additional minerals failed due to the request for increases in taxes to pay for the Korean War.[27]

In Truman's 1951 Revenue Act he again proposed reduction of the oil and gas depletion allowance from 27.5 per cent to 15 per cent as well as cuts in other depletion allowances. During the Ways and Means Hearings there were 59 pro-depletion witnesses and 32 other pro-depletion statements submitted for the record. The only opposing voices were the two labor witnesses, joined this time by representatives of the National Grange and the American Farm Bureau.

The Ways and Means Committee not only decided against reducing the depletion allowance but also doubled the allowance for coal from 5 to 10 per cent, and extended the 5 per cent rate to twenty-one new substances. The Senate Finance Committee granted the allowances to ten additional resources, ensuring that every Finance Committee member with any mineral production in his state was covered by the changes.[28]

The lonely battle against the depletion allowance was led by the freshman senator from Minnesota, Hubert Humphrey. Humphrey noted that the amount of percentage depletion bore no relation to the capital cost of property, but only to income, at a cost to the nation in 1947 of about $400 million annually. He also pointed out that depletion was of prime benefit to large producers with capital investment of $100 million or more, who were the recipients of 80 percent of the depletion benefits. The advocates of the depletion allowance included spokesmen for the large integrated oil companies and the API, the smaller independent oil producers, and Congressional and state representatives from oil-producing states. The most vocal supporters in the Senate included Lyndon Johnson along with John Carlson (R-KS), Matthew Neely (D-WV) and Robert Kerr (D-OK). Johnson emphasized national security, suggesting that encouragement of domestic production was required to provide excess production capacity equal to 15 percent of annual production for unforeseen emergencies. He also argued that a reduction in the depletion allowance would reduce production, diminish tax payments to the federal government, and force oil companies to raise prices, leading to lower consumption, and a possible reduction in economic growth. Nash concluded that the reason for the overwhelming defeat of the Humphrey amendment (a 71-9 vote) was satisfaction of legislators with the performance of the industry and their reluctance to tamper with established policy.[29]

The combination of little or no support in Ways and Means and the use of the closed rule when tax legislation went to the floor of the House precluded any attempts to reduce the depletion allowance in that chamber. Numerous minor attempts to lower the depletion allowance were undertaken in the Senate, largely in the form of floor amendments or in the form of applications and redefinitions of regulations by the Treasury Department. An amendment to cut the depletion allowance from 27.5 percent to 15 percent was introduced by John Williams (R-DE) in 1958, and one calling for a gradual cut over a three-year period from 27.5 percent to 20 percent in 1964. Paul Douglas (D-IL) offered a graduated form of depletion allowance to protect wildcatters in 1954, but was unable to obtain a roll call vote, though it was voted on and rejected in 1958, 1959, 1960, 1962 and 1964. The 1958 battle exposed the potential for a split between the larger companies and the independents. With thirty Senators voting for a cut in the depletion allowance, Russell Long told William Proxmire (D-WI) to concentrate on taxing the multinationals since oil imports would soon eliminate the domestic industry. The only serious threat until 1969 was an indirect administrative approach by Kennedy in the 1963 tax cut bill, proposing modest limits to both the depletion allowance and the foreign tax credit.[30]

Once reason for the failure to change the depletion allowance was the paucity of pressure from the Executive branch during the 1950s and 1960s. During the Eisenhower administration neither of his Treasury Secretaries were motivated to attack this indirect subsidy. George M. Humphrey had headed and later returned to M.A. Hanna and Company, whose interests included sizeable investments in oil. His successor, Robert B. Anderson, had been a member of the National Petroleum Council, a director of the American Petroleum Institute and the Independent Petroleum Institute of America, and president of the Texas Mid-Continent Oil and Gas Association.[31]

It was reported in 1970 that Anderson entered into an agreement with oil company officials, prior to his becoming Secretary of the Treasury under which Anderson would receive $450,000 over four years starting during his term in office. A second $450,000 was a contingent payment dependent on oil revenues (and thus oil prices) from properties whose lease was held by a subsidiary of Standard of Indiana. However, documents showed that Anderson only received $290,000 as of 1963. According to James Hill, an former associate of Sid Richardson, Anderson had advised that he would require an one million capital gain to continue in government, and Richardson proceeded to see what he could do.[32] The years of the Humphrey-Anderson reign at Treasury was the only time during the 1950-1969 period that there was no attempt on the part of the Department to investigate the depletion allowance. While Kennedy and Johnson (as President) were less friendly to the industry, reducing the depletion allowance was not a priority of their administrations.

The other important tax break worth expending political capital to defend was the foreign tax credit, which granted a credit toward domestic taxes equal to the taxes paid to oil producing countries. While an income tax credit was available to American firms doing business abroad since 1918, the credit for oil taxes was a post WWII development. The death of the dictator Gomez in Venezuela resulted in a surge of resentment against the foreign oil companies, and new concession laws increased the royalty tax to 16 percent in 1938. Taxes were increased again in 1941 and 1944, and finally 50-50 (50 percent) profit sharing was instituted in 1948. The Venezuelans, aware of the low cost of Middle East oil, sent a delegation to the Middle East to push the concept of the 50-50 profit split. King Saud had already increased demands on Aramco for additional revenue even before the arrival of this delegation. The State Department solution for providing more revenue for the Saudis was for Aramco to hand to hand back a large portion of its concession to the Saudi government which could then raise income by offering it to other companies. This suggestion was not received with enthusiasm by the company. A second alternative was for Aramco to pay a higher royalty to the Saudis, but this would cut deeply into company profits. The solution finally adopted was for Aramco to accept a tax on profits and, with US Treasury permission, have this tax offset against US corporate taxes.[33]

The adoption of the tax credit allowed the oil companies to shift a large proportion of the burden of providing revenue for the host country to the US taxpayer. It also provided the international oil companies a powerful motive for shifting profits upstream (to production) where they could take advantage of tax credits while keeping profits low in the refining and marketing sectors, discouraging entry of firms which would provide outlets for competitors' crude.

DEVELOPMENT OF PRESSURE FOR OIL IMPORT QUOTAS

Given the commitment of the Eisenhower administration to free trade, and the desire of both the Majors and the second-tier companies to import oil, there arises the question of how the domestic oil industry managed to obtain mandatory restrictions on oil imports and maintain them for fourteen years. I conclude that three factors interacted to make an oil import quota possible: the political power of small producers in the major oil producing states of Texas, Louisiana, and Oklahoma; the ability of quota supporters to obtain powerful allies (especially the coal industry) through log rolling and shared interests; and industry compromises which prevented a schism between the international oil companies and the domestic producers.

Domestic producers were wary of the threat of Middle East oil as early as 1943. The State Department had supported the extension of Lend-Lease to Saudi Arabia, a proposal which was encouraged by Harry Collier, president of Socal, and William Rodgers, president of Texaco. Collier and Rodgers offered to create an oil reserve in Saudi Arabia to sell oil below the market price to the US government. The State Department's Committee on International Petroleum Policy proposed the formation of a Petroleum Reserves Corporation (PRC) which would acquire option contracts on Arabian oil. Harold Ickes and the Navy then proposed that the PRC acquire reserves by purchasing Casoc stock. Rumors of negotiations between the government and Socal and Texaco led to intense opposition by other oil companies, especially Exxon and Mobil, who were opposed to the idea of governmental competition. The failure of the PRC scheme led to an attempt to negotiate an petroleum agreement with Great Britain. An Anglo-American Petroleum agreement was negotiated in 1944, but withdrawn by Roosevelt from Senate consideration as a result of strong industry opposition. The essential purpose of the agreement was to reduce the drain on Western Hemisphere oil reserves by developing Middle Eastern resources. The pact would have established an Anglo-American oil cartel, which was supported by the Majors, given the inclusion of an antitrust immunity clause. Opposition by domestic producers aware of the threat of cheap foreign oil, combined with lukewarm support by the international companies once denied the antitrust immunity clause, doomed the treaty.[34]

A severe shortage of marketable petroleum plagued economic recovery during the first three years after the war due to shortages of tankers, drilling equipment, pipelines, and construction materials. This delayed efforts by oil producers to repair and expand output that had been depleted by overproduction during the war. With the lifting of wartime price controls, oil prices doubled, stimulating an increase in exploratory drilling and development of domestic production capacity. Meanwhile, with the encouragement of the Truman administration the international oil companies increased their imports to the United States, entering into long-term tanker contracts and expanding east coast refineries. While these actions helped alleviate the heating oil crisis without hurting the domestic price structure, they left a residue of permanently expanded imports which increased protectionist sentiments among domestic producers.[35]

During the late 1940s there was a spirited dialogue within and without the executive branch concerning the security implications of petroleum supplies. In 1945 the Petroleum Industry War Council adopted a resolution saying that oil imports should be restricted so as not to disturb the domestic producing industry. The Special Committee Investigating Petroleum Resources issued a report in January 1947 concluding that nothing should be done to weaken the productive capacity of the domestic petroleum industry. The Interior Department, reflecting the views of the independent oilmen, felt that America was self-sufficient for the immediate future. The State and Defense departments disagreed with this position. Louis Johnson, the Secretary of Defense, in a letter to Lyndon Johnson stated that "Our national security in time of a major war, from a petroleum standpoint, is still dependent upon availability of large amounts of foreign oil." The National Security Resources Board conducted a staff study on national security implications in 1948 prepared by Wallace E. Pratt, a former Exxon executive. The study recommended that the rest of the world be supplied from the Middle East, with the US requirements filled by domestic and South American sources. Excess capacity up to 1 million bpd should be encouraged by retaining production controls and price supports, and imports should be encouraged at a level equal to this excess capacity. No mention was made of the use of storage facilities for development of a strategic reserve.[36]

Political pressure to restrict oil imports began to be felt during this period. The drop in demand for petroleum products in 1949, combined with the increase in imports (the US became a net importer in 1948 as can be seen in Table 2.3) caused state prorationing commissions to significantly restrict output for the first time since the war. This caused domestic producers to turn to Congress to alleviate their distress. The IPAA formed an Import Policy Committee to lobby the Senate for an amendment to the Reciprocal Trade Act which would limit oil imports to five percent of domestic consumption, using the justifications of national security and international monopoly. At the same time IPAA officials went to New York to discuss voluntary import reductions with the Majors. The international companies described the situation as a temporary problem, produced by a brief recession which caused domestic demand to level off, while European markets were absorbing less oil than expected. They defended the need for imports by the need to avoid long-term depletion of US reserves. Their fallback position was that if imports were excessive, the sensible way to control them was by voluntary action among importers, avoiding governmental intervention which would be sure to result in market distortions.[37]

TABLE 2.3 US OIL PRODUCTION AND IMPORTS: 1946-1960
(thousand bbls)

Year	Domestic Production	Net Imports Crude	Refined	Total
1946	1,851,748	43,630	-59,077	-15,447
1947	1,989,850	51,177	-56,265	-5,088
1948	2,167,264	89,357	-35,887	53,470
1949	1,999,215	120,617	-4,434	116,183
1950	2,155,693	142,891	56,064	198,955
1951	2,452,676	150,469	3,673	154,142
1952	2,513,733	182,895	7,424	190,319
1953	2,596,166	216,524	14,384	230,908
1954	2,567,628	225,880	28,342	254,222
1955	2,766,325	273,850	47,526	321,376
1956	2,910,514	313,209	54,996	368,205
1957	2,912,143	323,012	44,390	367,402
1958	2,744,152	343,661	176,290	519,951
1959	2,895,671	349,818	222,698	572,516
1960	2,915,365	368,488	221,717	590,205

SOURCE: API, *Petroleum Facts and Figures*, 1959, 1967, *Basic Petroleum Data Book*, 1984.

In March 1950 the coal industry began to articulate its concerns with imports in a report by the National Bituminous Coal Advisory Council to Interior Secretary Oscar Chapman. The Council report stated that the damage to the coal industry was due to imports of residual oil, especially from Venezuela. Residual oil is used as a boiler fuel, competing directly with bituminous coal. The report stated that the short-term losses were being felt by domestic coal, oil and railroad operators, but in the long run the cost would be the danger of dependence on foreign sources of energy.[38]

In March, 1950, Senator Thomas (D-OK) introduced a bill to limit oil imports to 5 percent of domestic demand. A Thomas proposal for an import limitation had been defeated the previous September by a single vote, 41-40. Thomas used both the national security argument, stated as a desire to avoid a dangerous dependence upon foreign sources, and the need to prevent a net loss of revenue by reason of contraction of the tax base, to justify quotas. By April pressure for action intensified as Senator Neely proposed raising the crude oil tariff from 10.5 cents to $1.05, rejected by the Ways and Means Committee in June. The IPAA presented three alternative ways of dealing with imports to a Senate Labor Subcommittee headed by Neely; a quota, an increase in the tariff to $1.05, or a requirement that imports and exports of oil be balanced. The coal industry also expressed its displeasure with residual imports to the committee, but the State Department supported the contention of the importing companies that the problems of the coal industry were not due to imports. Testimony by farm groups to the House Interstate Commerce Oil Subcommittee, headed by Oren Harris (D-AR) was divided; some groups expressed concern about price increases if independents were forced out of the industry, while the American Farm Bureau worried about the effect on agricultural trade of the imposition of import barriers.[39]

A memorandum from Interior to the White House is evidence that the executive branch was well aware of the various interest groups with a stake in oil import policy. It suggested that oil distributors and energy consumers were against import restrictions, domestic oil companies, oil producing states, coal operators and labor had an vested interest in maintaining high domestic prices through import restrictions, while the international American oil companies were split between the desire to increase imports and to maintain high prices for their domestic output. Federal policy was in favor of oil imports with regard to trade policy, maintaining competitive markets, and conservation of domestic resources. The reasons for the federal government to restrict imports would be to either protect the coal industry or smaller domestic oil firms against competition from the Majors.[40] The outbreak of hostilities in Korea that summer ended the pressure against imports as both military and domestic demand for liquid fuels was stimulated by the war. By August the Texas Railroad Commission was raising production allowables.

Excess capacity was on the rise again by 1952, and production allowables declined steadily during the next two years, generating new pressure by independent producers for restraints on imports. Net imports of crude oil increased from 184 million barrels in 1952 to 274 million barrels in 1955. This increase was primarily from smaller international firms, a surge in imports of Canadian crude and a dramatic rise in imports of petroleum products. The problem faced by the Majors was that by exercising restraint they merely left the field open for the second-tier companies to increase their share of imports. The dilemma for the Majors was if they didn't practice restraint three things could occur: the prorationing system could collapse, causing a significant decline in the US price and thus the world price; the independents would have a stronger political case for pressing for mandatory oil import controls; the political support of the independents for tax breaks could be jeopardized, threatening the depletion allowance, intangible expensing and favorable treatment of foreign taxation, at least as they applied to the Majors.

In 1953, the Reciprocal Trade Agreements Act came up for renewal, and it became the focus of a renewed effort to control oil imports. The major source of pressure for import controls was from the coal companies, due to the increase in the importation of residual oil from Venezuela. Crude oil imported from the Middle East did not directly compete for markets with coal since it was a lighter grade of oil which produced a greater proportion of higher quality products when refined. Venezuelan oil, for the most part, consisted of heavier crude which when refined provided a higher proportion of residual oil. Because of its low value relative to transportation costs it was not economic to ship residual oil to European markets, but fuel oil distributors on the east coast of the United States were an ideal market. Given that the market for coal had been declining at an annual rate of 6.6 percent since 1947 due to competition from cheap natural gas and the loss of railroad and shipping markets to dieselization, the increase in residual imports was competition the coal industry, plagued by overcapacity, was unwilling to tolerate. The loss of coal to residual imports was the equivalent of 31 million tons, compared to the 105 million tons lost to dieselization and gas home heating, between 1946 and 1952. However, there was little hope of regaining the these markets through political action.[41]

While the coal companies and the United Mine Workers (UMW) had an obvious interest in restricting imports the railroads were a bit ambivalent. Many of the railroads in the middle Atlantic states depended upon coal shipments as their most important source of revenue. These same railroads were rapidly dieselizing their own engines and therefore stood to benefit from low diesel-oil prices and thus from crude oil imports. Therefore the Eastern railroads found themselves balancing the gain from restrictions on imports of residual fuel oil with the losses from reduction in foreign trade and more expensive diesel fuel. The railroads were not firmly opposed to oil imports but chose to support restrictions as a favor for a customer. Their optimum strategy would be to work to restrict residual imports while supporting importation of crude oil.[42]

The domestic petroleum industry was divided in its stance on oil imports. Domestic producers agreed that imports were excessive, but also believed themselves committed to a "free enterprise system," in which direct intervention by the federal government was anathema. Domestic oil producers without foreign wells were organized in the IPAA which took the strongest stance for import controls. But even the petroleum independents were somewhat passive relative to the behavior of the coal interests. Bauer et. al. report that various sources hinted at pressures exerted by the big five of the oil industry, who were the major customers for the Texas and Oklahoma producers who wanted to stop the foreign imports, to discourage support for restrictions. There was at least a general consciousness that producers might be antagonizing their primary customers if they were too aggressive. The second tier companies began to find themselves with investments abroad, causing them to become ambivalent concerning oil imports. They also were reluctant to call for federal controls for imports in the fear that controls might be extended to the domestic sector as well. The international majors opposed any kind of legislated restrictions on oil imports, though they were careful not to antagonize other members of the industry, avoiding the appearance of pressuring the smaller firms The major oil companies were afraid to be overtly politically aggressive because of their problems with antitrust, their image as a symbol of big business in the public mind, as well as the need to preserve a coalition in support of tax breaks for the industry.[43]

Among the groups organized against oil import quotas was the Venezuelan government, which through its Chamber of Commerce hired the Washington law firm of Cleary, Gottlieb, Friendly, and Ball in 1951 to mobilize opposition. The Venezuelan business community, along with some oil companies, undertook an energetic and astute public-relations program in the United States. They focused on interests that would be recognized as legitimate by Congress. The Venezuelans made studies of the exports of various American industries to Venezuela, then sent letters based on this research to manufacturers in those fields. In New England the approach was to both businessmen who sold to Venezuela and to industries and utilities which burned residual fuel oil. This resulted in protests flooding into Congress from such industries as traditionally protectionist textiles, creating an image in Washington of violent New England opposition to the bill, and the New England congressmen took the lead in opposing it. Aiding in the efforts was the National Council of Importers, which had been waging a small-scale fight for free trade for a number of years.[44]

A number of groups opposed oil import quotas not because of a specific interest in oil but due to their position on the issue of free trade. Among the leaders for free trade were the automobile companies and the major banks. By the 1950s the National Association of Manufacturers had been converted from its protectionist stance to a supporter of free trade. However, some ambivalence concerning free trade existed among businessmen. While the US Chamber of Commerce supported the Trade Agreements, it also suggested excluding those imports which subject domestic producers and manufacturers to unfair or unreasonable disadvantage in competition. One factor weakening the ability of exporters to counter protectionist pressure is that, outside of agricultural interests, most comapnies during the 1950s were largely preoccupied with the much larger domestic market and did not place a high priority on developing and protecting foreign markets.[45]

When Eisenhower in the spring of 1953 asked the new Republican Congress for a one year extension of the Reciprocal Trade Act, the measure went automatically to the Committee on Ways and Means. Although Representative Kenneth Keating (R-NY) introduced a bill (HR 4594) to extend the Act for a year, the committee chose instead to hold hearings on HR 4294, introduced by Richard Simpson (R-PA), which imposed quotas or import taxes on lead, zinc and petroleum imports, and required the President to follow Tariff Commission recommendations on peril points and escape clauses. The IPAA supported the bill, which limited crude imports to 10 percent of domestic demand, while coal interests testified in favor of the clause limiting imports of residual oil to 5 percent of demand. Some independent oilmen opposed the quota on residual oil imports on the ground that it would lead to greater imports of crude. The importing companies and oil marketers testified against quotas and during the hearings Jersey Standard announced a cut in oil imports while Socony-Vacuum reduced imports for the second quarter of 1953.[46]

As the result of a compromise between protectionists and the Administration, Simpson introduced 2 bills: HR 5495, which contained two major amendments to existing law, expansion of the Tariff Commission from six to seven members and reduction of the period for recommending action on requests for tariff relief, and HR 5496, a far more protectionist bill. HR 5495 was reported favorably by a vote of 23-2 on June 9 and the bill sailed through the House after defeat of Democratic attempts to delete the Simpson amendments. On June 26, the Senate Finance Committee approved the bill after deleting the seventh member for the Tariff Commission and rejecting attempts to add oil quota amendments. The Senate passed the bill by voice vote on July 2, and after minor changes in conference passed both Houses in early August. The Trade Bill extended the Act for one year and established a 17 member Commission on Foreign Economic Policy to study the trade program.[47]

In July, the attention of the Ways and Means committee turned to HR 5894, introduced by Simpson to replace HR 5496, which was basically the Trade Bill with quotas on imported crude and residual oil, and protection for the lead and zinc industries. President Eisenhower, trapped by a question at a press conference from a reporter who had been primed by the opponents of the bill, was forced to oppose the bill. The Ways and Means Committee was split, with the Republican leadership opposed to the majority of the Committee's Republicans, but the bill was reported favorably 15-10. Led by Sam Rayburn and Majority Leader Charles A. Halleck (R-IN) a motion to recommit passed the House, 242 to 161.[48]

The Commission on Foreign Economic Policy, established by the extension of the Trade Bill, had a membership of seven Presidential appointees, five Senators, and five Representatives, chosen in August 1953. President Eisenhower chose Clarence Randall, chairman of Inland Steel, who was in favor of free trade, to head the commission (which became known as the Randall Commission). Included among the members were Senators Eugene Millikin and Harry Byrd (D-VA), and Representatives Reed and Simpson. Randall, in order to counter the dissent of Simpson and Reed, attempted to court Millikin, chair of the Senate Finance Committee. The process was one of successive surrender, point-by-point, on the assumption that if Millikin and others got virtually all that they asked, they would go along with the report. The draft report called for further tariff-cutting powers for the president, but also wrote in what were key points for the protectionists: the peril-point and escape clauses. The acceptance of the escape clause was a key protectionist victory, since due to the uncertainty of the consequences of import development both the Japanese textile industry and the big oil companies followed policies of voluntary market restriction, according to Bauer et al.[49]

A coalition of coal producers, railroads, and organized labor formed the Foreign Oil Policy Committee which presented a well-documented and broad-based brief against excessive importation of residual oil to the Randall Commission. The Committee asserted that a double standard for oil-product imports, (the tariff on gasoline was 51 cents per barrel, but only 5 cents for residual oil) caused imports of residual oil to increase 136 percent while imports of more valuable distillates declined 18 percent. The coal alliance claimed that residual imports had caused lost wages amounting to $340 million, with about the same amount lost in rail-freight revenues.[50]

The IPAA began 1954 by deciding not to press for congressional action to restrict imports. The reasons for this reticence including a desire to wait for the Randall Commission to report to the President, the need to demonstrate that they had pursued other avenues of relief before again asking Congress for protection, and uncertainty with regard to the attitude of the new Congress. The president of the IPAA met with representatives of the major importing companies to attempt to arrange a mutually satisfactory agreement to voluntarily reduce imports. A resolution calling for unspecified legislation to deal with the problem was passed in May, with the encouragement of both the President of Humble Oil and the Vice President of Union Oil. By November, the mood of the IPAA membership had hardened, and it decided to return to its policy of pushing for a quota or tariff.[51]

The IPAA joined together with Texas independent producer groups in December 1954 to avoid the perceived fiasco of the Simpson bill. While the IPAA had endorsed the Simpson bill and worked for its passage, a number of Texas producers attacked the quota plan as unfair, clumsy and full of bugs. The independent producers met in December and issued a joint resolution declaring that oil imports had become so excessive as to "undermine the security of the United States and do serious damage to the economies of the oil-producing states and the nation." While the IPAA and five of the Texas associations committed themselves to any course of action which would bring relief, the Texas Independent Producers and Royalty Owners (TIPRO) limited its support to only an equalization tariff; a tax on imported oil that would eliminate its price advantage. TIPRO felt that a quota system would place domestic producers in a preferential position instead of one of equality with foreign producers.[52]

The 1954 elections returned a Democratic majority to Congress. The new Ways and Means chairman was Jere Cooper (TN) who was committed to lower tariffs along with his right-hand man, Wilbur Mills (AR), third-ranking Democratic member of Ways and Means. The chairman of the Senate Finance Committee was Byrd of Virginia, a traditional Southern Democrat and therefore not a protectionist. Sam Rayburn, Speaker of the House, and the Senate majority leader, Lyndon Johnson, were long-time supporters of both reciprocal trade and the domestic oil industry. Rayburn labeled Cooper's

Reciprocal Trade Renewal Bill House Resolution 1, signaling that this bill was of top importance. The bill was ready on the first day of the session, and five days later, the president sent a special message to Congress supporting renewal legislation.[53]

The general expectation was that a Democratic Congress would be more sympathetic to liberal trade. However, Democrats from areas with high unemployment like Oklahoma, Appalachia and Rhode Island, and the Georgia and Alabama delegations, pressured by demands from the textile industry, came out for protectionist legislation. The Ways and Means Committee reported HR 1 on February 14 by a vote of 20-5, voting down a series of restrictive amendments including residual import controls. The House, with a liberal-trade leadership, managed to complete action on the three-year renewal on February 18, 1955, six weeks after the session began. The bill passed the House, 295 to 110, but only after a rule to bar amendments was adopted 193 to 192, and a move to send the bill back to committee was defeated 206 to 199.[54]

While the battle over protectionism raged in Congress the administration was examining the problems of the energy industries through a Cabinet Committee on Energy Supplies and Resources Policy. It was headed by the chairman of the Office of Defense Mobilization (ODM), Arthur S. Flemming, with the secretaries of Defense, Interior, Commerce, Labor, State, Treasury and the Justice Departments as its members, but excluding the Council of Economic Advisors (CEA). In its final draft the committee concluded that oil imports should be kept at the 1954 ratio of imports to production (gross crude imports = 10.3% of production, the production figure excluded production of natural gas liquids) by voluntary action to avoid the necessity for governmental action. The committee's case rested on the national defense rationale of maximizing domestic production, ignoring the alternative suggestion of purchasing low-cost foreign oil to store domestically, suggested by the President. Despite reservations, Eisenhower allowed distribution of the amended report on February 26, 1955, but downgraded its status to a report of a advisory committee rather than of a cabinet committee, avoiding the impression of a presidential endorsement for oil import restrictions.[55]

In the Finance Committee discussion focused on the details of oil quotas, centered around the Neely amendment. The Neely proposal would restrict oil imports to 10 percent of demand, but also restrict residual oil imports to 10 percent of residual demand. The immediate effect would be to reduce imports of crude by about 9.5 percent, but reduce imports of residual by 60 percent. As the price of residual rose and use declined, residual imports would be forced even lower, allowing an increase in crude imports. During hearings before the Committee Eugene Holman of Standard Jersey and Sidney Swensrud of Gulf endorsed the cabinet fuels committee's recommendation that imports be held at the 1954 ratio of domestic production but warned they would only practice restraint as long as other importers cooperated. The IPAA and TIPRO both supported the Neely amendment on the principle that imports should be controlled.[56]

With the Finance Committee evenly split Senator Milliken came to the support of the administration, and on April 26, the Committee adopted a substitute for the Neely amendment which gave the president authority to fix quotas on imports if in his opinion they injured domestic industries vital to national security. The deletion of the mention of specific commodities was construed as an administration victory, but it would be this amendment which would be used to establish quotas on oil in 1959. The escape clause was changed so that relief could be considered for any product which was suffering competition, even if the industry as a whole was thriving. The whole bill was favorably voted, thirteen to two, by the Finance committee the same day. Passage of the bill in the Senate depended on a statement by Frank Carlson, a friend of the President as well as a sponsor of the Neely amendment, that the administration would use the quota powers in the bill, committing the administration to a deal. The bill then passed the Senate, was adopted by the House, and passed on May 15.[57]

The mechanism for determining whether the level of imports threatened national security involved a two-step procedure. The director of the ODM determines whether there was a threat to the national security and informes the President. If the President agrees that there was reason for such a belief he then initiates an immediate investigation to determine the facts. If this investigation confirms the threat to national security the President takes such action as he deems necessary to adjust the level of imports to remove the threat.[58]

In late July 1955, twenty-seven senators wrote the ODM director, pointing out the increase in imports during the first 6 months of 1955 over the previous year. Flemming responded by requesting data from oil-importing companies on their current level of imports and their future plans, as well as a statement of policy. He found that there had been a 15 percent increase in crude oil imports during the first seven months of 1955 and a 23 percent increase in imports of residual oil. Flemming wrote letters to 18 importing companies warning that if they did not reduce imports the Government would have to take action. Both the NPC and the IPAA called for some sort of quota, with the latter suggesting a $1 per barrel tariff as an alternative. The Chairman of Phillips, K.S. Adams, expressed concern that the government might freeze imports for firms in relation to their 1954 imports, hurting firms which had recently invested abroad. The IPAA agreed that the big importers would have to make room for newcomers.[59]

The presidential advisory committee, aware of the desire of New England consumers to continue importing residual oil, did not react negatively to the increase in residual imports, informing the importing companies that the planned levels of imports were acceptable. State Department concern about import restrictions on Western Hemisphere exporters (specifically Canada and Venezuela) also influenced Flemming to state that the level of imports from this region was currently acceptable. He suggested that the importers from the Middle East should reduce their planned imports of crude oil for the last nine months of the year by 7 percent. With the exception of Texaco, the Majors accepted the new recommendations of the ODM director. However a half dozen importing companies reported plans that were significantly above the limits.[60]

Despite the cooperation of the Majors, oil imports continued to increase in 1956 for a number of reasons. New participants in the Iranian concession were trying to export more to the US while several third parties had initiated Middle East imports under their own logos. Major oil discoveries in Alberta fueled an increase in Canadian imports. Given the general awareness in the industry that there was widespread violation of the advisory committee's recommendations, the expectation of the imposition of quotas led to an increase in imports by companies in order to increase their base level as a protection against the imposition of involuntary quotas.[61]

The increase in the level of imports caused Flemming to request an additional 4 percent decrease in Middle East imports for the second half of 1956, which translated in a 25 percent drop from planned imports. In June Representatives Frank Ikard (D-TX) and John Jarman (D-OK) attacked the administration for surrendering to the oil importing companies, and warned that if action wasn't taken mandatory quotas would be pursued. Congressional pressure on the ODM surfaced again in July, when 31 senators formally submitted a letter to the director requesting information and action, while clear warnings of intent to create a mandatory program were issued. In August nineteen domestic trade associations presented a petition under the 1955 Trade Act requesting a restriction on oil imports, led by the IPAA. The association's president, Robert Wood, took its case to the White House but Eisenhower deferred action on the grounds that quotas might involve antitrust considerations as well as lead to price fixing, harming consumer interests.[62]

The closure of the Suez Canal in late October brought a temporary respite from pressure for import restrictions. The diversion of the oil flow to tankers around Africa halved the crude carrying capacity of the world's tanker fleet while action against Britain and France by other Arab nations resulted in an European oil shortage. By the end of November additional US sales had reached almost 600,000 bpd, much of this due to a reduction in stocks. Producers with access to markets petitioned state commissions for increased allowables, while those in West Texas and New Mexico, limited by the lack of pipeline capacity, opposed any increase in allowables, hoping that pressure on supplies would cause additional transport facilities to be developed. A split between the Majors and the domestic independent producers developed, as the former cooperated with the government in arranging increased Middle East exports to Europe, shutting the independents out of this market. By the end of March the crisis was over and allowables were cut back.[63] The refusal of the Texas Railroad Commission to allow increases in production during the crisis resulted in a rise in prices, first in Venezuela then in the Gulf of Mexico. The result of higher domestic prices would be to further motivate companies to import crude oil.

VOLUNTARY CONTROLS

In March 1957 Flemming notified oil importers of his attention to act on the complaint of the IPA and requested information of crude oil imports. Flemming's successor, Gordon Gray, reported to the President on April 23, 1957, that the level of oil imports threatened national security. This conclusion was based on import plans for the second half of 1957 which would amount to 17.4 percent of domestic production, 500,000 bpd above the 1954 target level. Appointment of the Special Committee to Investigate Crude Oil Imports was delayed until June 26, 1957, forced by the steady increase in political pressure emanating from Congress.[64] The Committee issued its report on July 29, 1957. The Committee considered and quickly rejected the idea of stockpiling oil, subsidizing exploration, or slowing depletion by use of foreign oil:

> We have concluded . . . that if we are to have enough oil to meet our national security needs, there must be a limitation on imports . . . .
>
> Before arriving at this conclusion, we considered and rejected, for the reasons noted below, the following three proposals . . .
>
> 1. Import foreign crude oil and store it in this country within depleted fields or elsewhere. The practical problems of cost and the physical problems connected with the storage of crude oil would make this solution impracticable . . .
>
> 2. Enlarge government participation in exploring for oil reserves which, when discovered, would not be put into production. Such a course would be costly to an already overburdened government and would be contrary to the principles of free enterprise which characterize American industry.
>
> 3. Encourage increased importations in order that our own natural resources might be conserved. Your committee has concluded that for the following reasons such a policy would be unsound.
>
> (a) It would result in a sharp decline in domestic exploration by private enterprise. . . .
>
> (d) In brief, such a policy of encouraging importation . . . would mean that in an emergency the Nation would be confronted with all of the liabilities inherent in a static, as contrasted with a dynamic mobilization base, including the delays, waste and inefficiency that accompany efforts to strengthen any part of the mobilization base on a "crash" basis.[65]

The Committee recommended that all importing companies, except those with imports under 20,000 bpd in 1954, should be requested to reduce their imports to 10 percent below their average for the 1954-1956 period. New importers should be allowed to enter and share in the market but crude and product purchase agreements by transfer of allotments would be prohibited. The ratio between imports and domestic production for Districts I-IV (east of the Rockies) should be kept to 12 percent or less, or around 750,000 bpd.[66]

It is obvious that the policy recommendations were designed to placate as many interests as possible, and not to protect national security. Eisenhower felt that an adequate supply of oil to Europe was almost as important a priority as insuring supply for the US. He concluded that since the Western Hemisphere was incapable of meeting the supply requirements of the free world for more than a short period, any crisis which cut off supplies of Middle East oil would necessitate the use of force. Given this conclusion there is no strategic value in reducing imports of Middle East oil by depleting domestic reserves, a policy justified on the grounds of national security. The national security argument was used by Secretary of State Dulles in rejecting a tariff due to his concern that it would keep out Venezuelan and Canadian oil while letting in Arabian oil.[67]

The real objectives of policy were to restrict importation of cheap crude oil, protecting domestic prices and production; allow residual imports, placating New England; spread the benefits of cheap imported crude oil among as many importers and domestic refineries as possible; and avoid alienating Venezuela and Canada, who were more important in terms of both diplomacy and trade than the Middle East oil producers. The losers were the Major oil companies, second-tier companies with interests in Middle East oil, refineries with lower proportions of imported feedstocks, marketers of oil products, and consumers.

The role of Robert Anderson in the deliberations of the Special Committee deserves additional attention. He was working as a consultant to the committee before actually taking office on July 29, 1957. Anderson was reported to be the spokesman for domestic oil interests on the committee, while Herbert Hoover, Jr., represented the international firms. It was reported that Anderson played a leading role, both because of his experience in the oil industry and his special standing with the President on economic matters. Given his connection with a subsidiary of Standard Indiana, which benefited greatly from the quota system as an inland refiner, it is interesting to note that Anderson stated that a tariff was rejected on the grounds that the Attorney General said it would violate the GATT.[68]

Twenty-two importers were given a quota, of which fourteen agreed to comply with the import limits. Sun Oil refused to comply because of management's aversion to government control as well as concern about possible antitrust violations resulting from voluntary industry cooperation. Seven other companies agreed in principle to the program, but requested adjustments in their quotas. Tidewater, which had borrowed $200 million for a new refinery, including $35 million in equipment designed specifically for foreign oil, was the most adamant protester against basing import quotas on an historical basis. Standard Indiana also attacked the program, calling it arbitrary and inequitable. John Swearingen, executive vice president of Indiana, estimated that the cost to the firm of using domestic oil was 70 cents per barrel, while a spokesman for Ohio Standard estimated the difference between foreign and domestic oil on the East Coast at 50 cents per barrel.[69]

In November Matthew Carson, the program administrator, reported that only three importers, Sun, Tidewater and Eastern States Petroleum, were in noncompliance with the program. By early 1958, however, the program was starting to fall apart. While the old line importers, the five Majors plus Atlantic and Sinclair, had kept well within the bounds of their allocations, and total imports had been within the program goal for three months, new importers had increased both in number and in their volume of imports. At the end of 1957 new importers were asking for 100,000 bpd of import allocations, increasing to 200,000 bpd in requests by February, 1958.[70] Table 2.4 demonstrates how attempts to placate newcomers consistently eroded the position of the existing importers.

TABLE 2.4 IMPACT OF OIL IMPORT CONTROLS
(thousand barrels per day)

	Allocation		Actual Imports	
	Voluntary	Mandatory	1954-56	July 1957
The five Majors	375.3	253.6	397.5	443.0
Atlantic & Sinclair	116.3	82.6	134.5	130.8
Newer Importers:				
Foreign Oil*	103.4	94.9	50.6	104.1
Domestic**	87.7	56.7	36.4	102.1
Others	202.5	152.7	64.0	74.3
New Importers	0.0	98.1	0.0	0.0

*Foreign production: Sun Oil, Standard Indiana, Phillips
**Cities Service, Tidewater, Standard Ohio
SOURCE: *O&GJ*, Aug. 5, p. 47, Sept. 23, 1957, p. 65, Mar. 23, 1959, p. 56.

Despite the success of the voluntary program in limiting the growth in imports, continued restrictions on domestic production caused domestic producers to attempt to legislate mandatory import controls as part of the extension of the Reciprocal Trade Act in 1958. The IPAA plan called for a 30 percent duty on imports in excess of quotas. While coal interests wanted a flat 10 percent restriction on both crude and products, TIPRO returned to its previous position calling for excise tariffs. In order to head off congressional action Eisenhower announced cuts in imports of foreign oil starting April 1, from 783 to 713 thousand bpd, with the entire cut in imports coming from the seven largest importers, while 43 thousand bpd were transferred from existing importers to new importers. Enforcement was limited to loss of government business by noncompliers. Despite this action a bill sponsered by Frank Ikard became the focus of efforts to legislate restrictions. The Ikard bill would have established a quota, with allocations determined by the sale of import licenses. In order to weaken support for the bill, Interior Secretary Seaton proposed subsidies for domestic metals whose producers had been seeking import quotas while Carson and Commerce Secretary Sinclair Weeks, chairman of the Cabinet Fuels Committee, lobbied for continuation of the current plan. The Ways and Means Committee rejected the Ikard plan and voted to send the Trade bill to the House.[71]

With the defeat of a bill offered by Richard Simpson, which would have extended quotas to 100 industries, and a move by Daniel Reed to recommit by a roll-call vote, the Trade Bill moved to the Senate in June. The bill increased pressure on the President by permitting an investigation by the ODM on the request of any interested party, requiring the findings of the mobilization director to be reported to the President for action, eliminating the executive's option of calling for a second investigation before making a decision. In the Senate, an attempt by Russell Long to add the Ikard bill as an amendment in the Finance Committee was defeated, but a stronger National Security clause was added, which included economic dislocation due to imports as a factor impairing national security. The conference retained these changes and extended the act for four years.[72]

An increase in petroleum products entering the country, often from companies shifting refining activities from domestic to foreign installations, added independent refiners to the voices calling for mandatory quotas. This was reflected in a call for the restriction of the importation of petroleum products by the Special Committee to Investigate Petroleum Imports in June, 1958. The Committee recommended the limitation of imports of unfinished gasoline and other unfinished oils to the prevailing level, but it did not find any necessity for restrictions on importation of residual oil.[73]

In September an attempt was made to reorganize the voluntary program by its new administrator, Captain Matthew Carson. Imports were to be restricted to 10.1 percent of estimated total demand, to be determined by the Bureau of Mines. Beginning in 1959, imports would be allocated according to refinery inputs, regardless of the refinery's import history. During the first 6 months of 1959 refiners would receive either their historical allocation, up to 37.5 percent of inputs, or a minimum of 6 percent of inputs, rising to 9 percent of refinery runs in the second half of the year with a corresponding reduction for refiners with the highest proportion of imported crude. In 1960 all allocations would be made on the basis of equal percentages of inputs. Inland refiners were given the right to import, and although the quotas were non-transferrable, the imported oil itself could be sold before actually being refined. Given the large transfers in wealth implicit in this program it had no chance of succeeding on a voluntary basis. Texaco and Gulf strongly attacked the program, while the Small Business Administration, along with the Justice department and Senate small business committee, asked for more for small refiners.[74]

By the end of 1958 a program of mandatory controls seemed inevitable, despite the efforts of supporters of free trade. Clarence Randall, chairman of the Council on Foreign Economic Policy, argued that quotas would penalize the consumer as well as weaken the General Agreement on Tariffs and Trade. He doubted that national security would be served by depleting domestic reserves. On the other side of the issue a Texas study commission called for restrictions on oil imports, while in February 1959, a National Coal Policy Conference, representing the coal operators, the United Mine Workers and the railroads, called for controls over all oil imports including residual oil. Senator O'Mahoney of Wyoming threatened to introduce a mandatory oil import quota bill while three Pennsylvania congressmen introduced bills to limit imports of residual oil. The director of the Office of Civil and Defense Mobilization (OCDM, formerly the ODM) submitted his findings on February 27 that the importation of crude oil and products posed a threat to national security, ignoring a staff finding that imports of residual oil posed no such threat. The Attorney General and the chairman of the Council of Economic Advisors both posed objections to the finding. Despite the President's discomfort at being pushed into mandatory controls by special interests he issued Presidential Proclamation 3279 on March 10, 1959, establishing mandatory controls, in order to avoid Congressional controls which would provide little or no latitude for administrative flexibility.[75]

MANDATORY OIL IMPORT QUOTAS

Under Presidential Proclamation 3279 the Secretary of the Interior was granted the authority to make allocations and issue licenses to importers. Imports of crude and unfinished oils were restricted as of March 11, and those of finished oils by April 1, 1959. In Districts I-IV the maximum level of imports of crude oil, unfinished oils, and finished products, exclusive of residual oil, was set at 9 percent of total demand in that region. Imports of residual oil were to be kept under review by the Secretary of the Interior and adjusted, on a monthly level if necessary, as he saw fit. Regulations for restricting oil imports were to provide for a fair and equitable distribution using as their basis refinery capacity in relation to refinery inputs during a period selected by the Secretary, avoiding drastic reductions below voluntary program allocations. Imported oil could be exchanged for domestic crude if effected on a current basis and reported in advance to the Secretary. Appeals and petitions were to be heard by an Appeal board comprised of one representative each from Interior, Defense and Commerce.[76]

Implicit in any quota program is the potential for economic rents. Any refiner with access to imports will be able to collect economic rent, equal to the difference between the domestic and world price (ignoring for now the rents collected by domestic producers). The argument that they would then be able to unfairly compete with other firms is inaccurate, since their marginal cost of obtaining additional oil would still be the domestic price, providing no additional motivation for penetrating the markets of less favored competitors.

Instead of seizing these rents through a tariff (rejected for diplomatic reasons; preferential treatment for Canada and Venezuela would violate the GATT as well as alienate Middle East producers) or auctioning off the right to import oil (which had been proposed in the Ikard bill and conveniently forgotten), the Eisenhower administration followed the lead of the Carson plan and spread the rents around as many firms as possible. The political advantages were obvious: the domestic producers didn't care which refiners received benefits; the Majors and second-tier firms with foreign oil were already opposed to import controls; the cost to consumers was unchanged; and spreading economic rents to numerous small inland refiners created a new constituency in favor of the program. The only additional loser was the taxpayer. The cost of the program was disguised by prohibiting purchases of entitlements, but limiting transactions to barter did not prevent side payments or a general understanding within the industry of the value of a quota allocation.

The allocation procedure was designed to spread the benefits of the quota among as many refiners as possible, especially smaller firms. Part of the allocation was based on a percentage of the historical level of imports. The rest of the allocation was based on a sliding scale, which established rates for each level of inputs. Thus a refinery processing 30,000 bpd of crude in January 1960 could import 13% of the first 10,000 barrels, 11.9% of the next 10,000, and 10.8% of the third 10,000, for a total of 3,570 bpd. So not only were import rights transferred to refiners who had never used imported oil but small, often inefficient or obsolete refineries were given a sizable subsidy.

Almost immediately, a conflict between mandatory import controls and foreign policy arose. While imports were to be restricted it was also necessary for diplomatic reasons to avoid imposing restrictions on Canadian and Venezuelan suppliers. Due to the restrictions on Canadian imports there were plans to construct a pipeline from the Canadian midwest to Montreal which would have displaced Venezuelan oil. To prevent this an amendment to Proclamation 3279 was issued on April 30, removing restrictions on oil entering overland. The State Department recommended that traditional importers of residual fuel oil be licensed to fill any unused storage capacity, effectively decontrolling residual imports without formally removing residual oil from the program. The motivation for this action was to influence Venezuelan politics by weakening those elements in favor of nationalistic oil policies who contended that the US oil policy disregarded Venezuelan interests. Opposition from Interior and Commerce scotched this plan, the former to protect coal interests, the latter because the State Department plan lacked a quid pro quo, the protection of private oil interests in Venezuela.[77]

While studies of Congress tend to criticize that institution for being overly influenced by special interests to the detriment of coherent policy formation, the actions of the Eisenhower administration give little confidence that one can expect the executive branch to demonstrate greater wisdom, or at least, a more consistent vision of the national interest. While imposition of some sort of oil import restrictions may have been inevitable due to pressures from Congress (though the Eisenhower administration did not expend a great deal of political capital resisting Congressional pressure) this can not excuse the strange structure of the mandatory quota system. In effect, to protect an economically irrational prorationing system an economically irrational import quota system was developed.

As domestic interests succeeded in reshaping government regulation of the domestic oil market to their desire the impact of this regulation was felt in the world oil market. The increase in posted prices in 1957 had set the stage for subsequent friction between the companies and host governments. Attacks on the existing price level came from two sources; the new Venezuelan producers and the Russians. The Venezuelan newcomers had bid for concessions there in 1955-6 to obtain low-cost uncontrolled supplies for their domestic refineries. The imposition of import restrictions in the late 1950s put pressure on them to find alternative markets just as their productive capacity was coming on line. They had paid large bonuses to the Venezuelan government in the expectation of a rapid pay-out of their investment. With quotas calculated on the basis of historical shipments, the independents were locked out of the US market. The only way for the independents to recoup their investments would be to take other markets away from the Majors.[78] The imposition of the import quota in 1959 had an inhibiting effect on the foreign exploration plans of some American independents. Superior tried to merge with Texaco in 1959 but was stopped by the Justice Department, so it sold its Venezuelan operations for Texaco stock. Sohio stopped the acquisition of new concessions after 1960 and signed long-term contracts for foreign crude in 1963. Other companies reacted by establishing processing relationships with local refiners, buying existing refineries and even building their own refineries.[79]

Around the same time as the oil began flowing in the wells of the independents in Venezuela the Russians commenced an escalation of exports to the noncommunist world. From less than 7 million barrels in 1952 exports rose to 24 million in 1955, 58 million in 1957, 139 million barrels in 1960, 328 million in in 1965 and 439 million in 1972. The Soviets were both willing to compete on price and by arranging for barter deals with countries whose currencies were soft. Between 1955 and 1960 the average export price to non-communist buyers was $2.20 per barrel. Oil shipped to Europe in 1961 averaged $1.70 per barrel, 30 to 50 percent below posted prices.[80]

While domestic producers discounted the importance of the voluntary program its impact was felt both in the existence domestically of import rights and internationally as a contributor to the downward pressure on price. The value of the right to import oil was about 80¢ per barrel, which suggests that voluntary quotas were having a significant effect in shifting imports to other markets.[81]

In any case discounting began in earnest in the end of 1957 when BP offered the equivalent of a 7 percent discount to Argentina. By early summer reports of Venezuelan crude at discounts of 75-80 cents a barrel were widespread. Second-tier companies operating in Venezuela were offering deep cuts, including a 1 million barrel sale to Petrobas at 45-50 cents under the posted price. Even US prices were under pressure with declines in the largest fields of about 7 cents and 13-20 cents elsewhere. The reopening of the Suez Canal depressed the tanker market, putting further pressure on crude prices. There were both direct and indirect discounts offered, the latter taking on a variety of forms, such as long-term financing of crude oil purchases, acceptance of soft currencies as partial payment by American suppliers, extra concessions for quality differences, "spiking" of crude with refined products, freight allowances on delivered prices, and recharters of vessels at favorable rates to f.o.b. buyers.[82]

The Major oil companies were faced with a serious dilemma. Constraints on markets combined with the entry of second-tier firms required competitive pricing to maintain markets. But competitive pricing meant discounting from posted prices, cutting into profit margins. Ironically, it was the Majors' past control of the market which trapped them--in a competitive market host governments would have accepted price movements as normal--but the existence of the oligopoly made any downward price movement triggered host government activity to combat what they perceived as exploitation by the Majors. The first attempt to adjust posted prices occurred in 1959, initiated by Shell lowering Venezuelan prices followed by BP in the Middle East. Aramco and other Middle East producers matched the BP price decision while Venezuelan prices declined once more for a total decline of about 25 cents per barrel. Since this merely dropped prices back to the pre-Suez level, the host governments, while not pleased with the drop in posted prices, were not surprised. The increase in output during the period kept revenues steady or increasing even after the price cut. Despite these moves real prices remained well below posted prices. A routine level of discount was considered to be 20-35 cents. Even the Majors were forced to discount, with Creole (Exxon) selling most of its oil at an average discount of 10 percent.[83]

Exxon made the decision to cut posted prices again in July 1960 despite warnings in both the industry press and from some of its directors that there would be serious political consequences. The original plan was to cut prices worldwide but pressure from the company's manager in Venezuela convinced the company to limit cuts to the Middle East. Exxon announced a cut of 10 cents a barrel, and the other Majors reacted negatively before following suit. Shell argued strongly that the cut should include Venezuela, but accepted the pattern of prices set by the others.[84]

The first Arab Petroleum Congress meet in Cairo, in April 1959, after the first price cut. Representatives from Venezuela and Iran were invited as observers since it was obvious that the large non-Arab exporters would need to cooperate with the Congress to have any impact on host country--oil company relations. Perez Alfonzo of Venezuela had been trying to convince the other exporters of the need for conservation, partially out of conviction but also because he realized that production restraint by Venezuela was useless unless the other countries did likewise. He was able to convince a kindred soul, the

American educated Saudi Arabian oil adviser, Sheikh Abdullah Tariki, of the need for an organization of producing countries. The two collaborated in the preparation of a secret agreement which was signed by the heads of delegations of the exporting countries.[85]

The second price decline galvanized the host governments into action. On September 14, 1960, the representatives of the governments of Iran, Iraq, Kuwait, Saudi Arabia and Venezuela met in Baghdad and created the Organization of Petroleum Exporting Countries (OPEC). The grievance that pushed them to this step was not so much the price cut but the fact that they had not been consulted beforehand. A month later OPEC held a meeting in Beirut, attended by oil company men. Despite accusations by Tariki that the oil companies were violating the 50-50 principles through deceptive practices the oil companies refused to assuage these suspicions by opening their books.[86] It is ironic that a policy to ensure the US against the threat of the loss of foreign oil would help lead to the development of the organization which would fulfill the worst fears of those concerned.

THE OIL IMPORT QUOTA: 1960-1970

The Mandatory Import program quickly drew opposition from a number of Senators, including William Proxmire, who introduced a bill to rescind the quotas, as well as George Aiken (R-VT), Wayne Morse (D-OR), S.L. Holland (D-FL) and John Pastore (D-RI). Charles Vanik (D-OH) attacked the program in the House. President Eisenhower at his press conference reassured critics that his administration plans to keep a close check on domestic oil prices, supporting the fear of Representative Tom Steed (D-OK) that an increase in prices could lead to an increase in import quotas.[87]

Despite the ban on selling entitlements a market for inland quotas developed as soon as the ink was dry on the Presidential proclamation. The average value of entitlements was estimated in the range of 80 to 85 cents at the beginning of the mandatory program. One effect of the trading system was to increase pressure on posted prices, since international oil companies, in order to find markets for their foreign production, would sell their own oil at prices below posted prices to companies with entitlements with whom they would then "swap" oil. If the company ships in its own tankers it could be profitable to offer more for entitlements in order to use its tanker fleet at close to full capacity. Companies with tanker fleets operating in the Middle East had an advantage in bargaining for entitlements since the actual landed cost of their oil was lower than that of Venezuelan oil. Companies without tankers fleets had significantly higher landed costs, requiring them to offer discounts from posted prices.[88]

The imposition of quotas was quickly followed by a struggle over the level of residual oil imports. Residual quotas were raised by 62,000 bbls daily for the first half of 1960, reduced from the 100,000 bpd recommended by the Interior Department due to strong protests from the coal industry. As the coal industry turned up the pressure Interior reacted by cutting residual imports. An additional cut of 60,000 bpd was planned for the first half of 1961, despite a rise in the price of residual oil from $2.02 to $2.25. Interior Department policy was determined by the conflict between importers, New England industrial users, residual oil marketers on one side and the coal industry, the UMW, and some independent oil producers on the other side.[89]

Despite cooperation between coal and oil producers to restrict oil imports basic differences between the industries surfaced. The National Coal Policy Conference, along with several members of Congress, attempted to establish a special Joint Congressional Committee to study the need for a national fuels policy in 1959 and 1960. The API attacked the resolutions calling for a fuels policy as an attempt to outlaw competition with coal, especially the call for "the optimal allocation of the various fuel and energy resources," which the API labelled as an attempt to implement end-use controls. The Mid-Continent Oil and Gas Association also opposed the national fuel policy proposal. A redrafted resolution, developed in the Senate Interior Committee, neutralized opposition from independent oil producers. The IPAA noted efforts to tone down the end use provision while TIPRO questioned the campaign being waged against the fuels policy proposal, stating that the petroleum industry needed the support of the coal industry to pass a gas decontrol bill, to continue the mandatory oil import program, and to retain percentage depletion.[90]

With the entrance of the Kennedy administration there were some shifts in executive branch attitudes toward the oil industry. Almost immediately after taking office the new Interior Secretary, Stewart L. Udall, allowed an temporary 100,000 bpd increase in residual imports due to the hard winter. A one day hearing on February 20 exposed the new Secretary to the interest groups on both sides of the residual import issue. On March 9, Udall increased the quota for residual oil as of April, revoking an earlier order issued by former Secretary Seaton. Quotas were increased about 39,000 bpd over 1960, and allocations were changed to favor smaller importers.[91]

Hearings on the oil import program were held by Interior in May. The IPAA asked that imports be limited to 10% of crude demand instead of 9% of total demand, while TIPRO suggested that imports be restricted to 11.8% of domestic crude production and for quotas to be sold by bid. East Coast refiners and foreign producers attacked the inland refiner allotments as creating "windfalls," while Standard Ohio and the Independent Refiners Association of America defended quotas for all refiners and, along with Standard Indiana, attacked the historical basis for assigning quotas. The sliding scale also came under attack, chiefly from the large companies.[92]

The formerly pro-control OCDM, now headed by Frank Ellis, was petitioned by the New England Congressional delegation to reappraise the national security rationale for residual controls. The Office of Emergency

Planning (OEP), which replaced the OCDM, came out in November 1961 in favor of removing restrictions on residual oil imports, stating that residual imports could have no effect on national security, a conclusion supported by the Defense Department. On the other hand, the Interior Department pushed for a tightening of import controls, arguing that the national security rationale should be extended to include the potential claims on the American industry to satisfy the free world's petroleum requirements. Interior also argued that imports would endanger the survival of the dispersed inland refiners, and opposed the OEP's position on residual imports. The Budget Bureau, noting the $2 billion dollar cost of quotas, pointed out that stockpiling of proved reserves might be achieved more efficiently by other means than quotas, suggesting the use of savings achieved by elimination of the depletion allowance to subsidize oil exploration. The Justice Department was concerned about the anti-competitive nature of the program, while the State Department felt that import restrictions were incompatible with the goal of reducing barriers to international trade. Given the conflicts between interest groups and within the administration, action was deferred by formation of the Petroleum Study Committee (PSC), an attempt to defuse conflict until after the passage of the 1962 Reciprocal Trade Bill.[93]

Passage of the 1962 bill became the priority of a Democratic administration committed to free trade for a number of reasons. Concern about the nation's negative balance of payments, acerbated by the nation's foreign military commitments, the sharp rise in overseas investment by American firms, fear that US companies might be shut out of the lucrative markets of the European Economic Community, and confidence in the superiority of American manufacturing all contributed to the desire to lower trade barriers. Active support for the bill came from the major multinational firms, the American Bankers Association, the Chamber of Commerce, the American Farm Bureau Federation and the AFL-CIO.[94]

The Committee for a National Trade Policy, which had been established in 1953 as a consequence of the battle over oil imports, served as coordinator for free trade groups and as liaison with the White House. The Administration's strategy included splitting the opposition by granting nonlegislative concessions to groups which sought some protection from the trade winds. A seven-point program to assist the textile industry was announced in early May 1961, and attempts were made to placate the Northwest timber industry, but the administration was not as responsive to the demands of the oil industry. The bill retained a national defense clause to protect other mineral industries as well as oil from attempts to lower trade barriers.

Kennedy sent the Trade Expansion Act with a special message on foreign trade policy to the Congress on January 25, 1962. The bill, delegated authority to the President to reduce tariffs by 50 percent, staged over five years, and elimination of tariffs below 5 percent. The peril-point provision was eliminated, but the Tariff Commission was required to advise the President as to the probable economic effect of tariff reductions. Adjustment assistance provisions were included in the bill to soften the impact of increased imports. Congress changed the vote needed to override a presidential decision against use of the escape clause to a majority, instead of the two-thirds vote as mandated in the 1958 law.[95]

The PSC report was finally sent to the President on September 4, 1962. The committee suggested that crude oil prices would be about $1 a barrel lower without controls (total demand for petroleum was around 3.5 billion barrels per year), and oil import licenses were estimated to be worth about $1 million a day to recipients. The report cited Treasury Department estimates that percentage depletion cost $1 billion per year. The PSC accepted the temporary need for oil import controls but recommended that the Secretary of the Interior initiate discussions with the Interstate Oil Compact Commission on steps to rationalize the prorationing system and lower domestic production costs. The PSC stated that tariffs were preferable to quotas due to simplicity of administration and the greater reliance on market forces, but if not acceptable then fees and auctions should be used for allocating import licenses. The report was never publicly released, the reason given being that the President did not wholly accept the report, though it was finally made public through the efforts of Representative Walter Rogers (D-TX) in June 1963.[96]

The PSC report was ignored because the Administration had already committed itself to continuation of the present program as part of its maneuvers over the trade bill, involving implicit agreements with Robert Kerr and Russell Long. The rise in excess production capacity to record levels (27.1% as of January 1962) motivated the administration to adopt Interior Department recommendations to tighten import controls. On November 30, President Kennedy issued a proclamation which implemented the Interior's proposals, setting the level of imports in Districts I-IV at 12.2 percent of domestic oil production rather than 9 percent of total demand. Exempt imports from Canada and Mexico would be subtracted from the total allowable imports. The impact was to lower imports about 70,000 bpd.[97]

While the PSC report was being ignored by the Administration the OEP report on residual imports received a more favorable reception. A rough draft was delivered on September 7, with the final report on February 13, 1963. Though the draft report stated that there was no reason for continuing controls on residual, the White House pressed the view that controls should be continued to reassure coal interests while expanding quotas to placate consumer interests. In the final report the OEP called for only a relaxation of controls. Much of the pressure for continuation of residual import controls came from the Secretary of Labor, as well as the desire to avoid alienating Senator Randolph (WV). The final administration decision was to continue controls but allow a ten percent increase in allocations as of April 1.[98]

The limited geographical impact of residual oil imports made it difficult for the coal coalition to generate sufficient political support to achieve its goals. Domestic refiners had little interest in producing additional residual oil since its price was low (due to competition from coal and natural gas) and domestic crudes could be converted to higher price products. Therefore the political struggle was limited to the importing states along the East Coast versus the coal producing states. This trend was illustrated by the full-page advertisement printed by the National Coal Policy Conference on April 2, 1962. It was signed by 17 Senators and 83 Representatives. 53 of the Representatives were from the six coal states of Illinois, Kentucky, Maryland, Pennsylvania, Virginia and West Virginia, but only 6 Senators from these states signed on, excluding both Pennsylvania Senators, who wisely chose a low profile considering that the largest city in the state, Philadelphia, was the leading refinery center on the Eastern seaboard.[99]

With the Trade Bill behind them the Kennedy administration chipped away at import restrictions, changing the basis for calculating quotas from current production to projected production, adding about 30,000 bpd to imports for the second half of 1963. The oil industry also faced challenges from the Kennedy administration on both the tax and prorationing front. Instead of attacking the depletion allowance directly, the administration proposed changes in the operation of the allowance which would have effectively reduced its rate from 27.5 percent to about 22 percent. The American Iron Ore Association, American Mining Congress, API, IPAA, National Coal Association, and the National Crushed Stone Association joined in defending depletion. TIPRO raised the question of equal tax treatment of foreign and domestic operations, calling for credits for state taxes. The Ways and Means Committee originally rejected the Treasury proposals on June 6, but Wilbur Mills decided to allow two minor reforms, valued at $50 million vs. the $200 million in revenue that the Treasury was attempting to gain. The first report by the Justice Department on the Interstate Compact to Conserve Oil and Gas since 1959 questioned the operation of the oil production control system. The Department expressed its concern about increased competitive pressure on the independent producer and refiner due to inconsistencies between states which they saw as favoring large integrated companies. In a related development, the Compact Commission undertook a survey of the domestic industry's production and conservation practices. The survey appeared to be motivated by the PSC study.[100]

When Lyndon B. Johnson succeeded John F. Kennedy in the White House, he tried to cut his ties with the oil industry and allowed Stewart Udall, continuing as Interior Secretary, a relatively free rein. Between 1965 and 1968, Udall condoned whatever changes and exemptions in the Mandatory Program which seemed helpful for achieving the Administration's broader goals. In addition his fellow cabinet members, congressmen, and oil industry managers sought to exploit the Program for various purposes. The Council of Economic Advisors wanted to liberalize imports to help hold down prices, the State Department favored imports from Canada and Venezuela, and Commerce supported exemptions for the petrochemicals industry to shore up the balance of payments. The congressional delegation from New England pressured to get residual decontrolled, and then for an exemption for a foreign trade zone. California senators tried to block imports of refined products from Puerto Rico, while the Texas and Oklahoma delegations pushed for a reduction of total imports. The Oil Import Appeals Board had a docket of small-company cases that required almost daily decisions. By 1968, the Interior Department had authorized special allocations for asphalt refineries, electric utilities, heating oil importers, petrochemical companies, and foreign trade zones from which refined products would presumably be exported.

In March 1964, Secretary Udall raised the limit on imports of residual oil, generating protests from the bituminous coal industry while failing to placate fuel oil users. In reaction to pressure by the coal industry President Johnson appointed a special cabinet committee on Appalachian coal. Johnson also instructed the US representative to the GATT to push for lower barriers to US coal exports. The report, never released, recommended that import controls on residual oil could be relaxed without effecting national security, and without serious impacts on the coal industry. Udall's initial reaction was to propose a new program for April 1965 which would eliminate quotas for New England and Florida, since coal did not compete with residual oil in these states. Pressure from White House aides, concerned with the political fallout from this decision, caused Udall to reverse his decision. In December 1965, Interior raised permissible imports of fuel oil by 50 percent, effective until April. In March 1966 Udall implemented a new program which expanded eligibility for quota licenses, effectively eliminating residual oil restrictions, while retaining nominal quotas as a sop to the coal industry.[101]

Canadian producers, taking advantage of free access to American markets due to the overland exemption, increased exports in the early 1960s, exceeding levels agreed upon in 1961. The Canadians agreed to limit imports to 125,000 bpd in 1963, but exceeded this limit in 1964. The Canadians pointed out that the imbalance in payments in current accounts would require a restriction on imports of US goods if oil exports were not allowed to increase. In 1963 historical quotas were lowered for Northern Tier refiners who were the major consumers of Canadian oil, and Canadian imports were counted against total imports. During the rest of the decade Canadian imports continued to rise, raising a dilemma for US regulators; an increase in Canadian imports increased Venezuelan exports to Canada but also increased Venezuelan perception of unequal treatment under the import control program. While it was tempting to use the national security exception of the GATT to allow special treatment of Venezuelan oil, there was concern that this would encourage trading partners to use it as a precedent for installing trade barriers. Most trading partners considered the national security rationale for the oil import program as specious, but it had never been challenged, a situation which might have changed if a Venezuelan preference was instituted.[102]

Diplomacy was also involved in a strange twist concerning Mexican imports. Since Mexico shares a border with the US it qualified for an overland exemption. However the lack of a pipeline from the Mexican oil fields to the US refineries resulted in Mexican oil being shipped by tanker to Brownsville, Texas, unloaded onto trucks, driven to Mexico and then back to the US, reloaded on tankers and shipped to East Coast ports. Imports reached 50,000 bpd in 1961 when an agreement limited Mexico to 30,000 bpd, but the "Brownsville shuffle" was continued until 1971.[103]

The chemical companies had argued from the beginning of the mandatory program that they should receive allocations along with refineries. They contended that the program gave an unfair advantage to domestic refineries, competitors in production of petrochemicals, and foreign chemical producers. The issue surfaced briefly in 1962 but did not become an important issue until an interagency task group in 1965 studied the oil import program as a means to promote exports to improve the balance of payments. Secretary Udall announced a policy to grant import allocations to petrochemical companies, starting with a 30,000 bpd allotment (part of the 12.2 percent limit) in 1966, and petrochemical allocations rose to 94,000 bpd by 1972.[104]

Beginning in late 1965 a number of special arrangements were made by the Interior Secretary. The principal refinery in Puerto Rico, which is in the customs territory of the United States, was dependent upon Venezuelan crude. A special allowance was established in 1959, permitting imports into Puerto Rico to meet local demand for petroleum products plus demand by foreign countries for exports of products from Puerto Rico. In 1965 Phillips Petroleum arranged with Interior to receive an allocation for importation of 50,000 bpd of unfinished oils in Puerto Rico, with the right to ship 24,800 bpd of motor fuel to the US, in return for constructing petrochemical facilities on the Island--in effect a $15 million annual subsidy for investing $100 million in Puerto Rico. Phillips cleverly switched its source of oil from Algeria to Venezuela, adding needed support to balance the opposition of other oil companies. The success of Phillips encouraged other companies to arrange similar deals. In 1968 Sun Oil received an import allocation of 66,000 bpd and an export allocation for 29,500 of products bpd in return for investing $125 million in facilities, building a $10 million harbor and payments to a conservation fund. Similar deals were struck with Union Carbide and Commonwealth Oil. These arrangements required oil imports from Western Hemisphere sources, thus meeting Venezuelan demands for special treatment.[105]

While the Phillips agreement in Puerto Rico was engineered behind the scenes by Abe Fortas and Oscar Chapman (former Interior Secretary) an agreement with Hess Oil in the Virgin Islands was the result of open lobbying. Leon Hess, a heavy contributor to the Democratic party, was able to overcome the opposition of the Oil Import Administration and the Assistant Secretary for Mineral Resources. Udall allowed Hess to build on the Virgin Islands while rejecting all other applications in order to preserve the natural beauty of the islands. Udall justified the special quota allowing Hess to ship 15,000 bpd in finished products to the mainland on the grounds of economic benefits to the island, including $2.7 million in annual payments to the Virgin Islands Government. The Hess quota was then subtracted from allowable imports for other refiners.[106]

The most ambitious proposal was made by Occidental oil in 1968, proposing to build a refinery in the coastal town of Machiasport that would process 300,000 bpd of Libyan crude oil, with the refinery constituted as a foreign trade zone. To make it profitable, Hammer sought an exemption to allow 100,000 bpd of the plant's products to enter US customs territory from the zone. The depressed state of the Maine economy, the prospect of cheaper product prices in the region, as well as a proposal to pay 20¢ per barrel to a marine foundation helped mobilize the entire New England congressional delegation and six governors in its support. The scheme also mobilized opposition throughout the oil industry and attracted the scrutiny of the press to the oil import program. President Johnson deferred to the next Administration a decision on the Machiasport project, dumping the problem of oil import controls on Richard Nixon.[107]

Udall, while granting these exceptions to the program, attempted to placate domestic oil producers, ordering an 18,000 bpd cut in imports for the second half of 1964. The IPAA continued to lobby for more significant cuts at its annual meeting in November, and exclusion of exceptions. The IPAA leaders rejected a tariff as politically impossible and decided to put quota bidding under study, while choosing to take no position on residual imports on the grounds that opposition created more ill will than the value to oil producers. The group took its case to Udall and asked for a 275,000 bpd reduction in imports. It complained that production by independents had declined slightly since mandatory controls were instituted while major companies experienced a 12 percent increase in output. Lower production rates and higher costs were resulting in increasing indebtedness for independents. The API, in its first official policy statement on imports since the mandatory program was established, endorsed restriction of oil imports. The Institute was careful to stress that controls should be designed to minimize market distortions.[108]

Hearings on the mandatory program were held by Interior in March 1965. Independent producers argued for an imports freeze or reductions up to 375,000 bpd. For the first time there was no stated opposition to crude import controls, though most major East Coast refiners asked for removal of residual oil import restrictions while four refiners joined the coal industry in defense of residual controls. Cities Service, Sun, and Texaco called for replacing the quota system with import duties, TIPRO asked for a quota bidding system, while Gulf, Indiana Standard, Ohio Standard, and Tidewater opposed tariffs or quota bidding. Most companies were opposed to exceptions, while the usual battles over historical and sliding scale allotments were fought.[109]

While the large international companies initially opposed the Mandatory Oil Import Program, as early as 1962 they had begun to appreciate the economics of the program. They realized that there was a trade-off between increased imports and the profitability of their domestic operations. Since state proration authorities limited domestic production, increasing imports would require maintaining costly spare productive capacity, while oil price erosion decreased revenues from domestic investments. On the basis of such analyses, together with a growing sensitivity to the political costs of continued opposition, the international majors began to support the Mandatory Program. By 1965, as evidenced at the Interior Department hearings, all the major companies had come to support the principle of mandatory controls, reflected in a 1966 statement by the National Petroleum Council in which it finally endorsed federal import controls.[110]

The major companies were also very aware of the politics of oil, especially when explained by Russell Long at a NPC meeting in March, 1964:

> . . . Senator Long said that those in Congress from the oil states "are especially interested in the domestic phases of the industry, because that is the part that gives employment to our people and means revenue to our state governments, and it is essential to our economy." But, added Long, "we would like you fellows that produce oil overseas to realize this, that when problems come up with regard to your tax credit overseas or even your depletion allowance overseas, or the special tax treatment to your employees that you have overseas, the fellows you are going to rely upon to protect your activities in that respect are the same people who are interested in the domestic production of oil here. . . . I believe your industry would make a great mistake not to realize that; that as far as the government is concerned, as far as the fair treatment you are entitled to expect from your government is concerned, the people who will be your advocates are people who are very much interested in domestic oil. . . . It is very much to your advantage to have a very healthy domestic industry and do everything within your power to cooperate to that end."[111]

The API contended that the depletion allowance is in the interest of the public and should be applied without discrimination between large and small operators and foreign and domestic production. The independents generally went along with this position for fear that an attack on foreign tax breaks could lead to challenges for domestic depletion provisions. In 1957 an independent refiner attacked the depletion allowance on foreign oil and the foreign tax credit. In 1962 Representative Tom Steed supported lowering domestic depletion to 15 percent for companies grossing more than $5 million, and proposed a bill to force companies to choose between foreign depletion or the foreign tax credit in retaliation for major company opposition to his amendment to the Trade bill restricting oil imports. Senator Mike Monroney (D-OK) in 1962 warned that: "While I am confident that if we work at the problem we can maintain the present depletion allowance with the support of a united industry, I am equally confident that we cannot maintain it if the industry is divided." While the IPAA in 1965 supported the united front, the executive committee of TIPRO voted to ask Congress to determine whether oil companies operating overseas enjoy any tax advantages not available to domestic producers. TIPRO had already questioned the fairness of the foreign income tax credit during 1961 Senate hearings on imports and small business and 1964 Senate hearings on import allocations.[112]

Concern with prices increased as the Vietnam War began to trigger inflationary pressures in the US economy in the mid-1960s. The Johnson administration began to pressure the Texas and Louisiana Commissions to increase production allowables during early 1966. A 1.5 cent price rise in the price of gasoline from 1964-1966 caused the administration to attempt to change the allocation pattern to increase competition. Gasoline's high visibility made a price rise more politically volatile than for most products. A one-cent rise in gasoline prices in early 1967 brought pressure upon the major companies to rescind the increase, with threats to increase imports of gasoline and stories leaked to the press. The Interior department was able to thwart the inflation fighters' desire for both more imports and reduction of the historical allocation, but it was less successful in fighting the pressure for exemptions.[113]

In 1967 both TIPRO and IPAA decided to take the legislative route to tighten import restrictions, and tried to include other industries in attempts to write the 12.2 percent of domestic production limit into law. The administration argued against the loss of flexibility inherent in an import control bill, and Secretary Udall promised not to use import controls to influence prices and to maintain the 12.2 percent limit east of the Rockies. He warned that a legislative campaign could backfire since consumer state congressmen were in the majority, raising the spectre of prices and tax treatment being brought under congressional scrutiny. Northeastern congressmen reacted to a bill authored by Russell Long to legislate the 12.2 formula by drafting an amendment to partially decontrol imports of No. 2 oil (diesel/heating oil). Opposition of Wilbur Mills to legislating import quotas killed the independents' proposal while a countermove to liberalize controls, led by Robert Kennedy, was thwarted by Interior.[114]

Import controls were much more successful than domestic producers were willing to admit. Crude and refined product imports (excluding residual) in PAD districts I-IV stayed below 9 percent of total demand for petroleum products. Imports as a percentage of non-residual consumption never rose significantly higher than the base period of 1954-56. The rise of imports as a percent of crude production was due to the fact that a larger proportion of domestic production was in the form of natural gas liquids, increasing domestic output of products with the benefits going to natural gas producers. The real problem faced by the independents was that the small fields on which they concentrated their efforts were becoming harder to find and more expensive to develop. Ironically, the more domestic exploration and production was stimulated by higher prices due to import controls, the faster costs rose as accessible deposits were depleted.

Both the majors and independent producers were in favor of changing the system to eliminate the proliferation of single-interest exemptions that were threatening both the effectiveness and viability of the oil import program. The advantage of an auction would be to allow the government to collect the rents generated by a quota and to simplify administration of the program, eliminate allocation decisions, and remove political pressure for exemptions. However, integrated companies and refiners were unanimous in their opposition to the auction proposal, with Interior receiving 100 negative comments and 1 positive comment (TIPRO). Small refiners protested the loss of import tickets, whose value had risen to $1.25 per barrel. Large companies expected any plan to be full of exemptions for small refiners, Caribbean plant operators, petrochemical manufacturers, etc.[115]

Controversy over the program continued to grow during 1968. The IPAA repeated its attempt to legislate controls by supporting efforts to amend the 1968 Trade Expansion Act extension. The National Consumers League joined the usual free trade opposition to quotas, while the API took a neutral stand. An April increase in home-heating oil prices increased political pressure to control oil prices, discouraging support for further limits on imports.[116]

THE RISING TIDE OF CRITICISM

In February 1969, President Nixon moved the authority over oil import policy to the White House from the Interior Department. A Cabinet Task Force on Oil Import Control was created in March to review the program, headed by Secretary of Labor George Schultz, and including the Secretaries of State, Defense, Treasury, Interior, Commerce and the Director of the Office of Emergency Preparedness (basically the old OEP). While responding to pressures from the oil industry to reorganize the import control program, President Nixon was also reacting to actions by some liberal Democratic senators, as the oil import program was becoming a controversial issue.

In March 1969, Senators Phillip Hart of Michigan and Edward Kennedy of Massachusetts commenced lengthy and intensive hearings before their Subcommittee on Antitrust and Monopoly, chaired by Hart, on the effects of governmental intervention in the petroleum industry. The hearings were the first systematic effort to examine and publicize the consumer costs of federal oil policy, involving testimony of academic economists as well as the usual industry spokesmen. The Subcommittee's chief economist, John Blair, was the author of the FTC's 1951 cartel report.

Although Franklin M. Fisher of MIT had been critical of the oil import program in his testimony in 1967 to the Senate Finance Committee,[117] the Hart hearings was the first time that oil policy had undergone intense, analytic criticism by an impressive collection of academic economists. During the hearings there was an enlightening exchange between Senator Roman Hruska (R-Neb) and Dr. M.A. Adelman of MIT:

> *Senator Hruska*. There is a sentence in the concluding paragraph of your statement that intrigues me very much . . . ‘ Import policy, offshore leasing policy, and the other areas mentioned deserve a look by Congress and the

> Executive Branch. They have never been seriously discussed and debated (least of all by the Cabinet Committees of the 1950s).'
>
> . . . Now, then, my question would be this. What explanation would you give to the many men, Senators, Congressmen, Cabinet officials, sub-Cabinet officials, the members of loyal, dedicated staffs that are many in number, what explanation would you give to them when they came to you and asked you what do you mean?
>
> *Dr. Adelman*. I think I would have to say that I have most of those hearings in my office and that is one reason why I have a space problem there, and that I looked at them before I wrote that sentence Senator. Sometimes the truth is inadvertently painful but I think I have to stand on that, that all of these voluminous writings and hearings were concerned with the preservation of one or another interest, special interests usually, the general interests sometimes, but they were not devoted to the basic facts that would permit anybody to make an intelligent choice in the national interest. And that was why, under questioning by one of your colleagues, I had to say repeatedly that I did not know, nobody else knew, and that it will take a good deal of time to find out. . . . With all the piles of paper we have, we have not gotten the knowledge or even an approach to the knowledge that we should have.[118]

The petroleum industry's privileged treatment came under attack by the economists, who examined the costs of prorationing, the percentage depletion allowance and import quotas, as well as challenging the national security argument. The debate which had been proceeding in academic journals since the early 1960s had surfaced in the political arena. At the same time the industry was coming under more frequent and rarely complimentary scrutiny by the nation's press. The Santa Barbara oil spill and other pollution episodes, as well as the growing perception of the depletion allowance as a major tax loophole, eroded public support for the industry.[119]

Unfortunately for the oil industry George Schultz took his mandate to study the program seriously, appointing Phillip Areeda, a Harvard law professor, to direct the study, and James McKie of Vanderbilt as chief economist. Along with the usual collection of testimony from interested parties the Task Force solicited position papers from academics, analyses from federal departments, and consulting reports. Instead of merely questioning the structure of the import control program the Task Force attempted to examine the rationale for and cost of the program.

The oil industry, in filing briefs with the Task Force, took its usual range of positions on oil imports. The API claimed that import controls were required to prevent the US from becoming dependent on foreign energy supplies, which would threaten the security of supply for the free world. The IPAA supported the 12.2 percent limit on imports, calling for elimination of exceptions and for imports to be limited to crude oil and residual oil, protecting domestic refiners. Jersey Standard supported TIPRO's proposal for auctioning of quota rights, but most companies opposed this option.[120]

Well before the Task Force report was sent to the President in February 1970 it had become controversial. The Charles Rivers Associates (CRA) study became the focus point for attack by the industry since it challenged the basic assumptions used to justify controlling oil imports. The report had been commissioned by the director of the energy policy staff of the Office of Science and Technology, S. David Freeman, in January 1969. The report, entitled *An Analytical Framework for Evaluating the Oil Import Quota Program*, was submitted to the Task Force in July. CRA estimated the cost to consumers of quotas at about $6.2 billion, and $3.9 billion to the nation as a whole, including inefficiencies due to prorationing. The annual cost of a sufficient stockpile was estimated between $1.8 and $2.7 billion dollars.[121]

The industry was quick to reply to the CRA study in its rebuttals submitted to the Task Force in August. Estimates of the cost of controls submitted by oil companies ranged between $0.6 billion and $1.1 billion. The companies assumed world prices would rise in the absence of controls and counted the $2.3 billion in lease bonuses, rents, royalties, and state taxes as credits to be counted against costs of the program. The companies also estimated that the cost of storing the 5 billion barrels called for by CRA would cost $4.5 to $5 billion a year instead of $2.7 billion. They agreed that storage programs could not secure the nation's long-term oil supplies. Interior filed a rebuttal to the CRA report, rejecting its analysis that national security could be better protected, and at a lower cost, by a strategic reserve instead of import controls. Ironically, an Interior Department staff report found that the cost of storage in salt domes was significantly less expensive than the above ground storage proposed by CRA. The Justice Department called for an end to the quota system due to the costs to consumers, and recommended that a competitive industry combined with a government program to assure the existence of sufficient reserves, through exploration and storage subsidies, could serve National Security interests more efficiently than the quota system.[122]

Louisiana Land and Exploration company, as part of the campaign against the CRA report, commissioned a study by the Stanford Research Institute (SRI). The SRI report challenged the $6.2 billion price tag placed on the import program by CRA, stating that there was no economic cost to import controls due to offsetting impacts on prices of other fuels. The Oil and Gas Journal criticized the Task Force staff, claiming that it was operating largely independent of its Cabinet bosses, and that other than Dr. McKie (whose expertise was referred to as "academic"), none of the staff possessed substantial knowledge of the oil industry. "The staff, however, seems skilled mainly in the art of conducting literature searches, collecting figures and opinions published by academics like themselves. . . ."[123]

The Task Force final report, despite being toned down due to political pressure, was still an unpleasant experience for the industry. It stated that the cost to consumers of oil import controls was about $5 billion in 1969. Oil import controls were estimated to save gas customers $0.5 to $1 billion due to the higher level of gas production, while inefficiencies in the industry, transfers to producing states, royalty owners, the federal government and oil companies accounted for the distribution of consumer losses. The Task Force considered a one-year cutoff of Arab or all Eastern Hemisphere oil (wartime threats to supplies judged to be of equal or lesser severity) to be the worst case security threat, involving a loss of 4-5 million bpd if import controls were not in place and Latin American oil was diverted to Europe.[124]

The existing import control program was not considered adequately responsive to security considerations while it was felt that gradual liberalization of import controls would not weaken the national economy. Although the probability of an extreme interruption was considered to be small, total abandonment of import controls was rejected. Instead a tariff system, involving some relaxation of import restraints was considered sufficient. Initially a tariff of $1.35 per barrel was proposed for 1971, followed by reductions in tariffs as long as reserves were sufficient to meet 1980 production targets. An initial 20¢ a barrel tariff preference for Western Hemisphere imports was recommended. Existing quotas would be phased out over three years. No mention of the alternative policy of using a strategic stockpile to protect against supply disruptions was included in the report's conclusions, despite cost data in appendix J which indicated that a 1.75 billion barrel stockpile (worst case scenario) would have an annual cost in the range of $750 million.[125]

The Secretaries of the Interior and Commerce and the Chairman of the Federal Power Commission rejected the findings of the Task Force and submitted their own report. Their reasons for opposition included: the substitution of a tariff for a quota, designed to produce a domestic price of $3.00 a barrel which they referred to as price-fixing; the adverse effect of the proposed program on exploration and development; and the economic loss to the oil industry which would weaken the national economy and thereby impair the national security. They proposed to improve the administration of the existing program and allow for an annual increase in oil imports, up to an additional 600,000 bpd by 1974.[126]

George Schultz had fought for a lower tariff of around 50¢ a barrel, which would have dropped domestic prices to $2.50, and he preferred the elimination of import controls, the position that an economist unencumbered by political maneuvering or national security considerations would choose. Due to the inability to get a single supporter among the other six Task Force members, he dropped this proposal. In the final report (after four drafts) he proposed a $1 tariff, to be implemented over a three to five year period, while the four members who accepted the report in effect, by their insistence on a tariff high enough to maintain existing prices, demonstrated that they merely wanted to replace quotas with a tariff.[127]

President Nixon released the task force report on February 20, 1970, but he was careful to announce that any decision concerning a change in the oil import program would be postponed until after congressional hearings and consultations with foreign governments. Nixon created a new Oil Policy Committee, headed by General George A. Lincoln, director of the OEP, with the members including the Secretaries of State, Treasury, Defense, Interior, and Commerce, the Attorney General and the Chairman of the Council of Economic Advisers. Although Schultz as Secretary of Labor was not included, he would join the committee later as Secretary of the Treasury. Day-to-day administrative functions would continue to be performed by the Oil Import Administration of the Department of Interior, but the policy direction of the program was to be provided by the Director of the OEP, acting with the advice of the Oil Policy Committee.[128]

Even before the scheduled congressional hearings on the report oil industry spokesmen were putting their criticisms on record. Walter Levy, a noted oil consultant, attacked a tariff on the grounds that it would encourage Middle East governments to raise taxes in order to capture the tariff revenues, instead of the present system where, he claimed, the benefits of oil quotas are passed on to consumers. TIPRO was in favor of the tariff but opposed attempts to use it to lower domestic oil prices.[129]

The major arena for discussion of the Task Force report was the hearings before the Subcommittee on Mines and Mining (House Interior Committee). General Lincoln was the first witness to testify. He stated that there was general agreement that oil imports were going to increase although there was disagreement as to the extent. He agreed with the Task Force's conclusion that the mechanism for control be changed from a quota system to a tariff system, although he did not feel committed to all the details of the majority report. Lincoln, fingered by Stanley Nehmer of Commerce as the only member of the Task Force to agree with Schultz's desire for a tariff and lower prices, was a target for pro-industry committee members.[130]

Industry spokesmen attacked the report before the committee, focusing on projections for US supply under the prices which would result from following policies recommended by the report. Standard Indiana claimed that an 80¢ reduction would lower domestic production by 4 million bpd (from 13.8 to 9.9 million bpd), and increase Eastern Hemisphere imports to 6.4 million bpd. The President of the IPAA claimed that the quota system resulted in a net gain of over $3 billion to the US economy. The estimate that there was 1.7 million bpd of excess capacity was challenged as too high, which was interesting since independents had consistently complained that imports were curtailing domestic production.[131]

The controversy over the report shifted to the Ways and Means Committee in June as Chairman Wilbur Mills held hearings on trade legislation. The committee voted to amend the Trade Expansion Act in July to specify that restrictions on commodities under the national security clause shall be by quota, a provision later voted down by the House. However there was an important difference between 1970 and previous years. East Coast congressmen, having won a victory over residual imports, concentrated on freeing heating oil imports from constraints during hearing before the Senate Banking and Currency Subcommittee on Small Business. The oil industry was able to prevent a tariff, as General Lincoln recommended to President Nixon in August that consideration of a tariff system be discontinued. On the other hand, Nixon authorized increased imports of No. 2 oil, and authorized increased oil imports for 1971, in response to November price increases for crude oil and gasoline.[132] The battle over tariffs could be considered the political high water mark for the industry--after this it would be every industry sector for itself.

NOTES

[1] E.E. Schattschneider, *Politics, Pressures and the Tariff*, (Hamden, Conn.: Archon Books, 1963).

[2] Robert A. Pastor, *Congress and the Politics of US Foreign Economic Policy: 1929-1976* (Berkeley: University of California Press, 1980), pp. 78-84.

[3] Pastor, *Politics of US Foreign Economic Policy*, pp. 85-93.

[4] John F. Manley, *The Politics of Finance* (Boston: Little, Brown and Company), 1970, p. 330-35.

[5] Pastor, *Politics of US Foreign Economic Policy*, pp. 99-100

[6] Manley, *The Politics of Finance*, pp. 187-89; Pastor, *Politics of US Foreign Economic Policy*, pp. 100-101.

[7] Raymond Bauer, Lewis Anthony Dexter and Ithiel De Sola Pool, *American Business and Public Policy*, 2nd ed. (Chicago: Aldine Atherton, Inc., 1972), p. 30; and Manley, *The Politics of Finance*, pp. 198-99.

[8] Don D. Humphrey, *American Imports* (New York: Twentieth Century Fund), 1955, p. 154.

[9] FTC-1952, pp. 351-54.

[10] FTC-1952, pp. 356-57.

[11] FTC-1952, pp. 361-67; M.A. Adelman, *The World Petroleum Market* (Baltimore: Johns Hopkins Press, 1972), pp. 134-35.

[12] Walter J. Levy, "The Past, Present and Likely Future Price Structure For the International Oil Trade," *Third World Petroleum Congress*, Volume 10, The Hague, 1951.

[13] Adelman, *World Petroleum Market*, pp. 144-48.

[14] Helmut J. Frank, *Crude Oil Prices in the Middle East* (New York: Federick A. Praeger, 1966), pp. 160-61; and Adelman, *World Petroleum Market*, pp. 150-52.

[15] Frank, *Crude Oil Prices*, pp. 61-70, 165-66; and Adelman, *World Petroleum Market*, pp. 156-57.

[16] Jean-Francois Landeau, *Strategies of US Independent Oil Companies Abroad* (Ann Arbor: UMI Research Press, 1977), p. 66.

[17] Landeau, *Strategies*, p. 20.

[18] McDonald, *Petroleum Conservation*, pp. 40-41.

[19] David Prindle, *Petroleum Politics and the Railroad Commission of Texas* (Austin: University of Texas Press, 1981), pp. 42-43, 49.

[20] Wallace Lovejoy and Paul Homan, *Economic Aspects of Oil Conservation Regulation* (Baltimore: Johns Hopkins Press, 1967), pp. 40-44.

[21] Lovejoy and Homan, *Oil Conservation Regulation*, pp. 129, 142.

[22] Lovejoy and Homan, *Oil Conservation Regulation*, pp. 71-75.

[23] Lovejoy and Homan, *Oil Conservation Regulation*, p. 122; Alfred Kahn, "The Depletion Allowance in the Context of Cartelization," *American Economic Review* 54 (June 1964): 310; James McKie and Stephen McDonald,

"Petroleum Conservation in Theory and Practice," *Quarterly Journal of Economics* 76 (Feb. 1962): 116; and M.A. Adelman, "Efficiency of Resource Use in Crude Petroleum," *Southern Economic Journal* 31 (1964): 105, 122.

[24] Nash, *United States Oil Policy*, pp. 34-35; and Stephen McDonald, *Federal Tax Treatment of Income From Oil And Gas* (Wash., D.C.: Brookings Institution, 1963), p. 12.

[25] Nash, *United States Oil Policy*, pp. 84-85; and Ronnie Dugger, "Oil and Politics," *Atlantic Monthly* 224 (Sept. 1969), pp. 70-71.

[26] Nash, *United States Oil Policy*, p. 165; Dugger, "Oil and Politics," p. 71; and McDonald, *Federal Tax Treatment*, p. 15.

[27] Nash, *United States Oil Policy*, pp. 197-98; and Bruce Ian Oppenheimer, *Oil And The Congressional Process* (Lexington, Mass.: D.C. Heath and Co., 1974), pp. 34, 101.

[28] Nash, *United States Oil Policy*, p. 198; and Oppenheimer, *Oil and Congress*, pp. 34-35, 101.

[29] Nash, *United States Oil Policy*, pp. 198-99.

[30] Oppenheimer, *Oil and Congress*, pp. 17, 22; and Richard Vietor, *Energy Policy in America Since 1945* (Cambridge: Cambridge University Press, 1984), pp. 224-25.

[31] Robert Engler, *The Politics of Oil* (Chicago: University of Chicago Press, 1961), pp. 159, 312.

[32] Bernard D. Nossiter, "Ex-Treasury Chief Received Oil Funds," *Washington Post*, July 16, 1970, p. 1.

[33] Tugwell, *Oil in Venezuela*, pp. 39-43; and Louis Turner, *Oil Companies in the International System*, 3rd ed. (London: George Allen & Unwin, 1983), pp. 47-48.

[34] Robert O. Keohane, "State Power and Industry Influence: American Foreign Oil Policy in the 1940s," *International Organization* 36 (Winter 1982): 168-78.

[35] Vietor, *Energy Policy In America*, pp. 92-93.

[36] Vietor, *Energy Policy In America*, pp. 92-93; Louis Johnson to Lyndon Johnson, in U.S. Congress, House, Select Committee on Small Business, *Hearings: Effects of Foreign Oil Imports on Independent Domestic Producers*, 81st Cong., 1st sess., May 1949, Part 1, p. 10; and Craufurd D. Goodwin, "Truman Administration Policies Toward Particular Energy Sources," in *Energy Policy in Perspective: Today's Problems, Yesterday's Solutions*, ed. Craufurd D. Goodwin (Wash. D.C.: Brookings Institution, 1981), pp. 106-7.

[37] Vietor, *Energy Policy In America*, p. 96; *Effects of Foreign Oil Imports on Domestic Producers*.

[38] Goodwin, "Truman Administration Policies," pp. 86-87.

[39] *O&GJ*, Mar. 23, p. 162, June 1, p. 30; June 15, p. 65, June 22, 1950, p. 117.

[40] "Memorandum Concerning Oil Imports," April 14, 1950, Oil Imports File, Joel D. Wolfshon Papers, HSLT, in Goodwin, "Truman Administration Policies," pp. 89-90.

[41] Bauer, Dexter and Pool, *American Business*, pp. 363-64.

[42] Bauer, Dexter and Pool, *American Business*, pp. 364-70.

[43] Vietor, *Energy Policy In America*, p. 102; and Bauer, Dexter and Pool, *American Business*, p. 365.

[44] Bauer, Dexter and Pool, *American Business*, pp. 325, 375.

[45] Humphrey, *American Imports*, pp. 157-69.

[46] *Congressional Quarterly Almanac*, 1953, p. 210-13; *O&GJ*, April 20, p. 98, May 18, 1953, p. 223.

[47] *Congressional Quarterly Almanac*, 1953, pp. 214-16; and Manley, *Politics of Finance*, p. 198.

[48] *Congressional Quarterly Almanac*, 1953, p. 215; and Manley, *Politics of Finance*, p. 199.

[49] *Congressional Quarterly Almanac*, 1953, p. 217; and Bauer, Dexter and Pool, *American Business*, pp. 40-48

[50] Vietor, *Energy Policy In America*, pp. 101-102.

[51] *O&GJ*, Mar. 1, p. 38, May 10, p. 74, and Nov. 1, 1954, p. 59.

[52] *O&GJ*, Nov. 29, p. 45 and Dec. 20, 1954, p. 63.

[53] Bauer, Dexter and Pool, *American Business*, p. 59

[54] Bauer, Dexter and Pool, *American Business*, pp. 66-67; and *O&GJ*, Mar. 7, 1955, p. 92.

[55] William J. Barber, "Eisenhower Energy Policy," in *Energy Policy in Perspective: Today's Problems, Yesterday's Solutions*, pp. 223-29.

[56] *O&GJ*, Mar. 21, pp. 122-24, Apr. 4, 1955, p. 105.

[57] Bauer, Dexter and Pool, *American Business*, pp. 70-72.

[58] Barber, "Eisenhower Energy Policy," p. 230

[59] *O&GJ*, Aug. 15, p. 99, Sept. 19, p. 88, Sept. 26, 1955, pp. 128-30.

[60] Barber, "Eisenhower Energy Policy," pp. 231-33

[61] Vietor, *Energy Policy In America*, pp. 104-5.

[62] Barber, "Eisenhower Energy Policy," pp. 233-34, Douglas Bohi and Milton Russell, *Limiting Oil Imports* (Baltimore: Johns Hopkins Press, 1978), p. 33; *O&GJ*, June 18, p. 133, July 2, 1956, p. 54.

[63] Bohi and Russell, *Limiting Oil Imports*, pp. 38-40.

[64] Barber, "Eisenhower Energy Policy," pp. 236-37; Bohi and Russell, *Limiting Oil Imports*, pp. 41-42.

[65] "Report to the President of the Special Cabinet Committee to Investigate Crude Oil Imports," reprinted in Cabinet Task Force on Oil Import Control, *The Oil Import Question* (Wash., D.C.: GPO, 1970), appendix C-2, p. 184.

[66] "Report to the President," pp. 186-187.

[67] Barber, "Eisenhower Energy Policy," p. 241.

[68] Nossiter, *Washington Post*, July 16, 1970, p. 14.

[69] *O&GJ*, Sept. 16, 1957, p. 101-103.

[70] *O&GJ*, Nov. 18, 1957, p. 137, Jan. 27, pp. 123-24 and Feb. 17, 1958, p. 65.

[71] *O&GJ*, Feb. 10, pp. 68-69, Mar. 31, pp. 50-51, Apr. 14, p. 66, May 5, p. 74 and May 26, 1958, p. 58.

[72] *O&GJ*, June 16, pp. 64-65, July 21, 1958, p. 51; and *Congressional Quarterly Almanac*, 1958, p. 170.

[73] "Report on Imports of Petroleum Products by the Special Committee to Investigate Crude Oil Imports," reprinted in *The Oil Import Question*, appendix C-2, p. 192.

[74] Bohi and Russell, *Limiting Oil Imports*, pp. 59-60; Vietor, *Energy Policy In America*, pp. 113-14; *O&GJ*, Sept. 22, p. 58, Oct. 13, 1958, p. 78.

[75] Barber, "Eisenhower Energy Policy," pp. 247-51.

[76] "Adjusting Imports of Petroleum and Petroleum Products Into the United States," Presidential Proclamation 3279 reprinted in *The Oil Import Question*, appendix C-2, pp. 198-199.

[77] Barber, "Eisenhower Energy Policy," pp. 255-56.

[78] Frank, *Oil Prices in the Middle East*, pp. 97, 135.

[79] Landeau, *Strategies*, pp. 74, 97-98.

[80] Landeau, *Strategies*, pp. 97-98.

[81] Yoram Barzel and Christopher Hall, *The Political Economy of the Oil Import Quota* (Stanford: Hoover Institution Press, 1977), p. 20.

[82] Frank, *Oil Prices in the Middle East*, pp. 96-105.

[83] Frank, *Oil Prices in the Middle East*, pp. 108-9.

[84] Anthony Sampson, *The Seven Sisters*, 3rd Ed. (New York: Bantam Books, 1983), pp. 186-89.

[85] Fuad Rouhani, *A History of OPEC* (New York: Praeger Publishers, 1971), p. 86.

[86] Sampson, *The Seven Sisters*, pp. 193-197.

[87] *O&GJ*, Mar. 30, 1959, p. 104.

[88] *O&GJ*, Apr. 27, 1959, pp. 83-85.

[89] *O&GJ*, Dec. 14, 1959, p. 43, June 6, p. 80; Aug. 22, p. 56, Dec. 5, 1960, p. 82, Apr. 24, 1961, p. 71.

[90] *CQ Weekly Report*, Aug. 26, 1960, pp. 1490-91.

[91] *CQ Weekly Report*, Mar. 31, 1961, pp. 541-42.

[92] *O&GJ*, May 15, 1961, pp. 95-96.

[93] William J. Barber, "Studied Inaction in the Kennedy Years," in *Energy Policy in Perspective: Today's Problems, Yesterday's Solutions*, pp. 295-99.

[94] Alan Stone, "Capitalism and Case Studies," in *The Political Economy*, ed. Thomas Ferguson and Joel Rogers (Armonk, New York: M.E. Sharpe, 1984), pp. 303-9; *CQ Weekly Report*, Mar. 9, 1962, pp. 403-4.

[95] Pastor, *Congress and Foreign Economic Policy*, pp. 107-14.

[96] Barber, "Studied Inaction in the Kennedy Years," pp. 305-8; and *O&GJ*, July 8, 1963, p. 82.

[97] Barber, "Studied Inaction in the Kennedy Years," pp. 304-5, 308-310; *O&GJ*, May 7, p. 70, Dec. 3, 1962, p. 93.

[98] Barber, "Studied Inaction in the Kennedy Years," pp. 313-16, and Bohi and Russell, *Limiting Oil Imports*, pp. 153-54.

[99] *New York Times*, April 2, 1962, IV, p. 7.

[100] *O&GJ*, June 17, 1963, p. 43; *CQ Weekly Report*, July 12, 1963, p. 1125; US Congress, House, Committee on Ways and Means, *President's 1963 Tax Message, Hearings before the Committee on Ways and Means*, 88th Cong., 1st sess., 1963, pp. 278-300.

[101] James L. Cochrane, "Energy Policy in the Johnson Administration: Logical Order versus Economic Pluralism," in *Energy Policy in Perspective: Today's Problems, Yesterday's Solutions*, pp. 380-92.

[102] G. Griffith Johnson to the Secretary of State, *Report of the United States Delegation to the United States-Canadian Discussions of Petroleum Policies and Programs*, Ottawa, Canada, December 13-14, 1962, declassified 11/26/79; William Roth to Lee White, June 1, 1965, Office of the Special Representative for Trade Negotiations, declassified 1/19/78.

[103] Bohi and Russell, *Limiting Oil Imports*, pp. 133-34.

[104] Bohi and Russell, *Limiting Oil Imports*, pp. 161-64.

[105] Kenneth W. Dam, "Implementation of Import Quotas: The Case of Oil," *Journal of Law and Economics* 14 (Apr. 1971), pp. 44-46; *O&GJ*, Aug. 14, 1964, p. 46, Jan. 18, p. 51, June 14, p. 77, Aug. 2, p. 67, Sept. 20, 1965, p. 92.

[106] Dam, "Implementation of Import Quotas," pp. 46-47; and *O&GJ*, Nov. 13, 1967, pp. 70-71.

[107] Dam, "Implementation of Import Quotas," pp. 47-48; and Vietor, *Energy Policy in America*, pp. 140-41.

[108] *O&GJ*, June 29, p. 32, Nov. 2, pp. 30-31, Nov. 16, p. 129, Nov. 23, 1964, pp. 28-29.

[109] *O&GJ*, Mar. 15, 1965, p. 43-47.

[110] Vietor, *Energy Policy in America*, pp. 133-34.

[111] Vietor, *Energy Policy*, p. 134.

[112] O&GJ, April 8, 1957, p. 71, June 18, pp. 78-79, Aug. 6, 1962, p. 99; *New York Times*, Aug. 1, 1965, III, 1:1; US Congress, House, Select Committee on Small Business, *Small Business Problems Created by Petroleum Imports, Hearings before Subcommittee No. 4*, 87th Cong., 1st sess., Part I, Nov. 20-24, 1961, pp. 559-60; and US Congress, Senate, Select Committee on Small Business, *Oil Import Allocations, Hearings before the Select Committee on Small Business* 88th Cong., 2nd sess., Aug. 10, 11, 1964, pp. 50, 65.

[113] Cochrane, "Energy Policy in the Johnson Administration," pp. 374-76; and *O&GJ*, Feb. 27, 1967, pp. 68-69.

[114] *O&GJ*, May 15, pp. 44-45, May 29, pp. 35-36, Sept. 25, pp. 76-77, Oct. 30, 1967, p. 53.

[115] *O&GJ*, Apr. 1, pp. 87-88, Apr. 22, 1968, p. 42.

[116] *CQ Weekly Report*, Feb. 2, 1968, p. 155; and *O&GJ*, Aug. 12, 1968, p. 42.

[117] US Congress, Senate, Committee on Finance, *Import Quotas Legislation, Hearings before the Committee on Finance*, 90th Cong., 1st sess., Part 1, Oct. 18, 19, 1967, pp. 374-75.

[118] US Congress, Senate, Committee on the Judiciary, *Governmental Intervention in the Market Mechanism, Hearings before the Subcommittee on Antitrust and Monopoly*, 91st Cong., 1st sess., The Petroleum Industry, Part 1, Economists' Views, Mar., Apr., 1969, pp. 30-31.

[119] *Governmental Intervention in the Market Mechanism*; and *New York Times*, May 18, 1969, III, p. 1.

[120] *O&GJ*, July 21, pp. 32-33, July 28, 1969, pp. 102-3.

[121] James Burrows and Thomas Domencich, *An Analysis of the United States Oil Import Quota*, (Lexington, Mass.: Heath Lexington Books), 1970.

[122] *O&GJ*, Aug. 25, pp. 58-59, Sept. 1, p. 82, Sept. 29, 1969, pp. 34-36; and *New York Times*, Aug. 14, 1969, p. 1.

[123] *New York Times*, Oct. 22, 1969; and *O&GJ*, Sept. 1, 1969, pp. 63-64.

[124] Cabinet Task Force on Oil Import Control, *The Oil Import Question*, February 1970 (Wash., D.C.: GPO, 1970), pp. 20-38.

[125] *The Oil Import Question*, pp. 124-30, 137-38, 299-303.

[126] *The Oil Import Question*, pp. 343-44.

[127] *The Oil Import Question*, pp. 131; *O&GJ*, Jan. 19, 1970, p. 23.

[128] Statement of the President, February 20, 1970, printed in U.S. Congress, House, Committee on Interior and Insular Affairs, *Oil Import Controls, Hearings before the Subcommittee on Mines and Mining*, 91st Cong., 2nd sess., Mar., 1970, pp. 10-11.

[129] *O&GJ*, Jan. 19, pp. 24-25, Feb. 6, 1970, p. 52.

[130] *O&GJ*, Mar. 16, 1970, pp. 104-5.

[131] *O&GJ*, Mar. 23, pp. 36-37, Apr. 13, 1970, pp. 56-57.

[132] *O&GJ*, June 22, p. 46, Aug. 3, p. 56, Aug. 24, pp. 30-31, Dec. 28, 1970, p. 67; *CQ Weekly Report*, Apr. 10, 1970, p. 959.

III

The Oil Crises

MARCHING SLOWLY TO DISASTER

The irrelevance of the domestic debate over control of oil imports began to become apparent in 1971. The need to protect the prorationing system was fading, as evidenced by an *Oil and Gas Journal* editorial in March which called for the end of market-demand proration. The editorial stated that rising demand and falling reserves made overproduction unlikely, while market-demand was a magnet for industry critics, who included Hendrik Houthakker, chairman of the Council of Economic Advisors (CEA). Opponents of import quotas were also becoming bolder and more prominent. A bill was introduced in May by Representative Silvio Conte (R-MA) to replace quotas with a shut-in reserve to protect against loss of imports. The plan was based on a paper by Walter Mead and Philip Sorensen. The two professors claimed that a shut-in reserve could be maintained at a cost of 32¢ per barrel of initial-year production vs. $1.04 per barrel for additional production resulting from import quotas. Industry calculations quoted in rebuttal put the cost of reserve capacity at $1.20 a barrel.[1]

While consumers and producers fought over oil policy the Nixon administration was taking tentative steps to develop some form of coherent energy policy, though the motivation may have been as much to defuse controversy as to develop policy. David Freeman argued for a Department of Natural Resources in submissions to the Advisory Council on Executive Organization, which endorsed his recommendations in its report to the President in late 1970. In July 1970, John Ehrlichman established a subcommittee of the Domestic Council to deal with energy problems, to be headed by the Chairman of the CEA. The Department of the Interior also pushed ahead with its own studies, one by the National Petroleum Council (NPC) and one directed by John J. McKetta, Jr., of the University of Texas.[2]

In June, 1971, President Nixon gave what he called the first comprehensive message on energy to the Congress. The message stressed an expanded nuclear power program, including development of the liquid metal fast breeder reactor, and a program for coal gasification. Other elements included accelerated outer continental shelf leasing, shale reserves leasing, and the centralization of energy functions in a Department of Natural Resources. The CEA and Office of Management and Budget opposed the nuclear and gasification programs on the grounds that they had not been justified economically, but strong lobbying by the American Gas Association and an understanding reached with nuclear proponent Representative Chet Hatfield (D-CA) kept them in the message. Both the administration and Congress then proceeded to ignore the message, since energy policy was still a low priority item on the political agenda.[3]

Congressional policy-making on energy was as disorganized as the executive branch. The Joint Atomic Energy Committee handled nuclear power, while the two Interior and Insular Affairs Committees dealt with the Interior Department's minerals policies. Public Works Committees covered pollution, the Judiciary Committees dealt with antitrust and Ways and Means and Finance were responsible for taxes and trade. In the 1971 session of Congress a number of investigations were initiated. The Senate Interior and Insular Affairs Committee was involved in a two-year study of national fuels and energy policy, which included participation of the Senate Public Works, Commerce and Joint Atomic Energy Committees. The Senate Select Small Business Subcommittee on Government Regulation began hearings on supply and cost of heating oil. A House Republican Task Force on Energy and Resources began a study of energy while the Democratic Study Group planned to establish its own task force on energy in the fall.[4] The result was a multitude of channels by which interested parties could made themselves heard, but no arena in which political compromise could lead to a coherent energy policy.

By August energy concerns had been submerged relative to other economic problems, especially that of inflation. Acting under authority of the Economic Stabilization Act of 1970, President Nixon proclaimed an immediate 90-day freeze on prices and wages on August 15, 1971. A cabinet-level Cost of Living Council (CLC) was established to administer the freeze. Two immediate problems occurred; independent marketers of gasoline experienced a profit squeeze because traditional discounts to bulk purchasers were eliminated by the large companies and the price of heating oil had been frozen at a depressed level due to choice of the base period. The resulting shortage of No. 2 oil increased pressure from marketers and northeastern Congressmen for imports of heating oil, which the President subsequently allowed in a proclamation on November 5.[5]

Phase 2 of the price control program was announced on October 7, and would last from November 15, 1971 to January 10, 1973. Decisions on price changes were to be handled by a price commission. Smaller firms were not required to report price increases consistent with the basic guidelines, intermediate firms were required to report increases, while large firms had to obtain advance approval for increases. Since allowable increases were based on profit margins, different companies were allowed to sell the same product at different prices. The low profitability of No. 2 oil relative to gasoline caused shortages of the former and excess production of the latter. This factor combined with unusually cold weather in the fall of 1972 would result in a shortage of heating oil, acerbated by the rule that allowed price pass-through of import costs only if foreign products were kept physically separate from domestic products, discouraging imports of No. 2 oil. An additional bizarre twist of regulations allowed oil companies to adjust prices of diesel fuel but not heating oil relative to other product prices, even though the two fuels are identical.[6]

One method of increasing pressure on oil companies to keep product prices down was by increasing oil import quotas. Authorized imports for 1972 were increased in January, May and again in the last quarter of 1972, reacting to tight heating oil supplies. In January 1973 Nixon suspended controls on imports of No. 2 fuel oil and raised imports of crude and refined products 915,000 bpd above 1972. Reacting to the steady increase in imports the IPAA and TIPRO asked for a share of import quotas to be granted to independent producers.[7]

In early 1973, with the election behind the administration, attention returned to energy policy. On January 5 Secretary of the Treasury George Schultz was given the additional title of assistant to the President for economic affairs, including jurisdiction over oil prices and oil imports. Control over the oil import program was transferred to the Treasury Department on February 2, and Deputy Secretary William Simon was named chairman of the Oil Policy Committee. The functions of the OEP were transferred to the President. On February 23 a White House Special Energy Committee, composed of Schultz, Ehrlichman and Kissinger, was announced. Charles J. DiBona, president of the Center for Naval Analyses, was appointed special consultant for energy, and his group would be known as the National Energy Office.[8]

One pressing reason for reorganization of energy responsibility in the executive branch were the rapid changes in the international oil market. The ascendance to power of Colonel Qadhafi in a coup in September 1969 in Libya started the process that would lead to a series of price increases, culminating in the 1973 price hike. The revolutionary government formed a committee to negotiate a revision of posted prices in December, and to put pressure on the oil companies cutbacks in production were ordered in May, 1970, beginning with Occidental. The oil companies, led by Exxon, took a firm line, offering only a 5¢ per barrel increase. With Occidental dependent on Libya for a majority of its crude, the company turned to its negotiating partner Exxon for alternative supplies, but Exxon refused to help. The result was the capitulation of Occidental, accepting in September 1970 a 30¢ per barrel increase in payments plus a change in profit sharing from 50-50 to 58 percent for the government. In a few weeks the other independents had settled for similar terms, followed by the Majors.[9]

Emboldened by the success of Libya other countries soon followed with similar demands. In the fall of 1970 the majors agreed to extend the 55 percent tax rate to Nigeria and the Gulf producing states and make some adjustments in posted prices. Then, in December, Venezuela increased its effective tax rate percent to 60 percent and declared that the government would unilaterally set tax reference prices, following the example that Algeria had set in July. The oil companies, who had tried to ignore OPEC for a decade, now wanted to obtain an agreement with the organization to prevent the leapfrogging in demands which was occurring in the Mediterranean. The Libyans made new demands in January, 1971, only three months after signing a five year agreement. To counter this threat the companies, having learned from their mistakes, arranged an oil-sharing arrangement. However, the companies' stance on comprehensive negotiations was undermined by American and British diplomats.[10]

In 1971 negotiations commenced in both Teheran and Tripoli. In Teheran the Gulf countries threatened an embargo if their demands were not meet, giving February 15 as a deadline. The companies surrendered a day before, giving the countries their requested terms. Posted prices increased 30¢ per barrel, in addition to an inflation allowance. The companies agreed to make the 55 percent tax uniform and to adjust the posted prices of particular crudes upward. This was followed in March by even more generous concessions to Libya equivalent to an escalation of 65¢ per barrel. Participation was the next issue raised, as Sheikh Yamani, the Saudi Arabian oil minister, felt that it was a more viable alternative than nationalization, because it would keep the companies interested in maintaining current price levels. In the next OPEC-company meeting in January 1972 participation was pushed along with a request for an 8.5 percent price increase to compensate for the devaluation of the dollar. Yamani turned the screws on the Aramco consortium, whose members after a short but bitter round of protests accepted participation.[11]

Along with the constant pressure on prices and ownership placed on the companies by OPEC, the Arab countries were beginning to test the potential of oil as a political lever. Libya had nationalized British Petroleum interests in retaliation for an Iranian seizure of Arab-owned islands in the Persian Gulf. The Libyan ambassador to Egypt suggested that the Arab states should employ their economic resources to pressure the Western countries late in 1971. The president of Iraq saw participation as a method to gain control of resources to be used in the struggle against imperialism and Zionism. Anwar el-Sadat, President of Egypt, said that US support of Israel was jeopardizing American oil interests in January, 1972.[12]

A perception that an energy crisis was upon the country emerged during the winter of 1972-73. The collision of environmental regulations, price controls, natural gas regulation, oil import constraints, and explosive demand growth resulted in apparent shortages of fuels. Air pollution controls restrained the use of coal and high sulfur residual, increasing demand for natural gas, whose price was controlled at artificially low levels by the Federal Power Commission (FPC). Shortages of natural gas shifted consumption to heating oil, whose supplies were limited due to price controls. Increased demand for oil products, combined with the leveling off of domestic production, resulted in a dramatic increase in oil imports. Projected share of imports from Middle East countries were as high as 60 percent of imports by 1980, raising fear of "political blackmail." While an October 1972 study by the OEP suggested that the United States could reduce energy demand by 1980 by as much as the equivalent of 7.3 million bpd of oil, Congressional legislative proposals excluded conservation, price increases or taxes, and decontrol of natural gas production.[13]

One problem in developing a consensus for action was the serious differences among oil analysts regarding the future of world petroleum markets and in the policy recommendations flowing from their prognoses. As Senator Henry Jackson pointed out during hearings on oil imports in 1973:

> Some experts emphasize the development of a "seller's market" based upon rapid growth of world energy demand and the group interests of the exporting nations in limiting production. Others emphasize the vastness and low costs of developing Middle Eastern petroleum resources, and the actual and potential rivalries among oil producers; they foresee a world "surplus" of oil persisting for at least two decades . . . the crux of the debate is whether the producing countries can collectively control production and prices without the active cooperation of the international oil companies and of consuming country governments. There is a corresponding lack of consensus whether the growth of United States imports would inexorably tend to increase the exporting countries' bargaining power, or would on the contrary help to restrain price increases by stimulating exporter competition for the US market.[14]

One issue on which economists agreed was the value of some form of strategic reserve. However, the administration, and especially the Interior Department was not enthusiastic about the proposal. The following testimony by Secretary Rogers Morton in front of the Senate Interior Committee illustrates the government's attitude:

> It is true we do not have, and it has not been required, policy in the United States to set up a great reserve inventory that will be used. I couldn't conceive of inventory going beyond 90 or 120 days and being carried by the Federal Government or the consumer. . . .
> . . . When you say use up the supply of other nations means becoming dependent on other nations, and this would offset the security you would gain by some short-term, 90 or 120 days of storage. After all, that time will slip by quickly, and when you find yourself exhausted, you find yourself at the mercy of other countries.[15]

While criticism of the administration's energy policy or lack of such were mounting in Congress, other factors were eroding support for the import quota program. World oil prices during 1972 had been rising toward domestic levels.

At the same time domestic excess capacity had disappeared, eliminating the ability of the Texas Railroad Commission to make supplies available for independent refiners by raising allowables. Thus independent refiners were caught in a double bind since the value of their import tickets had plummeted, lowering their subsidy, while crude oil supplies had become tight, forcing them to pay more for supplies. A number of alternatives to the program were considered before settling on an fee on imports above the quota level, combined with fees on products. To help independent refiners and marketers, the Oil Import Appeals Board was authorized to issue crude tickets to be used to exchange for products.[16]

On April 18 the President presented a second energy message to Congress, based on proposals put together by DiBona. Nixon suspended direct control over oil imports, tariffs were eliminated and holders of import licenses could import oil duty free, while imports in excess of the 1973 level would be charged an fee. Fees were set higher for products than crude to encourage domestic refinery construction. The President called for decontrol of natural gas from new wells, judicious use of provisions of the Clean Air Act where coal burning was concerned, and streamlining of licensing procedures for nuclear power plants. A seven percent tax credit on unsuccessful wells and a 12 percent credit on successful wells was proposed. Nixon also gave lip service to conservation, directing the establishment of an Office of Energy Conservation in the Department of the Interior.[17]

While the Nixon administration attempted to deal with oil product shortages by eliminating constraints on imports and encouraging expansion of domestic refinery capacity, its price control policies continued to disrupt the proper functioning of the market. Phase 3 price controls began January 11, and lasted until June 13, 1973. The rules promulgated under Phase 2 were extended in Phase 3, but were to be self-administered by firms. Shortly after the program started, fuel oil prices jumped 7.4 percent, prompting an outcry from the public and Congress. The Cost of Living Council concluded that the price increases were justified, but on March 6 the Council ruled that the twenty-three major companies would have to supply cost justification for price increases which caused revenues to rise more than 1 percent. Revenue increases greater than 1.5 percent were forbidden if profit margins were above the base period average. Since companies couldn't pass cost increases along, and world prices were at or above domestic prices, there was a lack of incentives to increase production or imports. There were also no incentives for major integrated companies to make oil available to inland refiners. However independent refiners could pass along product price increases to consumers, which encouraged them to bid up the price of available crude, and expand their market share at the expense of the large companies.[18]

The Oil Policy Committee felt itself under pressure from Congress to develop a plan to aid independent refiners, who were struggling to find crude supplies, and hurting due to the loss of subsidies formerly available under the oil import quota program. Senator Thomas Eagleton (D-MO) introduced an amendment to the extension of the Economic Stabilization Program which would give the government power to allocate crude oil supplies and refinery output. Senator Henry Jackson (D-WA) introduced another bill in May containing authority to allocate all fuels, down to the retail level. In the hearings chaired by Jackson the emergence of consumer interests was exemplified by the testimony of Lee White, former FPC Chairman, representing the Energy Task Force of the Consumer Federation of America (CFA). White asked for allocation of crude to independent refiners and the reservation of 75 percent of gasoline imports for independent marketers, on the grounds that their continued existence was vital to the national interest. A voluntary allocation program was established in May, but it basically consisted of scrambling to find supplies for companies who complained to their congressmen.[19]

Senator Jackson also opened hearings on May 30 on S 1586, to establish a system of petroleum reserves and limit oil and natural gas imports. John Lichtblau, of the Petroleum Industry Research Foundation, a research group representing oil importers, and the API testified in favor of a strategic reserve. DiBona and Love, however, testified against the bill, because of provisions to discriminate against insecure sources of imports, and the expense of developing a reserve. Instead they proposed further study and support of the administration's proposed energy supply incentives. In June the Senate Judiciary Subcommittee on Antitrust and Monopoly held hearings on competition in the oil industry. A major theme was criticism of the major oil companies by state attorneys general, accusing them of deliberately driving independents out of business. The voluntary allocation program was attacked for its inability to force compliance by the large companies. In July the House Commerce Committee held allocation hearings in which the large companies and the Nixon administration came under attack once again.[20]

On May 31, 1973, Senator Jackson requested that the Federal Trade Commission prepare a report within 30 days on the structure of the petroleum industry. A preliminary report was sent to Jackson on July 6, which he released for publication on July 13. Five days later, responding to this political pressure, the FTC filed a complaint charging the eight largest oil companies with maintaining and reinforcing a noncompetitive market structure in oil refining. The FTC relied on a dubious economic theory of "shared monopoly," especially in light of the relatively low level of concentration in the oil industry. The case was eventually dismissed in September 1981 after the Commission decided that further proceedings were not in the public interest and $100 million was expended by the various parties to the suit. However, the filing of the complaint fueled the suspicions of congressmen, convinced that "Big Oil" had engineered the energy crisis of 1973. These attitudes would be reinforced by the oil price increases and record oil company profits in 1974.[21]

Maneuvers by Roy Ash, Director of the OMB, to restructure energy organization in the White House and head off Senator Jackson, who was proposing to create an Energy Council, resulted in the formation of an Energy

Policy Office. Governor John Love of Colorado, with no energy experience, was chosen to head the new office, with DiBona as his assistant. Nixon in his June 29 statement establishing the office also announced a $10 billion, 5 year R&D program focusing on coal, a 7 percent conservation goal within the federal government, and he asked Congress to approve an Energy Research and Development Administration (ERDA) and a new Nuclear Energy Commission. One reason for suggesting an ERDA was the realization that the Department of Natural Resources proposal was in trouble in Congress due to battles over committee turf.[22]

William Simon, despite his belief that it would be futile, ended up supporting mandatory allocation after the attempt to institute a voluntary allocation plan failed. Love and DiBona initially resisted mandatory allocation, but by mid-September Love and Ash were advising the President that an allocation program was certain to emerge from Congress. The administration was split between those who wanted Congress to enact the program and take the blame for its inevitable failure, and those who wanted the Executive Branch to maintain control of energy policy. Ash and Love proposed allocating crude to small refiners and products to distributors along with a ten cent gasoline tax, or to merely allocate crude. By October, Schultz, Morton and Simon supported a general program of allocation, while Love, Ash and Laird wanted allocation restricted to propane and heating oil. The gasoline tax was quietly dropped. On October 2 Love announced the establishment of a mandatory allocation program for home heating oil and propane gas.[23]

Phase 4 price controls began August 12, 1973, following a 60 day price freeze, during which companies were allowed to sell domestic and imported products at different prices. The logic behind Phase 4 regulations was to impose ceilings on retail prices for gasoline and No. 2 oil, and thus put pressure backward onto costs, which caused a squeeze on profit margins for independent retailers as wholesale prices continued to rise. The problem was worsened by the choice of the base period, a period when retail prices were lower due to price wars. The FTC added fuel to the fire with a staff report accusing the major companies of manipulating the shortage to the disadvantage of independent refiners and marketers, a charge which the Director of the FTC's Bureau of Competition repeated in testimony before the Senate Judiciary and Commerce Committees.* Simon pointed out that the existence of these independent firms was due to surplus crude and refining capacity in the past, the disappearance of which could hardly be attributed to the large companies.[24]

Under Phase 4 controls domestic crude oil was subject to a two-tier price system, with the price of oil from existing wells, frozen (initially set at $4.25 a barrel) while new oil, defined as any increase in production from a property above the May 1972 level, could sell at the world level. For every barrel of new oil produced, one barrel of old oil was released from price controls. The two-tier system managed to alienate almost everyone in the oil industry. Independent producers wanted all oil decontrolled, while the major refiners would prefer that it all be controlled. Independent refiners, with limited access to old oil were fearful of price competition, while stripper well owners were especially vociferous about their need for exemption.[25]

THE OIL EMBARGO OF 1973

The participation agreements and nationalizations of 1972 had the effect of releasing a small but significant quantity of crude to the international oil spot market. The importance of these sales was not the revenue received but the price signals to the producing countries concerning increasing scarcity in the oil market. As markets tightened, the Majors limited sales to the second-tier firms in order to supply their own affiliates, forcing a scramble for crude supplies. In August 1973 Libya sold 40,000 bpd for between $6.25 and $7.00, while selling the bulk of its share of production for $4.90 a barrel. Prices for low-sulfur fuel from other sources was over $5.00 a barrel due to the impact of environmental regulations. By September 1973, aware of the prices that oil products were bringing on the spot market, OPEC members were demanding a revision of the Teheran/Tripoli agreements. Negotiations commenced in Vienna on October 8, 1973, (the day after the Yom Kippur war was initiated) and OPEC demanded a 100 percent increase in posted prices, a request to which the companies did not want to give their acquiescence in order to protect themselves politically in the consuming nations. Negotiations broke off on October 12 and four days later the Persian Gulf producers unilaterally raised posted prices by 70 percent to $5.11 per barrel.[26]

The Arab producers announced a five percent cutback on October 17, and three days later Saudi Arabia followed with a ten percent cutback plus a complete embargo of the United States and the Netherlands. By January 1974 it became evident to the Arab producers that the embargo, begun in October, was ineffective in terms of selectively punishing the nations which were its targets, and Arab production increased by 10 percent, raising it to 85 percent of the pre-embargo level. While the embargo may have been a failure as a political weapon it was quite successful in demonstrating that prices could increase substantially without quenching demand in the short-run. Tunisian oil in November sold for $12.64 and low-sulfur Nigerian for more than $16 a barrel. Six month contracts for Iranian oil in December brought bids of $9-17 per barrel, mostly from American independents. When OPEC met in Teheran in mid-December these bids encouraged the Shah of Iran to push for a price of at least $14 on the grounds that the nearest substitute for oil cost $17. The Saudis, concerned with the consequences of price increases on the world economy, fought for a lower figure, with the two sides eventually compromising on a price for equity oil (the share of oil production owned by the company operating in the oil field) of $7.11 and a posted price of $11.65 for benchmark crude.[27]

The initial response to the embargo was a message by Nixon to the nation on November 7 in which he warned that at least 10 percent of petroleum demand would not be met. He urged voluntary conservation, and asked Congress to allow opening of the Naval Petroleum Reserves, relaxation of environmental standards, and the power to prevent power plants from switching from coal. Nixon called for a commitment to a program, Project Independence, to develop domestic energy sources and eliminate dependence on foreign energy by the end of the decade. Despite the call for action the administration was divided on what action to take. Schultz and Laird opposed gasoline rationing, calling for a 30 to 40¢ per gallon gasoline tax, while CEA chairman Herbert Stein opted for gasoline price increases. OMB Director Ash preferred a modified rationing scheme. The NPC, David Rockefeller, and 25 senators supported rationing over a gasoline tax hike. With Love unable to make decisions and Nixon paralyzed by Watergate, presidential aide Alexander Haig and Counselor for Domestic Affairs, Melvin Laird, took over. Ash, Haig and Laird decided that Love's Energy Policy Office was inadequate for the tasks ahead and pushed for an new agency, the Federal Energy Office (FEO), created by executive order on December 4, 1973. Simon was designated FEO administrator, with John Sawhill of OMB appointed as his deputy.[28]

The first task of the new office was the administration of the Emergency Petroleum Allocation Act of 1973 (EPAA). Passed by Congress on November 27, the EPAA required the administration to prepare and implement allocation plans for crude oil and products within thirty days. First priority was given to consumers involved in food production, fuel production, health and public safety, and mass transit. Cutbacks on distillate and residual oil supplies for other classes of user were ordered, and gasoline production was reduced at the refinery level. The EPAA exempted stripper well production from price controls while transferring authority for price control over the oil industry to the Federal Energy Office in December 1973. Price control was complicated by the rise in the price of imported and uncontrolled oil which aggravated the conflicts between different sectors of the industry. The integrated oil companies, putting their needs first, failed to renew one-year supply contracts with independent refiners which expired on December 31. The FEO required crude oil suppliers to sell to existing customers unless a bona fide offer at lawful prices was refused. Crude supplies were reallocated among refiners so as to require refiners with supplies above the industry average to sell the excess to refiners with supplies that were below average, at the weighted average cost plus transportation costs and a 6 percent handling fee. Sellers were allowed to raise product prices and recover 84 cents per barrel of crude sold. The sellers objected because the weighted average didn't cover the replacement cost of oil; buyers complained that the crude was at a price which made their products noncompetitive, which didn't matter under the shortage conditions that existed until mid-1974, but was a problem once markets loosened.[29]

For the Treasury and the CEA the solution was to decontrol oil, establish a short-run price ceiling at about $12 a barrel, and use a windfall profit tax to lower returns to producers to about $6.50 a barrel. A coupon system was proposed to return part of the increased costs to consumers. This was prevented by fear of inflationary impacts, the threat of consumer and Congressional rebellion, and the fact that the EPAA prohibited price decontrol for petroleum products unless the President found there was no shortage of these products. Nixon instead raised the base price of old oil to $5.25 on December 19 and proposed a windfall profits tax as part of his request for emergency energy legislation. The disclosure that almost 10 percent of Nixon's 1972 campaign receipts had come from people associated with oil and gas companies killed all possibility of price decontrol.[30]

The National Energy Emergency Act (S 2589), which would have given the President power to impose gasoline rationing and implement energy conservation measures, was quickly reported by the Senate Interior Committee and passed on November 19 by a 78-6 vote. The House Commerce Committee, chaired by Harley Staggers, (D-WV) decided to write its own version, including a clause to provide for restrictions on the windfall profits of energy producers. Purchasers of energy products could petition for relief from excessive product charges to the Renegotiation Board, which could establish a lower sales price or order refunds. Windfall profits were defined as those in excess of the average profits obtained selling a product during the years 1967-1971. The windfall profits provision expressed the anti-oil sentiments of the House, triggered by the large increases in oil company profits during the first nine months of 1973.[31]

The bill passed through the conference with the windfall profits provision intact but then ran into trouble in the Senate. Oil state senators objected to the windfall profits provision and started a filibuster December 21, led by Russell Long and Paul Fannin (R-AZ). The House balked at a new compromise which deleted the windfall profits restrictions, and voted it down 228-36. Another attempt to pass the Energy Emergency Bill failed when it was recommitted to a House-Senate conference by a 57-37 vote on January 29. The coalition which defeated the bill included northern Democrats who thought the bill overly lenient in relaxing the 1970 Clean Air Act's standards, oil state senators who wanted to eliminate windfall profit provisions, and backers of the President who wanted to strip the bill of everything but authority to impose rationing and conservation and provisions easing clean air regulations.[32]

In January, controversy over oil company profits continued as a bill (S 2799, HR 11925) was offered by Senator George McGovern (D-SD) and Representative Les Aspin (D-WI) to levy an 85 percent tax on profits exceeding the average profit rate of the four previous years, excluding an amount equivalent to increases in a company's spending on oil supply. Gerard Brannon, a member of a OMB advisory task force, proposed as an alternative repeal of percentage depletion and intangible drilling tax credits. Senator Jackson introduced a bill in February which would have reduced the price of new oil, and roll back the price of old oil, a provision which House-Senate conferees decided to include as a partial substitute for the windfall profits provision of S

2589. Stripper and new oil would be sold for $7.09. The rationale for control of new oil was the statement of producers (IPAA in November, 1973) that $5 to $6 a barrel was sufficient to motivate exploration for new sources. After the Senate approved the conference report on February 19, the House passed it on February 27. Nixon then vetoed the bill in protest of the rollback provision, a move applauded by oil industry lobbyists, and supported by the Senate 40-58. The President also objected to the bill's unemployment compensation fund and the provision of low-interest conservation loans to homeowners.[33]

The House Ways and Means Committee adopted proposals by Al Ullman (D-OR) on March 19 which would eliminate the 22 percent depletion allowance in stages by 1980, tying declines in the tax to rises in prices. The Committee also approved Nixon's windfall excise tax but included a plowback provision and restricted use of the foreign tax credit. On April 1 the Committee approved a new plan backed by Chairman Wilbur Mills to eliminate depletion over three years, by the end of 1976. The final version included the depletion phase-out and the plowback provision on the windfall excise tax. The House Democratic Caucus, under the instigation of William Green (D-PA) and under the influence of the general hostility to the oil companies, in an unprecedented setback for Mills voted for a full House vote on amendments that would eliminate the foreign tax credit and depletion. Speaker Carl Albert (D-OK), in order to defuse the situation, announced that House action would be postponed until July. Mills then killed the Energy Tax Bill by merging it with a general tax reform bill, which the Rules Committee refused to report.[34]

With the embargo over and the supply situation easing, the FEA (successor to the FEO) adjusted its allocation system. The Buy/Sell list, implemented on June 1, 1974, modified the original system for evening out crude supplies among refiners by requiring the 15 largest integrated oil companies (excluding several that were classified as independents because of a low ratio of supplies to refining capacity) to sell crude oil to approximately 100 small and independent refiners. The sales requirement for sellers was based on the ratio of its refinery capacity to the total capacity of all 15 firms. Sellers could only charge their weighted average cost plus a 30¢ handling fee.[35]

The FEA, aware of the mood in Congress, sought a strategy which would avoid extension of the allocation act or a tightening of controls. A plan to issue tickets to refiners in relation to the percentage of capacity that could be supplied by low-cost oil, with a sliding scale for small refiners, was adopted and renamed the "entitlements scheme" by the FEA. The Treasury economists, supported by OMB Director Ash and CEA staff, opposed the plan because of fears that it would merely generate a new set of vested interests, who would then block decontrol of crude prices. On July 21, 1974, twenty-eight senators, led by Philip Hart, wrote to Sawhill urging adoption of the entitlements plan to preserve the viability of independent refiners and marketers. Sawhill was also influenced by estimates by his staff that decontrol could contribute 0.2-0.3 percent annually to inflation, a concern shared by the Commerce Department.[36]

It was announced after September hearings that the FEA would proceed with the entitlements program. On December 4, 1974, the FEA issued rules establishing what was known as the "entitlements" program. Each refiner received entitlements for old, price controlled crude, equal to the national average ratio of old crude to refinery runs. Refiners with less than 175,000 bpd of refinery runs received bonus entitlements on a sliding scale as under the import quota program. Refiners with less than the national average of old oil would sell entitlements to other refiners at prices fixed by the FEA, based on the spread between the estimated controlled ($5.25) and uncontrolled prices ($10.50 to $12). The initial price was set at $5 per barrel, with the price rising to $6 in a month.[37]

PRESIDENT FORD: PICKING UP THE PIECES

When Gerald Ford became President in August he moved quickly to centralize control over energy policy. An Energy Resources Council was created with Rogers Morton in command and Frank Zarb of OMB as executive director. Sawhill resigned from the FEA and was replaced by Zarb on November 25. Morton and Zarb put together a team which developed a new program with the goals of reducing reliance on insecure sources of energy, enhancing security against supply interruptions, and funding of energy supply, while maintaining consumer purchasing power. FEA staff analyses of proposals were conducted in mid-November, and legislative measures were discussed in December at meetings between representatives of OMB, CEA, the Energy Council and other parties. The President was briefed on the outcomes and settled on a legislative package on December 27.[38]

President Ford's January 15 State of the Union address included proposals for a strategic petroleum reserve (SPR) to be filled with production from the Elk Hills Naval Reserve, to lower consumption through energy taxes, removal of oil price controls, deregulation of natural gas production, and relaxation of pollution control standards. The tax plans included a $3 per barrel increase in import fees, imposed in $1 increments on the first of the month during February, March and April, a $2 per barrel excise tax on crude oil and products, a 37¢ per thousand cubic feet (equivalent to a $2.15 per barrel tax on a Btu basis) tax on natural gas, and a graduated windfall profits tax on the increase in the price of oil above base levels. The oil excise taxes and import fees were expected to raise about $9.5 billion annually, the natural gas tax about $8.5 billion and the windfall profits tax about $12 billion per year. $5 billion of these taxes would be paid by state and federal governments, and the other $25 billion would be returned to the economy through tax cuts. The objective of the program was to cut imports by 900,000 bpd in 1975 and 1.6 million bpd by 1977.[39]

Congress quickly reacted to the President's program by moving to block implementation of the oil import fees. The House Ways and Means Committee in a 19-15 vote on January 24 reported HR 1767, which postponed the President's proclamation raising oil import fees by at least 90 days. The fee delay was tied to legislation to increase the debt limit for six months to avoid a veto. This action came after Ford refused Chairman Ullman's request for a 60-day delay before imposing fees. In the Senate, Senators Kennedy and Jackson prepared a joint resolution that would prohibit import fees for 60 days, and require Congressional approval after that waiting period. The battle lines between the Democratic Congress and the new President were forming as Speaker Carl Albert stated that Congress should consider alternative policies, including gasoline rationing. Lowell Weicker (R-CN) and Majority Leader Mike Mansfield introduced a bill to require the President to establish a gasoline rationing program within 60 days. Ford responded at a January 21 press conference by threatening to veto any gas rationing bill sent to him and vowed to go ahead with his own program given the lack of alternatives offered by Congress.[40]

HR 1767 was passed on February 5 by a 309-114 vote after the Rules Committee separated it from the debt ceiling measure. The Senate passed the bill on February 19 by a vote of 66-28, sending the bill to the President who vetoed it on March 4, but accepted the suggestion of congressional leaders that he defer the second and third increases and his plan to decontrol old oil until May 1. The Senate Democratic leadership decided after the veto to postpone indefinitely a Senate vote to override the veto while in the House, leaders delayed a vote for a week. The reason for the overwhelming opposition to the import fee was the perception of Congressmen that the public would not tolerate additional increases in energy prices. A Roper poll found that 63 percent of respondents felt that increased energy costs were the primary cause of inflation. Hostility to the oil industry was also evident, with 60 percent feeling that the industry was making too much profit. Even the industry union charged that workers and consumers were being robbed blind.[41]

The House and Senate Democratic leadership developed their own energy plan, the product of two task forces; a Senate task force led by John Pastore (D-RI) and a House task force led by Jim Wright (D-TX). The Democratic plan included elimination of the depletion allowance (except for small producers) and foreign tax credits, and the establishment of tax credits and loan and grant programs to encourage residential and commercial energy conservation. It included a 5¢ gasoline tax and mandatory efficiency standards for automobiles, requiring a 50 percent improvement in gas mileage by 1980 and 100 percent by 1985, with taxes and credits based on mileage. The plan also provided standby authority for mandatory conservation measures and an energy trust fund, to finance an independent National Energy Production Board to mobilize public and private resources to increase domestic energy production and encourage conversion to abundant fuels. The Energy Board would have the power to limit oil imports and set up a central purchasing mechanism. A strategic reserve would be established, holding 500 million barrels of oil by 1980, and 1 billion by 1985. Liberal Democratic congressmen dissented from the Democratic leadership, calling for lowering energy prices to cost-based levels via a price rollback, with old oil controlled at $4.25 a barrel, and prices for new oil ranging up to around $7 a barrel. They asked for tough, mandatory fuel-efficiency standards rather than increases in consumer prices and taxes to institutionalize the conservation ethic.[42]

The Democrats on the Ways and Means Committee formulated a comprehensive energy plan of their own in February. They suggested gradual import cutbacks by phasing in import quotas, reducing imports by a million bpd over a two to three year period, with the long range goal of limiting imports to 25% of domestic consumption by the early 1980s, making a federal agency responsible for purchasing foreign oil by auction. A strategic reserve, extension of mandatory oil allocation authority, and stand-by rationing authority were included in the program. Gasoline consumption would be reduced by a phased-in increase in federal gasoline taxes starting at 5¢ a gallon, eventually rising to 40¢ in five years. Energy related taxes, including gasoline mileage taxes on automobiles, would be used to finance an energy trust fund to increase energy supplies and promote conservation. Decontrol, if implemented, would be put into effect gradually by increasing the controlled price of old oil $1 per year over a 5-year or 6-year period or by gradually reducing the quantity of old oil subject to price control, with a windfall excise tax ranging up to a maximum of 85 percent on windfall profits over $2.00 per barrel (as proposed by the Treasury Department), with plowback provisions. Percentage depletion would be repealed except for the first 3,000 bpd of production for two years, while foreign oil would lose both depletion and intangible expenses deductions, and use of the foreign tax credit would be restricted.[43]

William Simon, in his testimony in front of the Ways and Means Committee in March explained the administration's opposition to rationing, gasoline taxes, plowback provisions with a windfall profits tax and import quotas. He noted that both national leaders and public opinion supported rationing, but that this was probably due to ignorance of the costs and consequences of such a program. "Last year, when we considered the feasibility of rationing gasoline, we concluded that while it could be implemented, it would take four to six months to set up, employ about 15 to 20,000 full-time people, incur $2 billion in federal costs, use 40,000 post offices for distribution, and require 3,000 state and local boards to handle exceptions. . . ." Simon stated that a price increase in crude oil was more effective in reducing consumption than a gasoline price increase, since there is just as much opportunity to conserve other petroleum products as there is to conserve gasoline. Plowback was not justified because oil producers will retain sufficient price incentives after a windfall tax. He argued against quotas, pointing out that an import fee and a quota have identical price implications unless price controls are instituted, which then requires a bureaucracy to allocate products, and would result in gasoline lines.[44]

The administration's policies were opposed by a number of groups which purported to represent the public interest. Groups like the National Council of Catholic Charities, the National Retired Teachers Association and the American Association of Retired Persons, testified against policies which would increase energy costs for consumers, supporting quotas, gasoline rationing or rebatable gasoline taxes, automobile mileage taxes and mandatory petroleum allocation. Gary DeLoss of the Public Interest Research Group attacked FEA pricing policies for allowing the old oil price to rise to $5.25 a barrel and permitting the price of new oil to be determined in a so-called "free market" on the grounds that these pricing policies failed to satisfy the Allocation Act standard for equitable pricing. He proposed a oil price rollback plan setting old oil at $4.25 per barrel, eliminating the stripper exemption and pricing new oil at $7 per barrel. It was claimed that rationing energy through allocation would have economic benefits including shifting consumer spending from energy to other goods, reduction of disruptions due to curtailments of natural gas, reduction of the trade deficit, and reduction of health and environmental costs associated with the production and consumption of energy. Lee White, speaking for the CFA, supported controls and endorsed the judgments presented by DeLoss.[45]

While various groups struggled to develop an energy policy the oil depletion allowance came under attack again, first in Ways and Means, and then on the House floor. The provision came up during the writing of the Emergency Tax Cut bill (HR 2166) in the Ways and Means Committee. The committee twice rejected efforts to add an amendment to the bill to eliminate depletion, and when the bill went to the Rules Committee, Ullman and other senior Democrats urged a closed rule. But seven committee members, led by Sam Gibbons (D-FL) and William Green, took the issue to the Democratic Caucus to allow consideration of depletion on the House floor, after the Rules Committee voted 23-12 against a proposal allowing floor action. Green 's request was approved by the Democratic Caucus February 25 by a 153-98 vote, and the House voted 248-163 against depletion on February 27. Charles Wilson (D-TX) then offered an amendment to keep depletion for the first 3,000 bpd of production, which although it was struck down by a vote of 216-197, signaled there was more support for independents than for the major oil companies.[46]

The Senate Finance Committee, chaired by Russell Long, dropped the depletion repeal provision from the bill on March 4. During maneuvers on the Senate floor Lloyd Bentsen suggested a compromise which would retain depletion for the first 3,000 bpd of production by independent producers. The Bentsen amendment along with a provision by Vance Hartke (D-IN) to tighten tax treatment of foreign oil income were added to HR 2166, but the Finance Committee countered by offering a substitute for the House bill. After an attempt to invoke cloture on HR 2166 failed by one vote, Majority Leader Mansfield maneuvered to have the bill recommitted to the Finance Committee with compromise provisions. Ernest Hollings (D-SC) and Kennedy then offered an amendment to eliminate depletion except for the first 1,000 bpd of production by independents, and included Hartke's foreign tax provisions. A number of attempts by oil state senators to eliminate or ease the depletion repeal were rejected, before acceptance by one vote of a proposal from Dewey Bartlett (D-OK) to raise the limit for independents to 2,000 bpd. The final version of HR 2166, hammered out in conference, retained the foreign tax credit though limiting its application. Depletion was eliminated retroactive to the first of the year except for small producers.[47]

The House Ways and Means Committee began consideration of a draft measure of an energy conservation bill March 18, but it would take two months to report a bill. Oil state Democrats joined Republicans in opposing some elements while the committee tried to avoid inclusion of provisions unacceptable to the House Democratic caucus. The gasoline tax, for example, began as a 40¢ over five years proposal in the original Committee plan, but was quickly reduced to 3¢ with an option for an additional 20¢ conditional on the rate of consumption. Import restrictions began with a quota and retention of a $1 import fee, a ban on additional fees, extension of mandatory petroleum allocation through 1980, and a petroleum reserve holding six months of imports. A goal of reducing imports to 25 percent of consumption by 1985 was added, but the fee was changed to a 2 percent duty on crude and 5 percent on products, with a $1 limit. To appease Northeastern Democrats 2 million bpd of imports, out of 5.5 to 6.5 million bpd of imports set as goals for 1977-1979, were reserved for distillate and residual fuel oil for three years. Finally, import goals were retained but the President was given authority to raise quotas to levels that would avoid shortages. The Ways and Means bill, the centerpiece of the House Democrats' energy program, was gutted as soon as it hit the House floor as the gasoline tax was eliminated June 11. The measure was passed in mid-June and sent to the Senate Finance Committee where it died after a few hearings.[48]

The one area where Congress demonstrated decisiveness was in opposition to the Ford Administration's attempts to decontrol oil prices. In March the Senate Interior Committee approved a bill to extend the veto period for Congress to reverse decontrol to a month while requiring the President to set a price ceiling for new oil. The House Commerce Committee reported a measure requiring congressional review of any price increase of more than 50 cents. The Senate Interior Committee reported a bill to extend the EPAA price controls in June (S 1849) which passed both Houses, the Senate by 62-29 July 15 and the House by 303-117 on July 30, and was sent to the President. Ford vetoed S 1849 on September 9 and the Senate sustained his veto by a 39-61 vote, with the margin provided by five Republicans and three oil state Democrats who had voted for the bill, switching after a strong lobbying effort by the White House. The President and Congress agreed to extend controls until November 15 while Congress attempted to write a bill acceptable to the President.[49]

President Ford had first proposed decontrol on April 30 with a plan to phase out controls over twenty-five months. His next plan, announced July 14, called for a thirty month phaseout and a ceiling of $13.50. This was followed by a 39 month plan with a $11.50 price ceiling. The thirty month plan was defeated 262-167 on July 22, and the thirty-nine month plan was shot down by a 228-189 vote on July 30. The shape of price control became part of the discussion for HR 7014, the Energy Policy and Conservation Act (EPCA). Debate in the House centered on two alternatives, one from Robert Krueger (D-TX) which was basically the President's proposal conditional on passage of a windfall profits tax, and one offered by Harley Staggers and Bob Eckhardt (D-TX), which retained the $5.25 price ceiling on old oil, rolled back new oil to $7.50 a barrel (from $12-13), and allowed $10 per barrel for high cost oil. The House voted 218-207 to substitute the Staggers-Eckhardt decontrol system for the Krueger language. The bill passed the House in late September and went to conference, where it had to be resolved with four Senate bills covering the same topics. In conference it was decided to base price controls around a weighted composite price. The administration asked for an $8 price and 15 percent annual price increases, while the conferees settled for $7.66 and a maximum increase of 10 percent per year. The final compromise kept the price of old oil at $5.25 and capped new oil at $11.28. Ford reluctantly signed the act into law on December 22, 1975.[50]

The FEA found no guidance in the EPCA to develop regulations to meet its statutory objectives to both protect inefficient producers and refiners while promoting competition and maximizing economic efficiency. Due to natural decline, the rules on releasing old oil were irrelevant, so the FEA decided to change the base period to 1975 and use a decline rate to encourage additional production from old fields. The FEA then had to split the 10 percent price increase in 1976 between old and new oil, due to controls being based on the composite price. Congress passed the Energy Conservation and Production Act in August, 1976, amending the pricing provisions of EPCA so as to exempt stripper wells from price controls.[51]

The EPCA permitted the President to exempt up to 2 million bpd of oil transported through the Alaska pipeline from the computation of the composite price, but Carter would include it in the program in April 1977 to increase flexibility. Since the wellhead price of Alaskan oil, $6.80 per barrel, once transportation costs were subtracted ($7 per barrel) was less than the composite average, now around $8.50, this allowed an increase in the ceiling price of other types of oil. While the "release" program (releasing one barrel of old oil from controls for every barrel of new production) was eliminated, the FEA was authorized to take in account declining production from old fields. The conservative calculation of decline formulas, and stringent standards on enhanced recovery projects discouraged domestic producers from investing in existing properties. It made more sense to delay investment until the field declined to the point where it qualified for stripper status and became exempt from price controls.[52]

Oil company divestiture was another issue that raised emotions in Congress. Senator Philip Hart was a leader of the forces to break up the oil companies, having pursued this course of action for a decade. The first vote came as a retaliatory move by Senators James Abourezk (D-SD) and Gary Hart (D-CO) during consideration of a natural gas pricing bill. Their amendment, co-sponsered by Hart who left his hospital bed to vote on the bill, would have prohibited integrated operations for companies producing 100,000 bpd after 1980. It was rejected 45-54. Two weeks later they tried again, with a 350,000 bpd limit, losing 40-49. An amendment to force horizontal divestiture, proposed by Edward Kennedy, was also rejected. The API formed a special committee with a budget authorization of $1 million to fight the break-up forces. Divestiture was supported by labor unions, consumer groups, and some oil companies, and politicians aware that attacking "big oil" would sell during an election year.[53]

Two years after the embargo energy was no longer an important issue, having little effect on the 1976 Presidential race. Supplies were plentiful, consumers had become accustomed to higher prices, and the EPCA prevented prices from rising quickly. Although it was obvious that the nation was becoming more dependent on oil imports, there was no political will to do anything to prevent this from occurring. The split between the parties, regions and different interest groups made forming a consensus impossible. The Republican candidates, as would be expected, were in favor of market solutions while the Democrats supported government intervention. Ford campaigned on the position that the nation could reach self-sufficiency by 1985 if his plan for massive spending ($100 billion) for new projects was approved while Reagan said that self-sufficiency could be achieved by removing government constraints on production. Democrat candidates endorsed price controls, more conservation, concern for environmental impacts of energy production, and except for Carter, all came out in favor of divestiture to break up the major integrated oil companies.[54]

THE CARTER ENERGY PLAN

Even before the inauguration the Carter team was soliciting studies and advice on energy policy. The General Accounting Office weighed in with a paper identifying eight critical energy issues for the incoming Congress to consider. The Twentieth Century Fund released a report from a task force which included Walter Levy, Morris Adelman, and John Love. The study recommended encouraging development of domestic energy sources by providing price incentives for oil and gas, supported increased use of coal and nuclear power, and stressed the need for speedy development of an oil stockpile. The National Research Council released an interim report in January urging greater stress on conservation. Carter gave his people three months to resolve the differences between the two approaches, having promised the country a plan by that time in his inauguration address.[55]

The energy group in the Carter White House was centered around James Schlesinger from the start. John F. O'Leary, the new FEA head, had worked with Scheslinger at the Atomic Energy Commission. Five other members of his team had worked with or under him before. The energy group also included Alvin Alm from the Environmental Protection Agency and S. David Freeman, who had just finished directing the Ford Foundation's *A Time To Choose* energy study. There was no industry representation in the group, and while an attempt was made to gather advice from government, industry, and even the general public, policy was basically made by a small group under pressure to complete their work by April 20. The Secretary of Transportation, OMB, CEA, and the assistant secretary of the Treasury for tax policy were all left out of energy policy formulation.[56]

Carter tried to prepare the nation for his proposals with a televised talk on April 18 in which he stressed the seriousness of the energy crisis, pointing out that imports of oil had increased steadily despite the embargo. The dramatic nature of his talk was an indirect acknowledgment of the difficulty of obtaining support for the sacrifices implicit in any comprehensive energy program:

> Many of these proposals will be unpopular. Some will cause you to put up with inconveniences and to make sacrifices. The most important thing about these proposals is that the alternative may be a national catastrophe. Further delay can affect our strength and our power as a nation.
>
> Our decision about energy will test the character of the American people and the ability of the President and the Congress to govern this nation. This difficult effort will be the "moral equivalent of war"—except that we will be uniting our efforts to build and not to destroy.
>
> I know that some of you may doubt that we face real energy shortages. The 1973 gas lines are gone, and with the springtime weather our homes are warm again.
>
> But our energy problem is worse tonight than it was in 1973 or a few weeks ago in the dead of winter. It is worse because more waste has occurred, and more time has passed by without our planning for the future. And it will get worse every day until we act.[57]

The speech was followed two days later by a televised address to a joint session of Congress. The objectives of the energy program included reducing the annual growth rate of energy demand to 3 percent, reducing gasoline consumption by 10 percent, lowering oil imports to 6 million bpd, and establishing a one billion barrel strategic reserve. Carter proposed a gasoline tax which would rise 5¢ per year until conservation goals were meet, as well as ending price controls on gasoline; tax credits and grants to encourage conservation; decontrol of the price of new oil and natural gas combined with a wellhead tax on old oil; conversion of industries and utilities to coal and support for development of solar and geothermal energy.[58]

Carter's proposals were similar in many ways to those of President Ford, with an emphasis on conservation instead of increasing supply. Decontrol of oil prices was attempted by Ford, but under Carter's plan the economic rent from old oil would be captured by the government with a crude equalization tax, instead of a substantial share accruing to the oil companies. Ford had attempted to end controls on gasoline and natural gas, while Carter would retain controls on natural gas, though increasing the price for new gas. The gasoline tax was a major difference between the two Presidents; Carter proposed an escalating tax whereas one of the factors leading to the firing of John Sawhill was his support for a gasoline tax. Carter was also more willing to regulate automobile mileage and apply taxes on gas guzzlers, where Ford relied on voluntary efforts by automakers. While both Presidents pushed for the conversion of power plants to coal, Carter also included taxes on industrial use of oil and gas to encourage conversion. Carter, unlike the Republican Presidents who preceded him, did not see relaxation of environmental standards as a solution to the energy crisis, and was willing to support a much greater range of conservation incentives than the Ford administration.

The response to the Carter proposals was mixed. The big three automobile companies were opposed to the gas guzzler tax and mileage regulations. AFL-CIO President George Meany criticized the plan for eliminating the breeder reactor while its Executive Council was against the crude equalization tax due to objections to using price to encourage conservation. Construction union leader Robert Georgine said the entire program was flawed and that the proposed gasoline tax was regressive. The IPAA complained that no new funds would be made available for exploration and asked for additional tax incentives as well as decontrol, while the API merely stated that the program would not increase supplies. Environmentalists and tax reform groups supported the Carter program, though consumer groups criticized the program's emphasis on higher prices.[59]

In late 1976 Speaker Tip O'Neill (D-Mass) tried to establish a new energy committee to resolve the problems associated with numerous committees claiming jurisdiction over energy matters. Affected House committee chairmen resisted his plan, and a compromise was reached to set up an Ad Hoc Committee on Energy which would receive the mark-up versions of legislation. The Ad Hoc Committee would not have authority to change the recommendations of the standing committees, but it could recommend amendments for consideration on the floor. One day after Carter's speech to Congress the House by voice vote authorized the committee. Carter's energy package was introduced on May 2 as the National Energy Act, with about half of the bill's provisions referred to the Interstate and Foreign Commerce Committee. All the tax provisions went to Ways and Means, while minor parts of the plan were sent to three other committees.[60]

O'Neill chose Thomas Ashley (D-OH) to chair the Ad Hoc Committee, presumably for his lack of past involvement in energy policy controversies. O'Neill was also careful to stack the 40 man committee (27 Democrats, 13 Republicans) with Democrats who would be more receptive to the President's proposals than party members in general, having concluded that passage of the bill would require a consensus among moderate and liberal Democrats, writing off Republicans and conservative Democrats. He took care to placate committee chairmen by allowing them to manage their parts of the package on the floor, as well as seeking their approval before selecting the Ad Hoc Committee's Democrats. O'Neill assigned a deadline of July 13 for legislative committees to complete hearings and markups on the Energy Act. The bill would then go to the Ad Hoc Committee for polishing, and hopefully be passed before the August recess.[61]

All the elements of the bill finally arrived at the Ad Hoc Committee on July 19 when the Commerce Committee filed its report. The Ways and Means Committee killed the conditional 5¢ annual gasoline tax as well as a flat tax of 3¢ per gallon, voting merely to extend the current 4¢ tax. The mileage limit for application of the gas guzzler tax was lowered from 18 miles per gallon (mpg) to 15 mpg in 1979, and for 1985 the limit was lowered from 27.5 to 23.5 mpg. Most of the other provisions of the program were retained, though natural gas deregulation was narrowly defeated in the Commerce Committee by a 22-21 vote on June 29 and then a 23-20 vote on July 14. Given the need to get the bill to the floor by August 1 to meet the Speaker's deadline, the Ad Hoc Committee had only five days to mark up the 113 provisions in the bill.[62]

In a tribute to Speaker O'Neill's management of the bill the Ad Hoc committee got the bill to the House floor on schedule. Dan Rostenkowski (D-IL) added an amendment providing for a 2¢ increase in the gasoline tax in 1978, and an additional 2¢ in 1979, which passed by a vote of 22-18. The Administration compromised with oil and gas state representatives, supporting an amendment authored by two Texas Democrats, Bob Eckhardt and Charles Wilson, preserving regulation of natural gas while raising the price for new gas from $1.45 to $1.75 and expanding the definition of new gas. Part of the deal with Majority Leader Wright was a promise that the committee would refrain from attacking the intangible drilling allowance, though the oilmen expected the issue to be brought up on the floor.[63] The bill went to the Rules Committee where O'Neill and Richard Bolling (D-MO) wrangled a modified closed rule for floor debate.

Gas deregulation became a major issue on the House floor. The compromise amendment was a target for criticism from both pro- and anti-deregulation elements, and a combination of Republicans and deregulation Democrats threatened to overturn the compromise, leading O'Neill to make a rare speech blasting oil lobbyists, and the deregulation amendment was defeated 227-199. Attempts to add a gasoline tax increase to the bill were overwhelmingly rejected, by votes of 339-82 for a 5¢ increase and 370-52 for a 4¢ increase. The House also rejected a proposal for a plowback deduction from the crude oil equalization tax by a vote of 223-198. The final bill (HR 8444) was passed 244-177 after a Republican attempt to kill the crude oil equalization tax failed by a 219-203 vote. Passage of the bill with the equalization tax intact was credited to O'Neill's leadership and restraint on the part of consumer groups who choose not to attack the plan for tactical reasons.[64]

In the Senate the Carter energy program was split into two bills, with tax related sections in a bill (S 1472) assigned to the Finance Committee, and the other sections in a bill (S 1469) assigned to the Energy and Natural Resources Committee. Russell Long, chairman of the Finance Committee, was a strong supporter of the oil industry. Henry Jackson, chairman of the Energy Committee, was a strong proponent of price controls. Long attacked Carter's plan as a disaster on the supply side, providing no incentives for producers. Jackson opposed allowing domestic oil to be sold at world prices, and objected to the proposed gasoline tax, but he did support the coal conversion provisions of the bill. To further complicate matters Senator Kennedy proposed adding a divestiture amendment to the coal conversion bill, which mobilized the oil industry but also raised the threat of Carter support for vertical divestiture if the energy plan was gutted.[65]

September saw the energy bill flounder in the Senate, with the administration wheeling and dealing in an attempt to salvage the remains of the plan. The only provisions to pass the Senate were a weakened version of the coal conversion section and an energy conservation bill, setting minimum fuel efficiency standards for 1980 cars. The Jackson committee, unable to reach agreement on natural gas deregulation, sent the issue to the Senate floor. Russell Long attacked the administration for relying on energy conservation rather than accelerated production. Senate Majority Leader Robert Byrd (D-WV) allied himself with Long, supporting his proposal to use crude oil equalization tax revenues to subsidize energy projects. However, given the lack of support for the tax in the House, siphoning off proceeds to finance energy industry projects would doom the tax in conference.[66]

The Carter administration, which in April was praised for its speed in presenting a comprehensive energy program, was being criticized in September for having moved too hastily. Criticism centered around the haste in which the program was drafted, the lack of consultation with key committee chairmen, and poor strategic decisions in maneuvering the program through Congress. The Schlesinger group had excluded members of Congress and key lobbying groups from participating in the drafting process, forfeiting the opportunity to obtain commitments to the President's program. Consumer lobbyists faulted the administration for not consulting with Long and Bentsen before raising the natural gas price ceiling to $1.75, leaving them free to push for deregulation. Failure to involve the Senate Finance Committee in preparation of the tax provisions left opponents to the gas guzzler tax, the crude oil equalization tax and the industrial energy users' tax free to work against their passage. Another problem that arose were sloppy analyses, which lowered the administration's credibility as they came under attack. The administration's lobbying effort was also denigrated, with some congressmen and staffers contemptuous of the level of expertise displayed by the administration group in working the hill. Carter was criticized for not following up on his April speeches for four months, allowing popular support to dissipate.[67]

A serious problem faced by the Carter administration in pushing their energy plan through Congress was the ignorance and apathy of the American people. Despite the President's speeches in April, a Gallup poll at the end of the month found one-third of respondents believing that the country was self-sufficient in oil. In August 1977 a *New York Times/CBS News* poll found only one-third of respondents believed the nation's energy problems were as bad as the President said. 38 percent thought that energy shortages were real, but 49 percent felt energy shortages were a scare tactic by the oil companies to justify charging higher prices. Only 48 percent of respondents were aware that the country needed to import oil. 59 percent were against allowing natural gas prices to increase to finance additional exploration, and 66 percent felt it was unfair to raise gasoline taxes to reduce driving. Support for the President's handling of the energy problem declined from 58 percent in April to 47 percent in August.[68]

Secretary of Energy James Schlesinger warned Congress in October that failure to adopt the crude equalization tax would result in the imposition of $15 billion in import duties on imported oil, requiring a $5 a barrel fee. Four days after his warning the Senate Finance Committee tentatively agreed to eliminate all of the major taxes proposed by Carter and voted 10-6 for a proposal, offered by Bob Dole (R-KS) to prohibit increases in duties on oil imports except in wartime. The oil state Democrats and Republicans on the committee had a working majority, eliminating any possibility of Chairman Long pushing through the President's tax program even if he was so inclined. Long's strategy was to take a skeleton energy tax bill to the full Senate which was expected to pass it, allowing Long a pivotal role in the House-Senate Conference.[69]

While the criticism of the Carter administration's handling of its program may have had some merit, the complaints voiced in the Senate suggest that no energy plan which significantly impacted consumption or production could have been enacted by Congress. Liberals and consumer advocates disliked the program's emphasis on using price increases to encourage conservation. Oil state Senators and Republicans thought the program lacked sufficient incentives for production. Coal conversion for utilities and large industrial plants was weakened in the Energy Subcommittee on Energy Production and Supply, chaired by Floyd Haskell (D-CO). Haskell objected to the effect on industry, air quality, and the cost of electricity if plants were forced to convert. The oil industry through an intense lobbying campaign obtained deregulation of new natural gas on October 4 by a vote of 50-46. The Senate Finance Committee killed the crude oil equalization tax as the CFA and labor groups attacked provisions added by Senator Long to give part of the funds collected to energy producers. On the Senate floor the industrial oil and gas user tax was gutted, turning a net $40 billion gain for the Treasury into a $3 billion deficit by 1985. While the Senate voted for $40 billion in tax credits, the only revenue item which passed was an extension of the 4¢ gasoline tax through 1985, estimated to bring in $20 billion. The Senate gave Long a mandate going into conference with the House that no taxes would be allowed unless they were combined with tax incentives for increased production, conservation and conversion to alternate fuels.[70]

The threat of oil import fees was raised again in February 1978 when William Miller, the new chairman of the Federal Reserve Board, called for an import fee of $5 or $6 dollars a barrel if the crude equalization tax was not adopted in the next month. Besides the need for leverage with Congress, the fee was also attractive due to pressure from allies to slow oil imports, address the trade deficit, and strengthen the dollar. The Senate responded by removing the President's authority to impose oil import fees or quotas in an amendment to an unrelated appropriations bill in June 1978. The vote followed the rejection of an attempt to table the amendment, led by Herman Talmadge (D-GA) and Henry Jackson. A similar amendment, sponsored by Dole and passed by the Senate in late 1977, had been held up in the conference committee handling the energy tax bill. The House narrowly rejected the provision which was dropped in conference, though it did serve as a signal of the determination of the Senate to avoid energy taxes.[71]

The conferees slowly moved toward agreement over the next year, painfully working out compromises motivated by fatigue. Senate and House negotiators announced an agreement on a gas deregulation plan on April 21, 1978. The coal conversion bill was the first piece of the package that was approved by the Senate in July. The conferees finally reported a natural gas bill in August, which passed the Senate September 27, 57-42, after two attempts to recommit the measure. The Senate accepted the energy conservation report of the conference on October 9 and the energy tax conference report October 15, the last day of the session. The House then approved the entire package. The final energy program included little of the original plan. The crude oil equalization tax, the gasoline tax, and the industrial user tax were all rejected, and the gas guzzler tax was reduced, but conservation tax credits survived almost completely unscathed. For natural gas, instead of continued regulation to be extended to intrastate markets with a $1.75 ceiling, Congress settled for a phased deregulation plan, ending controls by 1985 and raising the ceiling price to $2 per MCF.[72]

DECONTROL AND THE WINDFALL PROFITS TAX

Just when legislators thought they had finished with energy as an issue the next stage of the policy controversy arrived, as the EPCA of 1975 made oil price controls discretionary as of June 1, 1979. The dilemma for the Carter administration was aggravated by the 14.5 percent OPEC price increase announced for 1979, adding 0.7 percent to an already intolerable inflation rate. While Carter wanted to raise the price of domestic oil to the world level in order to stop subsidizing imports (note the gap between domestic and foreign prices in Table 3.1), the equalization tax proposal had been brutally murdered by Congress. Decontrol without a windfall profits tax would possibly generate sufficient opposition by labor and consumer groups to extend mandatory price controls, but a windfall profits tax would have to run the gauntlet of Ways and Means and the Senate Finance Committee. Given the unpopularity of the oil companies and the problems with inflation, the Carter solution was to gradually decontrol prices in combination with a windfall tax. The White House deliberately chose to decontrol prices before asking for a windfall tax, avoiding a situation where liberals and oilmen would combine to defeat a windfall tax linked to decontrol, one group because of the tax, the other to prevent higher prices.[73]

TABLE 3.1 DOMESTIC OIL PRICES 1974-1980 ($ per barrel)

Type	1974	1975	1976	1977	1978	1979	1980
Old Oil	5.03	5.03	5.13	5.19	5.38	---	---
New Oil	10.13	12.03	11.71	11.22	11.98	---	---
Stripper	---	---	12.16	13.59	13.93	---	---
Alaskan	---	---	---	6.32	5.21	10.57	16.87
Domestic*	6.74	7.56	8.14	8.57	8.96	12.51	21.59
Refiner Acquisition Cost							
Domestic	7.18	8.39	8.84	9.55	10.61	14.27	24.23
Foreign	12.52	13.93	13.48	14.53	14.57	21.67	33.89

*Domestic is average wellhead price.
SOURCE: API, *Basic Petroleum Data Book*, 1985, *CQ Weekly Report*, Dec. 23, 1978, p. 3470.

Motivation for decontrol of oil prices were provided by the events in Iran and Harrisburg, Pennsylvania. The cessation of Iranian exports led the International Energy Agency to ask its members to reduce consumption of oil by 5 percent on March 2, with the understanding that the US share would be equal to a reduction of imports of 1 million bpd. The accident at Three Mile Island at the end of March created doubt about the future of nuclear energy, and increased the desire to stimulate additional production and conservation of fossil fuels. An interagency group, headed by domestic policy adviser Stuart Eizenstat and including representatives from OMB, the NSC, Treasury, State, and the Department of Energy (DOE), developed a plan in response to the cutoff of Iranian oil. Along with the usual Carter administration menu of programs and proposals the plan's main theme was to allow the domestic price of oil to rise to world levels.[74]

On April 5, 1979, Carter announced his decision to decontrol oil prices gradually over a 28 month period, starting June 1, and asked for a windfall profits tax of 50 percent. Revenues from the tax would be used for an energy security fund to finance development of new energy sources, from synfuels to gasohol. Henry Jackson, the leading decontrol opponent in the Senate, admitted that there were insufficient votes to stop decontrol. In the House opponents of decontrol were more numerous. An attempt by Toby Moffet (D-CT) in the Commerce Committee on May 2 to add an oil price control amendment to the annual authorization for the Energy Department was stopped by a 21-21 vote. The Democratic caucus then voted 153-82 to reject decontrol on May 23, but the vote did not bind party members to support such legislation on the House floor. Among those opposing decontrol in the caucus were the Speaker, Tip O'Neill, the Majority Whip, John Brademas (D-IN), and the chairman of the Subcommittee on Energy and Power, John Dingell (D-MI). The Democratic leadership kept Moffet's amendment off the House floor until October, by which time the momentum of oil price control advocates had dissipated.[75]

The President was bucking the tide of public opinion in decontrolling oil. A poll taken in April 1979 by the Associated Press and NBC showed 50 percent opposed to decontrol and only 37 percent in favor. Given decontrol, 66 favored the windfall tax and 24 percent opposed it. A New York Times poll in early June found that 62 percent favored gasoline rationing over allowing price increases. This poll also disclosed that while 70 percent disapproved of Carter's handling of the energy problem, only one-third of respondents felt that he deserved "a lot of blame" for it, compared to the 69 percent who blamed the oil companies, Congress (59%), people who waste energy (58%), and Middle East oil producers (47%). The public was still undecided whether there was a problem, as 36 percent thought "we produce as much as we need," with only half of those polled aware that the US had to import oil. Antagonism to the oil companies was fueled by gasoline shortages in the spring of 1979, followed closely by rising prices and record company profits. The shortages hit California in April and spread east in May and June. A study by the GAO found that a decline in crude imports in the first four months of 1979, combined with DOE's allocation and price regulations caused the regional shortages of gasoline.[76]

Labor unions and consumer groups opposed decontrol due to its impact on prices. Environmental organizations endorsed decontrol for the same reason but protested that the windfall tax was too mild and the proposed Energy Security Fund would subsidize the wrong kind of projects. The API and the IPAA, not unexpectedly, endorsed decontrol but objected to the windfall tax. Mobil broke from the other large oil companies by proposing continued controls on old oil in return for freeing future discoveries from controls and special taxation. The motivation for this proposal seemed to be the company's desire to defuse the inflammatory rhetoric surrounding the decontrol issue, and its belief that oil prices would rise sufficiently to allow profits on new oil to outweigh those lost by maintaining controls.[77]

The development of a consensus for decontrol, or at least the acceptance that decontrol of oil prices was inevitable, shifted the focus of conflict to the structure of the windfall tax which would be levied on the difference between controlled and uncontrolled prices. President Carter sent Congress the details of his proposed tax April 26, warning that: "The congressional battle over the windfall profits tax . . . will be a classic confrontation pitting the common and public good against the enormous power of a well-organized special interest." Decontrol was expected to produce $15.4 billion through 1981 for the oil companies, of which the windfall tax would collect $3.3 billion, with $5.1 billion to be paid as federal income tax, $1 billion to the states, and $6 billion would be retained by the oil companies. The money collected by the federal government would go into an Energy Security Fund to help low income families and mass transit.[78]

Decontrol was to operate by gradually shifting oil from lower price tiers to higher price tiers to decontrolled prices. Newly discovered oil, and most marginal and enhanced recovery oil was to be decontrolled as of June 1979. Lower and upper tier oil were to be phased out beginning January 1980. Producers could decrease the quantity of their production classified as lower tier oil by 1.5% a month starting in June 1979 (becoming upper tier oil), and by 3% a month in 1980. Upper tier oil would be decontrolled by gradually increasing the controlled price by 4.6% a month, starting in January 1980, until it reached the world price on October 1, 1981. A 50 percent tax rate would be imposed on the difference between the oil's actual price and its "tier price." For windfall profit tax purposes, oil moved from lower to higher tiers at a slower rate than under decontrol; thus some "old" oil might sell at $13, but have a tax price of $6, and be charged 50% of $7 as the windfall tax. Uncontrolled oil would be in "tier 3," and its windfall tax would be calculated on the difference between the December 1978 OPEC price of $16 and its market price. Tier 2 oil would be taxed on the difference between $13 and its sales price, which would be the market price after 1981. All "tier 1" oil would be "tier 2" oil by 1983, and "tier 3" oil by 1986 for windfall profit tax purposes.[79]

The main issues to be fought over were the tax rates and the inclusion of a plowback provision. Carter, aligned with the liberals in Congress, opposed a plowback provision, but oil industry interests led by Russell Long were adamant in their support for such a clause. Liberals, frustrated over their inability to stop price decontrol, shifted their energy toward increasing the rate of taxation in the windfall tax proposal. James Shannon (D-MA), a protégé of O'Neill, co-sponsored a bill to impose an 85 percent excise tax, while Pete Stark (D-CA) introduced another bill which not only imposed an 85 percent rate but would make Alaskan oil liable to the tax. In response to these proposals Ways and Means Republicans introduced a bill with a 25 percent tax rate and a plowback exemption.[80]

While windfall profits tax bills were working their way through Congress the world oil market was going through its second convulsion in six years. Anticipation of the OPEC price increase, announced at the meeting in Abu

Dhabi in late 1978, resulted in increased stockpiling of crude, a trend which intensified when political disturbances in Iran extended to the oilfields. By the end of December exports from Iran had ceased and the effects of the shutdown began moving through the market. Before the Iranian trouble the quantity of crude at the Majors' disposal was well in excess of the requirements of their refining affiliates, allowing 3.5 million bpd for sale to outsiders. The loss of Iranian output eliminated this surplus, and the resulting tightness in spot crude markets drove prices up for the small amount of oil traded, less than 10 percent of the world's oil trade. Arab Light was sold for $23 a barrel, almost $10 above the official selling price. The price increase for the marker crude (Arab Light) to $14.54 planned by year's end was moved forward to April 1, 1979, at the March meeting of OPEC. The willingness of the Saudis to increase production was the only restraint to further price increases.[81]

Prices in the spot market continued to rise in April and May 1979 as independent refiners scrambled for supplies, from the $20-25 range of March to $35-38 in May. In response most of the OPEC members began leapfrogging prices while the Saudis vainly tried to maintain some stability in the market. At the OPEC meeting in Geneva in June Saudi Arabia accepted the verdict of the spot market and raised the official price for Arab Light to $18 a barrel, while the other members added a market premium of $2, making the actual marker price $20 a barrel. Spot market prices remained firm during the next few months, hovering around $32-35 due to stockpiling caused by uncertainty. Prices began creeping upward before the OPEC meeting in Caracas in December 1979, by which time Arab Light was bringing around $39-40 on the spot market despite a 6-7 percent increase in stocks in the OECD nations. One reason for the rise in price was the shift by producing countries to the spot market for selling their oil; it was estimated by Walter Levy that spot transactions, which had accounted 3-5 percent of sales in January, were 15-20 percent of sales in December 1979.[82]

Saudi Arabia announced in mid-December that it would raise the marker price to $24 a barrel, back-dated to November 1, followed immediately by Libya announcing a $4 increase and the other leading price hawk, Iran, declaring a $5 increase. With the hawks pushing for a $30-35 basic price the Conference broke up without any agreement, due to the determination of Yamani to hold the price line. Saudi Arabia raised prices by $2 in January as a concession to the general drift in OPEC but Kuwait, Abu Dhabi, Qatar and Iraq proceeded to raise their prices by the same amount. During the first part of 1980 spot prices for Arab Light dropped to the $35-37 range and the volume traded on the spot market dropped from 3 million bpd at the end of 1979 to well under 1 million bpd in April. Attempts by Saudi Arabia to close the OPEC pricing gap failed as other OPEC producers added $2 to their prices in April and June. The combination of recession and price-induced conservation lowered demand sufficiently to drive down the spot price of Arab Light to $31 at the end of August, while the Saudi Arabian increase of $2 to $30 per barrel for Arab Light narrowed the price gap within OPEC as other producers struggled to maintain their premiums. However, the outbreak of hostilities between Iraq and Iran caused a virtual cessation of Iranian exports, producing another jump in spot prices with Arab Light reaching $38-42 at the end of October. The December 1980 OPEC Conference resulted in increases of $2-4 a barrel to a weighted average of $35, despite the refusal of Yamani to accept an increase in the price of Arab Light.[83]

This pattern of increases had an substantial impact on the politics of the windfall tax, as it became evident that the energy problem would not fade away, while the estimates of the revenues generated by price decontrol and recovered through the windfall profits tax grew at an exponential rate. In May 1979 the CBO estimated that decontrol would result in $63 billion in additional revenues by 1985, with $21.5 recovered through the windfall profits tax and $14.7 paid in other taxes. The House Commerce Subcommittee on Energy and Power estimated $50.7 billion in revenues by 1985, with $12.8 billion to the windfall tax and $15.1 billion in other taxes. Carter administration estimates gave $59.7-83.8 billion in revenues by 1985, with $28.9-34 billion recovered through taxes (the higher estimate assumed 3 percent real annual increases in the price of oil). In July the administration's no price increase estimate for 1990 was $142.8 billion in tax revenues if HR 3919 was enacted. In midsummer the House estimated $104 billion for windfall tax revenues (based on $20 oil and 1 percent annual price increases), while by October the Senate estimates of revenues (based on $30 oil and 2 percent annual price increases) through 1990 had increased to $1.1 trillion, with $207 billion going for state taxes and royalty payments, $388 billion to federal income taxes, and $138 to $273 billion in windfall profit taxes.[84]

The strategy used by the administration, on the advice of Al Ullman, chairman of Ways and Means, was to separate the decision on the tax rates from the decision on the ultimate destination of the revenues. The tax bill was the arena in which the Democratic liberals took their stand, attempting to eliminate the overseas tax credit for the oil companies. They also pushed for a higher rate, trying for 85 percent which was rejected, but defeating a proposal in Ways and Means for 60 percent. The committee settled on an amendment offered by Dan Rostenkowski for a 70 percent rate by a vote of 22-14. The committee also slowed the rate at which old oil moved up tiers for tax purposes, increasing the effective tax, and applied stricter standards to defining newly discovered oil. The House rejected the Ways and Means bill, an attempt to restore the White House proposal, and a plowback provision before voting 236-183 for an alternative measure (HR 3919) offered by James Jones (D-OK) and W. Henson Moore (R-LA) lowering the rate to 60 percent. The Jones-Moore measure reduced the tax for marginal wells, phased out the tax by 1990, and treated tertiary oil as new oil. Before the House voted, Senator Long in a letter to Moore warned: "I wish to suggest very strongly that no House member should vote for a tax which is any higher than he wants to see become law. When the Senate votes on this measure, the mood at that time might cause the Senate to vote for an even higher tax than the one passed by the House."[85]

As oil prices rose and gas lines continued, Carter called a 10-day domestic summit at Camp David, following that with a televised speech on July 15 in which he described the nation as suffering a "crisis of confidence." Carter announced that he was setting import quotas to limit imports to 8.2 million bpd, following notification by the Treasury Secretary Michael Blumenthal that oil imports were a security threat. Carter had pledged to limit oil imports to 8.5 million bpd by 1985 at the June 29, 1979 economic summit meeting in Tokyo. The President proposed using import quotas to meet this limit. He was challenged by Charles Vanik who cited a Library of Congress study that showed demand exceeding the quota as early as 1982, resulting in a shortage of 400,000 to 1 million bpd and import rights selling for $3 to $8 per barrel if auctioned by the government. The Senate showed its disapproval by voting 70-23 on October 30 for an amendment which would allow Congress to block quotas.[86]

The President asked Congress to authorize a special Energy Security Corporation to finance the production of substitutes for imported oil, with a goal of 2.5 million bpd by 1990. These substitutes included fuels from coal, shale oil, biomass and unconventional natural gas. The program would be financed through $83 billion to be raised by the windfall profits tax and $5 billion from the sale of special energy bonds with the money going to price guarantees, direct loans and loan guarantees.[87]

In the Senate both the windfall tax and the Energy Security Corporation proposals came under attack. The synfuels crash program was criticized as being overly ambitious by both Jackson and Long. The Senate Energy Committee decided to limit the first stage of the program to $20 billion, with a Congressional review before committing more funds. The Banking Committee also was critical of the program, sharing jurisdiction over the bill with Energy. In the Finance Committee questions were raised concerning the cost-effectiveness of various energy projects, with aides admitting that both Finance and Energy lacked the expertise to make such judgments. Due to the uncertainty over the size and shape of the Energy Security program, Long decided to separate the issues of the windfall tax and use of the resultant revenues.[88]

The arguments over the windfall tax became a battle of numbers, with costs and benefits of various exemptions and tax rates tossed around. The Carter administration, in defense of its windfall tax, gave estimates of the costs of exceptions, based on the estimate of $142.8 billion in revenues from HR 3919 by 1990. A stripper well exemption would cost $32.7 billion, $41.4 billion to exempt independent producers, and $18.6 billion to exempt newly discovered oil. Bob Dole proposed exempting oil discovered after 1978 from the tax, a move which the Department of Energy estimated would increase production 150,000 bpd by 1985 and 225,000 bpd by 1990, but which Lloyd Bentsen claimed would result in double those numbers. The Finance Committee approved the Dole proposal 15-1, with Senator Gaylord Nelson (D-WI) the only dissenter. The Senators felt that encouraging increased domestic production of oil made more sense than accumulating revenue for the Government to invest in synfuels. The Committee also exempted stripper oil, tertiary oil, and oil owned by governments and non-profit institutions. A 75 percent tax rate was set on old oil, and a 60 percent rate on new oil. The total expected revenue from this bill was estimated by the Committee at $138 billion, compared to $273 billion expected from the House bill by 1990.[89]

On the Senate floor numerous amendments to strengthen and weaken the Finance Committee bill were proposed. Dale Bumpers attempted to substitute the House bill but his motioned was tabled November 26, 50-32, though the margin was due as much to a refusal to vote for a House measure as opposition to tougher tax provisions. Lloyd Bentsen slipped an amendment to exempt the first 1,000 bpd of independent producers (worth $10 billion) through the Senate, while a week later a proposal to eliminate part of the depletion allowance for independents was defeated. Howard Metzenbaum failed in his attempt to submit an amendment making the windfall tax permanent, while Daniel Moynihan was more successful with a proposal to delay the phase-out of the tax until 90 percent of $210 billion (instead of $128 billion) was collected. An amendment to raise the tax on oil discovered between 1973 and 1978 from 60 to 75 percent (adding $22.5 billion) survived the threat of a filibuster by oil state Senators. The filibuster was used to block an attempt to extend the tax to new oil, and after three attempts to invoke cloture failed a compromise was reached, lowering the tax rate on new oil from 20 to 10 percent and increasing the base price. The total revenue raised by the Senate bill was expected to reach $178 billion by 1990.[90]

The House and Senate conferees on the windfall tax agreed to draft a bill to raise $227 billion by 1990, splitting the difference between the two bills. The battle shifted to the question of which Senate credits would be eliminated to raise the additional revenue. The major companies and the 200 large independents (producing >1000 bpd) supported the House view that taxes should be kept low on new, tertiary and heavy oil, opposed to the 12,000 smaller independents, supporting Senate proposals to provide them with an exemption. Residential conservation credits were eliminated, and alternative energy credits greatly reduced, primarily because they lacked economic justification. The final agreement, after considering 30 proposals, lowered rates on the first 1,000 bpd on both old oil and stripper oil for independents. Oil owned by state and local governments, Indians, and non-profit institutions was exempt form the tax. The tax would be phased out over a 33-month period that would begin in January 1988 conditional on raising $227.3 in revenues. Regardless of revenues collected the tax would phase out beginning in 1991. Residential and business tax credits were included in the bill, along with low-income energy assistance. The bill easily passed the House on March 13 and the Senate on March 27, 1980.[91]

A provision of the windfall profits tax bill which drew little attention at the time was an amendment to the Trade Expansion Act of 1962 to limit the president's authority to impose oil import quotas. It allowed Congress to block quotas by passing a resolution in each House, which if vetoed, must then be passed by a two-thirds vote. This provision had been added by the Senate, and survived the conference. With decontrol accomplished Carter was no longer interested in using quotas or import fees to coerce Congress, and he resisted pressures from his economic advisers to push for a 50¢ gasoline tax, gasoline price decontrol or an import fee to encourage additional conservation in December. Both Vice President Mondale and domestic policy adviser Stuart Eizenstat were strongly opposed to any measures that would hurt consumers.[92]

Oil fees were mentioned again during discussions over balancing the 1981 budget, and this time the pain of budget cutting balanced the reluctance to anger consumers. A $13 billion budget cut and a $4.62 oil import fee were announced March 14 as part of Carter's anti-inflation package, though the presidential proclamation establishing the fee wasn't signed until April 2, and the fee wouldn't go into effect until May 15. The fee was designed to raise the price of gasoline by 10¢, using the system of refiner entitlements to attempt to prevent increases in other product prices. The fee would be eliminated if Congress passed a 14¢ Federal gasoline tax.[93]

The oil fee was disliked by the oil companies, but more important, it was unpopular among the voters. A *New York Times/CBS* poll found that the oil import fee was opposed by more than two-thirds of respondents, while over one-third felt that the rise in gasoline prices was the most painful aspect of inflation. Opposition in Congress quickly arose, both on grounds that the tax was inflationary and because it was perceived as a roundabout way of raising revenues in order to circumvent the need for tax legislation. The issue quickly emerged during the primary campaign, as supporters of Edward Kennedy used it to attack the President. An attempt to tie the oil fee to a tax cut failed to assuage opponents, as only the Democratic leadership stood with the President. On May 14, both the House and Senate voted to overturn the President's order, and in early June the House voted 367-30 against the fee, while the Senate vote was 73-16. Part of the reason for the overwhelming defeat was Carter's objection to a compromise budget reached in Congress, which antagonized Congressmen who were resentful of the President's intervention in the Congressional budget making process. A veto was then overridden by votes of 334-34 in the House and 68-10 in the Senate.[94]

Expectation of future oil price shocks were commonplace in 1980, as a report of the House Intelligence Oversight Committee predicted a world oil shortage within the next six years. Wharton Econometrics forecasted a 21% increase in oil prices in 1981. In October the Office of Technology Assessment said world oil production would show little or no increase over the next 20 years. Robert Pindyck argued that the Iran-Iraq could result in an additional 20-35% increase in oil prices.[95] A general consensus that the prices would continue to increase was held by the vast majority of economists.

THE REAGAN REVOLUTION

The policy of the Reagan administration toward energy became evident in the composition of his Energy Department transition team. Michael T. Halbouty, a Houston oilman, reflected Reagan's attitudes in his belief that there was plenty of oil to be found and that conservation was synonymous with uncomfortable. Two other oil company executives were also part of the transition team, but some balance was provided by Richard Fairbanks, a lawyer who had served at both the EPA and the Domestic Council during the Nixon administration.[96] The Reagan choice for DOE Secretary demonstrated the low priority of energy in his administration; James Edwards, former governor of South Carolina, had no relevant experience and took the job on the assumption that it would be for a limited tenure since elimination of the department was one of the Reagan campaign promises.

The administration plan was to split DOE between the Commerce and Interior Departments, the result of a battle between Interior Secretary James Watt and Commerce Secretary Malcolm Baldrige. About 70 percent of DOE would go to Commerce in the form of a new Energy Research and Technology Administration which would handle weapons, nuclear programs and high-tech research. The rest would revert to Interior, reversing the administration reorganizations of the previous decade. The realization that the dismantling of DOE would not result in significant savings doomed any chance of obtaining congressional approval. Two influential Senators who had voted against forming DOE were now opposed to dismantlement. James McClure (R-ID), chairman of the Senate Energy Committee, was concerned that the influence of his committee would be diminished while John Tower (R-TX) didn't want nuclear weapons programs run by lower level officials. Jack Brooks (D-TX), chairman of the Government Operations Committee in the House, stated quite bluntly that the proposed reorganization made no sense.[97]

While the Reagan administration was unable to dismantle DOE it did manage to institute its free market philosophy with respect to energy policy. Like many Reagan policies, this was more an extension of proposals instituted by Carter than a dramatic shift in direction. Oil decontrol was already in place when Reagan took office, while natural gas prices had been allowed to climb closer to free market levels. Congress had also shifted in its attitude toward control, as former pro-control members such as John Dingell accepted the need to allow price to dictate behavior. The election of Northeast Democrats such as Bill Bradley (D-NJ) and Paul Tsongas (D-MA), both serving on the Senate Energy Committee and favoring higher prices encourage conservation, illustrate the shift in attitudes toward market solutions. The inability of government to efficiently regulate energy markets had become apparent during the 1970s, as the FPC, FEA, FERC, and the Economic Regulatory Administration found themselves buried by the thousands of cases for which they were required to make decisions. The GAO found the DOE allocation program to be chaotic. Studies by the Energy and Justice Departments exonerating the oil companies for the gasoline shortages of 1979, and blaming the Government's own pricing and allocation rules increased skepticism toward regulation.[98]

Reagan ended price controls on oil January 28 by executive order, eight months before they were scheduled to expire, freeing the 25 percent of domestic production still subject to price controls. Decontrol also eliminated, as of March 31, the requirement that large oil companies sell crude oil to small refiners. Edwards made it clear that small refiners would receive no help from the Reagan administration. Small refiners turned to Congress but were unable to achieve their usual success in maintaining subsidies. Senator Howard Metzenbaum put decontrol to a vote for the record, losing 68-24. In the National Energy Plan (NEP-III) of 1981, the President took credit for a movement towards market solutions and blamed the Carter administration for intervening in the market, but implicitly recognized political realities by omitting a call for decontrol of natural gas prices. NEP-III stated that nuclear power was safe, ignored nuclear industry subsidies and continued a commitment to the breeder reactor. Government support was to be withdrawn from synfuels, solar and conservation since high energy prices were sufficient to motivate action in these areas.[99]

Reagan was no more effective than Carter had been in obtaining natural gas price decontrol. The Natural Gas Policy Act of 1978 had initiated gradual decontrol of new gas (discovered after April 1977), to be completed by January 1985. By fall 1982 the complex regulations governing natural gas prices which controlled old gas but allowed high prices for certain new gas and imports, combined with the widespread prevalence of "take-or-pay" provisions in contracts signed during the shortage years, had resulted in prices in certain areas equivalent to that of heating oil. This meant that deregulation would have little effect on consumers, but would cause a significant shift of earnings between owners of old and new gas. By March 1983 rising natural gas prices had increased political pressure for stricter controls even among Republicans. Reagan submitted a bill on February 28 which phased in decontrol of old gas over three years, but would also end automatic pass-through of costs by pipeline companies to consumers, and permit pipelines to break contracts for expensive gas. After eighteen months of vicious battles in both houses of Congress the bill was killed due to the unwillingness of Democrats to vote on a bill which would cost them support among either producer or consumer states during a Presidential election.[100]

Reagan was more successful in reducing the windfall tax on the oil industry, a campaign commitment, though his desire to limit taxation had to be balanced with the need to reduce the massive budget deficits resulting from the attempt to implement supply side economics. Originally the only break for oil in the 1981 tax bill was an increase in the tax credit allowed small royalty owners from $1,000 to $2,500. Dole, the new Finance Committee Chairman, usually a supporter of the oil industry, found himself in the awkward position of resisting pressure from Bentsen and David Boren to add tax breaks for oil producers. Boren finally pushed through an amendment to reduce the windfall tax on newly discovered oil. Rostenkowski, the new Chairman of Ways and Means, realized that the Southern Democrats were the key to control of the House and accepted exemption of stripper oil from the windfall tax and reduction of windfall tax rates on new oil. A depletion provision favorable to independents was added in the House, but many of the oil tax breaks passed in the House were eliminated in conference. The Economic Recovery Tax Act of 1981 increased the tax credit allowed small royalty owners from $1,000 to $2,500, provided an exemption from windfall taxes on oil production of two barrels per day for 1982-1984 and three barrels for 1985 onward. The bill also exempted oil from stripper wells and reduced the windfall tax on newly discovered oil by 2.5 percent a year until 1985, with the tax dropping to 15 percent in 1986.[101]

The Treasury plan in 1984 proposed phasing out the oil depletion allowance, which applied only to independents, over a five year period except for oil from stripper wells. In return it would phase out windfall profits tax by 1991. In the Ways and Means Committee Rostenkowski proposed repealing the depletion allowance but the committee retained the stripper provision due to pressure by Treasury Secretary James Baker and Vice President George Bush, though shortening the phase out period to three years. Baker said that oil and gas incentives needed to be preserved because of national security concerns (though he was quite willing to eliminate tax credits for solar energy).[102] The battle in the House became irrelevant when the depletion allowance was retained in conference with the Senate.

With the crash of oil prices in 1986 the windfall profits tax became meaningless, at least temporarily, since prices fell below the base level above which taxes were implemented. Despite this Howard Metzenbaum, Bill Bradley and Majority Leader Robert Byrd prevented the Senate from coming to a vote on repeal during most of 1987. Their reasoning was repeal would not help the industry, that independents had already been given sufficient protection from the tax, and when prices rose again the large companies would be the beneficiaries of the repeal. Phil Gramm (R-TX) argued that the tax cost over $100 million for record keeping even if no revenues were being collected, and that repeal would give producers a psychological lift. Finally on July 16 the Senate voted to repeal the windfall tax, 58-40, as an amendment to the comprehensive trade bill, and later added it to the deficit-reduction bill. However, Dan Rostenkowski barred consideration of the windfall tax in conference, due to a ban on tax-losing provisions in the reconciliation bill reached in the budget summit agreement.[103] The tax was finally eliminated when a repeal provision was attached to the omnibus trade bill in the Senate and accepted by the House in 1988.

An oil import fee was proposed in the first year of the Reagan administration, primarily as a source of revenue. Influential Republicans, including Herbert Stein, chairman of the CEA under Nixon, as well as the current chairman, Murray Weidenbaum, Treasury Secretary Regan and David Stockman were quietly promoting the fee, while Energy Secretary Edwards was opposed. Both the White House and Congress gave the fee some consideration during 1981 and 1982, but opposition by oil companies, consumer groups and the Saudis prevented implementation. Dole was the biggest supporter of a fee in Congress, but the Ways and Means committee, influenced by the objections of Northeastern congressmen, had little enthusiasm for the proposal. Chairman of the House Energy Committee, John Dingell, denounced the fee, proposing a tax on refined products as an alternative. In addition, the administration was committed to lowering inflation in order to bring interest rates down and an oil import fee or gasoline tax would have the opposite effect. Analyses by the Energy Information Agency also cast doubt upon the effectiveness of the fee in reducing budget deficits. They stated that a $5 import fee would raise $8 billion in net revenues the first year, $7 billion the second and $1 billion the third, declining due to its negative impact on the economy. The CBO projected an annual increase in net revenues of $8-9 billion.[104]

One reason for the increased attractiveness of oil import fees was the break in world oil prices. In the spring of 1981 excess supplies began building on the oil market, fueled by the Saudi Arabia decision to maintain high production levels. Spot prices fell to slightly below the official price of $32 a barrel by July. In August, despite a Saudi cut in production of 1 million bpd, the glut continued to grow. A Saudi increase of $2 in the price of marker crude in October unified the price structure only in the sense that all OPEC producers were pricing themselves out of the market. By the API meeting in November oil executives were predicting that oil prices would stagnate for a year or two. Spot prices remained firm through most of 1982, sustained by the willingness of Saudi Arabia to cut production to 5-5.5 million bpd, but prices began to slip by the end of the year, with Arab Light bringing only $30.25-$31 in November. The shakeout in the world oil market came in the spring of 1983 after OPEC failed to agree on production allocations in January. Spot market prices slipped to $28 a barrel as the UK and Nigeria announced price cuts. At the March meeting in London OPEC finally succeeded in coming to some sort of price and production allocation agreement. The marker crude price was reduced from $34 to $29 a barrel, individual production quotas were assigned to members, and Saudi Arabia accepted the role of the swing producer.[105]

A gasoline tax of 5¢ a gallon, raising the federal tax to 9¢, was approved by the House Ways and Means Committee to finance highway and mass transit systems in December. The House quickly passed the bill, but it ran into opposition by both liberals and conservatives in the Senate where it was temporarily blocked by a filibuster. The filibuster was finally broken on December 18 and the tax passed the Senate, 54-34, to days later. A filibuster by Jesse Helms over the conference bill was overcome with the aid of President Reagan, and the tax passed on December 23.[106]

In Reagan's 1983 State of the Union message he called for a conditional $5 per barrel excise tax on oil to be imposed in 1986-1988 to reduce the federal budget deficit if the deficit was more than 2.5 percent of GNP. The API opposed the conditional tax on the grounds it would increase uncertainty in the industry and raise consumer prices. The Democratic leaders in both Houses were also dubious concerning the value of the standby excise tax, preferring to repeal indexing of federal income tax rates.[107]

In March of 1983 the oil import fee issue rose again as Senator Pete Domenici (R-NM) proposed a levy a $10 a barrel tax on imported petroleum. Paul Volcker, chairman of the Federal Reserve, threw his support behind an oil excise tax to take effect if oil prices declined sharply, suggesting a price around $25 per barrel as the level which should trigger such a tax. The IPAA came out against a tax on imported oil at its midyear meeting, citing harm to the interests of organizations and industries that have supported IPAA on tax, natural gas, and related issues. The organization was also convinced that a tax on imported oil would likely result in a similar tax on domestic production, while the windfall tax limited the benefit from higher prices. The issue was mute since President Reagan, facing re-election the next year, had no interest in taxes which would raise prices or endanger the economic recovery.[108]

Suggestions of the possibility of a price collapse in the world oil market began to be voiced during 1982, and continued in 1983, though intermixed with claims that the glut was temporary and prices would rise once more. Despite the weakness in world oil markets, OPEC managed to maintain prices near the $28 per barrel official price level into 1985, though in real terms the price continued to decline.[109]

STRATEGIC PETROLEUM RESERVE

The strategic petroleum reserve (SPR) was authorized in the Energy Policy and Conservation Act of 1975, four years after it was proposed by Sorenson and Mead. The original target was 1 billion barrels, but it was lowered by Carter to 750 million barrels for budgetary reasons in 1977. Construction and filling of the reserve lagged far behind plans, with only about 72 million barrels in storage at the end of 1978, compared to the 250 million barrel interim target set under the Carter plan (see Table 3.2). Estimated storage costs for the first 500 million bbl of reserves increased from $765 million to $1.5 billion, due to initial underestimation of costs and poor management. The fact that systems for removing oil from the reserve were also not in place at this time was an additional embarrassment. By 1979 most of the construction problems had been solved but during the Iranian emergency of 1979 finding oil became a problem as suppliers broke contracts due to market conditions. Given the tight oil market the Carter administration decided to stop filling the reserve in April

1979 since the effect was to increase demand for imported oil.[110]

At the end of 1979, as supplies became available on the world market once again, the administration refused to resume filling of the reserve. The decision was made by an informal committee influenced by the opposition of the Europeans and the Saudis to filling the reserve. DOE Secretary Duncan made a trip to Saudi Arabia in February 1980 where he was advised that filling the reserves would encourage the radicals in OPEC as well as putting pressure on the Saudi government to cut production by a million bpd. When Duncan returned to Washington he decided that filling the reserve should be delayed until mid-1981. During 1979, Bob Dole and Bill Bradley, concerned about the rate of fill of the reserve, proposed an amendment to S 932, the Energy Security Act, to require the Government to resume filling the reserve at a rate of 100,000 bpd. Carter early in 1980 cut $600 million in outlays for oil purchases in his revised budget requests. John Dingell, during the conference on the synfuels bill, suggested that the government be prevented from producing oil from naval petroleum reserves, losing $2 billion in revenues, unless it continued filling the reserve up to the 500 million barrel level, language included in the final bill. Bradley, after a trip to Saudi Arabia in April 1980, said that bowing to Saudi pressure on the reserve actually made the Saudis feel more insecure, since it is difficult to have confidence in a protector who is so easily bullied.[111]

TABLE 3.2 FILLING UP THE RESERVE
(millions of barrels at the year's end)

Year	1978	1979	1980	1982	1984	1985	1986
Ford Plan	----	----	----	500	----	1,000	----
Carter Plan	----	250	500	----	750	750	750
Actual	67	91	108	293	451	493	515

SOURCE: *National Journal*, Feb. 24, 1979, p. 304, *Energy Statistics Sourcebook* (Penwell Publishing Co., 1986), p. 174.

Filling of the reserve resumed in September 1980 at a rate around 130,000 bpd. In December the fiscal 1981 Interior appropriations bill included funding for a 300,000 bpd fill rate, though the Reagan administration expected the average rate for 1981 to be 200,000 bpd. The 1982 budget request for the reserve allowed for a 230,000 bpd fill rate, which was still too slow for some members of Congress. The $3.8 billion budgeted for SPR in fiscal 1982 was a tempting target for budget cutters. David Stockman, along with Representative Phil Gramm proposed selling oil bonds to finance the SPR, a plan that was rejected by the administration due to opposition from Treasury, but accepted by the House. The Senate Budget Committee adopted a proposal of Nancy Kassebaum (R-KS) by a vote of 13-4, to cut energy funding by $3 billion by funding purchases for only the first quarter. She introduced a bill (S 707) to require importers to contribute to the SPR by selling oil to the government at 10 percent of the purchase price, with ownership reverting back to the companies after 11 years, a variation on a theme first proposed by David Weimer of the DOE to impose a 3 percent tariff on imports to fill the reserve. While Energy Chairman James McClure was agreeable to off-budget methods of financing the SPR, both Republicans and Democrats on the Committee objected to removing funding from the budget. The Senate ordered 1982 SPR funding out of the budget four times (the intent being to disguise the expenditure), and the House finally agreed to declare oil purchases an investment, making it an off-budget item.[112]

After the budget follies of 1981 focus on SPR shifted back to the fill rate. The limit on storage space in 1982 caused the Administration to slow the rate of SPR fill while delaying the construction of additional storage space. In the House, members from steel-producing districts pushed for construction of steel tanks to hold SPR oil, despite the fact that tank storage is a far more expensive method than the salt domes used in the program. The Senate wanted to lease storage space from private industry, either tanks or tankers, until space was available in salt domes. While Congress authorized $3.7 billion in 1983 for oil purchases, the Reagan budget called for only $2.1 billion, due to lack of storage space. The administration also deferred $400 million in spending on a new SPR cavern. Thus the fill rate, which had reached 292,000 bpd in 1981, dropped to 185,000 bpd in 1982 due to capacity restrictions, with a planned rate of 208,000 bpd in 1983. Congress ordered a fill rate of 300,000 bpd, but allowed a minimum rate of 220,000 bpd. The administration also developed a plan to auction SPR oil during an emergency at the President's discretion.[113]

In December 1984 Energy Secretary Donald Hodel stated that the government was considering an early halt in filling the SPR, stopping at 500 million barrels. His reasoning was the government was mandated to keep a reserve equal to 90 days of imports. With imports declining from the peak of 8,787,000 bpd in 1977 to 5,240,000 bpd in 1984, there was no point in spending the additional money filling the reserve. The president of the IPAA endorsed Hodel's proposal in order to save money and to jolt Americans out of their false sense of security. President Reagan's proposed fiscal 1986 budget asked for no new appropriations and capped the stockpile at the end of fiscal 1985. Representative Philip Sharp (D-IN), chairman of the Energy and Commerce Fossil Fuels Subcommittee, also favored slashing appropriations for oil purchases as well as halting construction of new storage facilities. Senator James McClure, chairman of the Energy Committee, while supporting additional reserves, was receptive to the idea of delaying filling the SPR for budget reasons. Congressional Democrats were skeptical that filling the reserve would resume once stopped, and felt a smaller SPR wouldn't provide sufficient security. The 1985 appropriation had financed a fill rate of 159,000 bpd, and before the administration freeze the 1986 planned fill rate was 145,000 bpd.[114]

The deficit reduction bill of 1986 authorized appropriations to fill the SPR at a minimum rate of 35,000 bpd until there was at least 527 million barrels in the reserve, and authorized funds for construction of the Big Hill Texas reservoir, increasing storage capacity to 750 million barrels. This followed an attempt by the Senate Republican leadership to incorporate a moratorium on filling the SPR in the bill. With imports rising to over 7 million bpd in August of 1986 President Reagan overruled his budget advisors and ordered continued purchases of crude oil for the SPR. This was seen as a victory for Energy Secretary John Herrington, who had argued at a Domestic Policy Council meeting two days earlier that prices were too low to pass up filling the reserve. James Miller, the Budget Director, had urged halting purchases as current contracts expired. The President's decision reflected a new resolve in Congress to fill the reserve, as the 1987 Fiscal Reconciliation Bill called for increasing the fill rate from 35,000 bpd to 75,000 bpd until the SPR reaches 750 million barrels.[115]

During 1987 concern with dependence on oil imports increased, as a DOE report in March stressed that dependence on Middle Eastern oil would pose a security threat for the rest of the century. Administration thinking shifted toward supporting a faster rate of filling the SPR, up to 100,000 bpd. In May Reagan told Congress that he supported raising the fill rate to 100,000 bpd providing savings could be made elsewhere in the budget to pay for this increase above the current rate of 75,000 bpd.[116]

THE OIL PRICE COLLAPSE

World oil prices had managed to stay fairly steady despite an increasing glut on the market through 1985, but the same forces that had led to the abrupt increases in prices in 1979-1980 were also responsible for the sudden decline in the beginning of 1986. The structural changes which occurred in the world oil market during the 1970s, in the shift of control of production from the Majors to the producing countries, made the Rotterdam spot market a key source of information and even a determinant of prices. The "Rotterdam market" is not a market in the conventional sense, a place where people meet to match bids and offer. Instead there is a trading activity which takes place in central Europe covering two distinct types of trade, international cargo trading, and the barge trade. Prices in the market are limited on the low side by the willingness of refiners to run crude at a loss; on the high side by the consumers' desire for products. Prices on the market are available only on a second-hand basis from sources such as Petroleum Argus, Petroleum Intelligence Weekly and, most importantly, Platt's Oilgram Price Service. Platt's price reports are obtained from a variety of sources and depend on the company's subjective judgment, not a statistical record of transactions. Platt's takes great care to check its sources, even making it a policy to move prices by small amounts in order to test the changes.[117]

The spot market increased in importance because of the change in the control of crude that had been occurring in the world oil market. Producer government ownership of OPEC oil increased from 2 percent in 1970 to 80 percent in 1980 while direct exports by OPEC national companies went from zero in 1970 to 36 percent in 1978, and exceeded 50 percent in 1980, with much of this oil diverted to intergovernmental supply contracts. As spot market prices increased governments cut supplies to customers or lowered volumes to free up crude for spot market related sales. This added to the pressures on the spot market as companies which lost supplies scrambled to replace them while other firms, though temporarily possessing secure supplies, built up inventories to protect themselves from future disruptions in supply. The movement toward spot trading, begun by OPEC producers in a tight market, continued in the weak post-crisis market as buyers preferred lower short-term prices over insecure long-term contracts. The quantity of internationally traded spot crude edged up from 1-2 percent in early 1979 to above 6 percent in early 1982.[118]

The growing importance of the spot market led to an increasingly volatile market in which the most important short-term factor was the movement of oil inventories. During periods of uncertainty or before an expected price increase stockpiling occurred, increasing demand for crude and driving spot prices upward; during periods of oversupply and expectations of falling prices stocks were liquidated putting increased downward pressure on prices. In contrast to the immediate impact of stock movements on the world petroleum market the only other short-term response is to lower consumption by recession or increase consumption through unanticipated growth. Since demand for petroleum is relatively inelastic in the short-run, inventory swings can cause large changes in prices. In the long-run changes in the technical composition of capital result in permanent changes in consumption patterns. OPEC attempted to control output, but lack of control over fringe producers made its market control tenuous, even when members accepted the required discipline.

The crash of 1986 was a surprise only in its abruptness and timing. The decline in oil prices was held off by a combination of OPEC production cuts and expectations within the industry that consumption would resume its historical growth. Unfortunately for oil producers, as Table 3.3 so clearly illustrates, declines in imports of the OECD countries, combined with increased third party production caused OPEC production to drop by 5.4 billion barrels between 1979 and 1985. Saudi Arabia, the swing producer, saw its production drop from 9.9 million bpd in 1981 to less than 3.4 million bpd in 1985. By July 1985 Saudi Arabia had lowered output to about 2.2 million bpd against its quota of 4.35 million bpd. With spot prices below official prices OPEC members were making long-term deals with various forms of discounts.[119]

TABLE 3.3 PRODUCTION AND PRICES
(Thousand Barrels/Day)

Year	*OPEC Output*	*Non-OPEC Output*	*US & WE Net Imports*	*Import Cost* Nominal*	*($1990)*
1972	26,711	3,433	22,865	3.22	9.37
1973	31,002	3,470	26,285	4.08	11.15
1974	30,729	3,973	25,555	12.52	31.48
1975	27,186	4,222	23,170	13.93	31.97
1976	30,743	4,455	25,930	13.48	29.10
1977	31,269	4,817	26,950	14.53	29.35
1978	29,880	5,384	25,939	14.57	27.28
1979	31,000	5,822	26,235	21.67	37.35
1980	26,967	6,504	22,715	33.89	53.36
1981	22,694	6,947	19,450	37.05	53.02
1982	19,287	7,549	16,675	33.55	45.20
1983	17,759	8,126	15,965	29.30	37.94
1984	17,529	8,627	16,370	28.88	35.83
1985	16,365	9,127	15,480	26.99	32.28
1986	18,694	9,299	16,925	14.00	16.31
1987	18,299	9,660	19,062	18.13	20.47
1988	20,426	10,042	20,221	14.56	15.82
1989	22,648	10,355	21.264	18.08	18.83
1990	23,868	10,867	21,044	21.76	21.76
1991	24,188	11,130	21,674	18.70	18.04
1992	25,293	11,401	22,047	18.20	17.12
1993	25,725	11,887	NA	16.14	14.81
1994	25,951	12,394	NA	15.07	13.55

* US Refiner average acquisition cost, 1980 dollars calculated using the GDP implicit price deflator.
SOURCES: *Petroleum Economist*, Mar. 1986, p. 112, Feb. 1995, p. 64; *API Basic Petroleum Data Book*, Jan. 1995, Section 6, Table 3; *BP Statistical Review of World Energy*, June 1985; International Petroleum Encyclopedia, Penwell Publishing, 1991, p. 320, 1994, p. 294. US & WE includes Canada and Japan. Non-OPEC includes only countries outside of both the OECD and OPEC.

In September 1985 the Saudi Oil Minister Sheik Yamani began to negotiate contracts on a netback basis. Under a netback contract the f.o.b. price of the oil is directly related to the selling prices of the products refined from it. This was in effect a continuous discounting method in a weak market. The purpose was to put pressure on both fringe producers and other OPEC members who were flagrantly cheating on their quotas. Immediately, analysts expected oil prices to decline to about $23 by spring once the winter price runup ended. In December OPEC agreed to abandon the policy of defending the $28 price through cuts in output. The impact of netback pricing combined with an increase in OPEC output was to lower the effective price of Saudi Light to $21 at the end of January and to $18.50 in February. This forced fringe producers such as the UK and Mexico to drastically slash prices to hold on to their market share, and by mid-February Mexico had lowered prices to $15 a barrel. Uncertainty with regard to Saudi Arabian production levels kept the oil market in turmoil as unofficial industry estimates put Saudi Arabia January output at 5 million bpd while the Saudis claimed production of 4.1 million bpd. The world oil price was on a roller coaster in 1986, with the average OPEC price dropping from $27.76 at the end of January to $9.83 in July, rising to $14.36 at the end of September, dropping back to $13.33 in November. In 1987 average OPEC prices became more steady, beginning and ending the year at a little over $17, and never rising above $18.[120]

The import fee continued to be mentioned as a possible source of funds to reduce the deficit during 1985, but Reagan stiffened his opposition to new taxes, including an import fee. Gary Hart offered a proposal for a $10 import fee, rejected by a 78-18 vote by the Senate in November 1985. Apprehension concerning oil prices prompted a proposal by Senator Malcolm Wallop (R-WY) in December to set a $22 price floor for US oil, with a tax to raise the cost of foreign oil if its price falls below that level. David Boren proposed a $5 fee on oil and a $10 fee on product imports, with an exemption for heating oil. In January Reagan made it clear in a meeting with Bob Packwood (R-OR), chairman of the Senate Finance Committee, that no new taxes would be acceptable. As prices dropped President Reagan said on February 3 that he might accept an import fee if tax reform legislation remained revenue neutral. Beryl Sprinkle, chairman of the CEA and the president's chief of staff, Donald Regan, opposed the fee because it would damage the economy. Oil state Congressmen supported a fee, but Eastern and Midwestern Senators were strongly opposed. The API said most of its members were against the fee, and the IPAA continued its opposition because of fear of government intervention and taxation of the independent's gains. Support for the fee came from bankers in the Southwest with large energy loans and state governments who lost millions in tax revenue from declining oil prices.[121]

Pressures for a import fee continued to be felt in March 1986 during Senate Finance Committee hearings. TIPRO argued that a sound national energy policy required an import fee on crude oil and products, along with incentives for drilling. The Independent Refiners Coalition said that a crude oil fee without a fee on products or an exemption for heating oil would be devastating to the US refining industry. Testifying against a fee were Citizens for a Sound Economy, the Society of Independent Gasoline Marketers, and the Chemical Manufacturers Association. Reagan reiterated his opposition to the oil import fee during telephone calls to the Oklahoma Senators and in meetings with five

Republican oil state congressmen in April. He expressed support for measures such as eliminating net income restrictions on depletion credits, increasing the depletion allowance on stripper wells, decontrol of old natural gas prices, and elimination of the windfall profit tax.[122]

In September Reagan ordered a study of the effect of lower oil prices on national security. The study group included officials from Energy, Interior, State, Defense, Commerce, Treasury, the National Security Council, OMB and the EPA. This followed publication of an API study in July warning of impending dependence on foreign oil in the 1990's, and a study by Chase Econometrics which stated that $12 oil would guarantee another oil shock. Lloyd Bentsen, charging that the Reagan administration had no energy policy, introduced a bill requiring an import ceiling at 50 percent of US consumption. The bill would have required the President to send a report to Congress each year forecasting imports, and to formulate a policy for preventing the ceiling from being exceeded. Chances of an import fee were dashed in August when the Senate rejected a rider by Gary Hart to the debt-limit bill providing a $10 fee on imported oil by a vote of 82-15.[123]

The success of Republicans running in the oil patch in 1986 gave implicit support to the Reagan administration policies on oil. The Democratic capture of the House put oil state Senators in influential positions, with Lloyd Bentsen slated for the chair of the Finance Committee, David Boren the chair of the Energy and Agricultural taxation subcommittee, Bennett Johnston, chairman of the Energy and Natural Resources Committee, and Jim Wright, Speaker of the House. By November 1986, a consensus was forming in the oil industry for the need for action, with the API calling for intervention by the federal government, suggesting a number of options including an import fee. Chevron Chairman George Keller suggested establishing a minimum floor price for domestic oil between $14 and $17, though independent producers were calling for a price of $25-32. Harry Broadman and William Hogan, in a study by Harvard's Energy and Environment Policy Center, called for imposition of an $10-11 import fee to reverse increasing dependence on foreign oil. Interior Secretary Hodel expressed concern regarding the national security threat of increased oil imports, sentiments echoed by a report by the National Defense Council Foundation.[124]

As support for an import quota increased in early 1987, so did the activity of opponents. Senators Claiborne Pell (D-RI), John Chafee (R-RI), George Mitchell (D-ME) and John Heinz (R-PA) pledged to work for the defeat of an import fee. They were joined by the Consumer Federation of America, Chemical Manufacturers Association, Petroleum Marketers Association of American and the New England Council, along with Dow Chemical and Exxon. A DOE report in March supported the contentions of oil state congressmen that oil imports threatened national security, but it also criticized an oil import fee as a solution. Secretary Herrington urged repeal of the windfall tax and the Fuel Use Act, and decontrol of old natural gas prices. He was joined by the IPAA in calling for restoration of the 27.5 percent depletion allowance for new and enhanced recovery production, and expensing of geological and geophysical costs. President Reagan supported most of these recommendations in a letter to Congress, but the depletion allowance suggestion was shelved by the administration. A coalition of 18 environmental, consumer and religious groups assailed the White House recommendations, accusing the administration of ignoring the potential for conservation and renewable energy, while the Independent Refiners Coalition decried the lack of attention to increased imports of refined products.[125]

Bennett Johnston in early April admitted that additional tax breaks for the oil industry were unlikely, but he hoped for an oil import fee combined with a gasoline tax as revenue enhancers. Since gasoline taxes fall heavier on the West, this would balance the Eastern objection to measures which increased the cost of heating oil. During the conference on the fiscal 1988 budget resolution the possibility of an oil fee or other energy taxes was raised once more as congressmen looked for politically acceptable revenue sources. The Ways and Means Committee also considered gasoline taxes and an import fee in June and July, and Reagan said a gasoline tax would be acceptable as part of a deficit reduction package. However, resistance to energy taxes doomed any effort to use them for revenue enhancement.[126]

A second attempt by Bentsen to limit US imports was shot down in the Senate by a 55-41 vote in July. Bentsen had tried to slide his 50 percent import ceiling proposal into the omnibus trade bill, but a coalition of senators backed by the administration opposed his move. Among the groups which fought the provision were the US Chamber of Commerce and the National Association of Manufacturers. A split in the oil industry hindered attempts to obtain a fee. The IPAA decided not to support an oil import fee, as the organization said that given the conflict in the industry, it would not be possible to obtain an import fee, and a united industry was required to lobby for other issues such as tax credits for exploration and drilling, increased depletion, and restoration of the investment tax credit.[127]

THE OIL CRISIS THAT NEVER HAPPENED

Prices in the world oil market stabilized in 1988 in a range between $13 and $17 as OPEC struggled to come to some accord that would restore discipline to the group and predictability to the oil market. By 1989 prices moved into the $17-$19 range as OPEC managed to hold on to a tentative agreement despite friction between Kuwait and the other OPEC members. The spot price of West Texas intermediate climbed above the $20 mark in March for the first time since February 1986, and fluctuated around this level for the remainder of the year.[128]

The Bush Administration made noises about developing a new energy strategy based around environmental concerns, but Bush's credibility was limited given his close ties with the Texas oil industry. The administration's energy credibility was further damaged by the appointment of former Louisiana Representative W. Henson Moore as deputy energy secretary in charge of Conservation and Renewable Fuels, after spending a decade in Congress voting against such policies. When Bush declared that increasing oil imports were an unacceptable threat to national security shortly after taking office, he was following a twenty year Presidential tradition. Unfortunately, Bush had no answer to the two problems which plagued any attempt at forming an energy policy: the desire of consumers for plentiful, cheap energy, and the regional nature of energy politics. Reducing oil imports would require lowering consumption which meant higher prices or end-use controls, neither of which would be politically popular. The regional nature of energy politics made it impossible to make policy, with high sulfur coal states, low sulfur coal states, and natural gas producing states unable to agree on any policy involving air pollution, and energy producing states and energy consuming states opposed on pricing issues, and everyone against energy taxes.[129]

Congress, having made it clear that gasoline taxes and oil import fees were unacceptable, turned to corporate average fuel efficiency (CAFE) standards as a substitute for energy policy in 1989. Spurred by concern with global warming, surging oil imports, and pressure to open the Alaska wildlife refuge to oil drilling, CAFE standards were seen as a painless way of reducing dependency on foreign oil. The EPCA in 1975 had set CAFE standards at 18 mpg for 1978 and 27.5 mpg by 1985. Responding to petitions from automakers, the Department of Transportation rolled back the target to 26 mpg until 1988, delaying the 27.5 mpg target until 1990. Global warming bills introduced by Senators Al Gore (D-TN), Tim Wirth (D-CO) and Representative Claudine Schneider (R-RI) in 1989 would have established a CAFE standard of 45 mpg by the year 2000. The automakers responded by supporting a "carbon fee" on all fuels.[130]

The Exxon Valdez tanker spill in March 1989 had little impact on world oil prices, but it strengthened the hand of environmentalists who wanted to prevent exploration in environmentally sensitive areas. For nearly a decade Congress had used the appropriations bill to impose temporary moratoriums on offshore exploration, and in June the House Appropriations Committee challenged the Bush Administration by voting to impose another one-year ban on off shore drilling along much of the nation's coastline. The Interior Department, in September, announced a delay in the sale of Atlantic oil and gas leases in response to pressure from Congress and state officials.[131]

In July 1990, Iraq President Saddam Hussein openly threatened to use force against Arab oil exporting nations if they did not curb their excess production, a threat aimed at Kuwait and the United Arab Emirates. Following this declaration OPEC reached an accord in Geneva to raise oil prices by $3 to $21 per barrel. Four days later, on August 1, Iraq invaded Kuwait on a pretext.

The industrialized world announced sanctions against Iraq, including an embargo on oil from Iraq and Kuwait.[132]

The immediate response in the oil market was an increase in the price of crude from $16 to $28 a barrel, as well as a rapid increase in gasoline prices of 4 to 14 cents a gallon, bringing calls for a probe of the industry. Congressional Democrats, led by Senator Joseph Lieberman (D-CT), urged President Bush to act against price-gouging by the oil industry. Bush responded by calling on the industry to show restraint in his August 8 address to the nation. The National Association of Attorneys General and the Justice Department's Antitrust Division announced investigations into oil company pricing decisions. Oil industry executives, fearing public and Congressional outcry, sought to hold down profits by limiting increases in gasoline prices and increasing funds set aside for future environmental expenses. Public opinion turned against the companies, with 72% viewing Big Oil unfavorably, and 65% disbelieving anything oil companies said about the Gulf War. Most Americans continued to believe that oil was more profitable than most businesses, despite an average return on equity of 12% since 1985 vs. an average of nearly 15% for manufacturing.[133]

In early 1990, the Bush Administration had proposed capping the strategic reserve at 750 million barrels and slowing the fill rate in order to cut costs, but Congressional leaders introduced bills calling for the reserve to reach a billion barrels. DOE officials warned Congress that they would advise Bush to veto bills expanding the reserve. With the invasion of Kuwait, support for an increase in the reserve becoming overwhelming in Congress, with the House proposing to store products as well as crude oil in light of the growing awareness of limited refining capacity. The Strategic Petroleum Reserve bill was unanimously approved by both chambers on September 13. The bill expanded the SPR to a billion barrels, created a three-year test reserve for refined products, and expanded the president's authority to sell off oil from the reserve, to anytime a domestic energy supply shortage exists or is imminent.[134]

Key members of Congress, as well as energy economists were dismayed by the Bush Administration's reluctance to release oil from the Strategic Reserve. With oil prices gyrating, mostly higher, Energy Department officials favored using the reserve to moderate oil prices, but Energy Secretary James Watkins failed to overcome the opposition of OMB director Richard Darman, Treasury Secretary Nicholas Brady and national security adviser Brent Scowcroft, to SPR oil sales because they would interfere with the market. Congressional attitudes were determined by regional interest, with oil consuming states in favor of sales while oil producing states were enjoying higher oil prices. Bush finally responded to increasing oil prices by announcing a five million barrel sale as a test on September 26. The House, after Secretary Watkins mentioned the next day that 15 million barrels would be a more realistic test, passed by voice vote a bill authorizing the president to withdraw this quantity.[135]

The Bush administration cautiously resisted any drastic moves to lower oil consumption, aware of the political dangers of asking Americans to sacrifice.

Pollsters found that about 70 percent oppose higher gasoline taxes to encourage conservation; and while 83 percent supported federal CAFE standards of 45 mpg, this was conditional on only raising the price of new cars by less than $500. Bush reiterated his support for increasing domestic energy production, including drilling in the Arctic Wildlife Refuge. DOE had no plans in the wake of the Iraq crisis, while Congress gave lukewarm consideration to gasoline taxes and import fees, as part of a budget accord, as well as CAFE standards. Northeast Congressmen opposed energy taxes and oil import fees, Western Congressmen gasoline taxes, coal states a Btu tax, the Administration and the auto industry higher CAFE standards, leaving few options.[136]

Despite the willingness of Saudi Arabia to increase production by two million bpd to offset the loss of Kuwaiti oil, world oil prices steadily increased during the fall of 1990, jumping to $31 in late August. Higher prices encouraged OPEC members to increase their production as the organization decided to prevent the kind of price escalation that choked off oil demand in the 1980s. The OPEC decision brought temporary stability to world oil markets, but prices on the futures market jumped back up at the end of September to close to $40 a barrel, and exceed that level in October when Hussein announces threat of a missile attack on Israel, in anticipation of an oil crunch developing from the Gulf Crisis. Oil prices would move sharply up and down the next few months, but gradually decline toward the end of the year as the threat of war receded, falling below $30 in December.[137]

Prices began to inch up as the deadline for Iraq to pull out of Kuwait neared, climbing back above $30 in early January, before collapsing when the success of the military intervention against Iraq became clear, with prices falling below $20 a barrel. For the next few years the cost of imported oil would fluctuate between $13 and $19 a barrel, evidence that the Iraq crisis was a mere blip in the long-term movement of the world oil market.[138]

In the wake of the Gulf War Congress attempted to write a new comprehensive energy bill. In the Senate Energy and Natural Resources Committee chairman Bennett Johnston, working with the committee's ranking Republican, Malcolm Wallop, put together a bill. Philip Sharp (D-IN), chairman of the Energy Subcommittee of the Energy and Power Committee, began hearings on energy policy. Johnston and Wallop planned to combine drilling in the Arctic Wildlife Refuge with higher CAFE standards, to be determined by the Transportation Secretary based on a set of criteria still up for discussion. Sharp also indicted that he would offer CAFE legislation as well as increasing the SPR to 1.5 billion barrels and allowing the president to draw down the reserve in response to rising prices.[139]

The proposed White House energy blueprint was close to the Johnston-Wallop plan. The report called for stimulating domestic oil production, including drilling in the Wildlife Refuge, consolidate nuclear licensing procedures. Under pressure from the OMB, tax credits for energy from renewable sources and a fund to pay for government conservation projects was eliminated, and CAFE standards were excised from the plan. The plan did not receive overwhelming applause on the hill. Environments criticized as a "drain America first" policy, with no provisions for increasing conservation. Sharp advocated keeping the administration proposal alive as a reference point for debate, and to let the Senate take the lead in steering an omnibus package through competing interests.[140]

In the House Sharp and Leon Panetta (D-CA), chairman of the House Budget Committee, introduced a variety of energy bills. Sharp's package was five bills, numbers HR 776 to HR 780, including guidelines and mandates to improve energy efficiency in consumer products; enlarge the SPR to 1.5 billion barrels, requiring oil importers to store 3 percent of their imports in the reserve and charging a 2¢ per barrel fee on imports to finance the reserve; and provide a tax credit for renewable energy. Panetta's bill, HR 560, provided a DOE fund to finance federal conservation projects and increased grants for weatherization projects for low-income families; increase the SPR to 2 billion barrels and establish a standby gasoline tax triggered by falling oil prices; mandate increasing mileage standards by 20% by 1996 and 40% by 2001, to 40 mpg, and increase federal research spending on conservation, and alternative fuels. Senator Wirth introduced his National Energy Efficiency And Development Act in the Senate on March 21, incorporating conservation and efficiency provisions, including Senator Bryan's (D-NV) 40 mpg CAFE standards.[141]

The Senate Energy and Natural Resources Committee Democrats opposed many of the Johnston-Wallop bill's provisions, including the opening of the Alaska wildlife refuge to drilling and easing regulation of the nuclear industry. Johnston tried to add a proposal that would impose CAFE standards of 34 mpg by the year 2001, but was blocked by a combination of Republicans, taking the marking orders from the White House to block the standards, and Democrats who favored Bryan's CAFE standards of 40 mpg. Johnston's original proposal for importers to set aside 9 percent of imports to fill the SPR was reduced to 3 percent. The bill was reported 17-3 after a final markup session on May 23, adding deregulation of the wholesale electricity business, banning oil drilling off the California coast until 2000 and dropping the 3 percent oil import set-aside. Two other Senate committees which claimed jurisdiction over parts of the bill, Environment and Public Works, and Commerce, Science and Transportation, planned to make major changes in the legislation.[142]

Developing a consensus on energy policy was especially difficult as concern with oil faded. A mere six months after the war only 12 percent of those surveyed in a Booz-Allen & Hamilton poll rated energy as one of the three most important issues, while the environment was rated as a top concern by 36 percent. The House Energy Subcommittee on Energy and Power didn't begin to mark up legislation until the middle of July, leaving the tougher debates until after the August recess. The slow pace meant that many of the tough ideological votes would be coming up in an election year.[143]

Philip Sharp's Energy and Power Subcommittee marked up a bill which would require all importers and domestic oil refiners to provide about one percent of their oil (or financial equivalent) for the SPR. Language in the bill broadened the president's authority to sell oil to include releases to relieve large jumps in oil prices. A clean coal technology was reauthorized, along with a proposal to promote alternative fuels, with a goal of displacing 10% of oil-based motor fuels by 2000, and 30% by 2010 was added. Electricity generation deregulation was included, with provisions to protect consumers and to require free access to transmission networks.[144]

In the Senate the Johnston bill was derailed by a filibuster lead by three freshmen Senators: Richard Bryan, Paul Wellstone (D-MN) and Tim Wirth. An attempt to shut off debate fell 10 votes short of the necessary two-thirds, 50-44. Opponents included those voting to stop Alaska wilderness oil drilling, electric utility deregulation and CAFE standards. Environmental groups disliked the proposals to aid coal and nuclear industries and the emphasis on production instead of conservation. Some committee chairman opposed the bill simply because they had been refused referral of parts of the program.[145]

Johnston tried again in January, offering a trimmed-down version of the energy bill, discarding the proposals to open the Arctic Wildlife Refuge and to increase CAFE standards. The environmentalists accepted that the economic climate did not encourage CAFE standards, while the Bush administration dropped its threat to veto the bill if it failed to include the Alaska drilling provision. Conservation and energy efficiency sections were strengthened and environmental protections restored, pro-nuclear provisions were retained while alternative fuel support was deleted. The revised bill passed the Senate on February 19 by a 94-4 vote. The only major accomplishment of the bill was the deregulation of wholesale electric power generation.[146]

The energy debate now shifted to the House, where energy policy was more disaggregated. The Energy Committee approved HR 776, similar to the Senate bill on March 11. The one significant difference was the retention of the 1% of oil imports to be stored in the reserve. The Science, Space and Technology Committee approved a five-year 17.2 billion energy research and development bill on April 1 and folded into the energy package bill. Interior was preparing a bill to stiffen regulation of nuclear power. The Ways and Means Committee added a tax package to promote renewable energy, conservation and alternative motor fuels, provided relief from the minimum tax for independent producers, and struck the import set-aside provision. Oil industry spokesmen claimed the import set-aside would wreck havoc on the industry, a contention that Sharp described as "That is the biggest bunch of bull you've ever heard."[147]

House leaders struggled to find an acceptable compromise, settling on a version on May 19. Among the provisions in this version were: streamlined licensing for nuclear plants, bans on offshore drilling, utility deregulation, and the Ways and Means tax provisions. The Rules Committee put the import set-aside provision back into the bill. A new provision, to prevent natural gas producing states from controlling output through prorationing was aimed directly at Texas, Oklahoma and Louisiana, who controlled 58 percent of the nation's gas supplies. The bill was passed by the House on May 27 by a 381-37 vote and sent to conference. The oil import set-aside provision was deleted by a

263-135 vote on a floor amendment offered by Dan Rostenkowski, with Republicans supporting the President's position by a 141-10 vote, and Southern Democrats by 62-19.[148]

On July 30 the Senate passed a revised version of its energy bill including energy-related taxes. The Bush Administration attempted to interfere with the conference negotiations which started September 10 by claiming that the proposals to restrict offshore oil and gas drilling and limit state regulation of natural gas production might draw a presidential veto. Johnston and Sharp made it clear that Congress would work out a final bill without giving these objections much weight. Energy efficiency measures were resolved by compromise, as were clean coal research programs, and ventures to promote renewable energy. The major point of contention was the lack of guaranteed access by producers to electric transmission systems in the Senate version. While tax relief for oil producers was in both bills, it was intertwined with a House proposal to provide urban aid. Another sticking point was the shape of mandates for government and private fleet owners to use alternative fuels.[149]

After a month of struggling over the bill the conference finally came up with a final product on October 1. The section dealing with off-shore drilling was dropped to be dealt with separately because agreement could not be reached. The conference also deleted the limitation on state power to prorate gas production. The tax incentives for conservation, renewable energy and alternative fuel cars survived the conference, as well as relief for independent oil and gas producers. The conference report flew through both Houses and was signed by the President on October 24, ending an eighteen month odyssey.[150]

The Clinton administration demonstrated its lack of knowledge of energy politics by admitting that it was considering energy taxes in early 1993. When Treasury Secretary Lloyd Bentsen mentioned on national television that the administration was considering broad-based energy taxes to raise revenue, intense lobbying was initiated by various interest groups before a proposal was even formalized. Groups both opposed taxes and if one was inevitable, wanted to impose it on someone else. The Gas industry suggested oil import taxes, gasoline taxes or carbon taxes; Coal said a gasoline tax would be best; the IPAA focused on the regressive nature of gasoline taxes and suggested an oil import fee; Airlines were against any tax on oil; Railroads were against any tax on coal; Truckers against gasoline taxes; the Petrochemical industry against oil import fees. Automakers supported a gasoline tax hike to head off tougher CAFE standards. The Northeast was against oil import taxes, the West and South against a gasoline tax, the Midwest and Mountain states against a carbon tax. New anti-carbon tax lobbies were raising money to fight the tax, while environmental groups pushed for both carbon and gasoline taxes.[151]

The Clinton administration finally proposed a 25.7 cent per million Btu tax to raise $73 billion over four years for deficit reduction. The tax was designed to avoid directly attacking the coal industry, which meant it was had little effect on either energy consumption or carbon dioxide emissions. Oil was taxed at double the rate of other fuels, less as a conservation measure than an easy way to raise revenue. Opponents made apocalyptic predictions of economic ruin, which given the small size of the tax were ridiculous. Even before the measure was sent to Capitol Hill Bentsen had begun making exceptions, beginning with heating oil and ethanol and methanol. Once in Congress pressure grew from various interest groups for additional exceptions and exemptions.[152]

The energy tax became the most controversial portion of the Clinton deficit reduction plan. The Clinton plan barely passed the House 219-213, with the energy tax intact, but the Senate was adamant about eliminating the Btu levy. Senator John Breaux (D-LA) proposed an 8 cent a gallon gasoline tax as an alternative, which would net about half as much revenue as the Btu tax, threatening the budget accord. Under extreme pressure from conservatives and Western Senators the Senate finally resolved on a 4.3 cent gasoline tax, which was reluctantly accepted by the House.[153]

After twenty years of attempts at formulating an energy policy which would reduce the nation's dependence on foreign oil the best that six presidents and Congress could accomplish was to raise gasoline taxes by 9.3 cents a gallon, establish CAFE standards requiring an average of 27.5 mpg for new cars, and a few inconsequential tax incentives and research programs for conservation and alternative fuels. The one fact which stood out after two decades was that energy taxes were a third rail of electoral politics, like Medicare and Social Security, touch it and face the possibility of electrocution. The combination of the Gulf War, Global Warming and never-ending deficits were insufficient to provide the political will to increase energy taxes.

NOTES

[1] *O&GJ*, Mar. 13, p. 19, Mar. 20, p. 54, May 31, 1971, pp. 20-21.

[2] Neil De Marchi, "Energy Policy Under Nixon, Mainly Putting Out Fires," in *Energy Policy in Perspective: Today's Problems, Yesterday's Solutions*, pp. 405-9.

[3] De Marchi, "Energy Policy Under Nixon," pp. 410-12.

[4] *CQ Weekly Report*, Sept. 18, 1971, p. 1940.

[5] Bohi and Russell, *Limiting Oil Imports*, pp. 213-14.

[6] Bohi and Russell, *Limiting Oil Imports*, pp. 215-16.

[7] *O&GJ*, Dec. 27, 1971, p. 58, Jan. 17, p. 50, May 15, p. 49, May 22, p. 22, Sept. 1972, pp. 59-60, Jan. 22, 1973, p. 45.

[8] De Marchi, "Energy Policy Under Nixon," pp. 418-20.

[9] Richard C. Weisberg, *The Politics of Crude Oil Pricing in the Middle East, 1970-1975* (Institute of International Studies, University of California at Berkeley, 1977), pp. 25-26, 42-44.

[10] Weisberg, *Oil Pricing in the Middle East*, pp. 50-55; and Anthony Sampson, *The Seven Sisters*, pp. 258-263.

[11] Weisberg, *Oil Pricing in the Middle East*, pp. 63-64, and Sampson, *The Seven Sisters*, pp. 272-283.

[12] *CQ Weekly Report*, Feb. 5, 1972, p. 248.

[13] *CQ Weekly Report*, Feb. 3, 1973, p. 201-6; and U.S. Congress, Senate, Committee on Interior and Insular Affairs, *Oil And Gas Imports Issues, Hearings before the Committee on Interior and Insular Affairs*, 93rd Cong., 1st sess., Jan. 11-22, 1973, pp. 3, 482.

[14] *Oil And Gas Imports Issues*, pp. 4-5.

[15] *Oil And Gas Imports Issues*, pp. 57, 66.

[16] *New York Times*, Mar. 17, 1972, 59:2; De Marchi, "Energy Policy Under Nixon," pp. 423-25.

[17] U.S. Congress, Senate, Committee on Interior and Insular Affairs, *The President's Energy Message And S. 1570, Hearing before the Committee on Interior and Insular Affairs*, 93rd Cong., 1st sess., May 1, 1973, pp. 5-15, 55; *CQ Weekly Report*, Apr. 21, 1973, p. 913.

[18] Bohi and Russell, *Limiting Oil Imports*, pp. 217-19.

[19] De Marchi, "Energy Policy Under Nixon," pp. 432-33; and *The President's Energy Message And S. 1570*, pp. 21-22, 367.

[20] *CQ Weekly Report*, June 16, 1973, p. 1536; U.S. Congress, Senate, Committee on Interior and Insular Affairs, *Strategic Petroleum Reserves, Hearings before the Committee on Interior and Insular Affairs*, 93rd Cong., 1st sess., May, July, 1973; U.S. Congress, House, Committee on Interstate and Foreign Commerce; *Mandatory Fuels Allocation, Hearings before the Committee on Interstate and Foreign Commerce*, 93rd Cong., 1st sess., July, 1973.

[21] Richard Porter, "The Federal Trade Commission v. the Oil Industry," *Antitrust Bulletin* 27 (Winter 1982): 753-763; *Is the Energy Crisis Contrived?* An AEI Round Table held on 22 July 1974, (Washington, D.C.: American Enterprise Institute, 1974).

[22] De Marchi, "Energy Policy Under Nixon," p. 434; and *CQ Weekly Report*, July 7, 1973, pp. 1812-16.

[23] De Marchi, "Energy Policy Under Nixon," pp. 433, 437-41.

*He undercut his own argument with the statement that the minimum size for an new efficient refinery would be at least 100,000 bpd—implying that most small refiners were probably inefficient. Halverson also admitted that price regulations had an anticompetitive effect, combined with the depletion allowance: "Second, it appears that a combination of high crude oil prices and low prices of refined products has kept margins which refiners can expect to earn at such a low level that new independent entry into refining has not been encouraged. Indeed, the oil depletion allowance itself creates for the majors an additional incentive to keep crude prices high and maintain small refinery margins since a dollar of refinery profit will always be taxed at a higher rate than a dollar of crude oil profit." *Mandatory Fuels Allocation*, p. 208.

[24] De Marchi, "Energy Policy Under Nixon," pp. 442-44; and Bohi and Russell, *Limiting Oil Imports*, pp. 221-22.

[25] Vietor, *Energy Policy*, pp. 242-43

[26] Weisberg, *Oil Pricing in the Middle East*, pp. 98-99; and Steven A. Schneider, *The Oil Price Revolution* (Baltimore: John Hopkins University Press, 1983), pp. 199-204.

[27] Schneider, *Oil Price Revolution*, pp. 236-37; and Weisberg, *Oil Pricing in the Middle East*, pp. 112-13.

[28] De Marchi, "Energy Policy Under Nixon," pp. 447-451; and *CQ Weekly Report*, December 1, 1973, pp. 3139-41.

[29] Bohi and Russell, *Limiting Oil Imports*, pp. 224-25

[30] De Marchi, "Energy Policy Under Nixon," pp. 466-68.

[31] Edward Morrison, "Energy Tax Legislation: The Failure of the 93rd Congress," *Harvard Journal on Legislation* 12 (1975): 384-86.

[32] *CQ Weekly Report*, Dec. 29, 1973, pp. 3458-59, Feb. 2, 1974, pp. 235-37.

[33] *CQ Weekly Report*, Jan 19, pp. 95-98, Feb. 9, pp. 311-13, Mar. 9, 1974, pp. 632-34.

[34] *CQ Weekly Report*, Mar. 23, p. 764, Apr. 6, p. 897, Apr. 13, 1974, p. 953; and Morrison, "Energy Tax Legislation," pp. 405-12.

[35] Vietor, *Energy Policy*, p. 245.

[36] De Marchi, "Energy Policy Under Nixon," pp. 468-72.

[37] Bohi and Russell, *Limiting Oil Imports*, pp. 226-27.

[38] Neil De Marchi, "The Ford Administration: Energy As a Political Good," in *Energy Policy in Perspective: Today's Problems, Yesterday's Solutions*, pp. 480-86.

[39] *CQ Weekly Report*, Jan. 18, 1975, p. 131-37; and De Marchi, "Ford Administration," pp. 486-88.

[40] *CQ Weekly Report*, Jan. 25, 1975, p. 171-73.

[41] *CQ Weekly Report*, Feb. 8, p. 257, Feb. 22, p. 356, Mar. 4, 1975, p. 475; *New York Times*, Feb. 3, 1975, p. 37:5.

[42] *CQ Weekly Report*, Mar. 8, 1975, p. 476-77; US Congress, House, Committee on Interstate and Foreign Commerce, *Energy Conservation And Oil Policy, Hearings before the Subcommittee on Energy and Power*, 94th Cong., 1st sess., Mar. 1975, pp. 79-107, 155-61.

[43] *CQ Weekly Report*, Mar. 8, 1975, p. 474-75; and U.S. Congress, House, Committee on Ways and Means, *The Energy Crisis And Proposed Solutions*, 94th Cong., 1st sess., Mar., 1975, pp. 7-21.

[44] *The Energy Crisis And Proposed Solutions*, pp. 27-46.

[45] *The Energy Crisis And Proposed Solutions*, pp. 375-77; *Energy Conservation And Oil Policy*, p. 626-42.

[46] *CQ Weekly Report*, Feb. 22, pp. 351-52, Mar. 1, 1975, pp. 419-422; *National Journal*, May 5, 1975, p. 514.

[47] *CQ Weekly Report*, Mar. 8, p. 471, Mar. 22, pp. 571-73, Mar. 29, 1975, pp. 633-37. An initial small producer exemption of 22 percent was retained on 2,000 bpd, declining by 200 bpd each year until 1980, when it would remain at 1,000 bpd. The depletion rate would decline to 20 percent in 1981, 18 percent in 1982, 16 percent in 1983, and a permanent rate of 15 percent in 1984. The rate would stay at 22 percent for oil produced by costly secondary or tertiary recovery methods. The small producer exemption was denied to any company that operated a refinery with a throughput more than 50,000 bpd. An additional exception was made for the first 1,000 bpd produced through secondary and tertiary techniques until 1983.

[48] *CQ Weekly Report*, Apr. 26, pp. 858-59, May 3, pp. 952-54, May 10, pp. 959-61, May 31, p. 1123, June 28, 1975, p. 1343.

[49] *CQ Weekly Report*, Mar. 8, pp. 475-77, July 5, p. 1438, Aug. 2, p. 1655, Sept. 13, pp. 1939-40, Oct. 4, 1975, pp. 2134-35.

[50] *CQ Weekly Report*, Aug. 2, p. 1655-57, Oct. 4, 1975, p. 2134; and De Marchi, *"Ford Administration,"* p. 503-5. It was suggested that Ford was motivated by three factors; the proximity of the New Hampshire primary which required the President to disassociate himself from the large oil companies, the accelerating inflation rate which encouraged restraint on price increases, and the need to have some sort of energy policy.

[51] Vietor, *Energy Policy*, pp. 253-55.

[52] Dennis Drapkin and Philip Verleger, "The Windfall Profit Tax: Origins, Development, Implications," *Boston College Law Review* (May 1981): 643-50.

[53] *National Journal*, Dec. 6, 1975, pp. 1657-63.

[54] *National Journal*, Apr. 3, 1976, pp. 440-45.

[55] James L. Cochrane, "Carter Energy Policy and the Ninety-fifth Congress," in *Energy Policy in Perspective: Today's Problems, Yesterday's Solutions*, pp. 549-51.

[56] Cochrane, p. "Carter Energy Policy," pp. 551-56.

[57] *Energy Policy*, pp. 251-52.

[58] *Energy Policy*, pp. 254-55.

[59] *CQ Weekly Report*, Apr. 23, 1977, p. 732; and U.S. Congress, Senate, Committee on Finance, *Energy Tax Act of 1977, Hearings before the Committee on Finance*, 95th Cong., 1st sess., Aug., Sept. 1977.

[60] *CQ Weekly Report*, July 16, 1977, p. 1435.

[61] *National Journal*, May 14, 1977, pp. 740-45; and *CQ Weekly Report*, May 21, 1977, p. 958.

[62] *CQ Weekly Report*, July 16, 1977, pp. 1438-39.

[63] *National Journal*, July 30, 1977, pp. 1196-99.

[64] *CQ Weekly Report*, Aug. 6, pp. 1623-1629; *National Journal*, Aug. 20, 1977, pp. 1303-5.

[65] *CQ Weekly Report*, May 21, 1977, pp. 957-59; *National Journal*, Aug. 20, 1977, pp. 1305-6.

[66] *National Journal*, Aug. 20, 1977, pp. 1305-6.

[67] *CQ Weekly Report*, Oct. 1, pp. 2062-63, Oct. 22, 1977, pp. 2233-35.

[68] *New York Times*, June 2, 17:5; Sept. 1, 1977, 1:3.

[69] *New York Times*, Oct. 3, 1977, 1:1; Oct. 7, 1977, 1:8.

[70] *CQ Weekly Report*, Oct. 22, pp. 2233-35, Nov. 5, 1977, pp. 2355-58.

[71] *CQ Weekly Report*, July 1, 1978, p. 1659-60; *New York Times*, Apr. 9, 2:3, June 28, 15:1, June 29, 1977, IV, 1:2.

[72] *National Journal*, Nov. 14, 1978, pp. 1760-68. MCF is industry shorthand for thousand cubic feet.

[73] *CQ Weekly Report*, Dec. 23, 1978, p. 3470; Apr. 14, 1979, p. 712.

[74] Joseph Yager, "The Energy Battles of 1979," in *Energy Policy in Perspective: Today's Problems, Yesterday's Solutions*, 1981, pp. 604-16.

[75] *Energy Policy*, pp. 256-57; and *New York Times*, May 23, 1979, 1:4.

[76] *New York Times*, May 8, 1:3; June 12, 1979, 1:5; Yager, "Energy Battles, " pp. 616-18.

[77] *National Journal*, May 26, 1979, p. 852; *New York Times*, May 3, 1979, IV, 1:4.

[78] *CQ Weekly Report*, Apr. 28, 1979, p. 772.

[79] Drapkin and Verleger, "Windfall Profit Tax," pp. 653-665.

[80] *New York Times*, Apr. 24, 1979, 1:2 *National Journal*, May 26, 1979, p. 857.

[81] *Petroleum Economist*, Jan., pp. 2-3, Feb., pp. 54-55, Mar., p. 132, Apr., pp. 129-30, Sept. 1979, pp. 350-51

[82] *Petroleum Economist*, June, p. 262, Aug., pp. 310-11, 346, Sept., p. 394, Oct., p. 442, Nov., p. 494, Dec. 1979, p. 538.

[83] *Petroleum Economist*, Jan., pp. 3-4, Feb., p. 90, May, p. 226, July, pp. 278-79, Aug., p. 362, Nov., p. 506, Dec. 1980.

[84] U.S. Congress, Senate, Committee on Finance, *Crude Oil Severance Tax, Hearings before the Subcommittee on Energy and Foundations*, 96th Cong., 1st sess., May 1979, pp. 49-50; *CQ Weekly Report*, June 2, p. 1040, Aug. 4, p. 1580, Oct. 27, 1979, p. 2416.

[85] *New York Times*, June 4, IV, 2:1; June 14, IV, 3:1; June 29, 1979, IV, 1:3; Drapkin and Verleger, "Windfall Profit Tax," pp. 671-75.

[86] *CQ Weekly Report*, July 21, pp. 1433-34, Oct. 13, p. 2265, Nov. 3, 1979, p. 2480; *Energy Policy*, p. 219.

[87] *Energy Policy*, p. 70, 210; and *CQ Weekly Report*, July 21, 1979, pp. 1433-39.

[88] *CQ Weekly Report*, Aug. 4, p. 1580, Sept. 8, 1979, p. 1915.

[89] *CQ Weekly Report*, Aug. 4, p. 1580, Oct. 27, 1979, pp. 2414-17; *New York Times*, Sept. 7, IV, 1:1, Sept. 26, 1979, IV, 1:1.

[90] *CQ Weekly Report*, Dec. 1, pp. 2707-8, Dec. 8, pp. 2759-60, Dec. 15, pp. 2818-19, Dec. 22, 1979, p. 2876.

[91] *New York Times*, Jan. 19, 1980, 31:1, Jan. 23, 1980, IV, 1:6; *CQ Weekly Report*, Dec. 22, 1979, p. 2875, Jan. 26, pp. 183-84, Mar. 8, pp. 668-69, Mar. 15, 1980, p. 843; Robert S. McIntyre, "Lessons for Tax Reformers From the History of the Energy Tax Incentives in the Windfall Profit Tax Act of 1980," *Boston College Law Review*, 22 (May 1981): 705-30. The compromise kept the 30 percent rate for new oil, but set the rate for pre-1979 oil at 70 percent of price above a $12.81 base (adjusted for inflation) except for the first 1,000 bpd of an independent's production, taxed at 50 percent. Stripper wells owned by major companies were taxed at 60 percent on a base of $15.20 but the first 1,000 bpd of independent stripper well production would be taxed at 30 percent. Actual revenues collected were $9.925 billion in 1980, $25.944 billion in 1981, $16.754 billion in 1982, $10.668 billion in 1983, $8.871 billion in 1984, and $6.53 billion in 1985, for a total of $77.757 billion. Collections ended early in 1986 due to the price decline. API, *Petroleum Facts and Figures*, Sept., 1987.

[92] *CQ Weekly Report*, Mar. 8, 1980, p. 669; *New York Times*, Dec. 7, IV, 4:1, Dec. 21, 1979, 1:5.

[93] *New York Times*, Mar. 11, 1:1, Mar. 15, 1:6, 36:5, Apr. 1, II, 12:1, Apr. 3, 1980, IV, 1:6.

[94] *CQ Weekly Report*, June 7, 1980, p. 1539-40; and *New York Times*, Mar. 18, 1:2, Apr. 17, IV 3:2, Apr. 17, IV, 11:6, May 14, 1:5, May 15, II, 12:1, June 5, IV, 6:5, June 6, 1:6, June 7, 1980, 1:4.

[95] *New York Times*, Jan. 17, IV, 4:4, June 16, IV, 6:6, Oct. 20, IV, 1:3, Dec. 9, 27:1.

[96] *National Journal*, Dec. 6, 1980, p. 2075.

[97] *CQ Weekly Report*, Apr. 10, 1982, p. 783-85.

[98] *CQ Weekly Report*, Oct. 25, 1980, p. 3208; Richard Pierce, "The Choice Betwen Adjudicating and Rulemaking for Formulating and Implementing Energy Policy," *Hastings Law Journal* 31 (1979): 1-3; *New York Times*, Apr. 24, IV, 5:1, July 18, 1980, 1:2.

[99] *CQ Weekly Report*, Jan. 31, p. 231-32, Mar. 14, p. 465, Aug.1, 1981, p. 1424; Drapkin and Verleger, "Windfall Profit Tax," pp. 659.

[100] *CQ Weekly Report*, Feb. 14, 1981, p. 313-14, Oct. 30, 1982, p. 2757-60 and Sept. 29, 1984, p. 2409.

[101] *National Journal*, Aug. 15, 1981; and *CQ Weekly Report*, Aug. 8, 1981, pp. 1431-32.

[102] *CQ Weekly Report*, Dec. 1, 1984, p. 3019, Nov. 23, pp. 2411-12, Nov. 30, 1985, pp. 2484-85; *New York Times*, Nov. 21, 1985, p. D 1:6.

[103] *CQ Weekly Report*, July 18, p. 1584, Nov. 7, pp. 2747-48, Nov. 28, 1987, p. 2932.

[104] *O&GJ*, Apr. 19, 1982, p. 60; *New York Times*, Oct. 5, 1981, IV 2:1, Mar. 15, IV 2:2, Apr. 8, IV 1:3, Apr.14, 1982, IV 1:5; and *CQ Weekly Report*, May 8, 1982, p. 1039.

[105] *New York Times*, May 11, IV, 1:3, May 13, IV, 2:1, Nov. 11, IV, 1:4; *Petroleum Economist*, July 1981, p. 319, Dec. 1982, p. 482, Mar., p. 111, Apr. 1983, p. 118.

[106] *New York Times*, Dec. 3, 1:5, Dec. 5, 40:1, Dec. 7, 1:6, Dec. 11, 13:3, Dec. 18, 1:2, Dec. 21, IV, 29:1, Dec. 22, 1:2, Dec. 24, 1982, II, 4:1.

[107] *O&GJ*, Jan. 31, p. 67, Feb. 14, 1983, p. 74.

[108] *New York Times*, Mar. 7, p. IV, 2:1, Mar. 9, 1983, IV 3:3; and *O&GJ*, June 6, 1983, p. 40.

[109] *New York Times*, Jan. 4, IV, 5:1, Mar. 8, 1:4, Feb. 22, IV, 12:6.

[110] *National Journal*, Feb. 24, 1979, p. 301; June 28, 1980, p. 1045; and *O&GJ*, Sept. 17, 1979, p. 46.

[111] *Energy Policy*, p. 231; *National Journal*, June 28, 1980, pp. 1044-49.

[112] *CQ Weekly Report*, Mar. 28, p. 553-56, May 16, p. 870, June 27, 1981, p. 1139; and *O&GJ*, Dec. 15, 1980, p. 46.

[113] *CQ Weekly Report*, June 26, 1982, p. 1549; *O&GJ*, Dec. 13, 1982, p. 28; and *Energy and the Environment, The Unfinished Business* (Wash., D.C.: CQ Press., 1985), p. 27.

[114] *CQ Weekly Report*. February 23, 1985, p. 355-56; *New York Times*, Dec. 12, 1984, p. D 23:1; and *O&GJ*, Dec. 31, 1984, p. 49. Hodel replaced Edwards in November 1982. Hodel was deputy administrator of the Bonneville Power Administration and president of the National Electric Reliability Council.

[115] *CQ Weekly Report*, Apr. 5, p. 757, Nov. 1, 1986, p. 2791; and *New York Times*, Aug. 7, 1986, p. D 3:1. Herrington, a California lawyer and a veteran of past Reagan campaigns, was assistant secretary of the Navy from 1981 to 1983 before becoming the White House assistant for personnel. He succeeded Hodel in January 1985 when the latter was appointed to replace

James Watt at Interior, and admitted during confirmation hearings his ignorance of energy policy. *Energy and the Environment*, p. 7.

[116] *New York Times*, Mar. 17, p. 1:3, May 7, p. D 1:1.

[117] Joe Roeber, "Dynamics of the Rotterdam Market," *Petroleum Economist*, Feb., Mar., Apr., 1979.

[118] Jochen H. Mohnfeld, "Changing Patterns of Trade," *Petroleum Economist*, Aug., 1980; idem, "Implications of Structural Change," *Petroleum Economist*, July 1982, pp. 269-72.

[119] *Petroleum Economist*, Mar., pp. 78-80, Aug., 1985, p. 270.

[120] *Petroleum Economist*, Mar., 1986 pp. 74-75; *O&GJ*, Feb. 24, pp. 26-27, Mar. 10, 1986, p. 24, various issues 1986, 1987; *New York Times*, Sept. 4, IV, 1:1, Sept. 17, IV, 1:6,.

[121] *O&GJ*, Jan. 13, p. 31, Feb. 10, 1986, p. 50; *New York Times*, Feb. 7, p. D 1:3, Feb. 10, 1986, p. D 2:1; and *CQ Weekly Report*, Feb. 15, 1986, pp. 295-97.

[122] *O&GJ*, Mar. 10, 1986 pp. 34-35, Apr. 28, 1986, pp. 50-51.

[123] *O&GJ*, Sept. 29, 1986, p. 40; and *Oil Daily*, July 23, p. 1, Aug. 4, p. 1, Aug. 26, 1986, p. 1.

[124] *Oil Daily*, Nov. 6, p. 1, Nov. 12, p. 1, Dec. 1, 1986, p. 2; and *O&GJ*, Nov. 17, p. 17, Dec. 1, 1986, pp. 27-28, Jan. 12, 1987, p. 26.

[125] *O&GJ*, Feb. 23, p. 38, Mar. 23, 1987, p. 22-23; *New York Times*, May 1, 1987, p. D 1:1; *Oil Daily*, Apr. 1, p. 1, Apr. 16, 1987, pp. 1-2.

[126] *Oil Daily*, Apr. 7, p. 1, June 29, p. 1, July 8, p. 1, Nov. 10, 1987, p. 1; and *CQ Weekly Report*, May 23, 1987, p. 1045.

[127] *O&GJ*, July 13, p. 23 and Oct.19, 1987, p. 17-18.

[128] *New York Times*, Sept. 27, IV, 5:3, Nov. 25, 1988, I, 1:6, Jan. 4, IV, 14:4, June 8, IV, 1:3, Nov. 29, 1989, IV, 1:6, API, *Basic Petroleum Data Book*, Jan. 1995, Sec. VI, Table 14.

[129] *CQ Weekly Report*, Jan. 20, 1990, pp. 157-62.

[130] *National Journal*, May 13, 1989, p. 1214; *CQ Weekly Report*, May 6, 1989, pp. 1039-40.

[131] *New York Times*, Apr. 12, I, 17:1, June 30, I, 1:5, Nov. 4, I, 1:5, Nov. 14, 1989, IV, 11:3; *CQ Weekly Report*, Jan. 20, 1990, p. 161.

[132] *New York Times*, July 18, D, 1:4, July 28, I, 1:6, Aug. 2, A, 1:6, Aug. 5, 1990, I, 1:4.

[133] *New York Times*, Aug. 8, A, 11:3, Sept. 16, 1990, I, 1:3; *National Journal*, Aug. 11, p. 1960, Sept. 1, 1990, p. 2075; *CQ Weekly Report*, Aug. 11, 1990, p. 2590; Richard Behar, "Big Oil's Bad Rap," *Time*, Feb. 25, 1991, p. 64.

[134] *New York Times*, Aug. 9, D, 1:3, Sept. 27, A, 1:5; *CQ Weekly Report*, Sept. 15, 1990, p. 2914.

[135] *New York Times*, Apr. 20, 1990, D, 1:3; *National Journal*, Nov. 17, 1990, p. 2798-99; *CQ Weekly Report*, Sept. 29, 1990, p. 3120.

[136] *National Journal*, Sept. 8, p. 2161, Sept. 15, 1990, pp. 2184-86; *CQ Weekly Report*, Sept. 22, 1990, pp. 3004-7.

[137] *New York Times*, Aug. 19, I, 1:3, Aug. 23, A, 1:5, Aug. 30, A, 1:6, Sept. 28, A, 1:4, Oct. 10, A, 1:5, Dec. 6, 1990, D, 1:1.

[138] *New York Times*, Jan. 15, D, 1:3, Feb. 16, I, 42:1; API, *Basic Petroleum Data Book*, Jan. 1995, Sec. VI, Table 3b.

[139] *CQ Weekly Report*, Feb. 9, 1991, p. 365.

[140] *CQ Weekly Report*, Feb. 16, p. 413, Feb. 23, pp. 464-65, Mar. 16, 1991, p. 669.

[141] *CQ Weekly Report*, Mar. 16, pp. 701-4, Mar. 30, 1991, p. 794.

[142] *CQ Weekly Report*, Apr. 20, pp. 971-72, May 18, pp. 1270-71, May 25, 1991, p. 1369-70; *National Journal*, June 1, 1991, p. 1312.

[143] *National Journal*, July 27, 1991, p. 1893.

[144] *CQ Weekly Report*, July 27, p. 2071, Aug. 3, p. 2157; Oct. 12, 1991, p. 2943.

[145] *CQ Weekly Report*, Nov. 2, 1991, pp. 3191-95; and *National Journal*, Nov. 9, 1991, p. 2767.

[146] *CQ Weekly Report*, Feb. 1, p. 248, Feb. 8, 1992, pp. 297-99

[147] *CQ Weekly Report*, Mar. 14, pp. 612-14, Apr. 4, pp. 870-71, May 2, 1992, pp. 1152-53.

[148] *CQ Weekly Report*, May 23, p. 1436-39, May 30, 1992, pp. 1530-32, 1564; *National Journal*, June 13, 1992, pp. 1412-13.

[149] *CQ Weekly Report*, Sept. 12, pp. 2710-11, Sept. 19, 1992, pp. 2802-3.

[150] *CQ Weekly Report*, Oct. 3, pp. 3030-33, Oct. 10, pp. 3141-43, Oct. 31, 1992, pp. 3464.

[151] *National Journal*, Jan. 30, p. 292, Feb. 13, 1993, pp. 403-4, 409-410.

[152] *CQ Weekly Report*, Jan. 30, pp. 214-15, Feb. 28, p. 450, Apr. 10, pp. 910-12, Apr. 24, 1993, p. 1007; *National Journal*, Apr. 17, 1993, 917-20.

[153] *CQ Weekly Report*, July 31, 1993, pp. 2023-28.

IV

Economists, Energy And Public Policy

Thinking on economic policy tends to be based on generally accepted economic theory, which in turn evolves over time as it moves from specialist publications to textbooks and becomes accepted knowledge. Therefore to understand how economic thinking influenced policy it is necessary to determine both what economists thought and when they thought it, as well as the disagreements among economists and the resolution (if any) of these disputes. Until the 1960s there were few articles written on oil, while the literature on resources was quite general, with limited application to the oil industry. This reflected the low priority given applied work in general, and energy economics in particular, in the profession. With the advent of energy crises of the 1970s a flood of articles on energy and oil appeared in the literature, many of which were highly abstract, with limited applicability to policy analysis.

The evolution of thought in the profession was gradual, with new concepts displacing existing beliefs over time. One reason for this slow acceptance of new concepts is that there has always been a split between theory and practice in economics. Theory is usually expressed in idealistic terms, often with a formal mathematical structure, while in practice economists are forced to deal with institutional constraints on optimal behavior, ambiguous circumstances which cannot be reduced to a simple theoretical structure, or a lack of data to permit estimation of crucial parameters of models. Thus one finds dissent toward imposition of compulsory unitization, questions raised about the effect of the depletion allowance, and even a challenge to generally accepted estimates of the costs of import quotas. However, a consensus can usually be observed, whether it is the belief that market demand prorationing establishes a government cartel while promoting inefficiency, that import quotas transfer wealth from consumers to producers without appreciably increasing national security, or that tax breaks for the oil industry resulted in unjustifiable subsidies.

EXHAUSTIBLE RESOURCE THEORY

The original exposition of the theory of exploitation of an exhaustible resource was presented by Lewis Gray in the *Quarterly Journal of Economics* in 1914. In the case of an exhaustible resource (one where prevention of exhaustion is noneconomical) the goal is to obtain maximum benefit from a limited supply of resources, taking in account the loss in present value of delaying production. Gray noted the impact of diminishing productivity (increasing marginal cost) and also mentioned the possibility of increasing costs of production as the stock of the resource declines. Gray attacked the distinction made between rent and royalty (defined as a return to cover the value of the mineral extracted), since the real economic rent comprises the entire net return including the royalty. A production tax will shift production between present and future, slowing the rate of exhaustion.[1]

While Gray covered the main points of the theory of exhaustible resources he did not provide a formal presentation, which awaited the 1931 article by Harold Hotelling. Hotelling applied the calculus of variations to determine the optimal path of production of an exhaustible resource. He began with the assumption that it is a matter of indifference to the mine owner whether he receives p_o at present or p_oe^{rt} at time t, and thus price will be a function of time such that $p = p_oe^{rt}$, where p is the net price received after paying the cost of extraction. Defining $q = f(p, t)$ as production at time t given price p, denoting T as the time of final exhaustion, and K as the total stock, then $\int^T qdt = \int^T f(p_oe^{rt}, t)dt = K$, and $f(p_oe^{rt}, T) = 0$, allowing determination of p_o. Hotelling showed that maximization of the social value of the resource, if future utility is discounted at the market rate of interest, will provide the same price path as under free competition. Under monopoly the problem changes, as the monopolist desires to choose output q so as to maximize net profit = $\int qp(q)e^{-rt} dt$. Hotelling demonstrated that production will be retarded under monopoly.[2] The influence of this article was limited for decades due to its highly technical nature, but it would be rediscovered in the late 1960s and be an important influence on economists during the 1970s.

F.W. Paish made a number of astute empirical observations on investment in gold mines in a 1938 article which unfortunately were ignored in resource theory. Paish suggested that uncertainty (both with regard to the quality of a deposit and future prices) would lower the rate of investment since overinvestment could not be rectified, while an underestimate of the optimal level of production could be corrected by further investment as information was gathered. This implies that a period of overoptimistic expectations would result in an extended period of production at a rate of output above the optimal level with perfect foresight. He observed that while selectively mining the richest ores (i.e. the deposits with the lowest marginal cost of production) is the economically optimal strategy, it is not the normal practice in mining. Paish suggested that a reason for ignoring selective mining was that by shortening the life of the mine the method would not be in the interest of the mine managers and employees.[3]

In 1944 William Pabst used Hotelling's work to explain the changing composition of an initially monopolistic resource market. Assuming that competitive firms faced a more elastic demand curve than that of a monopolist, pessimistic expectations of future prices would cause output to rise in the competitive sector as monopolistic firms sold off reserves at "bargain" prices. Expectations of increasing demand encouraged monopolistic firms to buy reserves, since the value of reserves was greater for them than for competitive firms.[4]

The economics of the common property problem, applicable to the economics of petroleum reservoirs under multiple ownership, was graphically depicted and analyzed by H. Scott Gordon in his 1954 article in the JPE. He constructed his argument around the fishery, applying it to various other examples of common property resources. Since fish are not private property is impossible to appropriate the rent from a fishery, and instead of producing where marginal cost equals marginal product, fisherman try to maximize average product to obtain the greatest total yield. Any fishing ground which yielded a positive economic rent would draw fisherman from other areas until the rent was dissipated. The problem is that each fisherman faces a situation where his marginal cost equals his average cost, although for the fishery as a whole marginal cost increases as the take increases. Without ownership rights the opportunity cost of producing one more unit is zero. Gordon used the petroleum industry as one example of the problem of the common.[5]

Orris Herfindahl extended the analysis of the mine to the industry, bringing in the rate of exploration and development in the long run. Given a constant cost of exploration, in long-run equilibrium there will be zero profits in both exploration and mining and rents will be dissipated. An increase in demand will result in higher prices and increased exploration and production until price declines to its former level. A decline in the cost of exploration will increase exploration activity and thus reserves, with prices declining until profits from exploration disappear, with both reserves and exploration activity permanently higher. Given a rising cost of exploration over time the life of mines will be lengthened while industry reserves will decline. The value of known resources will rise with the cost of exploration, which is dependent on cumulative discoveries, which in turn depends on the rate of price increases. If there are different regions the cost of exploration will equalize between regions, as will the value of the ore in the ground. Herfindahl then used data on petroleum exploration to show a relation between profits and the rate of exploration.[6]

Herfindahl extended his analysis in a 1964 paper, restating the Hotelling rule as the principle that the present value of the royalty (defined as $P_t - C_t$) must be the same at all times. He showed that a decline in the interest rate would both extend production of the resource (assuming a ceiling price) and increase the initial price. The effect of a change in costs and demand for the product were also explored. Herfindahl showed that deposits of different quality would not be exploited simultaneously, since the higher quality deposit would be used first, and exploitation of the lower grade occurring only after the lower cost resource has been exhausted (since the resource must be sold for the same price, the deposit yielding the higher discounted royalty will be exploited first). He extended the analysis to exploration for the resource when deposits of different qualities are found. Herfindahl noted that the large supplies of most minerals, including fossil fuels, makes scarcity unimportant in the determination of price.[7]

It would not be until 1967 that another formal mathematical formulation of the pure theory of exhaustible resources would make its way into the literature in a paper by Richard Gordon, nearly forty years after Hotelling's seminal work. By generalizing and clarifying Hotelling's presentation Gordon made the model more accessible to the profession. Gordon set up the problem as a constrained present-value maximization problem. Using this formulation he derived the basic Hotelling condition that scarcity rents will increase at the rate of interest. Increasing costs were incorporated as a function of the cumulative exhaustion of the resource. Gordon briefly touched upon models of competitive behavior, but as he observed, the problem quickly becomes difficult to describe without oversimplification.[8]

The combination of the controversy surrounding the publication of *The Limits To Growth* and the emerging energy crisis resulted in a flood of papers on the theory of exhaustible resources in the 1970s. Most of these articles were concerned with the question of scarcity and the relation of resources to growth, but a number dealt with price and production under various conditions. Practically all these papers accept the assumptions that supply and demand functions are continuous, uncertainty can be reduced to risk (i.e., a probability distribution can be postulated), and therefore the problem can be defined in a fashion so as to yield an optimal solution. While many of these articles are quite elegant, except insofar as they were used to construct models that were then estimated (Salant for example) they had little effect on the policy debate. Their primary impact was to provide intellectual legitimacy to attempts to model oil markets as markets for an exhaustible resource controlled by a cartel which maximizes future income. These models were often combined with game theoretical analysis of cartel behavior, primarily OPEC or part of OPEC and a competitive fringe. There were no attempts to determine whether oil markets were congruent with the implied assumptions of exhaustible resource theory, or to explain why the historical development of the industry contradicted the predictions of theory with numerous high cost deposits developed simultaneously with low cost deposits.

CONSERVATION

Parallel with the development of a formal theory of the economics of exhaustible resources was the articulation of a philosophy concerning use and abuse of natural resources. This dialogue had its roots in the Progressive era, when the concept of conservation entered the political arena in a major fashion in the form of the Conservation Movement between 1890 to 1920. To the early Conservationists, economic scarcity was defined by finite physical limits on the availability of natural resources. Waste, considered physical inefficiency in the use of resources, was emphasized since it was avoidable by wise policy. It was considered the product of ignorance or apathy; improper criteria built into the structure of consumer sovereignty, competition and laissez faire; inefficient government activity or inadequate government intervention. The most important type of waste from the perspective of petroleum policy was the mismanagement of resources resulting in a failure to obtain the maximum yield of extractive product from the physical resource.[9]

The Conservation movement was opposed to the exploitation associated with the rapid economic growth, irregardless of the nature of the actors. In fact, the movement often supported large corporations on the grounds that they could more readily afford to undertake conservation practices requiring efficiency, stability of operations and long-range planning. It was in effect the opposite of populism; instead of supporting decentralized control of the economy conservationists were in favor of the dominance of a technological elite which would bring scientific methods to the management of economic activity. They were opposed to big business only to the extent that it was felt that the central government was better positioned to promote conservation and efficiency. In many cases the conservation movement shared the aversion of corporate business to unrestrained competition, though for different reasons. The main conflict with business was in their vision of the decision-making process--conservationists felt that experts should make the decisions concerning proper use of resources, and a central authority should be responsible for enforcing these decisions.[10]

Erich Zimmerman critiqued the Conservationist position in his 1931 opus, *World Resources and Industries* He defined a resource as a purely subjective concept, dependent upon both preferences and technology, with availability determined by price in a market economy. Zimmerman felt that a serious error made by conservationists was in their refusal to draw the logical implications from the consequences of low prices, caused by unbridled competition, which resulted in overproduction and waste of resources. He quotes both L.C. Gray and R.T. Ely to illustrate the economist's concept of the problem of conservation as the trade-off between the well being of the present generation and that of future generations. Conservation was defined as a reduction of the rate of depletion or consumption, in contrast with economizing (increasing efficiency). Two ways of reducing exploitation of natural resources were postulated by Zimmerman; direct interference with the operation of the free market or the incidental consequence of monopolistic practices. He concluded that due to factors causing individuals to discount the future, the economic price system will cause overexploitation of resources, and that corporations are in a better position to conserve natural resources than individuals.[11]

Ciriacy-Wantrup developed a more rigorous definition of conservation by noting that it involves a shift in the intertemporal distribution of use of the resource toward the future; depletion then involves a shift toward the present. He also pointed out a number of factors necessary to make economically rational social policy with regard to resources: using the equating of marginal social revenues and costs as the basis for policy analysis; internalizing the costs and benefits associated with externalities; and consideration of the consequences of irreversibility when uncertainty exists with regard to the possibility of a catastrophic outcome. During the same year the Paley Commission Report was published, reflecting public concern with the question of availability of resources. Reviews by Colin Clark and Edward S. Mason, while taking issue with various estimates of population and resource consumption, agreed with the Commission's conclusion's that the United States would become increasingly dependent on foreign imports of raw materials.[12]

Anthony Scott challenged the distinction made by Ciriacy-Wantrup, Ely and others, between renewable and unrenewable resources. Exhaustion of natural resources is an nonissue, since what actually occurs is the consumption of higher grade stocks, replaced by lower grade stocks with higher costs of exploitation or refining. The consequent increase in expense of utilizing the resource will lead to substitution, increases in efficiency in use, and innovation in production. He also examined the relationship of conservation and national defense, starting with the imposition of duties on oak from the Baltic regions by Great Britain in the early 19th century, justified by Adam Smith's rationalization of protection for industries required for defense of the country. Scott pointed out the dilemma in relying on the market system for strategic materials, since the owner of natural resources in wartime is usually not allowed to make a wartime profit from sales, because high prices are incompatible with most wartime planned allocations. Thus, while expected demand for strategic materials would increase in anticipation of a conflict, the expectation of price restraints constrain entrepreneurs from investing in capacity to increase supply. Scott discussed stockpiling as an alternative, but pointed out that it is unpopular with industry, fearful of the release of stocks to the market.[13]

The publication of Scarcity and Growth by Barnett and Morse in 1963 would put the issue of depletion of exhaustible resources to rest until the doomsday predictions of the Meadows in the early 1970s. Barnett and Morse attempted to empirically test the hypothesis of increasing scarcity by examining the costs of producing resources over time. They found that extractive minerals exhibited increasing returns over time. They also found that the cost of extractive goods, relative to the unit costs of nonextractive goods, declined over time. They concluded that this trend was due to technological progress lowering costs, outweighing the effect of an increase in scarcity (in the sense of depletion of higher quality stocks). Barnett and Morse expressed optimism with regard to continued progress and felt that the problems to be solved would not be those concerned with diminishing returns or scarcity, but those associated with environmental issues, allocation of benefits and costs, and the form of the social production process.[14]

OIL POLICY PRE-WWII

Two major works on oil policy were published in the mid-1920s, one by George Stocking in 1925, and an especially influential book by John Ise in 1926. Both men felt that over-production was a chronic problem in the industry. Among the causes given by Ise for over-production of oil were: the small size of leases in many fields, forcing the drilling of too many wells and allowing entry of numerous small companies who were oblivious to the consequences of their actions; lease contracts requiring the drilling and operation of wells regardless of market conditions; and the refusal of state and national governments to sanction agreements among operators to limit production on the grounds that such agreements are in restraint of trade, and therefore contrary to the anti-trust laws. Stocking was more explicit about defining the problem as a mismatch between property rights and the natural unit of production, the oil reservoir.[15]

Ise defined waste as "any loss or use which fails to afford the maximum social utility or want satisfaction, for the present and for future generations." Ise granted the need to discount the consumption of future generations, and even touched upon the issue of present investment versus future consumption. He was pessimistic with regard to future supplies of oil and whether technological progress would result in development of substitutes. His basic contention was that since petroleum is a limited resource, improvements in the efficiency of finding and producing petroleum merely hastened the date when prices would have to increase rapidly. This concern with scarcity caused him to focus his attention on the waste of oil and oil products, including oil left in the ground (especially the abandonment of small wells due to low prices and the mismanagement of field development), losses in the storage, transportation, and refining of oil, but most importantly the immense waste associated with the consumption of oil products. "The greatest and most unpardonable waste of oil, however, has been its consumption for all kinds of low or unimportant uses, particularly its use as fuel, in place of coal, and its use for all sorts of unessential automotive purposes." Ise was skeptical about the value of consumer sovereignty in determining priorities of use, since he felt that consumers were unaware of the costs of their decisions.[16]

Stocking considered the overinvestment of resources in production as wasteful as the physical loss of oil associated with uncontrolled competition. He concurred with Ise that uncontrolled production resulted in wasteful utilization of the crude product, but Stocking felt that higher prices would help to eliminate waste. For him the problem was that low cost oil discouraged investment in refining technology which maximized the yield of higher valued products from a barrel of petroleum.[17]

Ise felt that the industry was a natural monopoly in many aspects, especially the transportation sector, and that the trend toward integration was inevitable. This was a trend which he applauded since the large corporations were more efficient and likely to keep prices higher. He observed that a large company could develop and operate a field far more efficiently than numerous small companies due to its control over drilling and production, but unlike Stocking he failed to draw the lesson that unitization was the route to efficiency. Both men recommended government intervention, in the form of ownership of as much oil as it could obtain, and regulation of the extraction of privately owned oil. This would permit government to control over drilling and over-production, and to enforce regulations to eliminate waste in reservoir development, oil transportation and refining. Since even a monopoly must rapidly discount future revenues, Ise wanted the government to restrict production to raise prices, and to capture monopoly profits through a severance tax and a gasoline tax.[18]

Writing in 1930, Leonard Logan agreed with the basic premise of Stocking and Ise that overproduction, in causing low prices, promoted waste of petroleum. Logan pointed out that Ise's criticism was from a "utility" viewpoint, and that maximizing social utility was a political and legal question. He focused on conservation as the elimination of economic waste, defined as occurring when the economic value of the good produced is less than the cost of production. For Logan, the objective of stabilization was to raise price to a level where the industry as a whole could obtain a reasonable return on capital, to reduce uncertainty in the industry, and to eliminate unessential uses of petroleum. He saw unitization as more scientific than prorationing, but noted that the latter had been more effective in practice. Logan preferred voluntary control to government regulation, and refused to see the inevitable resistance to unit operation as an insurmountable barrier to such a program.[19]

Stocking published two articles in 1933 examining the effectiveness of oil prorationing. He noted that economic characteristics of oil production included: capacity additions determined as much by chance as rational calculation; the rule of capture requiring rapid exploitation of a field once discovered; the large proportion of fixed investment and low marginal costs encouraging continued production at almost any price. The result was an industry recurrently plagued with over-production, compounded by the inelastic character of the demand for its major product, gasoline. In addition a surplus of refining capacity made refiners willing to process any crude oil they could obtain. Stocking concluded that prorationing had stabilized the price of gasoline, but also had a detrimental effect on the economy by transferring purchasing power from consumers to oil producers. Proration contributed to more efficient extraction of underground oil as an incidental result of a program whose main objective was stabilization of output and control of price. Stocking observed that a genuine conservation program would take as its point of departure the geological unit--the oil pool, requiring unified operation. Compulsory unit operation was the logical next step, with the elimination of tariff barriers and production restraints.[20]

Myron Watkins in 1937 was careful to distinguish between economic and technical efficiency, observing that conservation of physical resources was not necessarily economically optimal, since it was necessary to account for opportunity costs, both with respect to alternative investments in the present and the value of the resource in the future:

> Applying these concepts to the oil industry, it follows from all this that the enforced nonuse of available petroleum resources is not necessarily true conservation. Such a policy involves saddling the present generation with the definite handicap of going without something which might well yield now more than it cost to acquire, or more precisely might well yield more in proportion to its cost than efforts expended in alternative directions . . ., while preserving for future generations only a highly speculative opportunity for fruitful production and use.[21]

Watkins distinguished between conservation, the avoidance of waste in petroleum recovery or use, and stabilization, regulative efforts designed to assure a supposed improvement in economic performance regardless of the effects upon efficiency. Conservation had no relation to constraining production to market demand, but only involved use of the most efficient methods of petroleum recovery. Watkins also noted that social economy was not synonymous with private profitability. He was aware of the legal difficulties of devising a system of control which would promote technical efficiency without requiring economic distortions. He pointed out that if the goal of stabilization was concerned imperative as a policy measure, given the stated importance of a healthy petroleum industry, the government possessed the power of eminent domain allowing it to seize minerals, while state governments, following the example of draining districts, could force the unitization of oil pool development.[22]

POST WWII TO THE MANDATORY IMPORT PROGRAM

Paul H. Frankel argued in 1946 that the oil industry could not operate under competitive conditions in his *Essentials of Petroleum*. Because of the uncertain results of exploration, the high overhead costs at all stages of the industry, and a high inelasticity of demand in the short run, a fall in prices does not significantly choke off supply or stimulate demand. Uncertain results of exploration result in either too little or too much oil discovered in relation to the amount that can be sold at profitable prices. Crude-oil production, transportation, refining, storage, and distribution all require large amounts of capital, resulting in large fixed costs and relatively low marginal costs, causing production to continue indefinitely at low prices. Therefore the industry is subject to continuous crises in the absence of reasonably strong control over supply, with prices falling below the long-run cost of production due to relatively small surpluses, often for long periods. Some group of "leading interests" will then move to control the industry, the major companies and/or various governments.[23]

Frankel saw integration as a logical outcome of the structure of the industry. Production in foreign countries required large amounts of capital, which would not be risked unless there were assured outlets for the bulk of production. Refiners, faced on the one hand with recurring threats of shortages, and on the other hand the need for a steady demand for their products, were encouraged to integrate to assure supplies and markets. The large scale required for transportation also encouraged integration with respect to both producers and refiners. Horizontal integration was encouraged by the high level of risk in production, which could be alleviated by spreading investment over various regions and countries. Frankel concluded that oil is a big industry and any attempt to break it up will be self-defeating.[24]

The publication of *A National Policy for the Oil Industry* by Yale's Eugene Rostow in 1948 had a major impact on the debate concerning oil policy, and succeeded in raising the ire of industry participants. During the National Petroleum Council's meeting in April 1948, Rostow's book was the focus of vehement criticism. It was decided that the American Petroleum Institute be given the task of refuting Rostow, since it was beyond the NPC's charter to do so.[25] It is not surprising that the industry reacted so vehemently to his book, since it was an unabashed attack on the whole system of control.

Rostow began by claiming that the market for gasoline was characterized by the type of behavior expected in an oligopolistic market. Since the large sellers believed the demand for gasoline was inelastic they were reluctant to engage in price competition. Rostow asserted that both the demand for gasoline and the supply of oil were responsive to price, and that competition existed between petroleum products and other sources of energy, as well as between motor and other forms of transportation service. He claimed that price leadership and price discrimination were deeply established practices in the petroleum industry, with prevailing differences in prices bearing no relation to differences in cost, but representing the most profitable possible adaptation of monopolistic sellers to market forces.[26]

Rostow then assailed the prorationing system, pointing out that the Federal Government was an accessory, since the estimates of market demand by the Bureau of Mines were essential for operation of the system. He compared these estimates to trade association statistical services condemned in various anti-trust cases. He pointed out that the prevailing system of production control prohibited flush flows, but otherwise violated all the geologists' rules of conservation practice. Rostow remarked that if conserving the nation's oil supply was the goal of policy, then oil tariffs should be eliminated and foreign oil used in peace time, keeping American oil as a standby for defense purposes. Other steps to conserve oil could include discouraging consumption by taxation, prohibition of its use where substitutes exist, and mixing gasoline with alcohol. "But we are not really serious about oil conservation, and there isn't much objective ground for urging that we should be. However, our public policy is loyal indeed to the slogan of conservation. It is the supposed objective of our production control system, as it is the moving idea which has persuaded the Supreme Court and the public to condone our present monopoly methods of determining supply."[27]

Rostow called for imposition of unitary operations on the fields, requiring the organization of companies or cooperatives in which all surface owners would share on an equitable basis, either in proportion to their surface ownership or to the richness of underlying deposits. Rostow disagreed with Joe Bain's presumption that even if unitization were adopted, limitation of production to "market demand" would be required in order to prevent physical waste associated with "over-production." Rostow pointed out that other commodity industries survive without market control, and that neither experience nor analysis supported the view that the production of oil would be carried forward to socially uncompensatory levels. Given excessive oil producing capacity, likely to produce prices below long-run average costs under free competition, keeping prices at levels which would attract new investment would be self-defeating. Rostow explained the policy of prorationing on a flat per well basis as a way to restrict the production of the better wells, largely owned by the large companies, in favor of the smaller wells owned by independent and non-integrated producers, in order to improve the latter's negotiating position with the majors. Rostow also called for vertical and horizontal division for major companies of their four separate functions of production, refining, transportation and distribution, and the elimination of all quotas or other quantitative limits on imports.[28]

Joe Bain answered the Rostow diatribe, agreeing with Rostow's call for mandatory unitization, but he expressed concern with the effect on the industry of unrestrained production. Bain suggested that differences between producers and the general public would give rise to a divergence between the private and the social optimum intertemporal production pattern. He saw Rostow's opposition to curtailment stemming from a general distrust of the policy as monopolistic and suggested that a curtailment program could include taxation of excess profits. Bain agreed with Rostow on the issue of divestment and thought that there were no real benefits in efficiency from integration. However, he pointed out that there were significant economies of scale in refining for units of 60,000 bpd of capacity and larger, which would still allow room in the market for 50 or 60 refining firms.[29]

Rostow, in a reply, agreed with Bain that policy should not be dependent on the assumption that private and social costs were identical in oil production. He suggested that their differences were based on two issues; a different evaluation of the risk of oil shortages and a difference in the chosen means for curtailing use if necessary. Rostow felt that higher prices due to production constraints would merely be inflationary, especially in the case of gasoline where no substitute existed. An alternative would be to restrict use where substitutes existed, lowering demand without giving producers a subsidy.[30]

In response to Rostow's book, a research project, the Petroleum Monograph Series, was established by a grant from the American Petroleum Institute to Yale University in 1949. While the editorial board and the writers were to be free of pressure and censorship, of the six board members, one was a petroleum industry economist (John Boatwright) and another an oil company vice-president, three were academic economists and one was associated with the Brookings Institution. The first work published, in 1954, was a study of crude and product price formation by Ralph Cassady. He noted that the crude market was supply-controlled, but asserted that there was active competition for crude supplies. Cassady concluded that prices were stabilized at level which was probably higher than which would have existed without proration. Prices were posted by buyers, but premiums and discounts were used when posted prices differ from prices dictated by the market. He asserted that product pricing was basically competitive, though concerted activities exist sporadically in many markets, mostly at the retail level.[31]

Erich Zimmerman was chosen as the author of the work on conservation, published in 1957. He began by stating the traditional definition of conservation: "In the case of petroleum, therefore, the prevention of waste is unquestionably the principal form of feasible conservation. On that point there can be little disagreement." He went on to declare that the first priority of oil policy should be to encourage exploration to obtain information about the resource, followed by maximization of recovery from reservoirs, and then the prevention of waste. He contended that prorationing had lowered costs by reducing the need for exploration and lengthening the production horizons of reserves as well as limiting waste. In answer to criticism of production restraint, Zimmerman contended that while curtailment of output might raise prices in the short-run, it lowered costs in the long-run. He reasoned that in a free market the random nature of supply causes surpluses which require wasteful crude shortage, while erratic prices increased risk requiring a higher rate of return in the industry and thus higher prices. Zimmerman, as a logical outgrowth of his support for prorationing, felt that imports needed to be controlled to preserve the delicately balanced program of state and federal policies. He contended that national security required preservation of domestic productive capacity, even though the country would have to rely increasingly on imports in the long-run.[32]

Wayne Leeman offered an analysis of the domestic crude market in 1957, focusing on the determination of prices along the Gulf Coast. Leeman criticized state regulatory bodies for reducing production below Maximum Efficient Rate (MER) of production. He concluded that the argument that overproduction would result in wasteful storage was fallacious since the self interest of oilmen, once freed from competitive extraction, would ensure that oil is not produced in such quantities. Leeman described the Gulf Coast market as oligopolistic, with the three principal producing states and the five international majors as the members of the oligopoly. Leeman recognized that a constraint on state control of prices was the competition of rival states and importers for market shares. He felt that importers were not constrained by the threat of quotas and tariffs and that this competition had a significant impact on efforts to limit production and raise price. Frankel, while agreeing that crude oil prices in the Gulf were determined by an oligopolistic system, was skeptical concerning competitive pressure from either rivalry between states or importing companies. Though prices were not increasing, they were maintained at the cost of the marginal operator.[33]

James Nelson also examined the structure of the domestic industry in 1958. He pointed out the amusing use of doublespeak in the oil industry, where terms like "freely competitive market" for oil is defined to include output controls, "supply" means the amount of crude oil above the ground, and "demand" means the fixed amounts of oil products that consumers purchase regardless of price. Output was stimulated by high prices and tax breaks; to compensate for this, domestic output was cut by state control and foreign output by so-called 'voluntary' import quotas. Drilling of wells was stimulated by prices and tax laws and restricted by rules of thumb such as spacing requirements. Exploration was stimulated by prices and tax breaks and depressed by excess capacity and output limitations. Everything canceled out, except for high prices paid by consumers, and the after-tax incomes of "small independent oil men." Nelson recommended elimination of the depletion allowance, restriction of capital gains treatment to unitized properties, and limiting expensing to dry holes. Imports should be freely permitted and a tariff levied to pay for a national security reserve, allowing importers to hold the reserve oil in bond. If this reserve was insufficient the Federal government should purchase domestic reserves and hold them off the market.[34]

John Boatwright attacked Nelson's conclusions, citing estimates that regulation resulted in the recovery of 50 percent more oil from known fields and the drilling of one-quarter the number of development wells than under unconstrained production, as a justification of state proration, ignoring alternative regulatory schemes. Boatwright claimed that neither price nor maintenance of reserve producing capacity were the primary purposes of regulation, and the effect of regulation had been a lower long-run average price. He dismissed unitization by pointing to the failure of voluntary unitization and the inability to pass compulsory unitization through state legislatures. Bruce Netschert considered Nelson's program of changes for the oil industry to be so unrealistic that it was merely an academic exercise.[35]

The third book in the petroleum monograph series dealt with the issues of integration and competition, written by Alfred Kahn and Melvin de Chazeau in 1959. The authors flatly stated that regulation of output involved determination of price and that prorationing to market demand violated the principles of both economy and efficiency. They noted that there were two alternative methods of regulation to prorationing; restricting production controls to prevention of output above MER or mandatory unitization of all fields. They opposed restriction to MER since this would allow operation of the rule of capture, requiring continued regulation, but observed that mandatory unitization would stimulate the opposition of both the small firms and the state commissions and governments reluctant to surrender their authority. De Chazeau and Kahn saw imports as a challenge to long-established conservation policies and pointed out the role of special tax inducements in encouraging foreign development of oil. They noted that percentage depletion influenced the attitude of integrated firms toward crude prices. Oil price escalation would increase income after taxes when crude oil self-sufficiency (the percentage of oil an integrated company supplied its own refineries) was greater than 77 percent, and if 50 percent of crude price increases could be passed along to product prices, 38.5 percent self-sufficiency would make a higher crude price beneficial. Therefore most of the domestic integrated companies benefited from crude price increases.[36]

De Chazeau and Kahn called for the abolition of quota restrictions, recommending as the second best solution either a tariff or a subsidy to those producers who require support. They also called for a careful and critical review of defense requirements and their relation to the domestic industry. A suggestion to preserve a national defense capacity involved imposing a restriction of maximum output to a percentage of MER that would result in the required reserve capacity. Alternatively, if the current system was retained, transferable import quotas could be issued operators, equivalent to their cutbacks in allowable production below MER, over and above their normal allotments.[37]

There were a number of reviews of the three works in the monograph series. John McGee found the Zimmerman work disappointing and potentially misleading, especially the definition of conservation as maximizing ultimate recovery from the reservoir. He criticized Zimmerman for ignoring alternatives to prorationing, as well as assuming that unrestricted production with free imports would necessarily be a worse situation than the state run cartel. McGee also pointed out that the state run cartel was worse than a private cartel since it didn't try to minimize costs. McGee dismissed the Cassady work as lacking in serious economic analysis and found the De Chazeau and Kahn work to be poorly written, offering little new concerning integration, and lacking careful testing of theoretical propositions. He also felt that the model used to evaluate the attitude of integrated firms toward crude oil prices was too simple, and that they ignored the possible existence of monopoly power in crude markets.[38]

Harold Williamson and Ralph Andreano agreed with Stocking's criticism of Cassady's neglect of state programs of production control and their effect on prices, and Joe Bain's description of Zimmerman's work as an inadequate treatise on conservation problems which confused economic and engineering criteria of conservation. Williamson and Andreano felt that de Chazeau and Kahn left important questions unanswered, such as what they meant by "workable competition," the ease of entry into the industry, and the consequences of divestiture. Williamson and Andreano wondered if the authors' call for compulsory unitization ignored the possible consequences on market structure; and whether unitization would be as hard to implement as divestiture.[39]

THE GREAT DEPLETION ALLOWANCE DEBATE

The first blow in the percentage depletion debate was struck by the publication of an article by Douglas Eldridge of the Treasury Department in 1950. Eldridge conceded that special tax allowances encouraged investment in exploration and development of mineral deposits, resulting in augmented supplies and lower prices. Withdrawal of percentage depletion would not reduce production from producing properties, but the relative attraction of new investment would be diminished, resulting in gradually increasing prices and lower output. Eldridge concluded that a tax policy which offered incentives only for extraction neglected other means of improving the nation's mineral position such as increased efficiency in use or substitution of other inputs to production. He challenged the assertion that the types and degrees of risk in extractive industries were different from those in general manufacturing enterprises. In exploration, business arrangements had been developed which effectively reduced risk, by the large size of some corporations, royalty payments based on production, and farm outs and other methods of sharing risk. Eldridge pointed out that tax deductions represent costs to the government similar to payments of subsidies, but that these costs were hidden.[40]

The papers submitted by economists for the Subcommittee on Tax Policy in 1955 and the Tax Revision Compendium for Ways and Means in 1959 became part of the economic literature on depletion. The 1955 report included papers by Horace Gray and Arnold Harberger. Gray pointed out that percentage depletion was granted independently of any conservation enhancing moves by producers, and if there was a national defense interest it would be better served by direct subsidies or price supports, stockpiling or publicly financed research. Harberger argued that the corporate income tax should be designed so as to tax the return to capital equally in all industries. He noted that there are a number of tax advantages in the petroleum industry, lower capital gains taxation of successful discoveries if sold, deduction of intangible development costs, and the percentage depletion allowance. He rejected the argument that oil exploration was riskier and required a higher after-tax rate of return to attract capital; and noted if true instead of tax breaks higher prices should be charged so consumers pay the real price for oil. Harberger also pointed out that while no adequate study of efficient minerals policy for national defense existed, and that tax policies that encourage production of scarce minerals could not be a rational defense policy.[41]

Peter Steiner's assumed the justification of special tax treatment for the industry was an overriding public interest in having available petroleum supplies and the belief that such supply would not otherwise be forthcoming. He challenged the concept of ideal resource allocation because prices were not consistently related to costs across industries, and the market was unable to correctly allocate public goods. Steiner asked for more empirical research to determine the costs and benefits of depletion, to decide whether structural changes in the form rather than the level of depletion might be optimal. John Menge suggested that depletion allowed a firm to depreciate and amortize the same asset. Menge observed that the industry argument that depletion was required implied that the price mechanism didn't work for petroleum, and noted that when this occurs the government usually regulated or nationalized the industry. Even if one could show the need for standby oil reserves for national defense, there was no logical justification for applying depletion for foreign wells. Menge called for elimination of depletion on the grounds it led to misallocation of resources, suggesting that subsidies be used if industries need support.[42]

Stephen McDonald moved the depletion debate to the pages of the *National Tax Journal* in 1961. McDonald argued that a "neutral" tax policy was non-neutral with regard to its impacts on different industries. He assumed the rate of return on invested capital reflects real costs, principally risk which must be covered by output prices in the long run; and that the corporate income tax is shifted so as to restore the normal rate of return. Industries which had a higher rate of return on capital or higher levels of capital intensity would raise output prices relative to other industries to recover their normal return on capital, shifting resources away from these industries. McDonald estimated the penalty for the production sector of the petroleum industry, using a rate of return on capital of 20.2% vs. 10.3% for manufacturing as a whole, and concluded that allocative neutrality would require 21.9% deduction of gross income, which was close to the 22% effective benefit of percentage depletion and expensing.[43]

Richard Musgrave pointed out that a tax with perfect loss offset also reduces risk, balancing the effect of the tax on the rate of return on risk taking.

He suggested that if McDonald's argument was correct, the solution was to replace a tax on capital with one on all costs. If McDonald's implicit assumption of an infinite elastic supply of capital is replaced by assuming an inelastic capital supply, then there would be no long-run shifting and no change in relative prices. In this case the depletion adjustment would cause an uneconomic move of capital. Eldridge questioned McDonald's figures for returns in the production sector of the petroleum industry, estimating that returns on equity when companies with foreign production were eliminated only averaged 12.3% vs. 9.7% for all manufacturing. He saw no a priori reason to expect the rate of return to be higher in a risky industry, and doubted the contention that oil and gas production was riskier than other manufacturing, since a number of other industries had equivalent rates of return.[44]

McDonald responded by denying any intention to justify or rationalize percentage depletion, an intent he felt implied by Eldridge. McDonald asserted that given the existence of uncertainty rather than risk, and social costs of capital formation, then perfect financial loss offsets still would mean less than perfect economic loss offsets. He felt that inclusion of loss offsets would only slightly modify his conclusions. McDonald pointed out that existence of risk aversion would require a higher rate of return in risky industries. He conceded the relevance of the Eldridge adjusted calculations of the rate of return in domestic petroleum production, but pointed out that elimination of foreign operators left a group of companies heavily weighted with strippers and property operators who rarely engage in risky activities. Accepting Eldridge's figures would give an effective special deduction for the oil and gas industry of about 14.5%.[45]

Rene Manes added a new element to the debate by invoking the findings of De Chazeau and Kahn concerning the effect of integration on crude pricing policies. Since 51 percent of refinery runs in 1962 were made by majors who were 75 percent or more self sufficient, and additional refining was done by smaller firms and natural gas companies who would also benefit from price increases, it was not unreasonable to conclude that price leadership occurs in the industry. In this case the higher returns in the industry reflected market control and not greater risks, which would mean that the depletion allowance did result in the misallocation of resources.[46]

Peter Steiner struck the next blow, observing that the discussion on neutrality was limited to examining a subset of variables, and that choice of variables was part of the reason for the depletion debate. McDonald's challenge was based on measurement of neutrality from an initial assumption of a zero tax rate, and focused on neutrality at the level of prices of final outputs rather than at the value of investments. Steiner assumed that the oil industry would have both higher risk and longer effective asset life on productive ventures than the typical manufacturing industry. Steiner estimated that there would be a slight level of non-neutrality, but that it was very sensitive to the assumed values. Accepting McDonald's contention of non-neutrality, Steiner asserted that it made little difference due to the market's allowance for risk and because the risk disadvantage after taxes is offset by the loss-writeoff and expensing provisions of the tax laws.[47]

McDonald pointed out that the disagreement between himself and Harberger was that the latter felt the incidence of the tax was on capital, making the test of neutrality the change in relative values of marginal investments, while he assumed that the incidence of the tax was on consumers, making the test of neutrality the observed constancy of relative prices. However they were both concerned with resource allocation, an issue which McDonald felt that Steiner did not explicitly address. McDonald saw the problem as determining the incidence of the corporate income tax, which would then determine the proper test of neutrality. Steiner objected to McDonald's failure to distinguish between using a discount to the expected value of the present value to represent risk aversion vs. applying different rates of discount to net income streams according to the degree of riskiness. Steiner preferred the use of expected values since the rate of return criterion equates unequal length investments only if the investor can reinvest at this rate without risk.[48]

The depletion allowance came under renewed examination in a special 1970 issue of *Natural Resources Journal* on federal policy toward oil production. Davidson suggested the industry had an extremely high ratio of economic rents to sales revenues, and corporate taxation and depletion primarily impacted these economic rents, with only a minor effect on resource reallocation. He rejected the argument that the depletion allowance was an efficacious device to stimulate exploration, since this required a high elasticity of supply of oil lands combined with a low level of economic rents. McDonald restated his arguments, though tempered by inclusion of the effects of expensing, which he stated that combined with depletion was probably unneutral. He acknowledged that under assumptions less favorable to the industry but more consistent with the opinions of the majority of economists the tax provisions would be markedly unneutral, and that with respect to economic rents percentage depletion was economically perverse.[49]

OIL POLICY AND OIL IMPORT QUOTAS

William Thorp questioned the imposition of mandatory quotas by examining the application of the security clause of the Trade Agreements Acts. He noted that of the twenty-one cases filed with the Director of Office of Defense Mobilization only eight came to an actual decision, and in seven the petition was denied. In the case of petroleum, unlike the others which were turned down, there was no discussion of mobilization requirements, of supplies, or of alternative ways to encourage further development. No case of a present shortage of domestic oil or of the impact of the industry on the general economic welfare was made. He pointed out that: "The Congressional debates suggest a very close relationship in the minds of some members of Congress particularly interested in oil and their attitude toward the extension of the trade agreement legislation. This then was not a case which was considered in the light of some new national policy declaration but one in which the policy declaration was written with the particular case in mind."[50]

James McKie and Stephen McDonald launched the second wave of criticism of public policy toward the oil industry with their article in the *Quarterly Journal of Economics* in 1962. They noted that if firms in the industry developed each oil reservoir as a unit and chose optimum production rates and spacing patterns under conditions of perfect foresight, it was doubtful that there would be problems concerning conservation policy. McKie and McDonald then judged conservation regulation with regard to how well it approximated outcomes expected from a properly functioning market. They found that for the major prorationing states MER was subjugated to considerations of market demand, and the two factors used to determine production allowables were fair allocation of production and the individual well as the unit of production. This pattern of allocating production quotas encouraged overdrilling, resulting in excess costs of development well drilling of approximately one billion dollars a year. Conservation in leading oil-producing states was dominated by principles of equity, and that political pressures from producing interests caused regulatory authorities to impose regulations that led to inefficient methods of exploitation and wasteful expenditures on drilling. They concluded that present policies encouraged the waste of labor and capital in the name of conservation.[51]

Paul Davidson suggested that market demand prorationing diminished the effect of the rule of capture, but the social costs of inefficient wells was tolerable only if the administrative costs of MER prorationing outweighed its benefits. Compulsory unitization would be superior to prorationing, though price would fluctuate with discoveries and technological break-throughs. Unitization might increase the oligopsonist strength of the large integrated refiners, which could require the divestiture of production. Davidson asserted that the elimination of the depletion allowance would cause no reduction in incentives to explore for oil, merely lowering the income of landlords. Industry spokesmen claimed that reducing the depletion allowance would cause the short-run supply curve to shift upwards, and that the complete cost would be passed along to consumers, but this conclusion assumed no changes in administration of regulations controlling production and imports.[52]

Davidson's article initiated a dialogue in the *AER*. Duncan Campbell challenged Davidson's contention that the depletion allowance did not encourage exploration, rejecting the assumption that the supply of new oil land is inelastic. Campbell argued that land varies in its probability of oil bearing, there is free entry in the exploration sector, and therefore land within the extensive margin of exploration will be rented and explored. The depletion allowance moves the extensive margin outward, increasing exploration as well as rents on intramarginal land. He concluded that oil supply has an elasticity considerably greater than zero. Henry Steele criticized Davidson for ignoring political realities in proposing unitization as a policy measure. He contended that unitization would be politically difficult to implement, as states would continue prorationing along with unitization to protect marginal wells, and that state conservation regulations were approaching economic rationality over time. Steele pointed out that public officials had an inherent bias in favor of small business. He rejected the argument that integrated firms had a stake in higher crude prices, since with import quotas only domestic sufficiency mattered and only four firms had a domestic ratio above 77 percent. Pass-through of prices was limited because while total demand might be inelastic, individual firms faced an elastic demand curve.[53]

Davidson replied to Campbell by stressing that the depletion allowance does not result in a significant increase in exploration while providing a subsidy to mineral rights owners. Large rents earned on intramarginal lands were evidence that supply of oil lands is highly inelastic. Given uncertainty, the depletion allowance raised both landowners' and producers' estimates of the ex ante rent, and there is no reason that it should increase the amount of the ex post rent received by producers. He suggested government support of geological research as a more effective subsidy, if one is required, for exploration. Davidson answered Steele by pointing out that a system of lump sum taxes and subsidies would be a more efficient method of redistributing income to small producers than prorationing. Davidson noted that there were 8 refiners with domestic self sufficiency ratios greater than 77 percent, that worldwide sufficiency ratios were relevant, and that in the short-run crude prices could increase despite long-run forces which would lower prices.[54]

Rene Manes in 1963, was the first economist to explicitly address the impact of import quotas on the domestic oil market. The US price of crude depends on the quantity of imports, and the allocation of quotas has no impact on price, but acts to redistribute profits among firms. He concluded that the value of the quota was independent of the method of allocation but depended only on the level of allowable imports. Even if distribution of quota rights lowered refiners' average costs, their marginal cost of obtaining oil would stay the same.[55]

Two articles published in 1964 focused attention on the economic costs of the oil control system and helped produce public pressure outside the profession for change. Alfred Kahn balanced the rent and subsidy aspects of depletion, noting that economic rents amounted to between 25 to 30 percent of gross revenues, and that lease bonuses supported the conclusion that graduations between qualities of potential oil land was fairly steep. Kahn expected that withdrawal of percentage depletion would have little impact on supply and would probably cause states to push for price increases to protect marginal wells. Excess capacity due to regulation resulted in adequate returns on investment paired with inflated returns on sales. While the depletion allowance tended to reduce price relative to cost, production control increased costs while maintaining prices. Kahn noted that excess capacity consisted entirely of the ability to extract oil from discovered reserves at a more rapid rate due to intensive developmental drilling of known reservoirs. He questioned whether the security value of this short-run excess capacity was worth the cost, given a $500 million annual cost of excess development wells, and the cost of denying access to foreign oil priced $1.25 to $1.30 below domestic production along the East Coast.[56]

The paper which had the greatest impact outside the profession was the article by Morris Adelman, first presented at the 1963 meeting of the American Economic Association, abstracted in the *AER*, and published in the *Southern Economic Journal*. The paper was basically a discussion of the world oil industry, but he devoted one section to a description of the waste inherent in the domestic system of control. He felt that prorationing had a systematic bias in favor of overdrilling and more expensive wells, but worst of all it encouraged production from uneconomic strippers. Adelman referred to the Petroleum Study Committee's $3.6 billion estimate of the cost to consumers of import controls, and provided his own estimate of just over $4 billion per annum. The cost of imports was calculated by using a value of an import ticket of $1.25 per barrel minus 9¢ for inland transport costs. The discounted savings from eliminating marginal and stripper wells was estimated at $3.75 billion, including the value of lost reserves, providing sufficient funds for a buyout of marginal producers. An end to overdrilling would bring annual savings of $1.3 billion in development costs and $800 million in operating costs, which he considered a conservative estimate, and felt the total savings were probably around $2.5 billion annually. The $1.5 billion difference between import costs and the cost of domestic waste was the value of foreign oil, which should be weighed against the security benefit of domestic production, as the national security required a substantial domestic oil-producing industry.[57]

A major development in the intellectual ferment concerning oil policy was the advent of the consulting firm and the economic model. Two reports which had a significant impact were produced by CONSAD Research Corporation for the Office of Tax Analysis, and Charles River Associates (CRA) for the Office of Science and Technology (subsequently submitted to the Cabinet Task Force on Oil Import Control). CONSAD gathered an impressive crew of academic economists, including Robert Lucas, Dale Jorgenson and Arthur Wright, while CRA academic consultants included Edward Erickson, Wallace Lovejoy and George Heitman.[58]

The CRA report used rudimentary econometric modeling data and techniques. The base estimates of domestic supply (from Fisher and Erickson) and demand elasticities with respect to price were 1.0 and -0.5 (estimates for demand elasticities ranged between -0.26 to -0.65). Given free trade the wellhead price would be $1.75 per barrel, domestic consumption would be 5.9 instead of 4.9 billion barrels per year, and domestic production would be 2.3 instead of 3.9 billion barrels per year. Under these conditions the annual cost to consumers of an import quota was $6.2 billion per year, minus an portion of the transfer to refiners passed on to consumers. Of this sum, $1.6 was due to inefficient resource allocation, transfers to producers accounted for $3.9 billion, transfers to refiners $0.5 billion, and $0.2 billion to Canadian producers. If prorationing was eliminated production would drop to 3.4 billion barrels a year under free trade and the social cost of the quota would increase to $3.9 billion under the base case elasticity estimates. The cost of stockpiling oil aboveground was estimated at 54¢ a year per barrel at 8 percent interest and 62¢ per barrel at 10 percent interest. A three year stockpile would require 5 billion barrels at an annual cost of $2.7 to $3.1 billion a year.[59]

The CONSAD report relied on a neoclassical model of production and a model of the individual producer, as an attempt to develop an industry simulation model met with failure. The researchers failed to find any significant relationship between rate of return and total expenditures for exploration and development using industry data. They then used a single equation CES function to estimate a reserve-reaction forecasting model, and the simulation model to test their results. They had a problem with econometric estimation due to poor data. They estimated that a 3.1 percent decline in reserves would occur if percentage depletion was eliminated, with a $1.2 billion increase in tax revenue, a 4 percent decline if expensing was eliminated, and a 7.1 percent reserve decline if both were eliminated.[60]

In the aftermath of the Cabinet Committee's report additional attention was paid to oil import controls. George Hay showed that a tariff shifted the total value of the quota to the government, with refiners losing the value of their tickets. A follow-up to Hay by Amacher, Tollison and Willett examined the effects of tariffs and quotas on resource allocation. Modeling the domestic industry as a monopolistic producer, they showed that quotas reduced but didn't eliminate the market power of the domestic monopoly by shifting the monopolist's demand curve to the left, while a tariff would eliminate its market power. Richard Mancke calculated the social costs of import quotas between $1.38 and $1.77 billion for 1969. Charles Cicchetti and William Gillen, on the eve of the embargo, estimated the social cost, consisting of the decrease in costs of consumption and increase in consumer surplus, at $3.5 billion in 1970 and the subsidy to the industry at $7.4 billion per year.[61]

Kaj Areskoug assumed that demand for crude was perfectly inelastic, income elasticity was 0.8, and supply elasticities gradually increased over time. The estimated social cost of the import program was estimated at $1.5 billion, excluding transfers. He estimated losses due to an import cutoff, and calculated that the expected net benefit of the quota program given a one year cutoff of all imports with a probability of 10 percent would be $0.7 billion annually. George Heitman criticized Areskoug for assuming elasticity of demand is zero, the dubious value of his supply elasticity estimates, and the assumption that domestic production would fall by over 70 percent in the absence of import controls as price fell to 2 dollars.[62]

THE WORLD OIL MARKET

Adelman presented his vision of the operation of the world industry in three papers in 1964, the *SEJ* paper, one in the *Journal of Business* (a revised translation of an article originally published in a French petroleum journal), and an essay in a book of papers on natural resources. Adelman bluntly stated that petroleum was not a natural monopoly since incremental cost exceeds average cost. He also rejected the proposition that short-run operating cost was the competitive floor to price, since the relevant variable cost is development plus operating cost as most contracts are for at least a year. Therefore the only relevant fixed costs were finding costs, which in most regions of the world are a minor fraction of variable costs. Given the large available reserves in the Middle East the present value of an undeveloped barrel was zero, especially under the concession system where companies had a deadline on ownership rights. He suggested that the World Oil Cartel had been inoperative since WWII, and that producing countries would have problems colluding because each had different rates of discount, and therefore different ideas of the optimal market price. Thus he expected prices to gravitate toward costs of production, a trend which he felt would be acerbated by the opening of the American market, since this would provide an opportunity for sellers to increase net revenues. After 1975, the rate of discovery might fall enough that replacement cost will become positive, development costs rise, and supply price increase.[63]

A number of works dealing with the world oil market appeared in the 1960s, including books by Charles Issawi and Mohammed Yeganeh, Helmut Frank, Edith Penrose, Paul Bradley and J.E. Hartshorn. Issawi and Yeganeh presented some sketchy but novel data concerning relative costs of discovery, production and transportation, as well as details on the financial arrangements between the oil companies and the Middle East countries. Frank presented an historical description of price formation, including an account of the erosion of control in the world market which had been occurring since the mid-1950s. Bradley provided the first rigorous and systematic attempt to determine costs of production for different regions. Penrose focused on the activities of the major oil companies and their relations with both consuming and producing countries.[64]

The definitive work on the world oil market was published by Adelman in 1972, and ironically his work would be immediately outdated by the events of the following year. On the other hand many of his observations were validated by the longer view available from the vantage point of the 1990s:

> This, aside, Sheik Yamani's warning is correct and of utmost permanent importance. If the oil companies become buyers of crude oil (even if their producing subsidiaries remain in place as contractors) and the producing nations sell, there is no longer a means to detect cheating. The known agreed-

upon price floor has disappeared, and far below it there is only the ground of cost. . . .

The higher crude oil and product prices have no connection with world supply and demand for crude oil. They reflect no scarcity of crude oil present or foreseen. . . .

Supply and demand are as irrelevant to the future price as to the past. A price set by supply equaling demand means a price in the neighborhood of incremental cost, which in crude oil is chiefly the return on the new investment. . . .

The important new fact is the active cooperation of the producing nations. Nobody foresaw the extent to which they would act together and threaten to withhold supply if their terms were not met. They could do this because the industry was integrated. The profits of higher product prices went, not to independent transporters and refiners-marketers, but to the producing companies whose operations and properties were within their reach.

Nobody knows, even approximately, how high the OPEC nations can raise taxes, and thereby prices, before sales drop off so rapidly as to reduce their total take. But clearly taxes can be increased several times without reducing sales, because oil consumers have few alternatives. . . .

This is an odd, unstable condition. The OPEC nations still have great unexploited power because price can be greatly increased without provoking a loss in sales and in total revenue. Therefore they are likely to try again soon. Yet, with no prorationing system in sight they cannot control the level of output and allocate markets. Their market power can only be exercised through brinkmanship.

It is even possible that this schooling in real security problems will drive out the obsession with 'access' to petroleum. . . . Rarely has a word been so compact of error and confusion. Nobody has ever been denied access to oil: anyone willing to pay the current price could have more than he wanted.[65]

OIL POLICY AND NATIONAL SECURITY

The dialogue on oil and national security was originally played out by noneconomists, primarily in *Foreign Affairs*. As early as 1951 the need for an oil stockpile was broached by Halford Hoskins: "In another war requirements would mount far higher. To provide a fair margin of safety against the possibility that supplies form the Middle East would be cut off, and the possibility that those from the Caribbean might be temporarily interrupted, an oil stockpile of the order of a billion barrels of crude would not be over-large. . . ." He considered above and below ground storage but rejected them because of cost or technical difficulties, and suggested development of the naval reserves along with other shut-in reserves as a stockpile.[66]

The Paley Commission stated that the security problem in oil was a matter of total free world demand and supply. Severe rationing of nonmilitary use would be insufficient and therefore an emergency cushion of oil production, refining, and transportation capacity in the Western Hemisphere was imperative. The wise course would be to maintain a quickly expandable output from an established reserve of oil, and therefore the Commission recommended that the Federal Government encourage immediate exploration for oil on publicly owned off-shore lands, that it should require conservative well spacing and withdrawal rates designed to prolong the life of pools, and conservation agencies of oil-producing States likewise give heavy weight to this security objective in their regulation of well spacing and production rates. It also called for encouragement of expansion of crude reserves elsewhere in the Western Hemisphere and maintenance of standby refining and transportation capacity.[67]

Herbert Feis recommended in 1954 that the nation should encourage conservation, stimulate discovery of new domestic supplies, including depletion allowances and expensing, and maintenance of excess capacity as a margin of reserve. He supported the system of oil regulation, including import restrictions if the volume of imports was large enough to cause a lag in domestic production and development. He suggested that the government should acquire proven domestic reserves for emergency use. James Schlesinger on the other hand criticized the 1957 Cabinet Committee recommendation of voluntary import quotas. He saw the policy of rapidly draining domestic reserves as irrational given the nation's relatively small share of world oil resources, and felt that in the event of war that a substantial proportion of oil consumption could be diverted. He suggested development of an "oil bank," a shut-in reserve, along with subsidization for stripper wells to protect them from foreign competition.[68]

An issue of the Quarterly of the Colorado School of Mines in 1969 was devoted to the problems of oil policy and national security. McDonald proposed that the federal government purchase partially developed petroleum reservoirs and set them aside as a reserve. He estimated the cost of buying capacity equal to 25 percent of current demand (about 6 million barrels) at $7.5 billion and an implicit interest cost of $400 million annually along with maintenance costs. However, since the government would be purchasing excess capacity the social cost would be close to zero or even positive, and elimination of excess capacity would reduce political pressure opposed to rationalization of the industry. Lovejoy defined the emergency to be meet as an interruption of supply for six months to one year, resulting in a deficit of 4.4 million bpd for the Western Alliance after rationing, substitution, and use of idle capacity. Stock reductions could then meet this deficit for about 166 days. Given Adelman's estimate of stockpiling costs for Western Europe of about 35 cents per barrel in 1967, and allowing for 40-45 cents for the US, a one-year supply would cost $1.9 to $2.2 billion annually, while with a 56 cents per barrel figure, suggested to the author, the cost would be $2.7 billion.[69]

A more formal approach toward stockpiling was provided by Walter Mead and Philip Sorensen. They pointed out that in its determination of a threat to national security the ODM neglected, both in 1957 and 1959, to estimate the costs of oil import restrictions or alternative means of satisfying the national security requirement. Following the example of McDonald they choose an partially developed reservoir, the Elk Hills Reserve, and determined that the annual social costs would be $146 million per year for emergency production of 453 million barrels and total reserves of 1.36 billion barrels, at a cost for one year of capacity of 32 cents per barrel. They compared this to costs provided to the Task Force on Oil Import Control from Interior which gave a cost of 39¢ per barrel for salt dome storage, 98¢ for steel tank storage, or $1.04 per barrel for incentives under the existing system.[70]

Walter Levy gave the oil company view of the impending supply problem in 1971, calling for restraint on import volumes, adequate incentives to explore and develop new energy sources, and development of stockpiles.[71]

Adelman blamed the American State Department for the rise in prices since 1970, alleging that its tacit approval of OPEC actions encouraged the price increases. He rejected the explanation that increased demand was responsible since the increase in consumption in 1970 and 1971 was below the average of the previous decade. The existence of overcapacity in the world market and the tremendous low cost Middle East reserves should have pushed prices lower if there had been no collusion in the market. The key element in maintaining the system was the multinational oil companies, acting as tax collectors for the OPEC countries. He recommended stockpiling and detailed plans for rationing supplemented by higher excise taxes, and elimination of the oil company role in crude marketing. Adelman pointed out that by inaction the Western countries had allowed OPEC nations the opportunity to build up financial reserves that would permit them to withhold oil for extended periods.[72]

James Akins warned of the threat of an Arab embargo in 1973, pointing out that imports from the Middle East could reach 8 million bpd by 1980. The elimination of US spare capacity made the world vulnerable to the loss of production from one or two countries. He also pointed out that Middle East producers were receiving a quantity of funds which would make it hard to find adequate investment opportunities, tempting them to constrain production, and eliminating incentives to cheat. Political perceptions of the value of future income would be as important as economic calculations using discount rates for these countries. He called for cooperation between consuming countries (ignoring that this is as difficult to arrange as prevention of cartel cheating) and gave lip service to stockpiling.[73]

SUPPLY AND DEMAND ESTIMATES

The analyses of economists were based on theory but dependent on the information available concerning costs, demand and supply. Since the source of much of the required data were oil companies who either did not gather, or did not wish to release, the necessary information, economists were constrained in their ability to determine key variables describing demand and supply relations. Also, since many of the economists were not primarily petroleum specialists (industrial organization and microeconomics were the areas of expertise of the majority of contributors) they depended on data and projections available in the literature to make their judgments. Size of resources, costs, projections of supply and demand, and estimates of supply and demand elasticities were the parameters which influenced the judgments of policy analysts.

One problem with making policy decisions was that the technical experts (geological, not economic) were so diverse in their opinions of the size of available reserves and future production. Netschert concluded from his examination of the estimates of future production and reserves that there was no resource limitation on continued growth in capacity through 1975, and technology would continue to expand the ability to increase capacity. He thought that the total domestic availability of crude oil in 1975, at the same constant dollar costs, would be 6 billion barrels per year (11 million bpd primary production and 5.4 million bpd secondary production).[74]

Lovejoy, Homan and Galvin reviewed the literature on oil supply and cost a few years after Netschert. They unfavorably compare the 6 billion barrel oil and 1 billion barrel natural gas liquids projection of Schurr and Netschert, expected in 1975 at present costs, with the lower estimates of future availability and predictions of rising real costs prevalent in the industry. Another projection mentioned is the study by Milton F. Searl for the US Atomic Energy Commission, Fossil Fuels in the Future. Searl predicted that the real cost of producing fluid fuels in the year 2000 should be less than 50 percent above 1958 costs. Low and medium costs reserves were expected to be consumed by 1980 without imports, while high-cost reserves would last an additional decade, but medium price production from oil shale could make up the difference between supply and demand. The Department of the Interior estimated the future availability of oil and gas in 1960 to be around 3.3 billion bbls equivalent in 1975 and 3 billion in 2000 if prices are at $2.25 (in 1959$, or $5.94 in 1980$) up to 5.3 billion in 1975 and 4.5 billion in 1990 if prices rise to $4.50 ($11.88 in 1980$).[75]

The first econometric study of petroleum supply was performed by Franklin Fisher in the early 1960s, in which he concluded that price affected supply by both encouraging additional exploration and shifting exploration from less risky to more risky prospects. He found that the elasticity of supply of new reserves was about 0.31, the sum of offsetting factors; a 1 percent increase in the real price of oil increased drilling by 2.85 percent but decreased the success ratio by 0.36 percent and the average size of discoveries by 2.18 percent. Fisher explained this trend by assuming that a rise in the success ratio for wildcats tended to be associated with a depletion of the inventory of relatively small, relatively certain prospects. Fisher's estimation technique was a bit crude (by present day standards), using a log linear form for his equations and estimating each equation separately. Edward Erickson used the same basic model, corrected for errors in the price series and with slight changes in the model's specification, and obtained a supply elasticity of 0.9 using data through 1959.[76]

Fisher and a follow-up study by Erickson were very influential as the basis of calculations for the CRA report, but they were later criticized as techniques improved. Mancke rejected the sum of the component costs method of estimating the long-run supply curve, substituting a residual supply curve method. Mancke subtracted expected rents from the expected price to determine the industry's ex ante long-run marginal cost curve. He calculated that an unknown quantity of "lower 48 crude" paid no rent hence $2.51 must be the expected long-run real cost of this oil in 1969; for offshore oil, 16 percent of production had expected costs of $1.42 a barrel; and expected real costs of other crude was between these two costs.[77]

Dennis Epple stated: "The most troublesome aspect of Fisher's work is that his econometric specification is not derived from an economic model. . . . No attempt was made to anticipate the signs of the coefficients prior to estimation. Instead, the results were rationalized after the equations were estimated as Fisher explicitly acknowledges." He also pointed out that Fisher's formulation of the problem made it impossible to determine short and long-run elasticities, while the petroleum districts used in the study exhibit extreme variance in terms of size, shape of supply curves, and therefore should have been weighted by production in each region. Fisher also failed to include a variable reflecting the shifting of supply curves as deposits are depleted. Epple found that the supply curve was shifting to the left about 6.7 percent per year due to the effects of exhaustion, and to maintain discoveries at a constant level price would have to increase by 2.3 percent per year if costs did not change.[78]

Jeffrey Kolb also criticized the Fisher-Erickson model, suggesting that the estimated negative price elasticity for average size of oil finds arose not from a deterioration in prospects accepted for drilling, but from an inappropriate calculation of the average size oil find, and the misuse of the technique of pooling cross-sections of time series data. By excluding the East Coast district the estimated price elasticity of discovery of oil reserves became 2.4 (2.65 in other calculations) implying that oil discoveries were price elastic, rather than inelastic as the conventional wisdom held. Kolb's regression results implied that increases in wellhead oil prices raised the average size of oil find without substantially affecting the success ratio. Although the portfolio of prospects held by firms may have more small than large prospects, large relatively uncertain prospects were accepted for drilling only at higher prices.[79]

A number of studies of supply and demand were reported in the pages of *Oil and Gas Journal*, the bible of the industry, and unlike many self-serving analyses produced by the industry, these were free of intentional bias insofar as this observer can determine. A 1961 study by Texas Eastern projected demand for oil at 12.27 million bpd in 1970, 13.69 million bpd (5 billion per year) in 1975 and 15.26 million bpd in 1980, in the same range as the Schurr et al. estimates for Resources For the Future (RFF). In 1963 BP saw the North American demand (US and Canada) as 14 million bpd in 1970 and 17 million bpd in 1980. The 1966 forecast by Edward Symonds, petroleum economist for the first National City Bank of New York saw 1969 demand in the US at 12.66 million bpd. In 1967 Citibank forecast US demand in 1970 at 13.16 million bpd and production at 10.68 million bpd. By 1970 Chase Manhattan Bank saw 1980 consumption of oil reaching 23.5 million bpd if there were no supply limitations.[80]

The unexpected jump in consumption is more clearly illustrated by examining projections made in the 1950s and early 1960s for 1975.

TABLE 4.1 Projections of Petroleum Demand for 1975
(millions of barrels per year)

Product	*Paley*	*Schurr*	*Mobil**	*Actual*	
	(1952)	*(1960)*	*(1963)*	*(1975)*	*(1975e)*
Gasoline	2085	2228	2405	2450	2702
Distillates	1180	1124	1354	1041	1295
Residual	1110	1004	836	898	1245
Other	625	798	1096	1568	1884
Total	5000	5154	5691	5957	7126
(million bpd)	13.7	14.1	15.6	16.3	19.5

"1975e" is an extrapolation of consumption from the early 1970s to 1975 given no embargo.
*North America
SOURCES: *O&GJ, Resources For Freedom, Energy in the American Economy, API Data Book.*

Up until the late 1960s demand increased no more than would be expected if one used historical forecasts as a guide. The sudden explosion in demand in the early 1970s took everyone by surprise, as illustrated by the column in Table 4.1 for the no embargo (projected consumption in 1975 had the drop in consumption initiated by the embargo not occurred) 1975 consumption figures. Ironically, the effect of the drastic jump in prices, combined with the impact of the resultant recession, brought consumption in the range of the long-run forecasts for that year. Given the long lag time involved with major investment decisions in the industry, this suggests that the major companies had acted rationally given the available information. The problem was that consumers unexpectedly increased their demand for petroleum at existing prices, due to unpredictable factors, such as environmental regulations, continued price control of natural gas and demand arising from other uses such as aviation fuels and petrochemical feedstocks.[81]

MODELING POST EMBARGO OIL MARKET BEHAVIOR

The aftermath of the oil embargo and price hikes stimulated a deluge of articles explaining the operation of the world oil market. Michael Kennedy published an article in 1974 in which he presented a regional multi-commodity linear programming model of the oil market. The OPEC core supplies residual demand, the fringe supplies a constant quantity and other producers use the current price as the long-run price. He concluded that the price hikes of 1973 were not likely to persist, and that the most likely long-run price was between $3.50 and $5 a barrel.[82]

Morris Adelman was the chief proponent of the school of thought which claimed cartel status for OPEC. In a 1976 article he argued that scarcity could not be a reason for the increase in oil prices, using a simple calculation. Assume the average production costs of Persian Gulf oil rise to $2.50 (1975$) in the year 2000, assume $16 a barrel shale oil, therefore crude oil will provide $13.50 in rents. Discounted at 10%, the after tax return on equity from 1957-1965, the 1975 price per barrel should be $1.50, $1.25 for rent and $0.25 for production cost (Note however if a discount rate of 2% is used the 1975 price would be about $8.50). Adelman claimed that using the multinational oil companies to clear the market allowed the cartel to limit output and allocate supplies without resorting to collusion. He suggested that this process was aided by the existence of backward bending supply curves for some OPEC members.[83]

A number of papers combined oligopoly theory with intertemporal optimization of profits from production of an exhaustible resource to explain oil market behavior. Jacques Cremer and Martin Weitzman presented a dynamic model of OPEC viewed as a monopolist sharing the oil market with a competitive sector, and assumed perfect-certainty rational-expectations. The cartel attempts to maximize profits given the reaction function of the competitive fringe, with the latter facing a cost function that increases with cumulative production and a production constraint. They concluded that real oil prices would not change much over the next twenty years. Stephen Salant modeled the world oil industry as a collection of independent firms, participants in a Cournot noncooperative game, each maximizing its own discounted profits while taking as given the sales paths of the other firms. The cartel is in effect a firm owning larger reserves than other participants. He concluded that formation of the cartel raises profits of the competitive fringe by more than those of the cartel; after the initial jump in prices the cartel will raise the real price of oil monotonically over time; the cartel restricts its sales to take over the market after the fringe has exhausted its reserves. Other authors who used this approach were Gilbert, Lewis and Schmalensee, Ulph and Folie, and D.M.G. Newberry.[84] These models share a high level of abstraction and dependency on assumptions such as full information, rational expectations, and intertemporal optimization of profits with regard to the behavior of OPEC and the competitive fringe. The general consensus was that oil prices would continue to increase.

Esteban Hnyilicza and Robert Pindyck modeled OPEC as a cartel with two blocks; a spender group with a high discount rate and a saver group with little immediate need for cash. They showed that with fixed output shares the price path approximates the monopoly price path, but if shares are subject to control, price paths depend on the relative bargaining power of the blocks. The Pindyck and Hnyilicza model was extended in a paper by Pindyck two years later. Unlike almost every other commentator on OPEC behavior and cartel behavior in general, Pindyck neither saw monopolistic control causing prices to rise monotonically over time or the sudden collapse of the cartel due to fringe production. He made the observation that adjustment lags and large potential monopoly profits determined behavior. The large monopoly profits immediately available to the cartel encourages formation, and the lag in the adjustment in demand suggests an initial high price, which then decays as customers and fringe producers adjust to the new price. This holds especially when the competitive price is low, reflecting the absence of scarcity, and therefore higher monopoly profits relative to the competitive solution.[85]

Ezzati in 1976 modeled OPEC under the assumption that the member countries were constrained in their pursuit of profit maximization because of

TABLE 4.2 PROJECTED VS. ACTUAL OPEC BEHAVIOR
(output in billion barrels/year)

Year	*Projected Price*	*Projected Output*	*Actual Price*	*OPEC Production*
1975	20.34	9.94	21.33	9.92
1976	17.19	9.23	19.51	11.25
1977	15.76	8.94	19.76	11.41
1978	15.21	8.87	18.40	11.88
1979	15.09	8.91	24.60	11.29
1980	15.18	9.00	33.89	9.84
1985	16.65	9.67	27.03	5.85

"Actual price" is weighted cost of imports, subtract about $2 to determine world price.
SOURCE: Hnyilicza and Pindyck, "Pricing Policies," p. 143, API, *Basic Petroleum Data Book*, Jan. 1995

economic limits and political objections. Domestic investment is limited by absorptive capacity while foreign investment is restrained by fears of devaluation, mismanagement and possible expropriation. He rejected the thesis that the cartel acts as a monopolistic producer since it assumes the cartel is a cohesive unit. Instead Ezzati developed an intertemporal linear programming model which determined the minimum production requirements of OPEC countries. If demand for OPEC oil is greater than the production required for export by OPEC countries the cartel is stable. Using relatively low values of world demand and non-OPEC supply elasticities (-0.2, 0.25) he concluded that the cartel would be stable to 1980 if the OPEC price was maintained around $22 (1980$) and that the cartel would be in the position to ask for higher prices.[86]

The behavior of OPEC countries during the 1970s caused some economists to question optimal cartel behavior theories, though attributing some market influence to the organization. William Nordhaus decided that the production targets of many of the financially unconstrained OPEC countries as well as some fringe producers are determined by noneconomic considerations. He concluded that production targets (and capacity decisions) of individual countries are set noncooperatively, and therefore the path of output may deviate markedly from either a monopolistic or competitive production path. Nordhaus saw the role of the cartel as slowing the adjustment of oil prices in an asymmetrical fashion. In periods of excess demand the official price spiraled upwards chasing the spot market price; in periods of slack demand when spot prices fell below official prices OPEC restrained price cutting.[87]

Theodore Moran saw Saudi Arabia acting as a price leader in restraining prices due to security considerations. He questioned the ability of OPEC to follow an optimal economic strategy even if this was desired, since there had been a great degree of uncertainty with regard to both supply and demand forecasts. Hossein Razavi concluded that uncertainty leads to the choice of a more conservative strategy with regard to both foreign assets and production rates. David Teece also rejected the concept of a wealth-maximizing cartel, since the lack of increased supply by OPEC in the face of large price hikes (expected in a Pindyck world) and policies supporting inefficient domestic investments support political considerations as a more important motivation than income maximization.[88]

Philip Verleger built on the Nordhaus observation of the importance of the spot market in price formation. He tested the hypothesis that the price of oil was determined by the weighted value of the spot prices of petroleum products transmitted through a geometric lag process. He found that the relation between spot prices and official prices was significant for the period 1975-1980, and that producers of Arab heavy crude responded more quickly to changes in the market, while producers of Arab light were the slowest to respond. Verleger rejected the contention that price moderates in OPEC were moderates; they merely adjusted prices at a slower rate.[89]

There was also a literature that questioned the whole concept of OPEC as a cartel, suggesting that the oil price increase was due to other factors instead of collusion. There was no evidence that OPEC attempted to set production quotas until March 1982, and since that time the group has been unable to prevent the decline in the price of oil. Most analysts ignored the dramatic shift in expectations of ultimate oil reserves which occurred in the early 1970s, during which oil companies changed their expectations from an implied level of 4 trillion to the 2 trillion barrel figure supplied by Nehring. Johanny contended that production cutbacks (and price increases) by OPEC countries were due to the de-facto nationalization which accompanied the 1973 embargo, which resulted in a substantial decline in discount rates for producers of oil. He felt the 1979-1980 price shock was the result of price expectations and stockpiling driving up spot prices, not OPEC production decisions. Walter Mead's study of OPEC behavior during the seventies tended to collaborate Johanny's assertions, as he found that the countries usually considered to be residual suppliers in oligopoly models expanded their share of OPEC output between 1974 and 1977, a period when they would be expected to accept production cutbacks to support prices if OPEC was a cartel.[90]

Adelman challenged the lower discount rate hypothesis by pointing out that the OPEC countries had desired higher production rates than the oil companies during the period from 1950-1970. The embargo was obviously a collusive act, and not the gradual tightening of production schedules one would expect from a change in discount rates. He rejects the absorptive capacity argument since the countries managed to increase spending in proportion with increases in income, and credits the Saudis with the 1980 price hike, due to their production cutback in January. Adelman contended that OPEC strategy consisted of an ad-hoc policy to err on the side of higher prices, due to lags in information needed to set policy. Since everyone in the market is aware of this propensity, actors in the oil market are slow to sell when a glut appears, but quick to build up stocks when shortages are apparent. The fear of a shortage can then trigger a shortage and a jump in the spot price.[91]

POST 1980 PRICE SHOCK LITERATURE

In the early 1980s the conventional wisdom saw oil prices rising steadily for the rest of the century. The Energy Modeling Forum at Stanford had conducted a controlled comparison for ten participating models in 1981, using the following assumptions for the Reference Case scenario: long-run demand elasticities of 1.0 for income, and -0.6 for price, with world GNP growth at 3 percent a year. Short-run demand adjustment was specified to be consistent with a Koyck-lag structure with a coefficient for lagged demand of 0.9. The amount of oil reserves remaining at $18 (all prices in this section are given as

1990$) was estimated at 1,075 billion barrels, with $36 bringing forth another 15 percent, and $72 an additional 6 percent, while unconventional oil became available in 1995 at $109 a barrel. Of these reserves two-thirds are in OPEC. Seven of the ten models in the EMF study assumed OPEC was an imperfectly disciplined cartel which increased price when the market was tight and eased off when the market was sluggish. The median model results gave an oil price in 2000 double the 1980 level in real terms, and the range of predictions were for a price increase of 50-150 percent by the end of the decade.[92]

The International Energy Workshop, an informally organized group, polled its membership with regard to their expectations for international oil prices in 1983. The median oil-price projection for the year 2000 declined from 75 percent above 1980 in 1981 to a 48 percent increase predicted in 1983. The median 1990 price level was expected to be 10 percent above the 1980 in the latter poll. Most of the participants projected a 1990 price level between 25 percent less and 50 percent more than 1980, and the end of the century price level projection ranged between the 1980 price and an 100 percent increase.[93]

In 1983 George Daly, James Griffin and Henry Steele used a similar approach to Ezzati (1976) in determining OPEC stability. They developed production response functions for the OPEC members based on reserves, absorptive capacity and political constraints. Demand was modeled as a function of income and a distributed lag on prices while non-OPEC supply is constrained by reserves, which are a function of a distributed lag on prices. OPEC was divided into three groups, output maximizers who act like fringe producers; price maximizers who possess a backward bending supply curve; and the cartel core which acts as the residual supplier. Threats to cartel stability were indicated by divergence from wealth-maximizing reserves to production ratios. They concluded that the 1980 price of $50 was not sustainable given a price elasticity of demand of -0.73. With a much lower elasticity of demand, instability would still occur if world economic growth rates fell 1 percent below historical trends, but the pre-revolution price of $24 was compatible with cartel stability under those conditions.[94]

An explanation for the emerging weakness in the market for oil was provided by Brian Levy in 1982. He found that by 1980 National Oil Companies had increasingly replaced international oil companies on the producing end of the industry, and downstream the share of traded oil that flowed through the multinational oil companies was down to 50 percent, from 90 percent in 1973. The rise in direct marketing brought to the forefront the widely divergent interests of the producing countries in both pricing policies and political goals.[95]

Arlon Tussing suggested that oil should be seen as a commodity, and like other commodities demand will expand unevenly over time concurrent with erratic shifts in supply; therefore normal behavior for commodity prices is to swing over wide ranges. Since durable goods have service lives on the order of twenty years it may require over a decade to complete adjustments to major price changes, and the present buyer's market should continue for a number of years. Verleger agreed with the commodity premise, crediting the withdraw of the multinationals from the distribution of OPEC crude, and the increase in the volatility of supply and demand for this change. He saw the creation of a futures market as an additional factor leading to instability in the market, as it provided an alternative source of supply to independent refiners and lowered the cost of speculation.[96]

Fred Singer observed that the combination of conservation due to high prices, and the shift in refinery output from fuel oil to higher distillates due to substitution of other energy sources for fuel oil, had contributed to a serious weakening of demand for crude. He expected a continued decline in demand for OPEC oil as OECD demand fell and non-OPEC production increased. He predicted that the price of oil could fall as low as $12 for a short period, but that the $24 range was most likely while the market stabilized during the 1990s. Adelman saw the future of oil prices in 1985 as uncertain, with both the possibility of a substantial price increase orchestrated by OPEC, or a small chance of a real price collapse.[97]

Gately reviewed the current opinions on the oil glut and the future of oil prices in an article written just before the price collapse of 1986. He found two basic views of the current situation; the oil glut as a temporary phenomenon or the permanent end of OPEC control. The temporary excess explanation saw the glut as the result of the decline in demand due to the price hike of 1980, and while prices would continue to decline for a few years, eventually consumption would increase and fringe production drop until demand for OPEC oil was high enough to prevent cheating on quotas and price cutting. At this point (around 1990 when OPEC demand reaches 24 million bpd) OPEC would have to choose to keep increasing production and hold prices steady or maintain output and allow prices to rise. The alternative view saw the decline in oil consumption as irreversible due to improvements in conservation and energy efficiency embodied in new capital stocks, while non-OPEC production would continue in the face of lower prices. He reported that the July 1985 poll of the International Energy Workshop gave a median projection of oil prices in the year 2000 at only 10 percent above the 1980 level, reflecting the adjustment in expectations that had occurred in two years.[98]

With the price collapse of early 1986 economists turned to determining the floor price of world oil and whether prices would stay at that level. James Griffin and Clifton Jones attempted to model the abandonment decision for Texas oil makers as an optimization problem with inclusion of the cost of irreversibility, the loss of expected future revenues. They also checked their results by estimating the cost function for Texas producers. The authors concluded that it would require a price of about $5 a barrel, for a length of time sufficient to lower producer's price expectations, to cause a significant decline in production (14-21 percent).[99] Adelman estimated that the floor price was about $8 in the short run, below which there would be large cutbacks in US production, and $5 in the long run, sufficient to maintain investment in new reserves in non-OPEC nations. He estimated that the ceiling for OPEC as a monopolist was in the range of $19-26.[100]

ENERGY MODELS

The most important problem faced by economists trying to model and explain oil market behavior during the last two decades has been the uncertainty concerning the true shape of the supply and demand curves for crude oil. Economists basically assume that microeconomic theory and econometric techniques allow determination of these parameters, but in practice, at least with energy in general and oil in particular, it has been difficult to obtain stable estimates of these relationships which possess predictive validity. The range of estimates for US gasoline consumption, between -0.1 and -0.4 for short-run elasticities, and -0.4 and -1.4 for long-run elasticities, illustrates the lack of consensus.[101]

Given the diversity of estimates of gasoline demand, which is a far simpler estimation problem than that of crude petroleum demand, or aggregate energy demand, it is easy to understand the lack of consensus in the profession. The reasons for conflicting results on price and income elasticities include the choice of sectoral breakdown, type of dynamic specification and functional forms used, the data interval used for estimation, and the time period over which the parameters are estimated.

The advent of energy modeling added an additional element of confusion into the energy policy debate. There were four basic types of energy demand models, engineering models which determine the technical requirements for energy, financial models that compare the costs of using different technologies to supply the same goods and services, microeconomic models motivated by the behavior of the consumer, and macroeconomic models which attempt to calculate the changes in demand at the aggregate level. Given the choice of model (or models, as different approaches were often combined) one then faces the problems of estimating the parameters. It is not surprising to notice modelers adjusting estimates rapidly as events cast doubt upon their predictions. Table 4.3 gives a selection of energy demand forecasts made by two oil companies and the government over the decade of the 1970s.[102]

The plenitude of energy demand studies during the 1970s have been reviewed elsewhere, particularly in Martin Greenberger's book on energy modeling.[103] Suffice it to say that most of this work dealt with energy policy as a whole, and oil as merely one sector, and rarely as the major important focus of the studies. These studies were often a strange amalgamation of economics, engineering and ad-hoc estimates, influenced by political pressures and ideological concerns.

TABLE 4.3 Forecasts of Energy Consumption (Quads per Year)

Year of Forecast	Source	1980	1985	1990	2000
1975	Interior	86-91	99-109	118-129	137-165
1976	SHELL	87	101	116	---
1976	Interior	82-87	99-104	---	163
1977	EXXON	86	98	114	---
1978	DOE/EIA	78-83	86-91	97-105	122
1978	EXXON	85	94	107	---
1979	DOE/EIA	78	81-83	88-93	104-113
1980	DOE/EIA	76.3	78-80	86-89	97-107
1980	SHELL	---	84	88	---
1980	EXXON	---	83	85	96

Source: *Energy Demand Forecasting*, p. 95.

ECONOMISTS AND POLICY

In an article in 1974, Paul Davidson, Laurence Falk and Hoesung Lee attacked the orthodox economic premise that the market would lead to welfare maximization in the case of oil. They base their dissent on the nonexistence of forward markets, the certainty of false trading, and the impossibility of specifying any time rate of exploitation of resources that will be efficient. The authors therefore called for policies to prevent redistribution of income from consumers to producers, and eventually to property owners as economic rents. They called for consuming nations to reduce the degree of monopoly in the energy products market by vigorous antitrust policy and to attempt to break up OPEC by encouraging domestic production of energy. The authors felt that a price of about $9 ($25 in 1990$) would allow for self-sufficiency in 1980, assuming price elasticity of supply of 1.6 and demand of -0.5, and a 1980 production of 16.9 million bpd.[104]

Bohi and Russell discussed a number of alternative scenarios and policies in their 1975 book on US Energy Policy. They estimated that if the domestic price of oil persisted at $14.60/bbl, ($37 in 1990) a midpoint estimate of 1985 oil imports would be about 8 million bpd, using the NPC projections. Complete self-sufficiency was attainable at domestic prices between $18 and $24 ($45-60 in 1990$). Given that foreign prices fell to $14 the cost to US consumers of self-sufficiency would reach $63 billion annually. The annual cost of standby capacity to cover 180 days of imports at that level would be $5.4 to $7 billion (based on $18 to $24/bbl oil) per year. Reducing energy consumption by 1.8 million bpd was estimated to require a tax of $2.40/bbl, increasing consumer expenditures by $17.5 billion, while a subsidy to produce that amount would cost about $10.8 billion. They defined energy security as protection against another Arab embargo, and recommended storage and shut-in capacity to cover that contingency. It would be less expense to establish interdependence with producers than independence, requiring protection against threats of embargo without going as far as self-sufficiency.[105]

James Cox and Arthur Wright proposed using a tariff to obtain independence from the threat of oil embargoes. They differentiated between effective independence, the ability to avoid or mitigate the consequences of foreign actions, and self-sufficiency. They pointed out that self-sufficiency in the present guaranteed greater dependency in the future. The authors' scheme was a rebatable tariff, where oil importers were offered rebates of the import duty provided they hold excess domestic oil capacity or oil in storage as an emergency reserve. Edward Renshaw suggested using an increase in the tax on gasoline as an alternative method to tariffs to reduce the monopoly profits of OPEC. He noted that this would both reduce OPEC profits and balance the misallocation of domestic resources due to distortions in the pricing of transport services. The regressive nature of the tax could be balanced by reductions in other regressive taxes. Theodore Bergstrom came to a similar conclusion six years later, estimating the optimal gasoline tax rate tax between 25¢ and 50¢ per gallon.[106]

A number of writers tried to estimate the true cost of imported oil, contending that the market price did not include the full social cost. One of the externalities associated with oil imports was the increased risk of an oil supply disruption. Disruption costs included both the direct effect of supply reductions and the macroeconomic impact of the disruption on aggregate demand. A second cost was the marginal cost of higher imports, which included the increase in cost of all imports due to a higher price resulting from increased demand. Estimates of the social costs of imports varied, with Stobaugh and Yergin giving a range from $16-61 per barrel, Nordhaus $13-46, Hogan $2-40, and the Energy Modeling Forum models, $8-18. The policy recommendation made by most analysts was to use a tariff to pay for strategic reserves and encourage conservation and domestic energy investment.[107]

Gilbert and Mork argued that any intervention in the market in the case of another disruption could lead to negative macroeconomic effects. Murphy, Toman and Weiss agreed that a disruption tariff (imposed to collect some of the rents accruing to OPEC) would hold little benefit, since the macroeconomic disruption costs would balance the wealth transfer benefit. On the other hand there may be benefits to imposing a tariff during normal conditions and lifting it during a crisis, lowering macroeconomic costs. Eaton and Eckstein suggested that in the face of a competitive world market the SPR would merely make the government a destabilizing speculator, but in facing a monopolistic seller a government inventory policy could raise both US and world welfare, by making US demand more elastic.[108]

Oil economists seemed to present policymakers with a wide range of estimates and suggestions. However, the disparity among these studies is reduced if we distinguish between qualitative and quantitative estimates. Economists proved themselves lousy prophets of detailed economic behavior, overestimating domestic supply elasticities and underestimating long-run demand elasticities. They also tended to overestimate the ability of OPEC to control production, while underestimating the response of fringe producers to higher prices. The relatively poor performance of oil market observers based in Stanford and Harvard suggest that ability of analysts may not be strongly correlated with university status.

Starting with Richard Nelson in 1958 and continuing over the next few decades, economists have been relatively consistent and accurate in many of their qualitative observations. The need for an oil stockpile, to alleviate the ill effects of dependence on imports, has been a consistent theme. Compare this with the silly insistence by the oil industry and many politicians that we could ever achieve "Energy Independence," or that it would be desirable, if possible. Economists tended to agree that an oil import fee or tariff was the logical method for financing an oil stockpile. For the most part, except for a liberal fringe which distrusted the market, most economists also agreed that higher prices were the most effective method of inducing conservation.

NOTES

[1] Lewis C. Gray, "Rent Under the Assumption of Exhaustibility," *Quarterly Journal of Economics* 28 (May 1914): 66-89.

[2] Harold Hotelling, "The Economics of Exhaustible Resources," *Journal of Political Economy* 39 (April 1931): 137-75.

[3] F.W. Paish, "Causes of Changes in Gold Supply," *Economica* 20 (November 1938): 379-407.

[4] William R. Pabst, "Unstable Conditions of Competition and Monopoly in Exhaustible Resource Industries," *Journal of Political Economy* 50 (Oct. 1942): 739-49.

[5] H. Scott Gordon, "The Economic Theory of a Common-Property Resource: The Fishery," *Journal of Political Economy* 62 (Apr. 1954).

[6] Orris C. Herfindahl, "Some Fundamentals of Mineral Economics," *Land Economics* 31, (May 1955): 131-38.

[7] Orris C. Herfindahl, "Depletion and Economic Theory," in *Extractive Resources and Taxation*, ed. Mason Gaffney (Madison: University of Wisconsin Press, 1967), pp. 63-90.

[8] Richard L. Gordon, "A Reinterpretation of the Pure Theory of Exhaustion," *Journal of Political Economy* 75 (June 1967): 274-86.

[9] Harold J. Barnett, and Chandler Morse, *Scarcity and Growth* (Baltimore: Johns Hopkins University Press, 1963), pp. 72-82.

[10] Samuel P. Hays, *Conservation and the Gospel of Efficiency* (Cambridge: Harvard University Press, 1959), pp. 262-72.

[11] Erich W. Zimmerman, *World Resources and Industries* (New York: Harper & Brothers, 1933), pp. 2-4, 780-800.

[12] S.V. Ciriacy-Wantrup, *Resource Conservation: Economics and Policies* (Berkeley, University of California Press, 1952); Colin Clark, "Afterthoughts on Paley," *Review of Economics and Statistics* 36 (August 1954): 267-73; and Edward S. Mason, "Comment," *Review of Economics and Statistics* 36: 273-78.

[13] Anthony Scott, *Natural Resources: Economics of Conservation* (Toronto: University of Toronto Press, 1955), pp. 13-14, 67-70.

[14] Barnett and Morse, *Scarcity and Growth*, pp. 3-12.

[15] John Ise, *The United States Oil Policy* (New Haven: Yale University Press, 1926), pp. 123-26; and George Ward Stocking, *The Oil Industry and the Competitive System* (Boston: Houghton Mifflin, 1925), pp. 310-11.

[16] Ise, *Oil Policy*, pp. 145-180.

[17] Stocking, *Oil Industry,* pp. 165, 264.

[18] Ise, *Oil Policy*, pp. 239, 496, 514-16; and Stocking, *Oil Industry,* pp. 311-12.

[19] Leonard M. Logan, *Stabilization of the Petroleum Industry* (Norman: University of Oklahoma Press, 1930).

[20] George Ward Stocking, "Stabilization of the Oil Industry: Its Economic and Legal Aspects," *American Economic Review* 23 (March 1933): 55-70; and idem, "Oil Proration: An Analysis and Evaluation," *Land Economics* 9 (May 1933): 135-44.

[21] Myron W. Watkins, *Oil: Stabilization or Conservation?* (New York: Harper & Brothers, 1937), pp. 32-34.

[22] Watkins, *Stabilization or Conservation?* pp. 34-35, 89, 145.

[23] Paul. H. Frankel, *Essentials of Petroleum* (London: Chapman & Hall, 1946).

[24] Idem, "Integration in the Oil Industry," *Journal of Industrial Economics* 1 (July 1953): 202-211.

[25] NPC, Meeting Transcript, April 15, 1948, p. 90-96 cited in Vietor, *Energy Policy In America*, p. 93fn.

[26] Eugene V. Rostow, *A National Policy For the Oil Industry*, (New Haven: Yale University Press, 1948), pp. 12-15.

[27] Rostow, *National Policy*, pp. 26-35.

[28] Rostow, *National Policy*, pp. 45-53, 145, 148.

[29] Joe S. Bain, "Rostow's Proposals for Petroleum Policy," *Journal of Political Economy* 57 (Feb. 1949): 55-60.

[30] Rostow, Eugene, "A Reply," *Journal of Political Economy* 57 (Feb. 1949): 60-68.

[31] Ralph Cassady, *Price Making and Price Behavior in the Petroleum Industry* (New Haven: Yale University Press, 1954), pp. 111-20, 337-39.

[32] Erich W. Zimmermann, *Conservation in the Production of Petroleum* (New Haven: Yale University Press, 1957), pp. 9-10, 272-75, 338-54.

[33] Wayne A. Leeman, "Crude Oil Prices in the United States at the Gulf Coast," *Journal of Industrial Economics* 5 (July 1957): 180-91; Paul H. Frankel, "Crude Oil Prices—A Comment," *Journal of Industrial Economics* 6 (October 1957).

[34] James R. Nelson, "Prices, Costs, and Conservation in Petroleum," *American Economic Review* (May 1958): 502-15.

[35] "Public Interest in Petroleum Products -- Discussion," *American Economic Review* (May 1958): 520-24.

[36] Melvin De Chazeau, and Alfred E. Kahn, *Integration and Competition in the Petroleum Industry* (New Haven: Yale University Press, 1959), pp. 206-45.

[37] De Chazeau and Kahn, *Integration and Competition*, pp. 252-54.

[38] John S. McGee, "Conservation, Integration, and Pricing in the Oil Industry of the United States: A Review Article," *Journal of Industrial Economics* 9 (Nov. 1960): 63-81.

[39] Harold F. Williamson, and Ralph L. Andreano, "Integration and Competition in the Oil Industry: A Review Article," *Journal of Political Economy* 69 (Aug. 1961): 381-85.

[40] Douglas H. Eldridge, "Tax Incentives for Mineral Enterprise," *Journal of Political Economy* 58 (June 1950): 222-240.

[41] Gray, Horace M., "Percentage Depletion, Conservation, and Economic Structure," in Joint Committee on the Economy Report, *Federal Tax Policy for Economic Growth and Stability*, 84th Cong., 1st sess., 1955, p. 430-39; Arnold C. Harberger, "The Taxation of Mineral Industries," in *Federal Tax Policy*, pp. 440-49.

[42] Peter Steiner, "Percentage Depletion and Resource Allocation," in *Tax Revision Compendium*, submitted to the Committee on Ways and Means, (Wash., D.C.: GPO, 1959), pp. 949-66; John A. Menge, "The Role of Taxation in Providing for Depletion of Mineral Reserves," *Tax Revision Compendium*, pp. 967-78.

[43] Stephen L. McDonald, "Percentage Depletion and the Allocation of Resources: The Case of Oil and Gas," *National Tax Journal* 14 (Dec. 1961): 323-36.

[44] Richard A. Musgrave, "Another Look at Depletion," *National Tax Journal* 15 (June 1962): 205-8; and Douglas H. Eldridge, "Rate of Return, Resource Allocation and Percentage Depletion," *National Tax Journal* 15 (June 1962): 209-17.

[45] Stephen L. McDonald, "Percentage Depletion and Tax Neutrality: A Reply to Messrs. Musgrave and Eldridge," *National Tax Journal* 15 (Sept. 1962): 314-26.

[46] Rene P. Manes, "More Fuel on the Depletion Fire," *National Tax Journal* 16 (Mar. 1963): 102-106.

[47] Peter O. Steiner, "The Non-Neutrality of Corporate Income Taxation--With and Without Depletion," *National Tax Journal* 16 (Sept. 1963): 238-51.

[48] Stephen McDonald, "On the Non-Neutrality of Corporate Income Taxation: A Reply to Steiner," *National Tax Journal* 17 (March 1964): 101-4; Peter O. Steiner, "Rejoiner to McDonald," *National Tax Journal* 17 (Mar. 1964): 105-7.

[49] Paul Davidson, "The Depletion Allowance Revisited," *Natural Resources Journal* 10 (Jan. 1970): 1-9; Stephen McDonald, "Distinctive Tax Treatment of Income From Oil and Gas Production," *Natural Resources Journal* 10 (Jan. 1970): 97-112.

[50] William L. Thorp, "Trade Barriers and National Security," *American Economic Review* 50 (May 1960): 433-42.

[51] James W. McKie and Stephen L. McDonald, "Petroleum Conservation in Theory and Practice," *Quarterly Journal of Economics* 76 (Feb. 1962): 98-121.

[52] Paul Davidson, "Public Policy Problems of the Domestic Crude Oil Industry," *American Economic Review* 53 (Mar. 1963): 85-108.

[53] Duncan R.G. Campbell, "Public Policy Problems of the Domestic Crude Oil Industry: Comment," *American Economic Review* 54 (March 1964): 114-

18; Henry Steele, "Public Policy Problems of the Domestic Crude Oil Industry: Comment," *American Economic Review* 54 (March 1964): 119-24.

[54] Paul Davidson, "Public Policy Problems of the Domestic Crude Oil Industry: A Reply," *American Economic Review* 54 (Mar. 1964): 125-34.

[55] Renee P. Manes, "Import Quotas, Prices and Profits in the Oil Industry," *Southern Economic Journal* (July 1963): 13-24.

[56] Alfred Kahn, "The Depletion Allowance in the Context of Cartelization," *American Economic Review* 54 (June 1964): 286-314.

[57] Morris Adelman, "Efficiency of Resource Use in Crude Petroleum," *Southern Economic Journal* 31 (1964): 101-22.

[58] James Burrows and Thomas Domencich, *An Analysis of the United States Oil Import Quota* (Lexington, Mass.: Heath Lexington Books, 1970); and CONSAD Research Corporation, "The Economic Factors Affecting the Level of Domestic Petroleum Reserves," December 1968, in US Congress, Joint Publication Committee on Ways and Means and Committee on Finance, *Tax Reform Studies and Proposals*, 91st Cong., 1st sess., 1969.

[59] Burrows and Domencich, *Oil Import Quota*, pp. 148-59, 204-7.

[60] CONSAD, "Economic Factors," chapters 2, 6, 7, 9.

[61] George Hays, "Import Controls on Foreign Oil: Tariff or Quota?" *American Economic Review* 61 (Sept. 1971): 688-91; Ryan Amacher, Robert Tollison and Thomas Willett, "Import Controls on Foreign Oil: Comment," *American Economic Review* 63 (Dec. 1973): 1031-34; Richard Mancke, "The Long-Run Supply Curve of Crude Oil Produced in the United States," *Anti-Trust Bulletin* 15 (Winter 1970): 727-56; Charles Cicchetti and William Gillen, "The Mandatory Oil Import Quota Program: A Consideration of Economic Efficiency and Equity," *Natural Resources Journal* 13 (July 1973): 399-430.

[62] Areskoug, Kaj, "US Oil Import Quotas and National Income," *Southern Economic Journal* 37 (Jan. 1971): 307-17; George Heitman, "US Oil Import Quotas and National Income: A Comment," *Southern Economic Journal* 38 (Jan. 1972): 30-432.

[63] M.A. Adelman, "Oil Prices in the Long Run (1963-1975)," *Journal of Business* 37 (1964): 143-61

[64] Charles Issawi and Mohammed Yeganeh, *The Economics of Middle East Oil* (New York: Praeger Publishers, 1962); Helmut J. Frank, *Crude Oil Prices in the Middle East* (New York: Frederick A. Praeger, 1966); Paul G. Bradley, *The Economics of Crude Petroleum Production* (Amsterdam: North-Holland Publishing Company, 1967); and Edith Penrose, *The Large International Firm in the Developing Countries* (Cambridge, Mass.: MIT Press, 1968).

[65] Morris A., Adelman, *The World Petroleum Market* (Baltimore: Johns Hopkins Press, 1972), pp. 215, 253-61.

[66] Halford L. Hoskins, "Needed: A Strategy For Oil," *Foreign Affairs* 29 (January 1951): 229-37.

[67] Report to the President by the President's Materials Policy Commission, *Resources For Freedom, Volume 1, Foundations for Growth and Security*, June 1952, pp. 109-111.

[68] Herbert Feis, "Oil for Peace Or War," *Foreign Affairs* 32 (April 1954): 416-29; James R. Schlesinger, *The Political Economy of National Security* (New York: Frederick A. Praeger, 1960): 145-48.

[69] Stephen McDonald, "Conservation Policies and National Security," *Quarterly of the Colorado School of Mines* 64 (Oct. 1969): 145-56; Wallace F. Lovejoy, "Oil Import Policies As They Relate To National Security," *Quarterly of the Colorado School of Mines* 64 (Oct. 1969): 157-80.

[70] Walter J. Mead and Philip E. Sorensen, "A National Defense Petroleum Reserve Alternative to Oil Import Quotas," *Land Economics* 47 (Aug. 1971): 211-24.

[71] Walter J. Levy, "Oil Power," *Foreign Affairs* 49 (July 1971): 652-68.

[72] Morris Adelman, "Is the Oil Shortage Real? Oil Companies As OPEC Tax-Collectors," *Foreign Policy* 9 (Winter 1972-1973): 69-107.

[73] James E. Akins, "The Oil Crisis: This Time the Wolf is Here," *Foreign Affairs* 51 (Apr. 1973): 462-90.

[74] Bruce C. Netschert, *The Future Supply of Oil and Gas* (Baltimore: John Hopkins Press, 1958), pp. 56-58.

[75] Wallace F. Lovejoy, Paul T. Homan, and Charles O. Galvin, *Problems of Cost Analysis In The Petroleum Industry* (Dallas: Southern Methodist University Press, 1964), pp. 26-29.

[76] Franklin M. Fisher, *Supply And Costs In The US Petroleum Industry* (Baltimore: Johns Hopkins Press, 1964): pp. 3-5, 21-23; and Edward W. Erickson, "Crude Oil Prices, Drilling Incentives and the Supply of New Discoveries," *Natural Resources Journal* 10 (Jan. 1970): 27-52.

[77] Mancke, "Supply Curve of Crude Oil," pp. 727-48.

[78] Dennis Epple, *Petroleum Discoveries And Government Policy* (Cambridge, Mass.: Ballinger Publishing, 1975), pp. 52-54, 100-1.

[79] Jeffrey Alan Kolb, *An Econometric Study of Discoveries of Natural Gas and Oil Reserves In the United States, 1948 to 1970*, (New York: ARNO Press, 1979), pp. 27-35, 54-58.

[80] *Oil and Gas Journal*, July 17, 1961, p. 46, Nov. 18, 1963, p. 113, May 2, 1966, p. 58, May 1, 1967, p. 106.

[81] Report to the President by the President's Materials Policy Commission, *Resources For Freedom, Volume 3, The Outlook For Energy Sources*, June 1952, pp. 5-10; Sam H. Schurr and Bruce C. Netschert, *Energy in the American Economy 1850-1975* (Baltimore: Johns Hopkins Press, 1960), p. 251; *Oil and Gas Journal*, July 1, 1963, p. 99; and API, *Basic Petroleum Data Book 1984*, Jan. 1984.

[82] Michael Kennedy, "An Economic Model of the World Oil Market," *Bell Journal of Economics* 5 (Aug. 1974): 540-77.

[83] Morris Adelman, "The World Oil Cartel: Scarcity, Economics, and Politics," *Quarterly Journal of Economics and Business* 16 (Summer 1976): 7-18.

[84] Jacques Cremer and Martin Weitzman, "OPEC and the Monopoly Price of World Oil," *European Economic Review* 8 (1976): 155-64; Stephen Salant, "Exhaustible Resources and Industrial Structure: A Nash-Cournot Approach to the World Oil Market," *Journal of Political Economy* 84 (1976): 1079-93; Richard Gilbert, "Dominant Firm Pricing Policy in a Market For an Exhaustible Resource," *Bell Journal of Economics* 9 (Autumn 1978): 385-95; Tracy Lewis and Richard Schmalensee, "On Oligopolistic Markets For Nonrenewable Natural Resources," *Quarterly Journal of Economics* 95 (Nov. 1980): 475-91, A.M. Ulph and G.M. Folie, "Exhaustible Resources And Cartels: An Intertemporal Nash-Cournot Model," *Canadian Journal of Economics* 13 (Nov. 1980): 645-58; and D.M.G. Newberry, "Oil Prices, Cartels, and the Problem of Dynamic Inconsistency," *Economic Journal* 91 (Sept. 1981): 617-46.

[85] Esteban Hnyilicza and Robert S. Pindyck, "Pricing Policies for a Two-Part Exhaustible Resource Cartel," *European Economic Review* 8 (1976): 149-54; Robert S. Pindyck, "Gains to Producers From the Cartelization of Exhaustible Resources," *Review of Economics and Statistics* 60 (May 1978): 238-51.

[86] Ali Ezzati, "Future OPEC Price and Production Strategies as Affected By Its Capacity to Absorb Oil Revenues," *European Economic Review* 8 (1976): 107-38.

[87] William D. Nordhaus, "Oil and Economic Performance in Industrial Countries," *Brookings Papers on Economic Activity* 2, (1980): 341-99.

[88] Theodore Moran, "Modeling OPEC Behavior: Economic and Political Alternatives," in *OPEC Behavior and World Oil Prices*, ed. James Griffen and David Teece (London: George Allen & Unwin, 1982), pp. 94-130; Hossein Ravazi, "Effects of Uncertainty on Oil Extraction Decisions," *Journal of Economic Dynamics and Control* 5 (1983): 359-70; David J. Teece, "OPEC Behavior: An Alternative View," in *OPEC Behavior and World Oil Prices*, pp. 64-93.

[89] Philip Verleger, "The Determinants of Official OPEC Crude Prices," *Review of Economics and Statistics* 64 (May 1982): 177-83.

[90] Ali Johanny, *Myth of the OPEC Cartel: The Role of Saudi Arabia* (Chichester: John Wiley & Sons, 1980); idem, "OPEC and the Price of Oil: Cartelization or Alteration of Property Rights," *Journal of Energy and Development* 5 (Autumn 1979): 72-80; Peter Odell and Kenneth Rosing, *The Future of Oil*, (London: Hogan Page, 1983); Walter Mead, "An Economic Analysis of Crude Oil Price Behavior in the 1970s," *Journal of Energy and Development* 4 (1979): 475-91.

[91] Morris Adelman, "OPEC As a Cartel," in *OPEC Behavior and World Oil Prices*, pp. 37-63.

[92] Dermot Gately, "A Ten Year Retrospective: OPEC and the World Oil Market," *Journal of Economic Literature* 22 (Sept. 1984): 1100-14.

[93] Gately, "Ten Year Retrospective," p. 1105-06; Alan Manne and Leo Schrattenholzer, "International Energy Workshop: A Summary of the 1983 Poll Responses," *Energy Journal* 5 (1984): 45-64.

[94] George Daly, James M. Griffin, and Henry B. Steele, "The Future of OPEC: Price Level and Cartel Stability," *Energy Journal* 4 (1983): 65-77.

[95] Brian Levy, "World Oil Marketing In Transition," *International Organization* 36 (Winter 1982): 113-33.

[96] Arlon R. Tussing, "Oil Prices Are Still Too High," *Energy Journal* 6 (1985): 9-12; Philip Verleger, "The Evolution of Oil as a Commodity," in *Energy: Markets and Regulation*, ed. Richard Gordon, Henry Jacoby, and Martin Zimmerman (Cambridge: MIT Press, 1987), pp. 161-86.

[97] Fred Singer, "The Price of Oil," *Annual Review of Energy* 8 (1983): 451-508; Morris Adelman, "An Unstable World Oil Market," *Energy Journal* 6 (1985): 17-22.

[98] Dermot Gately, "The Prospects For Oil Prices, Revisited," *Annual Review of Energy* 11 (1986): 513-38.

[99] James M. Griffin, and Clifton T. Jones, "Falling Oil Prices: Where Is the Floor?," *Energy Journal* 7 (1986): 37-50.

[100] Morris Adelman, "The Competitive Floor to World Oil Prices," *Energy Journal* 7 (1986): 9-31.

[101] George Kouris, "Energy Demand Elasticities in Industrialized Countries: A Survey," *Energy Journal* 4 (1983): 73-94; and William C. Wheaton, "The Long-run Structure of Transportation and Gasoline Demand," *Bell Journal of Economics* 13 (Autumn 1982): 439-54.

[102] U.S. Congress, House, Committee on Science and Technology, *Energy Demand Forecasting, Hearings before the Subcommittee on Investigations and Oversight*, 97th Cong., 1st sess., pp. 95-96.

[103] Martin Greenberger, *Caught Unawares: The Energy Decade in Retrospect* (Cambridge, Mass.: Ballinger Publishing Co., 1983).

[104] Paul Davidson, Laurence H. Falk and Hoesung Lee, "Oil: Its Time Allocation and Project Independence," *Brookings Papers on Economic Activity* 2 (1974): 411-48.

[105] Douglas R. Bohi and Milton Russell, *US Energy Policy: Alternatives For Security* (Baltimore: Johns Hopkins Press, 1975).

[106] James Cox and Arthur Wright, "A Tariff Policy For Independence From Oil Embargoes," *National Tax Journal* 28 (Mar. 1975): 29-41; Edward Renshaw, "A Note On Petroleum Taxes To Increase Domestic Welfare And Reduce The Monopoly Profits Of The International Oil Cartel," *Public Finance* 31 (1976): 132-42; Theodore Berstrom, "On Capturing Oil Rents with a National Excise Tax," *American Economic Review* 72 (Mar. 1982): 194-201.

[107] Harry G. Broadman, "The Social Cost of Imported Oil," *Energy Policy* 14 (June 1986): 242-252.

[108] Richard Gilbert and Knut A. Mork, "Efficient Pricing During Oil Supply Disruptions," *Energy Journal* 7 (1986): 51-68; Frederic Murphy, Michael Toman, and Howard Weiss, "An Integrated Analysis of US Oil Security Policies," *Energy Journal* 7 (1986): 67-82; Jonathon Eaton and Zvi Eckstein, "The US Strategic Petroleum Reserve: An Analytic Framework," in *The Structure and Evolution of Recent US Trade Policy*, ed. Robert E. Baldwin and Anne Krueger (Chicago: University of Chicago Press, 1984), pp. 237-76.

V

Public Opinion And Oil Politics

Public opinion had a dramatic effect on the development of energy policy in a number of ways. Until 1970, its effect was felt by its absence in deliberations concerning oil policies. During the 1970s, even before the oil embargo, public attitudes toward oil and energy prices paralyzed the government, preventing decisive action to deal with impending "shortages." In the aftermath of the oil crises, the threat of reactivating a quiescent public has constrained Congress from passage of an oil import fee or gasoline taxes. In effect, policy made before energy became a public issue was determined by the interaction of "special interests"; afterwards policy would be determined by fear of public reaction instead of consideration of the "public interest." This argument will be further developed in Chapter 7.

PUBLIC OPINION AND POLITICS

Political scientists have offered a number of definitions of public opinion, including the one proposed by V.O. Key: "Public opinion . . . may simply be taken to mean those opinions held by private persons which governments find it prudent to heed. Governments may be compelled toward action or inaction by such opinion; in other instances they may ignore it, perhaps at their peril; they may attempt to alter it; or they may divert or pacify it. So defined, opinion may be shared by many or by few people. . ."[1]

David Truman considered public opinion as consisting of the expressions of the attitudes on an issue of the aggregate of individuals making up the public under discussion, which does not necessary include either all the opinions held by this set of individuals, or all the members of the public at large. The existence of a majority opinion on an issue is a characteristic of the public opinion. With regard to political effectiveness intensity of opinion will matter, especially when expressed as a willingness to act, as well as the size of a segment of public opinion. In this respect public opinion results from individuals seeking expression through mass communication channels to participate meaningfully in political and cultural affairs. Opinions are the result of value and belief calculations which enable people to find a sense of congruence between their own experiences and the state of the world. Public opinion then channels individual views into common expressions that can influence the development of a situation.[2]

Two major influences of public opinion are elites and the mass media. Elites affect public opinion both through their direct influence on the people they come in connect with, for example as leaders of their communities or of organized groups, and through their access to the mass media. Politicians have numerous opportunities to express their attitudes on issues since this is considered a legitimate component of news. Business leaders are often cited with regard to economic and social issues and journalists have direct access to forums in which to state their views on public issues. Social scientists and "intellectuals" through professional journals and magazines of opinion can influence other elites, and over time their ideas permeate to the mass media and subsequently to the public.

Page, Shapiro, and Dempsey give five conditions for new information to be effective in changing perceived utility of policies for citizens: (1) the information is actually received, (2) understood, (3) clearly relevant to evaluating policies, (4) discrepant with past beliefs, and (5) credible. They presume that people rely heavily upon the cheapest and most accessible sources of information; newspapers, radio and television. Television news meets the exposure condition, and the simplicity in which its messages are molded also fits the criteria for easy comprehension. Interestingly enough, polls show that television is by far the most credible of the three mentioned media for most people. The audience for television news compared to newsmagazines tends to be older, poorer, and less educated. Magazine reading is strongly related to education, while women of all ages, older viewers and the poorly educated are most likely to watch TV news on a regular basis.[3]

Before TV news is given the dominant role in determining the political agenda one should note the influences on decision-makers in the television newsroom. Both TV news and the newsmagazines use the *New York Times* and to a lesser extent the *Washington Post* as the standard setter. The wire services are also important as a source of material for all types of media news. A study of elites in 1974 found they are exposed to a select group of sources: the *New York Times* was the most widely read, with the *Wall Street Journal* popular among businessmen and labor leaders, and the Washington Post among the government establishment, labor, media and voluntary association leadership. Heavy magazine readership was headed by *Newsweek*, *Time*, and *US News and World Report*, with economic leaders reading *Business Week* and *Fortune*.[4]

Despite the fact that television journalists are influenced by the higher quality media, time, audience and economic constraints shape TV news. Stories which require lengthy description or explanation are often dropped from story lists, a procedure which caused economic stories to be omitted in the 1960s, though not in the 1970s, when they became too important to be killed. TV producers disliked their "talkiness" and cringed when unexciting films of oil fields or supermarket cash registers had to accompany them. Television's reluctance to use long and complicated stories results from fear of losing their audience (newsmagazines also emphasize short articles, out of respect for the audience's unwillingness to read longer stories). The most basic format consideration, shared by all news media, is that news becomes suitable only when it is transformed into a story, which must always include a lead, a narrative, and a closer. This story format therefore makes the news into a morality play, an equivalent of traditional legend or myth. Journalists often resort to current symbols because these are already well known and therefore likely to interest the audience while saving valuable air time or print space. Consequently, there is both a reluctance to take on complex issues and a tendency to replace the details that could make the issue comprehensible with symbols which engage the viewer's attention.[5]

The actual impact of TV news and other media on public opinion is still under debate, though there is increasing evidence of media influence. A study which regressed attention paid to an issue in newsmagazines with public concern with the issue, found that both the public's experience and the media influenced public opinion on economic issues. The media played a substantial role in forming energy and inflation consciousness, although it was the character of events and their portrayal, and not the simple amount of news coverage allocated which determined the impact on the viewer. The citizen's view of what is politically important derives most clearly from an active consideration of his environment, both local and distant.[6]

Iyengar and Kinder are concerned with the agenda setting and priming affects on opinion of media selection of stories. Drawing on the information processing perspective in psychology they conclude that attention is highly selective; people only notice particular features of special consequence, and the impressions formed of issues and politicians tend to be organized around a few central themes. Under ordinary circumstances, people seem to follow heuristics, judgmental rules of thumb such as reliance upon information that is most easily accessible, rather than exhaustive analysis. Iyengar and Kinder conducted a series of experiments to measure the impact of exposure to television news stories on evaluations of presidential performance. They found that by ignoring some problems and attending to others, television news programs profoundly affect which problems viewers take seriously, especially among the politically naive. The media influenced evaluation through priming: problems prominently featured in television broadcasts loom large in evaluations of presidential performance. The effects were stronger for energy than for unemployment, reflecting a similar difference observed in other experiments. The stronger effect for energy might reflect the comparative novelty of national energy problems, causing citizens, faced with complexity and uncertainty, to be swayed by the way television broadcasts framed the news.[7]

Page et al. analyzed the relationship between variations in opinion change measured by surveys and changes in media content. The most important effect on opinion was estimated to be news commentary (from anchormen, reporters in the field or special commentators), with a single commentary associated with more than four percentage points of opinion change. This effect was found earlier by the authors in studies using the *New York Times* as a media source. A single story about experts or research studies is estimated to produce about three percentage points of opinion change. Stories using the President or opposition party statements as sources have a much smaller effect, about 0.3 percentage points, on opinion. Statements and actions by interest groups tend to move public opinion in the opposite direction by about the same amount, except for groups thought to be concerned with broadly defined public interests such as environmental groups and groups like Common Cause. Popular president (>50% approval rating) has a small (0.6% per statement) effect on opinion, while unpopular presidents have a negligible effect on public opinion.[8]

A dissenting opinion concerning the impact of the media on public opinion comes from William McGuire, who contends that there is a lack of conclusive evidence presented in the literature. The agreement among experts that mass media political campaigning has a powerful effect on voting is not supported by the empirical data which show only a modest effect size. Evaluation research of the relationship between ad exposure and voting choice within individual voters reports small to vanishing effects, largely confined to late deciders, minor offices, and obscure candidates. He accepts the proposition that media coverage performs an agenda-setting function, but suggests that it is equally plausible that how salient an issue is to the voter determines how much coverage the media gives it.[9]

PUBLIC OPINION AND VOTING BEHAVIOR

If there is a connection between public opinion and policy formation common sense suggests that the mechanism is through the expectation of reward and punishment at the voting booth for popular or unpopular actions. This relation can operate in a number of ways: politicians can choose actions which coincide with the general flow of public opinion to increase their popularity; they can avoid actions which are likely to generate a great deal of displeasure; they can engage in symbolic gestures to curry favor with the public without committing themselves to unpopular programs. A leader who wants to further what he perceives as the public interest without joining the unemployment line will have to balance unpopular actions by either persuading the public that the action is in their interest and/or by other moves designed to balance the public's displeasure with the specific course of action chosen.

The work of Campbell, Converse, Stokes and Miller in the 1950s cast doubt on whether the media had an important impact on elections, at least through the avenue of issue awareness. On most issues they found that the public was marginally familiar with government policies, with only 50-70% of respondents possessing both an opinion and issue familiarity. They also found that respondents were confused concerning party positions on issues. Only 40 to 60 percent of the informed segment of the population perceived party differences and could relate their attitudes to those of the political parties.[10]

Nie, Verba and Petrocik, updating the Campbell study, found that since the 1956 election ideological evaluation of candidates and parties increased steadily over the next two decades. The public demonstrated a major increase in the level of attitude consistency (attitudes in one issue area being related to attitudes in other issue areas) over the same period, and the use of issues to evaluate candidates increased substantially. They suggested that issues were becoming more salient to citizens, reflected in the increase in political items included among the hopes and fears of respondents over the years. The individual voter was tending to evaluate candidates on the basis of information and impressions conveyed by the mass media, voting on that basis instead of by party identification.[11]

Recent studies of voting suggest there is a substantial split in the political environment faced by Representatives, Senators and Presidents. Voters are far more aware of candidates in Presidential elections vs. Senate races, and possess still more information concerning Senators than Representatives. Even back in the 1950s and 1960s Representatives depended on their own personal standing in the eyes of the voters to insure re-election, independent of both party and issues, though voters tended to support candidates of the party with which they identified. Thus, it is not surprising that incumbents have a greater advantage in the House than the Senate. The rise in the success of incumbents has been attributed to both the increase in campaign resources available to incumbents and their active role in focusing on re-election, especially the provision of constituent services. The first allows them to take more credit for the positions which they want to acknowledge, increasing their recognition with voters, while service to constituents is an effective and noncontroversial method of developing a positive image. This explains why people have contempt for Congress as an institution but love their congressmen. As Fenno points out: "Members run for Congress by running against Congress."[12]

In the Senate, elections resemble those of the House with the exception that incumbency plays a smaller role, which Fiorina assumes is the result of reduced opportunity for constituency service. National conditions and presidential performance also have little effect on voting for Senators, though issue stands and ideology are more important in Senate elections. With the relative decline in party voting since 1956, the refutation of the presidential coattail hypothesis, and the limited impact of economic conditions, the candidate's image has grown in importance as the explanation for re-election success. A decline in party voting has also appeared in presidential elections, but three other factors, attitudes toward candidates, perception of general conditions in the country, and policies advocated by candidates, tend to vary in importance depending on the specific election.[13]

The opportunity to strengthen personal ties with constituents has not been an complete blessing for Congressmen. When members of Congress used the general decline in party loyalty among voters as an opportunity to strengthen personal, as opposed to partisan, ties with their constituents, a less partisan electorate presented them with new problems. This new electorate is more volatile, intensifying the uncertainties of electoral politics, as larger electoral margins no longer guaranteed a reduced probability of future defeat, a finding reflected in the growth of the standard deviation of the interelection vote shift in House districts from an average of 5.1 in the 1950s to an average of 8.8 percent in the 1970s. This factor ended the trend of increasing proportions of reelected incumbents. Voter fickleness stems from their openness to the influence of short-term political forces such as new personalities, issues, and persuasive messages, requiring more frequent and extensive attention on the part of incumbents.[14]

Jacobson states that incumbents lose when challengers are able to project a positive image of their own and to persuade voters that the incumbents have liabilities that outweigh their usual assets. Issues and images are evidently part of this, especially as new organizations such as PACs, direct-mail outfits and party committees are making funds available to challengers to publicize their stands while well-organized single-issue groups are able to focus attention on emotional, divisive issues. The resulting increased issue awareness allows national conditions to affect the substantive content of campaigns as both incumbents and challengers position themselves on issues. Since one key for a successful challenge is to create a negative perception of the incumbent in voters' minds, linking him/her with support for an unpopular administration or opposition to popular one is an important tactic. Incumbents follow an opposite tack, as demonstrated by many Democrats in 1980 taking care to avoid association with Carter and his policies.[15] Thus media presentation of issues, always important for Presidents and Senators, has become a factor in House races.

At the same time, because of Watergate, the Capitol Hill press has become more combative and oriented toward investigation, accelerating a process which had begun in the Kennedy administration and escalated during the war in Vietnam. The local press, based in the states or outside the established circles in Washington, has traditionally had a more cordial relationship with Congress than the national press. Incumbents fare well, receiving twice the coverage of their challengers and more positive coverage than challengers or Congress as an institution. Since the electronic media is the most important source of information for the public, an alternative explanation of the love the congressman, hate the Congress phenomenon is that constituents receive positive impressions of their congressmen on their local television news while the national news is criticizing Congress and the government. Members of the House control their press by maintaining a close relationship with the local press which needs their cooperation, while attracting little coverage from the national press so long as they stay unindicted. Senators have less intimate relationships with the local media and they also tend to attract national coverage, which is less controllable than the local press, raising the spectre of looking bad on the evening news.[16]

An examination of the 1976 presidential election found that symbolic attitudes (liberalism-conservatism, party identification, and racial prejudice) had strong effects on voters while self-interest had almost none. Voter sophistication actually increased the power of symbolic attitudes. The authors contend that with a few exceptions, the literature shows that self-interest to have little effect on policy attitudes and voting behavior on such issues as the economy, civil rights, the Vietnam War, busing, and the energy crisis. In the exceptions the personal impact of the issue was seen as more definite, concrete and immediate than with the case for most political issues. They suggest that "single-issue" publics are comprised principally of those with shared symbolic loyalties. For the politician the problem becomes to avoid violation of voters' symbolic attitudes when communicating with their potential supporters.[17]

The question remains whether public opinion will actually effect policy formation or merely the presentation of policy decisions. There is some evidence that policy does follow public opinion, especially when opinion changes are large and sustained, and issues are broad-based and salient. However in some instances policy seems to affect opinion, and other times there is no change in policy in response to shifts in opinion. Even when there is public support for a policy change, inertia in the political system and support for the status quo can delay change, sometimes until the issue has declined in salience. This is not a problem when a proposal is seen as a solution to a crisis, which would of course be an issue of high salience.[18]

A factor which complicates the relationship between public opinion and policy is Down's "issue-attention" cycle. To Downs, "Public perception of most 'crises' in American domestic life does not reflect changes in real conditions as much as it reflects the operation of a systematic cycle of heightening public interest and then increasing boredom with major issues." Downs saw the cycle following a five stage sequence: The Pre-problem stage when a problem exists, capturing the attention of experts and/or interest groups; Alarmed discovery and euphoric enthusiasm, resulting from the public becoming aware of the problem due to some dramatic series of events and demanding that something be done to resolve the crisis; Realization of the cost of significant progress; Gradual decline of intense public interest, due to the public realizing the intractability of the problem; and the Post-problem stage where the issue fades from the center of attention. Usually by the post-problem stage new institutions, programs, and policies have been created, and problems once through the cycle receive a higher average level of attention and general concern than those still in the pre-discovery stage.[19]

The rise and decline of environmental interest illustrates this cycle. Up until 1968 there was incremental growth in awareness of, and development of policy to deal with, air and water pollution and other environmental problems. Suddenly in 1969-70 concern with the quality of the environment became a major concern, reflected in a jump in people mentioning it as one of the top three problems requiring action (from 17% in 1965 to 53% in 1970), a Ralph Nader study, and a massive student-organized environmental event, "Earth Day." The result was the passage of the National Environmental Policy Act in 1969, the formation of the Environmental Protection Agency and the passage of the Clean Air Act and the Water Quality Improvement Act in 1970, and the Water Pollution control Act Amendments of 1972.[20] Since that point the issue of pollution has disappeared as a major concern of the American people, but substantial funds are still allocated to pollution control and the issue commands far more attention than it did in the 1960s.

This issue-attention cycle provides a special inducement for a president to take the initiative. In the case of an issue which has already grabbed the public's attention he does not want to be seen ignoring the crisis or failing to take sufficient action. Given the president's access to the mass media he also has the opportunity to engender public concern about a problem and to influence public opinion so as to encourage passage of legislation he favors in Congress. But presidents also rely on their own popularity to push their proposals through Congress, with passage related to both their popularity and the number of proposals included in their program.[21] Thus it is important for the president to conserve his popularity and clearly state his priorities when it comes to his legislative program.

For Kernell, the changes that have occurred in Washington over the last thirty years have made it imperative for the president to go public to support his policies. Under pluralist theory the president's role was that of the chief bargainer, but given a political community consisting of independent members with few group or institutional loyalties the limited goods and services available for barter to the bargaining president would be quickly exhausted. Given the vulnerability of the new Congress to the tides of public opinion, other avenues of presidential influence open up. A president who is effective in going public will find frequent opportunities for leadership due to the sensitivity of self-reliant politicians to public opinion. By campaigning around the country, the president increases the costs of opposition and the benefits of acquiescence for Congressmen. A president will be most successful when he can present his policies as defending the public interest from private greed, following a tradition established by the Progressives.[22]

PUBLIC OPINION TOWARD OIL ISSUES

The energy issue followed the Downsian issue-attention cycle almost perfectly. While experts became concerned about impending shortages in 1971 it was not until 1973 that public consciousness began to focus on the problem of energy supplies. In fact Gallup did not even include energy issues as a separate question until after the oil embargo, while respondents did not include it among the most important problems facing the nation until 1974. The first poll on energy issues which I was able to find was a Harris survey taken in April 1973. 47 percent felt the energy crisis was very serious and an additional 30 percent felt it was somewhat serious. The top reason volunteered for the energy crisis was that "large companies are conspiring to raise prices through scaring the public," volunteered by 20 percent of the respondents. 75 percent surveyed agreed that the seriousness of the crisis was due to the expectation of sharp rises in the cost of gasoline and heating and air conditioning. 51 percent agreed that the US was running out of oil and natural gas and would become more dependent on foreign sources. On the other hand, 52 percent felt that "US know-how is so good we'll find enough energy to meet our needs without a lot of trouble." Only 35 percent of respondents expressed confidence in the ability of the Federal government to manage the energy crisis.[23]

The Gallup series which asks respondents their opinion concerning the most important problem gives a clear picture of the fluctuation of interest in energy issues. Before May 1973 and after January 1982 energy wasn't even included among the major topics listed. The highest level of interest in energy was in early 1974, with significant concern evidenced in 1978 and 1979 when it was considered the second most important problem after inflation. By 1980 however, energy had already faded in perceived importance.[24] While the 1974 and 1979 peaks were due to the effect of higher prices and gasoline shortages, the 1977-78 peak can only be explained by the attention given the issue due to President Carter's efforts to gather support for his energy program. The dominance of inflation as an issue in the 1970s illustrates an important aspect of energy politics during the decade; the political sensitivity toward price increases meant that the normal economic solution of allowing higher prices to clear the market was extremely painful to implement during this period.

For the period from 1973 to 1978 there is a fairly complete set of polls concerning the perceived seriousness of the energy situation. The emerging concern with energy in early 1973 followed a winter of publicized fuel oil shortages and projected gasoline shortages for the summer. During the embargo there was a peak in the number of people perceiving the energy situation as a serious problem, with 50 percent seeing it as very serious and 37 percent as fairly serious, but this focus on energy quickly declined once the crisis had ended. Concern with energy gradually shifted upward during 1975-1976, and for the rest of the decade about 40 percent of respondents saw it as a very serious issue, with another 40 percent viewing it as fairly serious.[25] It is interesting to note that Carter's televised speeches in 1977 (April and November) had little effect on the perceived seriousness of energy as an issue.

One problem in mobilizing public support for policies which required sacrifice was the widespread belief that the energy problem, while it might be an important problem and certainly worth serious consideration, was not the result of a change in the economic environment but merely a ploy by the oil companies in order to charge higher prices. This can be seen in the responses in Table 5.1 to the loaded questions posed by the Roper poll:

> "Some people say there is a real shortage of gasoline and fuel oil because demand has outrun the supply. Others say there really isn't a shortage of gasoline and fuel oil and the big companies are holding it back for their own advantage. What do you think. . . ."

and by the CBS/New York Times poll:

> "Do you think the shortage of energy we hear about is real or are we just being told there are shortages so oil and gas companies can charge higher prices?"[26]

Carter managed to create a temporary impression in the public mind that the shortage was real after both April and November 1977 speeches but this effect quickly dissipated. A much smaller increase in people thinking the shortage was real occurred after his April 1979 speech, suggesting that public cynicism concerning the energy situation was so deeply embedded that no amount of persuasion could shift public opinion for long.

This cynicism was reflected in a series of questions concerned with the seriousness of the oil shortage as a long term problem. The public seemed to be split into two groups; those who felt the crisis was genuine tended to feel that the energy shortage was a real problem that would worsen, while the largest group of responses were from those who felt the oil shortage was contrived. Cynicism peaked during 1974, with those seeing the shortage as contrived reaching 53 percent in May vs. only 21 percent who saw the problem as real, followed by a brief period of acceptance with the proportion of cynics falling to 33 percent in May 1977, before rising to 51 percent in 1979.[27]

One reason why the public was not convinced that there was a real shortage of oil was that they were ignorant of the dependence on foreign oil. Carter managed to convince a majority in 1977 that the US did need to import oil, but even in 1979 a solid third of the respondents felt that the country did not need to import oil. The Roper question on this issue was more a measure of confidence in the country's potential for independence while the Gallup question directly asked whether the country is currently dependent on foreign oil. Thus the increase in Roper respondents who felt the country could be independent in 1975 and 1979 reflects both ignorance and optimism, while the

TABLE 5.1 IS THE ENERGY SHORTAGE FOR REAL?

Date of Poll		*Poll*	*Real*	*Just Being Told*	*No opinion*
1974	Feb.	(Roper)	18	73	9
1974	April	(ORC)	25	66	9
1974	May	(Roper)	41	53	6
1974	Nov.	(Roper)	39	54	7
1975	June	(Roper)	44	47	9
1976	Nov.	(Roper)	44	46	10
1977	May	(Roper)	61	33	6
1977	July	(CBS)	38	49	13
1977	Oct.	(CBS)	42	51	7
1977	Nov.	(Roper)	56	39	5
1978	Jan.	(CBS)	43	47	10
1978	April	(CBS)	42	50	8
1979	March	(CBS)	20	69	11
1979	June	(CBS)	30	62	8
1979	Aug.	(Roper)	27	65	8
1979	Oct.	(Cambridge)	36	54	10
1980	Jan.	(Cambridge)	28	61	11
1980	April	(Cambridge)	27	59	14
1980	July	(Cambridge)	34	58	8
1980	Oct.	(Cambridge)	28	60	12
1981	Jan.	(Cambridge)	26	64	10
1981	July	(Cambridge)	25	65	10

CBS is CBS/New York Times

SOURCE: Farhar et al., *Public Opinion About Energy*, p. 63; Al Richman, "The Polls: Public Attitudes Toward the Energy Crisis," *Public Opinion Quarterly*, 43, Winter 1979, p. 577; Edward Byers and Thomas Fitzpatrick, "Americans and the Oil Companies: Tentative Tolerance in a Time of Plenty," *Public Opinion* 8 (Dec./Jan. 1986): 44

Gallup poll was a direct measure of ignorance. Even so, the consensus of all polls was that by the end of the decade only half the American people understood that the country depended on foreign oil for a significant proportion of its oil consumption requirements.[28]

The appeal of the conspiracy theory explanation for oil shortages, reflected in the "it's an oil company plot" responses, is demonstrated in the next group of polls, given in Table 5.2. Oil companies received far more blame for the energy crisis than OPEC, with the latter given a minor role until 1979, when a slim majority decided that OPEC deserved the credit for the massive price increases. Both Congress and

TABLE 5.2 WHO DESERVES BLAME FOR ENERGY CRISIS

Group	Jan. 1974	Feb. 1975	Jan. 1976	June 1977	June 1979
Oil companies	56	57	57	55	72
Electric Companies	15	26	29	31	25
Environmentalists	10	11	9	13	15
Congress	26	26	26	28	34
Administration	39	28	28	24	36
Consumers	18	20	18	31	23
Arab / OPEC	22	38	37	32	51

SOURCE: Richman, "The Polls," p. 580, from Roper polls

the various administrations received more blame than consumers and environmentalists, suggesting that people expected the government to solve the problem, and did not perceive a conflict between interest groups pushing for low prices and environmental protection and development of effective policies to deal with the crisis.

The biggest problem facing politicians in trying to formulate policy was that the public either did not understand, or refused to accept, basic economic truths. They wanted a plentiful supply of low-cost gasoline. During 1974 and 1975, when Ford was trying to develop an energy policy public opinion opposed gas rationing, 56-34 percent (January 1974, Gallup), and rejected a tariff on imported oil, 53-27 percent (January 1975, Harris), decontrol of gasoline prices by 88-9 percent (February 1974, Harris), and higher gasoline taxes 78-17 percent (December 1973, Harris). Respondents did prefer mandatory rationing to an oil import tax 60-25 percent (January 1975, Harris).[29]

Carter faced the same problem in trying to develop a policy for dealing with energy which would have sufficient support to get through Congress. A standby tax on gasoline, rising three cents a year for five years was opposed 57-32 percent, though if the alternative was permanent gasoline rationing the tax won out by 48-33 percent. The next year a five cent gas tax was preferred to rationing by 44-42 percent (August-September 1977, Harris). With regard to various policy proposals, people were in favor of increased spending to develop new energy sources, 77-11 percent, merging all federal agencies dealing with energy into a Department of Energy, 76-10 percent, imposing an energy conservation program on both consumers and industry, 70-18 percent, establishing a strategic petroleum reserve, 69-13 percent, but narrowly opposed deregulation of oil and gas prices, 40-38 percent, and strongly opposed raising the price of gasoline to a $1 a gallon, 77-15 percent. (February 1977, Harris).[30]

Thus, despite Carter's appeals to the American people it is not surprising that his program ran into rough sledding in Congress. Carter's proposal for a gas-guzzler tax was disliked by 50-42 percent and a five cent standby gasoline tax was rejected, 63-27 percent, by the general public (June 1977, Harris). The House rejection of the proposed five cent gasoline tax was approved by 71-21 percent (August 1977, Harris). His overall energy program was disliked by a margin of 57-33 percent. The House bill which was passed in the fall also inspired little confidence, with only 11 percent of respondents feeling that it would be very effective (57 percent felt it would be somewhat effective) in getting the country to conserve energy, and 12 percent believing that it would be very effective (55 percent somewhat effective) in increasing supplies of energy. (September 1977, Harris).[31]

As the second energy crisis unfolded, the preferred policy of the public shifted toward increasing energy production, from 41 percent in May 1977 to 57 percent in December 1978, with the gain coming from the ranks of those who had felt that both conservation and increased production should be pursued as policy. However, this was due to a perception that supply could be increased without higher prices but that conservation would require taxes or rationing. The public opposed permitting oil and gas prices to increase in order to encourage development of additional supplies by 76-16 percent, (March 1979, Roper), opposed price decontrol for crude oil by 50-37 percent (April 1979, NBC), and only 32 percent thought that higher prices would cause the oil companies to put out a greater effort to discover oil. (April 1979, CBS/Times). Two months later only 16 percent supported allowing the price of gasoline to rise to provide more money for drilling wells (June 1979, Gallup). Carter's windfall profits tax proposal was supported by only 37 percent, but 41 percent had either not heard of it or had no opinion (August 1979, Gallup).[32]

While Reaganites liked to interpret public dissatisfaction with Carter's handling of the energy crisis as a reflection of dislike of government meddling in the market place, the evidence pointed the opposite way: people wanted more government interference. In an October 1979 Gallup poll people felt that Edward Kennedy would do a better job of dealing with the energy problem than Carter by 50-27 percent. They also felt that Kennedy would be more likely to reduce inflation, 48-23 percent, and unemployment, 49-24 percent. No one ever accused Kennedy of being a disciple of the Chicago School of Economic Policy. The approval rating for Reagan's handling of the energy situation in March 1981 was only 37 percent, which rose to 43 percent in June, then steadily declined to 33 percent in August of 1982, compared to Carter's 33-37 percent rating in the Fall of 1977.[33]

The reason for the decline in the popularity of Reagan's energy policy is found in the dislike of his decision to lift federal controls on domestic crude oil. A *NBC News/Associated Press* poll in February 1981 found that 39 percent disapproved of the action, compared to 31 percent who supported the removal of controls. Roper found in a June 1981 poll that 46 percent opposed and 38 percent were in favor of deregulation of oil prices, allowing the price of domestically produced oil to rise to the level of foreign oil.[34]

A different kind of survey was conducted in April 1979 by two investigators from the University of Michigan. Brunner and Vivian asked 101 citizens nationwide to express their viewpoints on energy policy. Through a procedure known as Q-technique each sorted a sample of 48 statements culled from public discourse by agreement, disagreement or relative indifference. Cluster analysis was then used to develop groups of respondents who share in some degree the same viewpoint. The technique is specialized to the identification of shared viewpoints and qualitative questions of intensity, direction, and meaning. The results revealed that the predominant themes were resistance to energy price increases, widespread distrust of the oil and gas industry, local utilities, the President, and energy experts, and undertones of moral outrage. The core viewpoint had an selective emphasis on those aspects of the energy situation that pertained to their immediate circumstances, ignoring the national implications of issues. The authors suggested that the uncertainty and anxiety felt by individuals increased their demand for emotionally satisfying rather than realistic solutions. One result was scapegoating, in which personal insecurities arising from the energy situation were displaced on various national actors such as industry and government, especially since they were promoting price increases as a solution to the problem when most individuals saw price increases as the problem. Hopes for solutions were centered on consumer and solar energy organizations, who were not associated with either price increases or established actors in the energy debate.[35]

THE MEDIA AND PUBLIC OPINION ON ENERGY

I examined the occurrence of coverage of energy issues since World War II in both *Newsweek* and *Time*, and conducted an overview of the articles in *Newsweek*, under the assumption that it is representative of media coverage, especially since it is considered to be slightly more liberal than *Time*, and therefore more likely to slant its coverage against the major companies. While the magazine did publish a number of articles which cast suspicion on the oil companies role in the energy crises, this was due to the fact that so many newsworthy spokesmen were echoing those sentiments. The White House was also a target for accusations of mismanagement, and blame tended to shift from the companies to Congress as the decade wore on.

Not surprisingly, coverage rose during the Suez crisis in 1956, stayed at a low level during the 1960s, and didn't begin to rise until 1972. Coverage stayed high through 1975, but was actually limited during the Carter attempts to influence policy, though Newsweek did make energy the cover story twice in

1977. 1979-80 saw another peak in coverage, and while a decline occurred in 1981-83, it was less than after the first crisis, suggesting that energy had a higher level of latent importance as news. Since 1984, coverage has declined to levels not seen since the embargo of 1974, but stories tend to be much longer than during the 1960s.

TABLE 5.3 Rate of Occurrence of Articles on Energy Issues (average per year)

Year	*TIME*	*NEWSWEEK*	*Main Themes*
1947-48	5	6	Possibility of spot shortages
1949-50	2	4	Tidelands, Imports
1951-55	3	4	Iran, Tidelands, FTC report
1956-57	9	9	Suez, price boost
1958-66	2	3	World oil glut, oil import quotas
1967-69	3	4	Oil pollution, 1967 War
1970-71	4	6	Alaskan Pipeline, Tight oil market
1972	9	8	Pipeline, Oil Squeeze, Iraq
1973	30	29	Energy Crisis/ Arab Oil Squeeze
1974	29	35	Oil crisis
1975	24	20	Oil politics, OPEC policy
1976-78	7	8	Carter energy politics, oil glut
1979	38	37	Oil crisis pt. 2
1980	15	23	Oil price hike
1981	9	15	Oil cartel on the skids
1982-83	10	8	Oil glut
1984-87	5	5	OPEC's troubles, price crash

The tone of the articles changed over time, reflecting changes in American attitudes toward business and government. For example, the coverage of the release of Federal Trade Commission report on the international oil cartel was decidedly slanted toward the oil companies:

> The FTC report is the result of two years of study under the direction of Dr. Corwin D. Edwards, head of the commission's Bureau of Industrial Economics, a person whose economic theories have sometimes shocked businessmen.
>
> Overseas operations of the oil companies unquestionably amount to big business, and the tendency to monopoly is such that it is virtually impossible for small independents to break in. . . . But such immense amounts of capital are needed that small operators would have virtually no chance anyway.
>
> The big companies are thus a stabilizing influence on international economy. Any move to break up the present holdings for the mere sake of

> division must be weighed against the danger of get-rich-and-get-out operators, a weakened bargaining position for American business, and uncertain supplies.[36]

One of the reasons the oil industry, especially the large companies, received negative coverage was that a Congressional attack was usually considered newsworthy, as when Senator Joseph C. O'Mahoney (D-WY), who had a reputation as a trust buster, launched a probe of the oil industry behavior during the Suez crisis, examining the price hikes. However, the same article also pointed out that many sources felt that the refusal of Texas to increase output, due to the fear of Texas oilmen that they'd be vulnerable to sharp cutbacks when the crisis was over, was responsible for the rise in prices.[37]

Until the mid-1960s, most of the policy issues affecting oil were covered in only an extremely superficial fashion, with little or no explanation or analysis. For example, a mention of the depletion allowance in 1959 merely noted that the majority in Congress accepted the oil industry argument that any cut in the rate would discourage investment in exploration and new production, cost the public more in higher prices than it would save in taxes, and make the US dangerously dependent on overseas oil supplies. The article failed to examine whether there was any merit to that argument, or to refer to spokesmen who might have a dissenting opinion. The only critical article which I came across until 1967 was a column by Henry Hazlitt in 1959 questioning the merits of the Mandatory Oil Import Quota. Hazlitt referred to a pamphlet written by William H. Peterson, an associate professor of economics at New York University, and published by the American Enterprise Association of Washington.[38]

Newsweek is to be commended on its decision to engage Milton Friedman and Paul Samuelson to write a regular column on economics in the 1960s, providing one of the few public platforms for the profession to transmit ideas to the general public. In 1967 Friedman took on the illusion that the oil industry was a bastion of "free enterprise," an attack which had additional credibility considering the conservative bias of the author. Friedman criticized percentage depletion, prorationing and oil import quotas as special favors for the industry. He pointed out that companies with access to import permits were receiving a subsidy of about $400 million a year, and that these various policies cost consumers $3.5 billion a year, according to Gilbert Burck in an April 1965 article in *Fortune*. Friedman suggested that a tariff should be used if import restrictions were needed, which he doubted; that oil from Venezuela should not be restricted at all; that prorationing was an incredibly expensive way to maintain excess capacity for national security; and that only World War III was likely to produce severe supply disruptions. If insurance against an emergency was required, he suggested stockpiling oil, paying for holding reserve wells in readiness, and preparations for limiting non-essential consumption, all of which would be far cheaper than present policies. Friedman bluntly stated that the political power of the oil industry, not national security, was the reason for these subsidies.[39]

One reason that the coverage of the industry turned more negative during the late 1960s was that it was easier to report isolated events than to place them in the context of the operation of the industry and the world market. The Hart hearings drew attention to the politics of depletion and oil quotas. Between 1967-1970 oil pollution received attention, especially the Torrey Canyon and Santa Barbara incidents, intermixed with essentially fluff articles about how oilmen were covering the planet and the seas to find more oil, as in the North Sea and Alaska. There was no attempt to examine how environmental concerns would affect the ability of energy companies to supply the rapidly increasing demand for fuel and electricity. It was not until the fall of 1970 that the first suggestion of possible shortages reached print, with regard to possible brownouts due to the lack of utility fuel.

In 1971 the Alaskan pipeline got most of the limited attention on energy, with OPEC's attempts to raise prices meriting one article. The first major article on an impending shortage of world energy supplies made the magazine in June of 1972, though an earlier article on the pipeline quoted Interior Secretary Morton that the US would be dependent on foreign oil by the mid-1980s. The June article did finally deal with the conflict between environmental requirements and energy supply, but blithely concluded that the solution would be to produce still more energy. By the end of the year recognition that oil supplies were becoming tight was apparent. The term "Energy Crisis" was first used in a cover story in January 1973, which was the first real attempt by the magazine to examine the causes of the energy situation:

> The trouble is the US has no experience with rising energy prices.
>
> Since the giant oil and gas companies will be the primary beneficiaries of the sharply higher prices the Administration wants to engineer, they will be operating in a political fishbowl as never before. Congressmen and small-businessmen were already charging last week that the big oil companies had manufactured the fuel shortages in order to raise prices. . . .
>
> No matter how the political wars over energy policy are finally decided, there will be one contingency that will please no one. No matter what course the nation takes, it can't help but be more and more dependent on foreign sources--meaning mainly the Mideast--for energy during the crunch period of the next dozen years. . . .
>
> It is abundantly clear that the energy crisis is gaining momentum, that the possible solutions are incredibly complex and interrelated--and that the importance of the whole situation extends far beyond the US borders. Yet there is still some concern that Americans aren't really ready to come to grips with the new energy realities.[40]

While the magazine never actually suggested that the oil companies were responsible for the energy crisis, by quoting government and private sources on the one hand and then occasionally presenting industry rebuttals, it put the issue in doubt without actually examining whether the evidence justified the accusations:

> Indeed, the Administration's general hands-off approach served to feed the growing suspicion among some lawmakers that the whole gasoline-fuel-oil shortage was orchestrated by the big companies to wring price concessions from the public and bury the independents, who often sell at cut-rate prices.[41]
>
> The move came amid mounting charges that the nation's 23 major oil companies--which control more than 80 percent of the US's refining and distributing capacity--were using the shortage to drive independent competitors out of business. . . .
>
> . . . "It's apparent that they are forcing out the independents and are planning to replace them with their own . . . stations," Illinois Senator Adlai Stevenson III charged last week. He joined fellow Democrat Hubert Humphrey in calling for "a vigorous" Justice Department antitrust investigation. . . . The oil companies, of course, denied the allegations. The independents are simply greedy, huffed one executive. . . .[42]
>
> With the gasoline shortage becoming worse each day, growing numbers of Federal and state officials are asking: is the shortage legitimate? Or is it due to a conspiracy by oil-industry giants to drive independent competitors out of business, boost their own profits and win major concessions--such as approval of the trans-Alaskan pipeline--from Congress?
>
> . . . Even the normally pro-big-oil Nixon Administration is now suspicious. Richard Fairbanks of the President's Domestic Council said last week that "it is an open question whether the majors are tying to drive independents out of business. They may very well be embarked on such a course."
>
> With the charges piling up, the majors have belatedly started to answer back. . . . Ikard [API-President] blames the shortage in large part on the industry's uncommonly bad luck with its forecasts.
>
> But many critics refuse to believe that the oilmen could have been that stupid. They remain convinced that industry executives saw shortages coming and--without resorting to anything as heavy-handed as secret, illegal meetings--welcomed them as a means to higher profits and a way around environmental restrictions. . . .[43]
>
> Recurrent allegations that the fuel pinch is actually the result of a conspiracy of some sort among the oil giants grew more intense. For example, Senator Henry Jackson, . . . released a damning study of the oil industry made by the Federal Trade Commission. In it, the FTC charged that the petroleum industry "operates much like a cartel, with 15 to 20 integrated firms being the beneficiaries of much Federal and State policy." The FTC charged that these firms control pipelines, make agreements to exchange oil among themselves and have reciprocal arrangements for processing oil, and that this endangers independents, makes it difficult or impossible for new firms to enter the field and causes "serious economic losses to American consumers."
>
> Predictably, the oil companies struck back at the charges of conspiracy.[44]

There were less accusatory articles published after the oil embargo than during the year proceeding it and these were more evenhanded, suggesting that the revelation of the nation's vulnerability to foreign manipulation of the oil market made the oil companies' defense more credible. In fact, there was definitely a shift in tone from blaming the oil companies to disgust with the "do-nothing" Congress.

> Sen. Henry Jackson's permanent investigations subcommittee will try to get some answers in hearings beginning this week. A few points are beyond dispute: the energy problem started long before the Arab oil embargo. And most of the blame can't properly be put on environmental pressures. The subcommittee is not fixing responsibility in advance, but a report by its investigators sketches a sorry picture of an Administration unable first to grasp the magnitude of the problem and then unable to cope with it, and of an oil industry that was guilty at least of poor planning and at worst of selfishly ignoring the public interest.[45]
>
> As charges of conspiracy mounted, hardly a day passed at the new Federal Energy Office without another rumor that the industry was hiding stocks of oil to exacerbate the shortage. Last week, the story was that tankers by the score were anchoring off the US waiting for higher prices before they would unload their cargoes. FEO chief William Simon quickly squelched that one. . . .
>
> But the suspicion of blue-collar workers and liberals toward big business dies hard, and there was more than phantom tankers to keep the critics busy. The major oil companies, for instance, raised their gasoline prices as much as 5.8 cents a gallon, blaming the move on the rising cost of foreign crude oil. The increases were undoubtedly legal, but there was widespread evidence that gasoline dealers were passing on far more than their extra costs. . . . True, most of the violations amounted to only a penny or so a gallon and were quickly corrected; and IRS officials tended to blame much of the illegal pricing on dealer ignorance of the latest rules. . . .
>
> Whatever the investigations eventually turn up, there has been little if any evidence so far the current shortages are trumped up. . . .[46]
>
> If the industry's estimated needs are anywhere near accurate, however, oil profits clearly will not be enough to finance its huge capital programs. The spectacular run-up in earnings last year was basically a one-shot affair, the result of inventory profits. . . .
>
> . . Although the oil industry is quick to argue the pitfalls of government involvement, it is nonetheless looking to Washington to set energy priorities. What it is getting instead is indecision and partisan squabbling. . . .[47]
>
> Exhaustive Congressional hearings and studies through the years have never really proved that the companies have monopolistic clout. . . . Yet bigness alone is justification enough for divestiture, in the view of some lawmakers. . . .
>
> Much of the new support for the divestiture may be prompted more by a desire to please the voters and by the senators' own ineffectuality in solving the energy crisis. . . .[48]

A study by the Media Institute, a pro-business organization which researches media treatment of issues, also found that television tended to slant its coverage against the oil companies and ignored economics when discussing the oil crisis. The study analyzed 39½ hours of evening news, including 1,462 stories spanning the two energy crises, covering the periods 1973-74 and 1978-79. Researchers found about four subjects in the average story, each subject accounting for an average of 25 seconds of news, for a total of 5,620 separate instances in which subjects were raised. During the first crisis there were 994 stories playing for a little more than 24 hours, covering 3365 subjects, while the second crisis coverage involved 468 stories, for a total airtime of over 15 hours and covering 2259 subjects. About 18 percent of the subjects came from unidentifiable sources, while the remaining sources used included the Federal Government (40%), the Oil Industry (17%), Congress (12%), Non-oil industry (7%), OPEC (6%), State and local government (4%), and "Experts." The subjects covered included Solutions (33%), Causes (18%), Effects (23%), other topics (26%).[49]

Network coverage of solutions tended to stress conservation (excluding price-induced conservation) and rationing. Conservation was recommended 9 percent of the time, seen as successful (3%) vs. unsuccessful (2%), and various schemes described (12%). Rationing was advanced as a solution 12 percent of the time, considered undesirable (4%), and otherwise mentioned (5%). Other solutions which received airtime were Regulation and Price Controls (11%), Deregulation & Decontrol (4%), Taxes (7%), Price Mechanism (9%), and Domestic Production (8%). There were some shifts in coverage between the two crises. In the first crisis, almost all discussion of regulation was in support of the government's regulation of the industry and price controls were discussed without any indication that controls might be harmful. In the second crisis, coverage of regulation as a solution (including price controls and their repeal) increased from 12 percent of all solutions to 21% percent. Coverage of deregulation received increased coverage only when the Administration endorsed decontrol, and opposition to decontrol was given equal time. During the first crisis coverage of domestic production focused on the tradeoff between environmental goals and increased production, while during the second crisis the question was whether decontrol could increase production enough to reduce dependence on foreign sources of oil. The availability and price of oil were depicted as twin problems as if they were unrelated, though price-induced conservation was mentioned.[50]

A serious barrier to objective coverage of energy problems was that the government dominated discussion of the issue. The media turned to the Federal Government for solutions 55 percent of the time, Congress (18%), Oil Industry (9%), "Experts" (1%), other sources (17%). The government's dominance of the discussion of solutions increased in the second crisis, due to the increase in the role of Congress as a source from 14 percent to 26 percent. On the other hand, the oil industry was the source of 11 percent of all solutions in the first crisis, and only 6 percent in the second. The networks' use of the government for information parallels the finding of a previous study on coverage of inflation, which found that government was the source of 77 percent of inflation stories, and that government was exonerated as a cause of inflation in 77 percent of stories broadcast.[51]

The same pattern was seen in the discussion of causes, with sources dominated by the Federal Government (33%) and Congress (16%), along with OPEC (15%), the Oil Industry (17%), Experts (3%), other (15%), and an additional unidentifiable 13 percent. In the first crisis, OPEC and the Arab oil embargo were the subject of 43 percent of all discussion of causes, and OPEC and the Iranian cutbacks accounted for 26 percent in the second crisis. The oil industry was considered 32 percent of the time in the first disruption, and discussed as a cause 42 percent of the time in the second crisis. The US government was a distant third in both crises accounting for 18 percent of the discussion in the first crisis and 19 percent in the second. Other causes receiving mention included consumers and unspecified hoaxes and finite supplies of oil. OPEC welcomed blame, with television rarely questioning OPEC assertions, accepting OPEC's claim that embargoes and the like were an effective political weapon. The networks depicted OPEC members as greedy and capricious, ignored the economic motivations behind cartel behavior, and failed to explain why "moderates" might wish to set a lower price for their crude than "price hawks" (ABC) or "hard-liners" (CBS).[52]

The coverage discussing industry blame was in general terms 31 percent of the time, involving hoax or conspiracy (17%), excessive profits and price-gouging (11%), holding back supplies (11%), supplying unreliable data (9%), and denial of blame (20%). When the government was mentioned with respect to causes, it was usually in general terms or with reference to the allocation system, with regulation and price controls a distant third. Government sources were volubly critical of the oil industry, with Congressional hearings a fertile sources of dramatic film footage as a number of congressmen were shown chewing out oil company executives for the benefit of the folks back home. Reports in both crises discussed the possible effects of oil company campaign contributions to President Nixon and members of Congress. Though denials of oil company blame came from outside the industry, the networks managed to cast doubt on their validity. ABC, for example, reported on August 14, 1979, that the Energy Department had released a 50-page report clearing the oil industry of hoarding supplies to hold out for higher prices, but then proceeded to quote claims that the report was "misleading" and "an exercise in propaganda," and insinuated that the report was based on information provided by the oil companies. Even when experts were consulted, the choice of someone like Christopher Rand, characterized by CBS as a Mideast oil expert and former oil company employee, raised questions of network objectivity.[53]

Kent MacDougall noted that television news coverage is far more damaging to business than books slamming industry for corruption and environmental irresponsibility. A successful book may reach only 20,000 buyers, but TV news is the major source of news for the public. He points out that the most striking aspect of TV coverage of business is its paucity, with business stories focusing on White House announcements, Congressional hearings, and other developments laden with audience-pleasing conflict and confrontation. Although these stories usually include a rebuttal by the business under attack, they leave a negative impression. The oil industry blamed the news media for the public's belief that the companies concocted the energy shortage to justify higher prices, because it provided a forum for the charges of politicians using the energy issue for their own political advantage. Since many of the politicians who made what the industry calls irresponsible statements were highly placed federal government officials, the press could hardly be expected to ignore their actions.[54]

However, there is some evidence that members of the press were a little irresponsible in their willingness to scapegoat the oil companies. An example was the coverage given the General Accounting Office report on its investigation into whether American oil producers had deliberately worsened the 1979 oil shortage. The *New York Times* story said the GAO found that "American oil companies had aggravated last spring's petroleum shortage by cutting production of crude oil within the United States last winter while imports from Iran were disrupted," while the *Wall Street Journal* said the GAO concluded "there isn't any evidence that the major oil companies created the US oil shortage that occurred after the closing of Iran's oil fields." *The Wall Street Journal* gave the correct version and the Times ran a lengthy correction two days later.[55]

The need for the media, especially the electronic media, to portray the news as a morality play, combined with the innate American predisposition to distrust businessmen in general and large corporations in particular, made the major oil companies the perfect villains. The incidence of immoral and unethical behavior by businessmen as characters on prime time TV expresses the media's underlying attitude that dishonesty is normal behavior in the business world. Since few reporters in any medium had sufficient training or knowledge to understand the oil market or resource economics, much less economics in general (economics is rarely a required subject for graduation from most universities for journalism or communication majors), it is not surprising that faced with a complex and difficult subject, they gravitated to simple explanations. The simplest and most powerful script was some sort of conspiracy involving the oil companies raising oil prices to take advantage of the American public.

The oil industry viewed television as often unfriendly, inaccurate, and superficial, especially in comparison to newspapers and newsmagazines. There is some truth to this accusation as television is better suited to capture high drama and transmit emotional experience than it is to present detailed information and explain complex issues. The actual content of a half-hour news show, reduced to type, would fill only half the front page of a standard-sized daily newspaper. The networks have hired business specialists and tried to make an effort to lift the level of business coverage. But the networks also employ investigative reporters whose coverage is adversarial. MacDougall gives an example of an investigation of purported improprieties in the oil industry conducted by Brian Ross of NBC, aired in five segments on successive evenings one week in 1979. He accused Exxon Corporation of reducing shipments of low-sulphur fuel oil to a Florida utility in favor of supplying more profitable petroleum products to customers in Europe, and the Shell Oil Company of possibly stranding heating oil customers in winter. The key points of rebuttals by Exxon and Shell officials never made the air. Exxon and Shell complained to the National News Council, an independent group that monitors media performance, which ruled in the companies' favor, and two National News Council members labeled the Ross report "strident and accusatory," biased and at times deceitful, and "irresponsible."[56]

Oil companies complained that the news media reported soaring oil profits unsympathetically and out of context, though the news media's most serious shortcoming is that they report profits simplistically. Journalists accept corporate financial statements at face value, rarely questioning the accounting conventions that enable companies to report high profits to investors and the public, while at the same time reporting lower profits to the tax collector. While some journalists suggested that the oil companies doctored their books to overstate foreign income and understate US income, reporters generalized about oil industry bookkeeping manipulation since they were unable to prove any individual company guilty of it. One way of making accusations without evidence was demonstrated by a *CBS Evening News* story on oil profits and bookkeeping gimmickry. The story did not specifically accuse Mobil, but correspondent Ray Brady clearly implicated the company by concluding his general accusations posed in front of a Mobil filling station sign. Mobil asked for air time to correct the impression left by the broadcast, but CBS rejected the request, which struck even journalists as high-handed and unfair, given the network's exclusion of industry comment from the story and the seriousness of the charges.[57]

The oil companies attempted to fight this negative exposure through public relations efforts. Among the oil companies, Mobil was the leader in fighting the public relations battle, with a public relations budget of $21 million and a staff of 73 in 1980. Well known for its sponsorship of quality television drama and its newspaper opinion ads, Mobil also fed special reports and interviews to television. Since 1976, Mobil produced at least a dozen televised special reports on energy issues, with each report featuring a TV journalist interviewing Mobil executives and other experts who supported the company's energy positions. The reports were sent for free to television stations, which aired them late at night and on weekends, with acknowledgment of Mobil's involvement only in a credit line that flashed briefly on the screen at the end. Mobil sent out individual interviews to local stations to be inserted into newscasts, usually without identifying the company as the source. Atlantic Richfield also produced energy programming, but the company's efforts were far more credible than Mobil, as many stories failed to mention ARCO and others included comment from oil industry critics. Oil executives appeared frequently on panel interview shows such as "Meet the Press" to present their side of the issues.[58]

There is also some evidence that the large oil companies moderated price increases in heating oil prices during 1979 and 1980 in response to media pressure. While Erfle attempts to attribute this to the threat of regulation I think this is a stretch.[59] It is doubtful that the major American oil companies really thought that delaying increases in heating oil prices would have a significant impact on the political climate. A simpler explanation would be that it was an investment in good will, or at least damage control, in limiting the pain caused by a price increase in a politically sensitive commodity. Large oil learned from experience over the years to avoid the price leadership role when it meant lowering prices to producers or raising prices to consumers. Given market pressures, prices would inevitably rise, triggering a regulatory response if politically feasible. All price restraint could have achieved was to avoid specific attention from the media.

These efforts by the companies to counter their negative image in the media were doomed to failure because information coming from clearly self-interested sources is distrusted by the public. Polls by Yankelovich, Skelly & White (YSW) in 1976 and 1978 with regard to the credibility of a number of opinion sources showed that the least believable sources were "public issue ads" (an average of 8 percent of the public always or usually believed such advertisements, while 51.5 percent seldom or never believed them). While corporate executives and union leaders were not considered credible, Ralph Nader was the source most likely to be accepted (44.5 percent found him always or usually believable, while 19 percent said they seldom or never believed him). Similar results were obtained by Cambridge Reports, Inc., in an August 1980 poll that asked whether each of five groups or institutions "represents your interests" on the issue of energy policy and environmental policy. "Scientific experts" and "Ralph Nader and other consumerists" were considered most likely to support people's interests. The influence of the media was demonstrated by the finding that television commentators and newspapers were also viewed positively in the YSW poll, with close to two fifths of respondents believing them most of the time, compared to less than one fifth who did not. In effect, people discounted the credibility of sources according to their perceived self-interest.[60]

PUBLIC OPINION AND ENERGY POLICY

A revealing dialogue occurred in the pages of *Public Opinion* between Pat Caddell and Kevin Phillips in 1978. Kevin Phillips critiqued the Carter Administration's energy plan for misreading the public mood. Phillips maintained that the White House miscalculated the degree of public willingness to rest the economic future of the Untied States on energy conservation instead of expanding energy supplies, making a valid point that people identified conservation with a lower standard of living, which President Carter implied with his emphasis on the need to cut back on consumption. He was also correct in pointing out that except for a brief period, Carter was unable to convince the American people that the energy crisis was serious enough to justify economic sacrifice. Phillips however, biased by his ideology, made the mistake of thinking that the American people preferred higher prices to higher taxes or that they trusted the oil industry to find new supplies. He was correct in stating that Americans wanted new energy supplies, but he didn't understood that they wanted them at the old energy prices.[61]

Caddell's reply in the next issue revealed the idealistic bent of the Carter administration, which was determined to force a semblance of a rational energy policy on the nation. Caddell denied that Carter's energy policy was based on survey data, pointing out that given the results of polls Carter should have done what his predecessors did: nothing. As he points out, "it would be next to impossible to construct a popular energy policy, since large segments of the public oppose most possible solutions if they cost anything, whether that cost is expressed in taxes, higher prices, or environmental risk." Caddell pointed out that the principal energy problem according to most people was high prices. Whether those prices are the result of taxes or decontrol doesn't matter; if price is the problem then raising prices will not be perceived as a solution. What support existed for deregulation was dependent on confidence that higher prices would result in increased supplies, which required trust in the oil and gas industry that was almost non-existent. Even if Carter had been able to mobilize support, since attitudes concerning the energy crisis and trust in government did not effect the willingness to conserve energy voluntarily, it is doubtful that increased public confidence in Carter would have made higher energy prices more acceptable.[62]

There were three problems which combined to make policy formation practically impossible: the ignorance of basic economic principles among most of the American public, including much of the elite; the innate populist prejudice against big business which found a perfect target in the oil companies; and the loss of confidence in government due to Watergate. Even before the embargo the oil companies were suspected of manipulating supplies to cause shortages, while the source of the accusations by the independents was a matter of chagrin at the loss of a needed subsidy. This is not to say that the oil companies were the best of citizens; during the period between 1973-1981 the companies were found guilty of massive overcharges for crude and refined oil in their attempts to circumvent Federal price controls. With the agreement with Texaco in 1987 the Energy Department managed to collect more than $7 billion in settlements and judgments against oil companies.[63] But the reason for the willingness of politicians and "public interest" spokesmen to assume that the companies were responsible for oil shortages lies with the traditional animosity toward big business.

The small, individually owned or family owned business has always been socially acceptable, but Americans tend be wary of "big business," which is associated with monopolies, excessive power and profits, misleading advertising, and ruthless competitive practices. When Americans are asked whether big business has "too much power" in American life, 82 percent say yes, compared to 35 percent when the question is asked with regard to business in general. This attitude can be traced back to the anti-business tone of the Progressive era, based on a desire to reinstate the business climate of pre-industrial America, when success presumably depended solely on individual talent and integrity. One of the leading Progressive activists against big business was Louis Brandeis, who went as far as supporting price fixing and cartels to protect small business from the competition of large, more efficient competitors. For Brandeis and many others, the very size of large firms made them anathema, a basically emotional attitude which required that efficiency be regulated to a secondary consideration.[64] That this attitude has persisted can be seen in the willingness to subsidize small, often inefficient, producers and refiners, whether through tax breaks, entitlements under the Oil Import Quota Program of the 1950s or the Petroleum Allocation Program of the 1970s.

Along with the traditional animosity to big business, there was a decline in confidence and trust in business during the 1960s and 1970s which was paralleled by a widespread loss of faith in the leadership of government, labor, and other private and public institutions at more or less the same time. Lipset and Schneider ascribe this lack of trust and confidence to a basic paradox in the American belief system:

> Americans hold their political and economic institutions in high esteem and realize that it is the exercise of competitive self-interest that makes those institutions work. But at the same time, they do not admire self-interested behavior and look with suspicion on anyone who has succeeded in acquiring political or economic power.

The result is that Americans tend to separate the basic legitimacy of the institutional order from the question of confidence in particular leaders. The public is predisposed to turn against their leaders when things go wrong and to assume that more honest or competent power-holders could make a fundamentally sound institutional structure work. Those who voice the severest attacks on government, business, and labor usually do so in terms of abuses of power and betrayals of the public trust. But even they, like most Americans, are well aware of the positive and indispensable contributions these institutions make to the society as a whole and to their own lives. In other words, criticisms are usually leveled at the behavior of power-holders and not at the underlying structure and functions of the institutions themselves.[65]

The decline in confidence in business in general has been a gradual trend since the 1950s. It seems that there has been a general acceptance of the necessity of large companies as instruments of economic growth, combined with a growing concern with the concentration of economic power in the hands of these companies. There is more willingness to consider measures to counter this power, though the proportion of people who are wary of big business is much larger than the numbers who would actually want to take action to counter business power, suggesting that the existence of large companies is considered a necessary evil by much of the public.[66]

Polls demonstrate that Americans do not accept the implications of a capitalist system even though they value the institution of competitive free enterprise. A CRI poll conducted in early 1979 found Americans valuing consumer and public interests more than those of stockholders and workers. When asked: "Whose interests do you think a business should put first...," 35 percent replied that a corporation should be most concerned with the interests of the American people, 32 percent emphasized consumer interests, while 9 percent placed the stockholders first, and 7 percent said workers. The nonideological nature of the public attitude toward business is shown by the fact that liberals, moderates, and conservatives were alike in their populist resentment of profits and moneymaking. Business is justified in terms of two types of values; individualistic values which legitimatize rewards for those who show talent, hard work, and achievement, and social values that expect business to produce goods, provide jobs, and serve the public interest. The public discomfort with "rational economic" behavior of business is reflected in attitudes toward profits. The public has traditionally had an exaggerated notion of the magnitude of profits. When asked the actual profit on a dollar of sales for the average manufacturer, estimates of profit margins in ORC polls taken from 1971 to 1981 averaged between 25 and 37 cents (37 cents in 1983), 4 to 6 cents higher than what people felt was reasonable, and far above the 4.9 cents per dollar of sales actually earned by companies. People thought that in 1981 oil companies made 62 cents per dollar of sales. When asked by Roper how much the price of gasoline would be reduced if profits were passed along people said 13 cents in 1975 and 32 cents in 1979.[67]

Most people do not have a very good understanding of what profits are, confusing markup with profits. This confusion about profits led people to some strange conceptions, as a 1975 Procter & Gamble poll found that business profits were blamed by 56 percent as a major cause of inflation, with only federal spending, cited by 69 percent, and oil prices, mentioned by 85 percent, more frequently named.[68] Given the high profits reported by the oil companies, the escalating price of oil, and the high level of concern with inflation the negative impression of the oil industry was inevitable. In effect, the oil companies were held accountable for shortages of heating oil and gasoline during periods when price controls distorted the market, and then held responsible for inflation when prices were allowed to rise and company profits climbed.

The resentment against the oil companies, concern with inflation, and a general acceptance of price controls made deregulation of oil prices an action akin to political suicide. The public was very sympathetic to price controls during the 1970s, especially since they saw excess profits as a contributing cause to inflation. The illusion, expounded by conservatives, that Americans are against regulation and interference with the market simply is not supported by the evidence of the polls. What determines public attitudes is how regulation is defined, who is being regulated, and for what purpose. People do not like regulations that interfere with "free enterprise" (which the American people seem to view as an idealized combination of the best of Adam Smith and Thomas Jefferson), telling innocent firms what to produce, what to pay workers, what to charge for products. On the other hand regulation that involves protecting the public interest from business people who charge high prices, pollute, endanger lives, and indulge in corruption is approved.

In 1974 the most commonly stated justifications for regulation were monopolies, tax loopholes, excessive profits, and concentration of power, and the need to stabilize or freeze prices. Opponents of regulation felt that regulation favors big business over small business, protects monopolies, controls or limits competition, interferes with production and free enterprise, allows for price-fixing, and creates too much red tape. Those who say that there is too much regulation defend their position by expressing the same values and priorities as those who say there is not enough regulation! Not surprisingly, given the populist tendencies of the American people, the consumer movement is seen in a favorable light, and it is credited with obtaining improvements in the area of consumer protection. 77 percent of respondents to a 1976 Harris poll agreed that "the consumer movement has kept industry and business on their toes" and 69 percent agreed that "the consumer movement has helped . . . to improve the quality and standards of the products and services people buy."[69]

The limits of public acceptance of governmental involvement with business is thus delimited by the desire to be protected from adverse impacts of the economy while preserving the illusion of laissez-faire. Thus, while the public was split over the issue of oil company divestiture (a slim and unstable majority in favor), as unpopular as the oil industry was there was never anything close to a majority in favor of nationalization. The debate over divestiture centered about whether it would lead to greater government involvement, increase competition, or decrease efficiency and raise prices. While people felt that divestiture would reduce the power of the large oil companies (72-14%), lower oil prices (48-36%) due to increased competition, and create more jobs (56-25%), while not lowering efficiency or raising prices (46-34%), they also felt that big companies were more efficient than small oil companies (52-27%), that divestiture would hinder competition with foreign oil companies (42-37%), and that oil companies did not deserve punishment (44-30%). However, when it came to nationalization, polls from 1974 to 1980 showed a near majority ranging between 43 and 49 percent opposed while support for nationalization oscillated between 20 to 35 percent. During this time there was no steady upward trend in favor of the nationalization of the oil companies, despite the rise in prices.[70]

Wildavsky and Tenenbaum described some basic differences between different groups involved in the development of energy policy:

> Those whose outlook is guided by quantity theories are similarly divided into consumerists (who want prices kept down) and preservationists (who call for reduced demand) . . .
>
> The most deep-seated distinctions between quantity theories and price theories are moral. . . .
>
> In a quantity-oriented model, decisions by those who decide for others are justified on the basis of their presumed merit. . . . The correct quantity is determined by intellectual argument and judgment; . . . Quantities can be "correct" or "incorrect," but prices can only be agreed upon.
>
> Preservationists think oil prices are unfair because the intrinsic value of oil is greater than its merely monetary price. Consumerists reject price, but for other reasons:
>
> (1) An uncontrolled price system is impervious to unequal distribution of income among participants and unequal treatment of regions. . . . The posted price seems unjust because it represents a massive redistribution of money from millions of small consumers to a handful of large producers. . . .
>
> (2) The second problem consumerists have with unregulated price is that they believe major oil companies, like monopolies, have the power to set prices and promote favorable legislation. . . . It follows . . . that the income of oil companies is believed too high, after allowing for costs of production plus a fair profit.
>
> And once it is believed that oil companies are monopolies and that competition cannot exist, quantity theories may be justified on the grounds that government will make decisions just as intelligently as a giant monopoly, and perhaps more equitably.[71]

Public opinion supported the policies of Wildavsky and Tenenbaum's quantity theorists, especially the consumerists with their promise of cheap, abundant energy. The basic contention of the consumerists, that the monopolistic structure of the oil industry had resulted in the exploitation of the public, was generally accepted. The views of free market advocates were rejected along with the concerns of the preservationists; people were neither willing to accept the verdict of the market that higher prices were needed to balance supply and demand nor the contention that the energy problem was the result of increasing scarcity which would require reduced consumption.

These attitudes were not limited to the general public; they were also held by many Congressmen, especially in the Democratic party and from energy consuming states. I do not think that these attitudes reflected a desire to parrot the sentiments of their constituents. There is a tendency to assume that political elites must be more sophisticated than the general public, but, given the background of most politicians one must doubt whether many Congressmen are any more capable of understanding economic issues than their constituents. While they have the resources to educate themselves on the issues, they can easily surround oneself with "experts" who are willing to support attitudes based on emotional prejudices. I would point out that David Stockman was praised for his economic acumen, yet, reading his account of Reagan administration economic policymaking, it is obvious that his knowledge of economics is extremely rudimentary and quite distorted.[72] Given the tendency of professional economists to point out the difficult choices involved in intelligent policy making it is not surprising that they do not dominate the debates on economic policy. One does not have to postulate a cynical, re-election focused politician to explain the willingness of Congressmen to scapegoat the oil companies. Faced with a problem beyond their comprehension they chose the same posture as their constituents; to believe in theories which provided easy solutions.

Public opinion polls should also be viewed with care, as the wording of questions and posing of issues can give a false impression of public acceptance of policy. A perfect example of the confusion which can result from allowing one's prejudices to interfere with one's judgment is found in an article postulating widespread support for alternative energy. The authors quoted a number of polls that show that the public strongly favored further development of renewable energy resources, fairly vigorous promotion of energy conservation, and was against further development of nuclear power. They noted that most people prefer increased use of solar and wind energy over any other source, and support programs promoting conservation as long as they do not raise prices or increase taxes. They then concluded that the Reagan administration's policies were incongruent with public opinion. The fallacy in their argument was that polls which ask people which energy source they prefer usually do not include the costs of supplying that energy source. Of course people prefer renewable energy and conservation programs which are costless! An example of this type of question was when people were asked if they approved the part of the House passed energy program in 1977 "Increasing the strip mining of coal, but also protecting the environment" which was approved by an 80-8 percent margin (September 1977, Harris).[73]

What public opinion polls can reveal are the issues over which political conflict will polarize. When the public is demonstrating what can only be called "political cognitive dissonance," for example, when people want to reduce budget deficits without cutting spending or raising taxes, or they would like cheap, abundant energy or to increase energy production without environmental damage or higher prices, then politicians are placed in a quandary. As Walter Mondale demonstrated so clearly, the one thing you cannot do is to be explicit about the impossibility of meeting these conflicting goals. Historically, the bearer of bad news has had a short life expectancy. The key is determine which points have higher saliency. Given an accurate estimate of the more important issue in the trade-off, the rational politician promises the public to do both, then when making policy sacrifices the less important objective and hopes that his achievement in producing progress on the more salient issue outweighs the damage caused by alienating supporters of the other issue.

NOTES

[1] V.O. Key, Jr., *Public Opinion and American Democracy* (New York: Alfred A. Knopf, 1961), p. 14.

[2] David Truman *The Governmental Process* (New York: Alfred A. Knopf, 1951), pp. 219-21; Lance Bennett, *Public Opinion In American Politics* (New York: Harcourt Brace Jovanovich, 1980), pp. 9, 15-18.

[3] Benjamin Page, Robert Shapiro and Glenn Dempsey, "What Moves Public Opinion?," *American Political Science Review* 81 (March 1987): 23-43; Harry Holloway and John George, *Public Opinion: Coalitions, Elites, and Masses* (New York: St. Martin's Press, 1979), p. 252; Herbert J. Gans, *Deciding What's News* (New York: Vintage Books, 1980), p. 224.

[4] Gans, *Deciding What's News*, p. 181; Holloway and George, *Public Opinion*, pp. 244-46; Carol H. Weiss, "What America's Leaders Read," *Public Opinion Quarterly*, 38, Spring 1974, pp. 1-22.

[5] Gans, *Deciding What's News*, p. 161-62.

[6] Michael B. MacKuen and Steven L. Coombs, *More Than News: Media Power In Public Affairs* (Beverly Hills: Sage Publications, 1981), pp. 98, 122.

[7] Shanto Iyengar and Donald R. Kinder, "More Than Meets The Eye: TV News, Priming, And Public Evaluations of the President," in *Public Communication And Behavior*, Volume 1, ed. George Comstock (Orlando, Fla.: Academic Press, 1986), pp. 136-71; Shanto Iyengar, Mark Peters, and Donald Kinder, "Experimental Demonstrations of the 'Not-So-Minimal' Consequences of Television News Programs," *American Political Science Review* 76 (1982): 848-58.

[8] Page et al., "What Moves Public Opinion?," p. 23-43.

[9] William J. McGuire, "The Myth of Massive Media Impact: Savagings And Salvagings," in *Public Communication And Behavior*, Volume 1, pp. 173-257.

[10] Angus Campbell, Philip Converse, Warren Miller and Donald Stokes, *The American Voter* (Chicago: University of Chicago Press, 1960), pp. 171-80.

[11] Norman Nie, Sidney Verba and John Petrocik, *The Changing American Voter* (Cambridge, Mass: Harvard University Press, 1979).

[12] Gary Jacobson, *Money in Congressional Elections* (New Haven: Yale University Press, 1980), pp. 11-15; Donald Stokes and Warren Miller, "Party Government and the Saliency of Congress," *Public Opinion Quarterly* 26 (1962): 531-46; Morris P. Fiorina, "The Case of the Vanishing Marginals: The Bureaucracy Did It," *American Political Science Review* 71 (Mar. 1977): 177-81; Richard F. Fenno, Jr., "If, As Ralph Nader Says, Congress Is 'The Broken Branch,' How Come We Love Our Congressmen So Much?" in *Congress in Change*, ed. Norman J. Ornstein (New York: Praeger Publishers, 1975), pp. 277-87.

[13] Morris P. Fiorina, "The Presidency and the Contemporary Electoral System," in *The Presidency and the Political System*, ed. Michael Nelson (Washington DC: CQ Press, 1984), pp. 207-12.

[14] Gary C. Jacobson, "Running Scared: Elections and Congressional Politics in the 1980s," in *Congress: Structure and Policy*, ed. Matthew McCubbins and Terry Sullivan (New York: Cambridge University Press, 1987), pp. 39-46.

[15] Jacobson, "Running Scared," pp. 65-70.

[16] Michael J. Robinson, "Three Faces of Congressional Media," in *The New Congress*, ed. Thomas E. Mann and Norman J. Ornstein (Washington D.C.: American Enterprise Institute, 1981), pp. 73-90.

[17] David Sears, Richard Lau, Tom Tyler and Harris Allen, "Self-Interest vs. Symbolic Politics in Policy Attitudes and Presidential Voting," *American Political Science Review* 74 (1980): 679-82.

[18] Benjamin Page and Robert Shapiro, "Effects of Public Opinion on Policy," *American Political Science Review* 72 (Mar. 1983): 182; Alan Monroe, "Consistency Between Public Preferences and National Policy Decisions," *American Politics Quarterly* 7 (Jan. 1979): 17.

[19] Anthony Downs, "Up and Down With Ecology--the "Issue-Attention Cycle," *Public Interest*, 28 (1972): 39-41.

[20] Charles O. Jones,"Speculative Augmentation in Federal Air Pollution Policy-Making," *Journal of Politics* 36 (1974): 445-53; J. Clarence Davies III and Barbara Davies, *The Politics of Pollution*, 2nd ed. (Indianapolis, Dobbs-Merrill, 1975), pp. 26-57.

[21] Douglas River and Nancy L. Rose, "Passing the President's Program: Public Opinion and Presidential Influence in Congress," *American Journal of Political Science* 29 (May 1985): 192-95.

[22] Samuel Kernell, *Going Public: New Strategies Of Presidential Leadership* (Washington DC: CQ Press, 1986), pp. 23-25, 225-26.

[23] Washington Post, July 26, 1973, reprinted in US Congress, House, Committee on Energy and Commerce, *Strategic Petroleum Reserve*, *Hearing before the Subcommittee on Fossil and Synthetic Fuels*, 99th Cong., 2nd sess., Mar. 4, 1986, p. 472.

[24] George Gallup Jr., *Gallup Poll* (Wilmington, Del.: Scholarly Resources, 1987).

[25] Gallup, various polls; and Barbara C. Farhar, Patricia Weis, Charles T. Unseld and Barbara A. Burns, *Public Opinion About Energy: A Literature Review*, SERI TR-53-155, June 1979, p. 58.

[26] Al Richman, "The Polls: Public Attitudes Toward the Energy Crisis," *Public Opinion Quarterly* 43 (Winter 1979): 577.

[27] Richman, "Public Attitudes Toward the Energy Crisis," p. 578.

[28] Gallup, various polls; and Richman, "Public Attitudes Toward the Energy Crisis," pp. 579-80.

[29] Thomas E. Fusso, "The Polls: The Energy Crisis in Perspective," *Public Opinion Quarterly* 42 (Spring 1978): 131-33.

[30] Fusso, "Energy Crisis in Perspective," p. 134.

[31] Fusso, "Energy Crisis in Perspective," pp. 133-35.

[32] Richman, "Public Attitudes Toward the Energy Crisis," p. 583; Gallup, pp. 201, 251.

[33] Fusso, "Energy Crisis in Perspective," p. 134; Gallup. p. 78.

[34] Seymour Lipset and William Schneider, *The Confidence Gap*, rev. ed. (Baltimore, Johns Hopkins University Press, 1987), p. 233.

[35] Ronald Brunner and Vivian Weston, "Citizen Viewpoints on Energy Policy," *Policy Sciences* 12 (1980): 147-174.

[36] "Oil Firms Face Hottest Fight," *Newsweek*, June 30, 1952, pp. 46-47.

[37] "Troubled Oil," *Newsweek*, February 18, 1957, pp. 86-88.

[38] "Black Magic," *Newsweek*, June 1, 1959, pp. 74-76; Henry Hazlitt, "Oil Import Quotas," *Newsweek*, September 21, 1959, p. 107.

[39] Milton Friedman, "Oil and the Middle East," *Newsweek*, June 29, 1967, p. 63; Gilbert Burck, "U.S. Oil: A Giant Caught in Its Own Web," *Fortune*, Apr. 1965, pp. 113-19. Burck's article is the only serious reference to the PSC study which I came across in the popular literature. Burck actually referred to a number of articles in the Economic literature on oil, including the 1962 McKie and McDonald paper and Adelman's 1964 SEJ article.

[40] America's Energy Crisis," *Newsweek*, Jan. 22, 1973, pp. 52-60.

[41] "Mr. Nixon Drills A Dry Hole," *Newsweek*, Apr. 30, 1973, pp. 58-59.

[42] "A Step Toward Rationing," *Newsweek*, May 21, 1973, p. 76.

[43] "Is the Big 'Shortage' Just Gas?" *Newsweek*, July 2, 1973, pp. 59-60.

[44] "No Shortage of Complaints," *Newsweek*, July 23, 1973, pp. 58-62.

[45] "Who's to Blame For the Crisis?" *Newsweek*, Dec. 3, 1973, p. 89.

[46] "Energy: No Shortage of Suspicions," *Newsweek*, Jan. 14, 1974, pp. 63-64.

[47] Big Oil's Barrel of Problems," *Newsweek*, August 11, 1975, pp. 63-64.

[48] "Big Trouble For Big Oil," *Newsweek*, Oct. 27, 1975, pp. 81-82.

[49] Leonard J. Theberge, ed., *TV Coverage Of The Oil Crises, Vol. 1., Qualitative Analysis* (Wash., D.C.: Media Institute, 1982), pp. 5-7.

[50] Theberge, *TV Coverage Of The Oil Crises*, pp. 13, 17, 19-22.

[51] Theberge, *TV Coverage Of The Oil Crises*, pp. 25-26.

[52] Theberge, *TV Coverage Of The Oil Crises*, pp. 27-31, 37.

[53] Theberge, *TV Coverage Of The Oil Crises*, pp. 30-33.

[54] A. Kent MacDougall, *Ninety Seconds To Tell It All* (Homewood, Ill: Dow Jones-Irwin, 1981), pp. 10-14.

[55] MacDougall, *Ninety Seconds To Tell It All*, pp. 21-22.

[56] MacDougall, *Ninety Seconds To Tell It All*, pp. 46-50.

[57] MacDougall, *Ninety Seconds To Tell It All*, pp. 70-75.

[58] MacDougall, *Ninety Seconds To Tell It All*, pp. 117-123.

[59] Stephen Erfle and Henry McMillan, 'Media, Political Pressure, and the Firm: The Case of Petroleum Pricing in the Late 1970s," *Quarterly Journal of Economics* 105 (Feb. 1990): 118, 133.

[60] Lipset and Schneider, *Confidence Gap*, pp. 246-47, 364-65.

[61] Kevin Phillips, "The Energy Battle: Why the White House Misfired," *Public Opinion* 1 (May/June 1978): 9-14.

[62] Patrick H. Caddell, "Why Kevin Phillips Misfired," *Public Opinion* 1 (July/Aug. 1978): 54-58; David Sears, Tom Tyler, Jack Citrin, and Donald Kinder, "Political System Support and Public Response to the Energy Crisis," *American Journal of Political Science* 22 (Feb. 1978), p. 76.

[63] "Texaco Will Pay $1.2 Billion to US For Overcharges," *New York Times*, Feb. 24, 1988, p. 1:8.

[64] Herbert McClosky and John Zaller, *The American Ethos* (Cambridge: Harvard University Press, 1984), pp. 102, 131-34; Thomas K. McCraw, *Prophets of Regulation* (Cambridge: Harvard University Press, 1984), pp. 110-11.

[65] Lipset and Schneider, *Confidence Gap*, pp. 5-6.

[66] Lipset and Schneider, *Confidence Gap*, p. 30

[67] Lipset and Schneider, *Confidence Gap*, pp. 175-77.

[68] Lipset and Schneider, *Confidence Gap*, pp. 180-82.

[69] Lipset and Schneider, *Confidence Gap*, pp. 235-46.

[70] Lipset and Schneider, *Confidence Gap*, pp. 263-67.

[71] Aaron Wildavsky, and Ellen Tenenbaum, *The Politics of Mistrust* (Beverly Hills: Sage Publications, 1981), pp. 23-41.

[72] David Stockman, *The Triumph of Politics* (New York: Harper and Row, 1986).

[73] Marvin Olsen, Eugene Rosa, Riley Dunlap, Robert Howell and Don Dillman, "Public Opinion Versus Governmental Policy on National Energy Issues,' *Research in Political Sociology*, Volume 1, pp. 189-210; Fusso, "Energy Crisis in Perspective," p. 135.

VI

Congressional Voting On Oil Issues

THEORIES OF CONGRESSIONAL VOTING

A number of works in political science have attempted to explain Congressional voting behavior. The cue-taking theory of Matthews and Stimson assumed that given the large numbers of votes which legislators must cast combined with their limited information resources, congressmen will rely heavily on simple cues in deciding how to vote. Nine different cues were suggested, including: voting with the majority of the House, voting with the majority of the member's party, voting with the administration, voting with the conservative coalition, voting with the party leadership, and voting with the state party delegation. Cherryholmes and Shapiro developed a simulation of legislative voting, with a first phase during which region, party, and other variables are used to give each legislator a predisposition for each bill, resulting in a vote if the sum of these influences is sufficient to motivate the legislator. Otherwise the legislator is assumed to be swayed by conversations with other congressmen. Aage Clausen used cluster analysis to show that legislators tend to demonstrate continuity in voting along policy dimensions, explained by long-term ideology, party, constituency, and presidential support considerations. John Kingdon interviewed selected congressmen immediately following their votes on specific motions and concluded that there was a "consensus mode" of congressional decision making, with the legislator influenced by fellow congressmen, party leadership, staff, constituency, administration, and interest groups.[1]

Morris Fiorina reviewed literature on the impact of constituency on Congressional voting, beginning with the seminal work by Julius Turner. Turner used crude surrogates for constituent interests (foreign/native born, urban/rural, section) to show that Congressmen divided along these lines, though not as deeply as along party lines during the post WWI period. In the 1950s research on constituency influence conducted by Duncan MacRae in

Massachusetts found that legislators from marginal, or competitive districts tend to be disloyal to their party, a trait shared by representatives from atypical districts as noted by Turner (districts with characteristics "atypical" of those normally associated with the legislator's party). Further research failed to grant conclusive support to either the marginal or atypical district hypotheses, while work by Warren Miller found that congressmen from safe districts were more likely to vote their constituents' interests. Fiorina, based on his own research (and in general agreement with Huntington), concludes that in marginal districts the legislator often must choose a segment of the constituency to please instead of trying to please all sectors of the constituency since an optimal voting strategy may not exist. At the Congressional level political parties are supported by a heterogeneous constituency requiring compromise in formulating the party position, and therefore marginal representatives are less likely to systematically deviate from the party position.[2]

Recently, there has been a spate of articles from economists (and public choice advocates) concerning Congressional voting, especially with regard to economic issues. Since economists tend to take the public choice route they are often uncomfortable with the existence of ideology as an influence on congressional voting and tend to dismiss its impact. Peltzman attempted to separate the characteristics of Senators' supporters from the general characteristics of the entire constituency by regressing votes on demographic characteristics at the county level. He then used constituency characteristics to explain Senators' ADA ratings, although the same result is achieved by using party identification. Peltzman added labor's share of campaign contributions to the model, ignoring the question of causality (do Congressmen vote in order to obtain contributions or are decisions to contribute the result of Congressional voting), but improving the model's fit. Peltzman concluded that economic variables were sufficient to explain voting on economic but not social issues. On the other hand a number of studies using ADA ratings as a proxy for ideology have concluded that ideology is important in determining votes by Congressmen on economic issues. Kau and Rubin used a measure for ideology which was the residual (RADA) from regressing ADA ratings on economic variables such as income, union membership, consumer reports subscriptions, etc. They found that RADA had significant explanatory value in explaining House votes. Kalt and Zupan also provide support for a role for ideology in their study of strip-mining legislation, using a specific ideological measuring with reference to environmental legislation.[3]

It is clear that, excluding an occasional aberration like Richard Schweiker, senators hold to a generally consistent position (as evaluated by interest groups) over time. While senators facing reelection tend to slightly change position and become more ambiguous there is no evidence that this shift has a significant impact on their reelection chances. The fact that politicians are ideologically consistent in their voting behavior does not answer the question of whether that consistency reflects their own beliefs or a stable constituency with well-defined interests. An attempt to separate personal ideology of politicians was attempted by Carson and Oppenheimer, providing a measure of the senator's variance from his expected (with respect to his constituency) position. While they demonstrated the existence of a personal component in Congressmen's voting behavior, the larger question of determining the ideological component of voting behavior remains muddled.[4] This would require a general equilibrium approach in order to deal with the self-selection process whereby a majority of constituents in an election district choose to be represented by people whose expressed beliefs (at least through votes in Congress) are fairly similar to their own. Therefore, one would still expect politicians to reflect the attitudes of constituents independent of constituent economic interests. When those attitudes are associated with demographic variables a measure of ideology would be expected to be correlated with such variables.

VOTING ON OIL ISSUES

A number of studies have attempted to identify significant influences on Congressional voting on energy issues. Edward Mitchell, along with Lopreato and Smoller, found that ideology dominated (as measured by ADA ratings) energy voting in Congress in the mid-1970s. Unfortunately both studies were done hastily at the height of the controversy over energy legislation and lack formal rigor. A few years later two articles appeared which covered energy votes during the decade of the 1970s. Riddlesperger and King studied Senate votes from 1973-80, focusing on divisive votes and aggregating a wide range of votes into an energy votes rating (EVR) which was a measure of the percentage of pro-producer votes cast. While they find that both party and economic interests influenced voting I cannot accept their method of regressing EVR on variables and using the change in R^2 as variables are added as a measure of relative significance. Bernstein and Horn attempted to explain votes in the House of Representatives on energy issues, using an index of opposition to major oil companies on 5 votes during 1975-76 as their dependent variable. Independent variables included oil production, party, and ideology (ADA). The authors suggested a significant role for ADA ratings in explaining energy votes.[5]

Joseph Kalt, in his book on oil price regulation, included an econometric study of voting in the Senate on energy issues. His dependent variable, PROCRUDE, was the logarithmic transformation of the frequency with which a senator voted for bills which Kalt judged favorable to the industry, and in which refiner and consumer interests differed from producer interests. PROADA was the frequency with which a senator voted the ADA position on nonenergy issues in 1975-76. CRUDE measured the potential value of production by multiplying the quantity produced by the difference between the state's average wellhead price and the world price of oil. REFINE measured the value of large refiner shipments and SMALLREF for the value of small refinery shipments. Consumers were represented by total state expenditures on energy. His first run regressed PROCRUDE on the independent variables. Kalt also regressed PROADA on a number of variables, with statistically significant results for nonwhite, urban (negative), farm (negative), manufacturing, blue collar (negative), age, votes for McGovern, and party (negative). Kalt's use of aggregate data to develop a general score obscured different interest group alignments on different bills. Specification error is a problem since PROADA is both a proxy for ideology and for signals a Senator might receive from his political party or other sources which were excluded.[6]

Regressions on energy votes were conducted in a few other studies as part of general analyses of Congressional voting. Kau and Rubin found the following variables significant with regard to the House vote on S 2589, the Anderson amendment to strike from the conference report provisions to control the price of crude oil: consumers (negative coefficient), RADA (negative), age (negative), education and oil production. Kau, Keenen, and Rubin, using a three equation model of congressional voting examined the votes for the 1979 Windfall profits tax (HR 3919), specifically the Ways and Means Committee amendments providing for a tax rate of 70 percent. Ideology was measured by the vote for Ford in the 1976 election. Significant variables include Ford (negative), labor contributions, central city, blacks (negative), coal production (negative), oil production (negative). Oil and coal production were determined at the state level, not the district level. Business contributions were insignificant in influencing votes on the windfall tax amendment.[7]

THE DATA

The data used to develop the vote analysis varied depends upon the years covered. Senate data is more consistent since state level data has been available for most variables for the entire period covered, except for ADA scores (adjusted to eliminate energy votes). House data had to be manufactured for almost all variables for the 1950s on a county by county basis and then converted to district values. This method was used for energy variables for all periods, involving some estimation using geological maps and location and size of oil fields and coal deposits. In certain areas this may be a source of error, the most likely possibilities being with regard to coal production when information by county was withheld and oil production in regions such as Los Angeles county where congressional districts share a small geographic region possessing numerous oil fields. Following is a list of data sources used in this research (various issues unless specific years are given):

City and County Data Book, Department of Commerce. Bureau of the Census
Congressional Directory, U.S. Government Printing Office
Congressional District Atlas, Department of Commerce, Bureau of the Census
Congressional District Data Book, Department of Commerce, Bureau of the Census
Mineral Yearbook, Department of the Interior, Bureau of Mines
Coal Production, 1980, 1982, Department of Energy, Energy Information Agency
Bituminous Coal and Lignite Production and Mine Operations, DOE/EIA
State Energy Price and Expenditure Report, 1984, 1986, DOE/EIA
Petroleum Facts and Figures, 1971, American Petroleum Institute
Basic Petroleum Data Book, 1987, American Petroleum Institute
International Oil and Gas Development, Vol. II, Production, International Oil Scouts Association, Austin Texas.
International Petroleum Encyclopedia, Petroleum Publishing Co., Tulsa, Oklahoma.

METHODOLOGY

The basic method followed throughout this section will be to describe the vote (or series of votes) taken, the politics and interest group alignments leading up to the vote, and to develop a model of voting using this information as well as observations from the literature. I assume that a congressman's vote will be determined by his constituency, his party and his own voting tendencies. I did not explicitly assume a vote maximizing or any other public choice type model, since I preferred to allow the data to eliminate potential variables than assume ex ante that they were irrelevant. I felt it was sufficient to merely assume that the legislator is influenced by characteristics of his constituency and his or her own predispositions. I feel that Occam's Razor suggests a simpler hypothesis than one requiring the unrealistic assumptions implied by a formal public choice model. My goal was to determine what sort of changes, if any, have occurred in oil politics over the time period covered.

The technique used to analyze the data is called the probit analysis model, first postulated by Goldberger. An dichotomous decision, such as a vote for a bill, is made if a certain threshold (economists would use expected utility as the quantity but this is not a necessary assumption) value is reached. Assume that this underlying response variable is y_i^* s.t. $y_i^* = \beta' x_i + \mu_i$. Since y_i^* is unobservable we observe a dummy variable which takes the value 1 if $y_i^* > 0$ and zero otherwise (i.e. the actual vote). Therefore the observed values of y are realizations of a binomial process. In the case of the probit model we assume that μ_i is distributed $N(0, \sigma^2)$. Since β and σ must be estimated together the problem is simplified by assuming $\sigma = 1$. The likelihood function is then written as

$$L = \Pi P_i y_i (1 - Pi)^{(1-yi)} = \Pi F(x_i'\beta) y_i [1 - F(x_i'\beta)](1-y_i)$$

where F(*) is the standard normal and $y_i = 1$ for a yes vote and zero otherwise. Since the first-order derivatives of the likelihood (or the log likelihood) function are highly nonlinear the likelihood function is maximized using numerical methods. The econometric package used for this work, a micro-computer version of RATS, used the Gauss-Newton method to solve for maximum likelihood. The parameter estimators are consistent and asymptotically efficient, as well as asymptotically normal, so that the regression t test can be applied.[8]

There were three major problems which had to be addressed in modeling energy votes. The first is that for Senate votes there are insufficient observations to be confident of the large sample properties of maximum likelihood techniques. The second problem is that the House and Senate data is highly collinear. This problem is more severe in the Senate because aggregation increases collinearity between variables. Collinearity is strongest between energy production variables, production and cost variables, and party, voting and demographic variables. The third problem is that many of the variables used are crude proxies for interest group influence, and there is some specification error.

The choice of functional form is almost always arbitrary, unless there is some prior data or theory which would indicate a certain structure. In this case the probit model is a nonlinear specification. The choice of variables to be included should be based on theory; however if one accepts either interest group theory or public choice theory one would expect relevant economic variables, such as energy production and consumption, to be significant factors in decision-making on votes. Since prior work on energy votes has found education, percentage of nonwhite and urbanization to give significant results I tested these variables to see if they were influential in voting decisions on oil issues. Political theories of congressional voting would predict that party identification, region and political culture should play a role in voting decisions.

Unfortunately many groups that might influence voting are hard to identify and obtaining data which would give an accurate measure of their size and potential resources is extremely difficult and expensive. Some researchers have used subscriptions to consumer publications as a measure of consumer political power, but this may merely be a measure of consumer sophistication. I used the per capita cost of petroleum products and natural gas since it seemed reasonable that consumers would react to the cost of energy. However this measure overlaps with industry consumption of energy as regions with energy intensive industry will have higher per capita costs. The same problem arose when trying to separate classes of producers; I used per capita production by stripper wells and nonstripper wells for the Senate, and since stripper data wasn't available at the House level, production from large fields and other fields as proxies for production by small producers and large producers. This is a very crude attempt at separation as many large companies owned interest in fields that were dominated by stripper wells, while some small producers operated highly prolific wells.

A problem posed by the proliferation of plausible explanations for voting behavior is that Probit analysis, since it produces parameters which are only asymptotically efficient, will be more reliable as the ratio of observations to independent variables increases. On the other hand, econometric theory suggests that in most cases left out variable error (exclusion of relevant variables) is a more serious error than inclusion of irrelevant variables. However, a serious pitfall for the poorly trained is to mistake increased explanatory value from inclusion of additional variables as an indicator of a superior model. Thus one must artfully balance the need for parsimony, application of prior knowledge and data limitations to achieve meaningful results.

When working with aggregate data, such as Congressional district data, the researcher should be wary of the effects of multicollinearity among the independent variables. Unfortunately, in the literature on congressional voting, diagnostics for multicollinearity are either rarely performed or not reported. This is a serious oversight since multicollinearity will cause a loss in the precision of the estimators. While the estimators will still be unbiased they will no longer have minimum variance. Correction for multicollinearity involves a tradeoff between bias and precision unless additional data is available (which is not an option with regard to my analyses). The effect of near exact linear dependencies on the finite sampling precision of estimators in a nonlinear model like Probit is unknown.[9]

I tested my data for multicollinearity by scaling the X'X matrix to unit length, producing the vector of eigenvalues for this symmetric matrix, and calculating the condition numbers, determined as the square root of the ratio of the eigenvalues to the smallest eigenvalue. These numbers are roughly equivalent to the measures used by Belsley, Kuh and Welch to locate potential linear relationships among the independent variables.[10] Since the econometric package I worked with lacked the capability to perform the variance decomposition used to locate such relationships an alternative method was employed. First the condition numbers were examined (a value over 30 being an indicator of the presence of multicollinearity) to determine the existence of a potential problem. Variables which theory would predict to be of lesser importance were then eliminated and the effect on the condition numbers noted. Finally, the independent variables were regressed upon one another to identify possible relationships.

Variables include percentage living in urban areas (Urban), percent black or Indian (Nwhite), median family income (Income), median years of education (Educ), per capita value of oil produced (Oilval), per capita value of oil from stripper wells (Stripval), per capita value of oil from nonstripper wells (Nonstrip) [for the House votes per capita value of oil from large fields (Larval), and per capita value of oil from non-large fields (Nonlar) was used], per capita value of natural gas production (Ngval), per capital value of coal production (Coalval), per capita yearly refinery capacity (Refcap) [split into three variables; Smallref for refineries with capacity less than 30,000 barrels per day, Medref for refineries between 30-100,000 bpd, and Larref], per capita residual oil cost (Rescost) for the 1950s regressions and per capita petroleum cost (Petcost) for the later regressions, per capita natural gas expenditures (Ngcost), political party (Party; Democrat = 1), and whether the state is southern (South).

My basic methodology was to run an initial regression with a full set of variables for the House votes, and a constricted set for the Senate votes. The initial results were then used to eliminate variables which did not seem to possess explanatory power (t-ratio <1.0), contradicted theory (i.e. energy cost should not be positively correlated with votes that lead to higher prices), or might be obscuring the influence of other variables with which it is collinear (Natural gas production and cost, and oil production variables exhibited this behavior in numerous votes). While one may object to my method of eliminating variables until an acceptable model is obtained as a form of data mining, I would point out that the high degree of collinearity in the data, necessitated this strategy. I traded increased bias in my coefficient values for increased precision (smaller sampling variability). It should be noted that the exact sampling properties of estimators developed in this manner are unknown.[11] Since the statistical properties of probit analysis with multicollinearity are unknown this is not a serious loss, but it does suggest caution in interpreting results. Despite the flaws in my methodology I contend that robustness of parameter estimates, across model runs and similar votes is sufficient to elicit confidence in my results. Robustness across various specifications is a more reliable indicator of significance than flawed statistical measures.

VOTE #1: TRADE AGREEMENT EXTENSION ACT OF 1949 (HR 1211)

The specific vote was on the Thomas (D-OK) amendment to the reciprocal trade act which would establish import quotas for petroleum and petroleum products. The amendment was rejected by a 40-41 vote on September 15, 1949. The close vote was a product of concern over the increase in oil imports which caused state prorationing commissions to restrict output, and the general struggle over trade policy. The IPAA had already testified in favor of restrictions in front of the Ways and Means Committee hearings in February.

Testifying against the reciprocal trade bill in the Senate Finance Committee hearings were representatives from petroleum commissions in Oklahoma, Kansas and Texas, along with North Texas and California oil producers. The Finance Committee vote to approve the trade bill was 7-6, split along party lines, with the Democrats in support including Tom Connally of Texas and Edwin Johnson of Colorado. Eugene Millikin (R-CO) led the opposition to the bill. The Thomas bill was introduced during the debate and supported on the floor by Andrew Schoeppel (R-KS). The trade bill passed the same day as the Thomas vote by a vote of 62-19.[12]

The results of my analysis of the voting for the Thomas amendment are given in Table 6.1. Since I was unable to find ADA ratings before 1960, and votes in the Presidential elections in this period were only weakly ideological I used a variable based on a measure of political culture (Polcul) as an alternative measure of the constituency's underlying ideology. Values for this variable were obtained from calculations made by Ira Sharkansky using Daniel Elazar's observations on the various political cultures found in the United States. Elazar describes three principal cultures of Moralist, Individualist and Traditionalist. A Moralist values professional government which is interventionist, the Individualist is ambivalent toward government and the Traditionalist opposes government action. The scale is lowest for Moralist and highest for Traditionalist states.[13]

The results of the runs for HR 1211 are given in Table 6.1. A positive coefficient value implies that the variable increased support for a quota. The t-test were used to determine whether a variable was significant at the 95 percent or 99 percent threshold, though it should be noted that when testing a

TABLE 6.1 Probit Runs for HR 1211 (Y=1, support for Quota)

Variable	Run 1	Run 2
Constant	0.96	0.78
Stripval	-0.014	
Nonstrip	0.0420**	0.039**
Coalval	0.021	0.019
South	-1.99**	-2.08**
Urban	-2.69*	-2.19*
Party	-1.57**	-1.51**
Polcul	0.18	0.20*
Rescost	0.018	
Significance Level	0.0058	0.0020
Pseudo R^2	.462	.455

Observations = 89

*significant at the 5% level; **significant at the 1% level.

succession of models statistics such as the t-test have unknown distributions, and there is a reduction of the degrees of freedom. Therefore these estimates of significance should be interpreted conservatively.

The results demonstrate the limitations of the data. While one would expect small producers to support a quota and large producers to oppose it, Stripval was insignificant while Nonstrip was a factor in favor of quotas. I suspect that this reflects the inability of these variables to act as proxies for these two groups in the 1950s, though Nonstrip at this time dominates stripper value, and therefore represents oil production in general. It should be remembered that coefficient values reflect the marginal impact of a change in a variable value on the probability of voting. Therefore when a variable such as party has a negative coefficient it does not necessarily mean that Democrats on average are more likely to vote against quotas (though at this point in time they were), but that taking in account the impact of the other variables membership in the Democratic party will increase the probability that a Senator will vote against quotas. Thus South, Urban and Party are all influences against voting for quotas while Nonstrip, Polcul and to a lesser extent, Coalval, increase the probability of voting for an oil import quota.

The performance of the Polcul parameter suggests that it serves as a real measure of underlying political attitudes during this period of time. I suspect it is a proxy for a populist undercurrent which opposes big government and big business, and which looks favorably upon protectionist measures which help small business. The weak level of opposition associated with the South dummy variable could be due to the traditional Southern support for free trade (due to the Southern dependence on agriculture) or the willingness of Southern democrats to support a democratic president. These conclusions are strengthened by their correspondence with anecdotal evidence concerning the politics of oil import and trade votes during this period.

Given this model of voting on oil quotas based on party and economic interest, it is interesting to note the outliers; that is the votes which contradict the predictions of this simple model. A check on party influence is the vote on the Millikin amendment to the trade bill for which voting split along party lines, with Democrats in opposition 3-43 and Republicans in favor 35-0. Two Democrats, McFarland of Arizona and Gillette of Iowa, who voted against Millikin and come from states without oil voted for the Thomas amendment. Twelve Republican Senators voted against the Thomas amendment (all of whom voted for the Millikin amendment), including both Michigan Senators, Ferguson and Vandenberg, and Bricker of Ohio. Since Michigan was a substantial oil producing state (over 16 million barrels in 1949) while Ohio was a minor oil producer and major coal producer these votes are hard to explain. More understandable was the decision of four New England and two Middle Atlantic Republicans to vote against the Thomas amendment.

VOTE #2: AMEND NATURAL GAS ACT OF 1938 (HR 1758)

A number of attempts were made in Congress during the 1950s to free independent natural gas producers from FPC regulation. In 1947 a Supreme Court ruling (Interstate Natural Gas Co.) implied that sales by independent producers to interstate pipeline companies could be regulated by the FPC. Although the FPC contended that it did not have this power, several Congressmen took action as early as 1947 to ensure that the FPC did not extend its authority over independent producers. A bill was passed by the House in 1949 but died in the Senate. In 1950 the Senate passed HR 1758 by 44-38 (D 28-16; R 16-22) on March 29, and the House, led by Speaker Rayburn, narrowly passed the bill on March 31 by 176-174 (D 97-116; R 79-57). The bill was then vetoed by President Truman. The bill was supported by legislators from gas producing states and opposed by Congressmen representing urban gas consuming areas.[14]

Unfortunately my data on the House of Representatives doesn't begin until 1953 (House data must be recalculated after every realignment of districts) so I can offer only an analysis of the Senate votes. The Senate data did not fall into the simple pattern that one would expect, with the coefficient for natural gas production positive and significant, while coefficients for urban and natural gas cost negative and significant. The first run found only natural gas cost to be significant, and with the wrong sign, while the hypothesis that the coefficients were jointly equal to zero was barely accepted. Interrelationships between independent variables distorted the results. Natural gas production and natural gas expenditures were strongly related during this period, and this obscured the relationship between production and voting. Of the sixteen states with the highest per capita natural gas consumption, Arizona and Mississippi were the only states which did not produce natural gas. Dropping Ngcost eliminated this problem, but at the cost of explanatory power of the model.

The poor performance of the model limits the confidence that can be placed in the results, but some conclusions can be drawn. Natural gas production seems to support deregulation votes, which is to be expected, while southern support could be explained by regional antipathy to government regulation. Urban opposition, though statistically insignificant, is in line with observed behavior. Seven of the 28 Senators from gas producing states voted against the bill, including Knowland (R-CA), Murray (D-MT), Chavez (D-NM), Myers (D-PA), Kilgore and Neely, and O'Mahoney. The pattern of House voting was more consistent, with all the representatives from Oklahoma, Texas, Wyoming, Kansas, New Mexico, Arkansas and Louisiana voting for, and those from NYC and Detroit voting against the bill. However, Philadelphia representatives split 4-2 in favor of deregulation, while Los Angeles representatives were 4-3 against (there is no gas production in Philadelphia while Los Angeles County was a major producer).

TABLE 6.2 Probit Runs for HR 1758 (Y=1, against FPC control)

Variable	Run 1	Run 2	Run 4
Constant	-0.50	-0.13	-0.37*
Stripval	0.0066	0.0095	
Nonstrip	-0.0033	0.0014	
Ngval	-0.048	0.012	0.049
South	1.35*	1.09*	1.35**
Urban	-1.25	-0.68	
Party	-0.30	-0.24	
Polcul	0.046	0.066	
Ngcost	0.22**		
Significance	0.019	0.025	0.0014
Pseudo R^2	.220	.166	.151
Observations = 94			

VOTE #3: TRADE AGREEMENT ACT OF 1953 (HR 5894)

President Eisenhower's request for an one year extension of the Reciprocal Trade Act in the spring of 1953 set the stage for another congressional battle over oil import quotas. A series of maneuvers resulted in the introduction of two bills: HR 5495, the Trade Agreements Extension Act of 1953, and HR 5496, a far more protectionist bill. HR 5495 was reported favorably from Ways and Means by a vote of 23-2 on June 9 and the bill sailed through the House on June 15, after Democratic attempts to eliminate two minor amendments failed. In July, the attention of the Ways and Means committee turned to HR 5894, introduced by Richard Simpson, which was basically the Trade Bill with quotas on imported crude and residual oil, protection for the lead and zinc industries, and a strengthened Tariff Commission. The Republican leadership joined the President in opposition to the bill, but the majority of the Ways and Means Committee's Republicans supported the bill, which was reported favorably by a 15-10 vote. There were two recorded votes on the bill, one on July 23 adopting a rule providing for consideration of the bill, by a 219-183 vote (D 81-110; R 138-72) and a motion to recommit the same day. Led by Sam Rayburn and Majority Leader Charles Halleck the motion to recommit passed the House, 242 to 161 (D 137-56; R 104-105).[15]

While it would be preferable to use district data, the energy consumption variables and Polcul were only available at the State level. A new variable was added for the House, Gold, the vote for Goldwater in the 1964 election, since the Polcul variable is only available at the state level. The 1964 election was felt to be a better indicator of district ideological sentiments than the 1956 or

TABLE 6.3 Probit Runs for HR 5894

	Consideration vote (Y=1)		*Recommittal vote (N=1)*	
Variable	Run 1	Run 3	Run 1	Run 3
Constant	1.98**	1.75**	1.29**	1.26**
Larval	0.0039*	0.0017	0.0019	0.0013
Nonlar	0.0055*	0.0020	0.0071**	0.0046**
Ngval	-0.012**		-0.0056*	
Coalval	0.0076*	0.0063*	0.012**	0.012**
South	-0.87**	-1.04**	-0.82**	-0.97**
Urban	-1.23**	-1.22**	-1.38**	-1.43**
Party	-0.38*	-0.39**	-0.31*	-0.36*
Polcul	-0.053		-0.057	
Gold	0.092	0.61	0.46	
Rescost	-.035**	-.032**	-0.025*	-.023**
LeadZinc	0.024	0.025	0.029*	0.030*
Sign.	0.000	0.000	0.000	0.000
Pseudo R^2	.189	.171	.234	.226

Observations = 431

1960 elections. A Leadzinc variable was included to measure the influence of production of these minerals on votes for the bill. The only serious relationship among the independent variables was between Larval, Nonlar and Ngval, which is to be expected.

It should be note that while the null hypothesis (coefficients are jointly equal to zero) was rejected, this is almost automatic given the large number of observations in House voting. The low value of the pseudo R^2 term is evidence of the limited predictive validity of the model. Variables which were important in explaining opposition to the oil quota bill include urbanization, expenditures on residual oil, membership in the Democratic party and residence in a Southern state. This reflects the opposition of the New England and Middle Atlantic states to import quotas, especially on residual oil, and the general support of free trade by the Democratic party in general, particularly Southern Democrats. Variables explaining support for quotas include value of coal, petroleum and lead and zinc production. Note that Nonlar dominates Larval on the recommittal vote, which probably reflects the willingness of representatives from districts with small producers to hold fast while those from districts dominated by major fields were willing to shift their votes. The coefficients for coal and petroleum production suggest that the influence of coal production was greater, supported by reports that while the coal industry was united in its opposition to imports, the oil industry was split, with the larger companies quietly applying pressure to limit support by smaller producers for quotas.

VOTE #4: NATURAL GAS BILLS (S 1853, HR 6645)

In June 1954 the Supreme Court in the Phillips Petroleum case ruled that the FPC had jurisdiction over independent producers of natural gas who sold to interstate pipelines. The FPC then proceeded to issue regulations to independent producers. The House passed a bill on July 28, 1955 (HR 6645) by a 209-203 vote (D 86-136; R 123-67) to exempt natural gas producers from federal public utility regulation. A similar bill was reported in the Senate but no action was taken that year. Meanwhile the oil industry formed a committee to work for the bills while mayors of large cities organized against the legislation. On February 3, 1956 Senator Francis Case announced he would vote against the gas bill (S 1853) because an oil company lawyer offered him a $2,500 campaign contribution. On February 6 the Senate passed S 1853 by a 53-38 (D 22-24; R 31-14). President Eisenhower then vetoed the bill on February 17, expressing his concern about the allegations of impropriety. No attempt was made to overturn the veto, and no bill to deregulate natural gas would even come to a vote during the next two decades.[16]

The regression results (Table 6.4) reflect the problems faced in choosing a correct model when confronted with severe correlation in the variable set. Ngcost has the wrong sign in both the House and Senate initial runs and therefore was eliminated, but at the cost of a sizable amount of explanatory value. The oil variables obscured the impact of gas production, especially in the Senate, and were dropped since it was felt that gas production was a more important variable for inclusion in a model explaining a vote on a natural gas deregulation bill. Although the impact was minimal for the House votes, dropping oil production significantly weakened the validity of the Senate model. An encouraging note is that the impact of these changes on the other variables seemed to be minimal, as those which seem to be significant influences had robust coefficient values over the various model runs.

Although Polcul and Gold are both proxies for constituent ideology, Gold dominated in the Senate while both were significant factors in support of deregulation in the House. This suggests that the issue encouraged support from congressmen with constituencies in favor of less government involvement in the economy (which would explain the sign on Polcul) and/or reflected a tendency to support business interests. South seems to be unrelated to deregulation in the Senate but significant in the House. While this may reflect a higher degree of collinearity in the Senate data an alternative explanation is that Senators are more likely to vote with the party than with the regional pro-business bias, while southern Representatives felt less bound by party discipline during this period. Given that southern states score high on both Gold and Polcul this requires assuming that the southern pro-business bias exists above and beyond its already conservative tendencies.

TABLE 6.4 Probit Runs For S 1853 and HR 7125

	Senate Vote		*House Vote*	
Variable	Run 1	Run 4	Run 1	Run 3
Constant	-4.28**	-1.56*	-3.11**	-0.91**
Stripval (Nonlar)	0.032		0.0017	
Nonstrip (Larval)	0.013		0.0015	
Ngval	-0.018	0.054**	0.037**	0.065**
South	0.65		0.94**	0.90**
Urban	-0.65		-0.50	
Party	-1.67**	-1.24**	-1.85**	-1.67**
Polcul	0.15		0.17**	0.10*
Gold	2.86	5.24**	1.55*	2.05**
Ngcost	0.071*		0.059**	
Sign.	0.0059	0.0004	0.000	0.000
Pseudo R^2	.400	.239	.398	.301

Observations = 96 (Senate), 408 (House).

VOTE #5: VOTES ON THE OIL DEPLETION ALLOWANCE: 1951-1969

President Truman attacked the oil depletion allowance in his annual tax message of January, 1950, but the House Ways and Means Committee was more impressed by the numerous witnesses who testified in favor of the allowance. Truman proposed a reduction of the oil and gas depletion allowance from 27.5 per cent to 15 per cent as well as cuts in other depletion allowances in 1951. The Ways and Means Committee responded by doubling the allowance for coal from 5 to 10 per cent, and extended the 5 per cent rate to twenty-one new substances, while the Senate Finance Committee granted the allowances to ten additional resources. Hubert Humphrey attempted to challenge the pro-depletion forces by introducing an amendment to the Revenue Act of 1951 to reduce percentage depletion allowances for oil, gas, sulfur and non-metallic minerals. The Senate defeated the Humphrey amendment by the vote of 71-9, with only John Williams, Paul Douglas (D-IL), Benton (D-CO), Moody (D-MN), Lehman (D-NY), William Green, John Pastore, Estes Kefauver (D-TN) and Humphrey voting for the amendment (Morse, a Republican from Oregon, declared for the amendment).[17]

Numerous minor attempts to lower the depletion allowance were undertaken in the Senate, largely in the form of floor amendments or applications and redefinitions of regulations by the Treasury Department. An amendment to the Excise Tax Technical Changes Act of 1958 (HR 7125) to cut the depletion allowance from 27.5 percent to 15 percent was introduced by John Williams in 1958. The amendment was defeated 26-63 (D 15-28; R 11-35). An amendment to the same bill offered by William Proxmire, fixing the depletion allowance at 27.5 percent for taxpayers with gross incomes of $1 million or less, 21 percent for income between $1 and $5 million and 15 percent for over $5 million was also defeated, 31-58 (D 21-22; R 10-36). Paul Douglas offered a graduated depletion allowance to protect wildcatters in 1954, but was unable to obtain a roll call vote. The Douglas amendment was offered to the Tax Rate Extension Act of 1959 (HR 7523), but was rejected 21-54 (D 16-33; R 5-21). In 1960 Douglas offered a copy of the Proxmire amendment to HR 12381, but it was shot down by a vote of 30-56 (D 22-31; R 8-25). In 1962 a Mansfield motion to table a Williams amendment to the Revenue Act of 1952 (HR 10650) to reduce the depletion allowance from 27.5 to 20 percent was adopted by a 57-30 vote (D 37-20; R 20-10). The Douglas amendment to the same bill was rejected 23-50 (D 18-29; R 5-21). Williams and Douglas also offered amendments in 1964 which were both rejected by a vote of 35-63. It should be noted that both Eisenhower and Kennedy avoided taking a position on any of the Senate votes on depletion.[18]

The six votes from 1958 to 1962 raised a problem in that ADA data was available only from 1960 onward. Runs performed on votes after 1960 found that Polcul was insignificant when regressed along with Gold and ADA, suggesting that the latter two variables contained the information that had been provided by Polcul in earlier runs. Since ADA and Gold are easier to interpret (one a measure of the Congressman's voting record, the other a measure of constituents' votes for a highly ideological candidate) they are preferable to Polcul, which is a product of a political scientist's judgment of the underlying political characteristics of different states. In all the votes on the depletion allowance during this period a negative coefficient indicates that the variable increases the probability of voting for maintaining the depletion allowance.

The votes on the depletion allowance during this period provide the expected result that producers supported the allowance. Despite provisions in most of the bills which preserved the allowance for small producers the model provides no evidence of a split in the oil industry ranks, with Stripval actually a stronger factor in opposition to the bills than Nonstrip. An interesting result is the change from opposition to neutrality (or even weak support) by Democrats, with the break occurring with the election of a Democratic president. This suggests that the Democrats saw the depletion allowance as a safe way to embarrass a Republican administration–safe because nonoil Democratic senators could portray themselves as the taxpayer's friend while the oil state Democrats could be confident of victory given Republican support for the tax break. The most consistent factors explaining votes on the allowance are the support by Goldwater voters and opposition by Senators with high ADA ratings. In the absence of an active taxpayers' interest group opposition was

TABLE 6.5 1958-1962 Depletion Allowance Votes

Variable	HR 7125A	HR 7125B	HR 7523	HR 12381	HR 10650A	HR 10650B
Constant	2.58**	4.33**	0.50	-0.54	0.29	-1.21**
Stripval	-0.078	-0.026*	-0.076*	-0.075*	-0.057	-0.055
Nstrip	-0.014			-0.0085	-0.024*	-0.21
South			0.91	1.53*	0.96	
Party	1.08**	2.46**			-1.15*	-0.85
Polcul	-0.20*	-0.29**				
Gold	-5.90*	-10.78**	-6.25*	-3.62	-3.58	
ADA			2.73**	3.01**	2.75**	3.13**
Pseudo R^2	.391	.450	.474	.457	.397	.311
Observ.	92	93	89	94	99	94

HR 7125A: Williams amendment to reduce the percentage depletion allowance from 27.5 percent to 15 percent (8/11/58). Y=1.
HR 7125B: Proxmire amendment to fix the depletion allowance at 27.5 percent for taxpayers with oil and gas incomes of $1 million or less; at 21 percent for incomes between $1 and $5 million; and at 15 percent for incomes over $5 million (8/11/58). Y=1.
HR 7523: Douglas amendment to reduce, on a graduated scale, the depletion allowance on oil and gas wells (6/25/59). Y=1.
HR 12381: Douglas amendment to establish the Proxmire schedule (6/20/60). Y=1.
HR 10650A: Mansfield motion to table Williams amendment to reduce the oil depletion allowance from 27.5 to 20 percent (9/5/62). N=1.
HR 10650B: Douglas amendment to reduce the depletion allowance to 15 percent when gross income exceeded $5 million and reduce the allowance to 21 percent when income was between $1 million and $5 million (9/5/62). Y=1.

based on ideological grounds, while the majority of Senators thought it best not to alienate a leadership that was heavily weighted toward oil producing states.

Although oil producing district representatives were not disproportionately represented on the Ways and Means Committee, until 1969 no measure which lowered the depletion allowance survived the committee, which under the usual closed rule for Ways and Means bills precluded any action in the House to attack the allowance. According to John Manley, the conventional wisdom was that the party leaders, led by Sam Rayburn, would not let a Democrat on Ways and Means unless he was for the trade program, against cutting the oil depletion allowance, and for Medicare. Manley suggests that orthodoxy on these issues was enforced in indirect, informal ways, as the general feeling of the members was that they had been in the House long enough so the leadership knew that they were "reliable" men. Since Democrats were elected to the Committee in the Democratic caucus, they could not get the votes of members from oil states as a candidate for Ways and Means and be a crusader on the issue. Republican appointments to Ways and Means were also opposed to cutting depletion, preventing a bi-partisan coalition against the allowance.[19]

During the 1960s anti-depletion representatives gradually obtained positions on the Committee. Six of the seven members who voted for the Gibbons motion to cut the allowance to 16.5 percent in 1969 joined the Committee after Rayburn's death, and the seventh only a few months earlier. For the first time in eighteen years witnesses appeared before the Committee to ask for a reduction of the depletion allowance, including five House members and two experts on mineral taxation. There was a general impression that the public was interested in tax reform despite the lack of organized support for changes in the tax code. The depletion allowance became indefensible, except for representatives from oil producing districts, as it symbolized the unfairness of the tax system. A split developed between the Majors and smaller independents due to the 50 percent of net income limitation on depletion, which prevented most small producers from collecting the entire 27.5 percent allowance. Wilbur Mills, during the Ways and Means Committee hearings on tax reform, brought out the fact that depletion was more important for the large operators and intangible expensing provisions for the small producer. Given the growing opposition to the depletion allowance, it was easy for many Congressmen to support cutting the depletion rate, hurting large companies, while retaining expensing provisions which favored small companies.[20]

There were five separate roll call votes on amendments in the Ways and Means Committee to a July 18 motion by Hale Boggs (D-LA). Bogg's original motion reduced the allowance to 20 percent, eliminated depletion on foreign operations while raising the limit for applying the allowance to 70 percent of taxable income from 50 percent. This would have had little impact on small producers while having a substantial impact on larger producers, especially the international oil companies. Opposition from depletion foes convinced Boggs to modify his own proposal at the start of the July 21 session, deleting the net income ceiling change and restriction of the allowance to domestic operations. George Bush (R-TX) offered an amendment to reduce the allowance to 23 percent while raising most other minerals to that level (note that major oil companies had begun acquiring coal companies at this time), which was defeated 9-16 (D 2-14; R 7-3). Sam Gibbons moved to cut the allowance to 16.5 percent but was defeated 7-18 (D 7-8; R 0-10). Rogers Morton (R-MD) proposed a reduction to 22 percent, an amendment that lost by a narrow margin of 12-13 (D 3-12; R 9-1) despite support by Wilbur Mills. An amendment from Charles Vanik, a leader of the depletion foes, to eliminate the oil depletion allowance for overseas operations passed 14-11 (D 12-3; R 2-8). Finally the Boggs motion carried by a vote of 18-7 (D 13-2; R 5-5). The tax reform bill was granted a closed rule and passed the House on August 7.[21]

TABLE 6.6 Probit Runs for the 1969 Depletion Allowance Votes

Variable	HR 13270A	HR 13270B	HR 13270C	HR 13270D
Constant	3.12**	0.13	2.14	-3.44**
Stripval	-0.26**	-0.19*		
Nstrip		-0.074**	-0.017**	-0.028
South	1.27*			
Urban			-1.74	2.83*
Gold	-9.19**		-6.04*	
ADA	1.01	6.69**	2.24**	2.82**
Party	0.58		0.60	
Sign.	0.0019	0.006	0.000	0.0001
Pseudo R^2	.495	.780	.484	.407
Observations	78	98	98	95
Oil State Dem.	12-2-0	1-13-0	0-13-1	2-11-1
Oil State Rep.	8-0-0	0-8-0	0-8-0	0-8-0

HR 13270A: Williams amendment eliminating the increase in the net income allowance from 50 to 65 percent for producers with gross income of up to $3 million from oil and gas wells (11/24/69). Y=1.
HR 13270B: Ellender amendment restoring the depletion allowance to 27.5 percent (12/1/69). A "yea" vote was in favor of the President's position. N=1.
HR 13270C: Williams amendment to reduce the depletion allowance from 23 percent to 20 percent (12/1/69). A "nay" vote was a vote in favor of the President's position. Y=1.
HR 13270D: Young amendment providing that the depletion allowance be applied only to production in the United States (12/4/69). Y=1.

The Treasury Department, aware that President Nixon had pledged to support the 27.5 percent allowance, proposed to maintain the existing level while restricting depletion and deduction of intangible drilling costs to domestic production, a proposal which Ways and Means summarily rejected. Senate action began in the Finance Committee but extended to amendments on the floor. Russell Long was defeated in an attempt to restore the allowance to 27.5 percent in the Finance Committee, when the vote resulted in an 8-8 tie. An attempt to provide a deduction for a plowback provision failed, and Clifford Hansen (R-WY) then proposed a 23 percent depletion rate, which carried by a vote of 12-3. Long attempted to add a sliding scale depletion rate, but when that proposal was rejected he offered a provision to increase the net limitation for producers with gross income under $3 million from 50 to 65 percent, passed by a 10-4 vote. The foreign depletion provision was also restored in Committee.[22]

Four amendments were fought over on the Senate floor. John Williams proposed eliminating the increase in the net income ceiling passed by the Finance Committee, a proposal rejected 26-34 (D 18-16; R 8-18) on November 24. Allen Ellender (D-LA) then offered an amendment to restore the 27.5 depletion allowance, rejected by a 30-62 (D 13-40; R 17-22) vote. Williams came back with an amendment to reduce the depletion allowance to 20 percent, rejected 38-52 (D 27-26; R 11-26). Finally Stephen Young (D-OH) offered an amendment limiting depletion to domestic production, rejected 25-68 (D 23-31; R 2-37).[23] In conference a compromise was reached which reduced the depletion allowance to 22 percent, while the House restriction on depletion on foreign operations and the Senate increase in the net income ceiling were both dropped.[24]

The Williams amendment challenging the increase of the net income allowance was a vote on small producer interests, since they were limited to an effective depletion allowance rate of 15-20 percent because of the 50 percent net income limitation. The Young amendment was a symbolic attack on multinationals since they didn't need a depletion allowance on foreign wells in most cases because of the foreign tax credit. The two other amendments were straight forward votes on the level of the allowance. Negative coefficient values indicate that the variable influences voting in favor of oil producers.

In the case of the income allowance amendment Stripval is significant but Nstrip is not, while the reverse occurs with the other amendments. The danger for the oil coalition was the realization by small independents that the net allowance was more important than maintaining the existing level of the depletion allowance. The net income vote was the only on in which ADA scores were not a significant factor in influencing votes, suggesting either that liberals were indifferent to aid to small producers or that some liberals wanted to split the industry coalition. There was an unusual amount of non-votes on this amendment. Gold was significant only on the income vote and the vote to maintain the 23 percent compromise level for the allowance. The most intriguing result is the unimportant role that Party plays in these votes, especially in light of a Republican president who was committed to support of the depletion allowance. At the bottom of Table 6.6 are the votes by oil state Senators stated as the number yes-no-undecided.

VOTE #6: VOTES ON FUEL ALLOCATION–1973

Even before the oil embargo distortions in petroleum product markets caused by the Nixon price controls had engendered pressure in Congress to alleviate shortages and protect the independent marketers and refiners. Senator Thomas Eagleton introduced an amendment to extend the life of the Economic Stabilization Program (S 398), giving the government the power to allocate crude oil supplies among refiners and products among marketers, which was signed into law on April 30, 1973. Senator Henry Jackson introduced S 1570 to extend allocation to the retail level. Frank Moss (D-UT) then offered an amendment in the form of a substitute bill authorizing the President to initiate a mandatory fuel allocation program. The Moss amendment required petroleum refiners and distributors to supply to each customer petroleum products at the same rate as during the period from October 1971 to September 1972. The amendment was adopted 44-12 (D 31-4; R 13-8) on June 1, 1973, with the president taking no position on the bill. A Ribicoff (D-CT) amendment to create an Office of Emergency Fuel Allocation passed 56-0 while a Humphrey amendment to prevent unfair competitive practices in refining and marketing, and requiring a FTC study of the industry was adopted 80-0. One reason for the overwhelming support for allocation were the strong anti-major company sentiments expressed by witnesses to the Senate Judiciary Subcommittee in June, including five state Attorneys General.[25]

Serious political conflict emerged on the next three amendments to S 1570 as the focus shifted from allocation to price constraints. The McIntyre (D-NH) amendment urging the President to take further action to stabilize prices on crude oil and petroleum products was adopted 63-27 (D 47-6; R 16-21) on June fifth. Dewey Bartlett followed with an amendment to exempt stripper wells from allocation or price controls, which was rejected 42-51 (D 13-41; R 29-10). Bartlett then offered an amendment which would have effectively removed price controls on oil, rejected 21-72 (D 1-52; R 20-20). The bill then passed 85-10. An attempt in the House by Torbert MacDonald (D-MA) to add an allocation amendment to the Alaskan Pipeline bill was rejected 179-233 (D 128-99; R 51-134) by teller vote on August 2. When the House dragged its feet on the bill the Senate passed it again in August (HR 5777). Bartlett offered another amendment exempting stripper wells from controls, which was defeated 43-50 (D 14-39; R 29-11) on July 30, and the bill passed 80-9 the next day. The House finally passed an allocation bill, HR 9681, despite objections by the Nixon administration that it was insufficiently flexible. Jake Pickle (D-TX) offered an amendment to prohibit allocation at the producer level, rejected 136-245 (D 54-154; R 82-91), along with an amendment to transfer the authority to monitor the program from the FTC to the GAO. The bill passed October 17 by a vote of 337-72 (D 200-27; R 137-45). The conference report cleared Congress on November 14 and was signed into law as PL 93-159, the Emergency Petroleum Allocation Act of 1973.[26]

A positive coefficient on Senate allocation vote runs (Table 6.7) means the variable contributed to opposition to allocation or price controls. S 1570A and S 1570C are basically price control votes, while the other two amendments involve small producer exemptions from controls. The runs for S 1570C are of questionable validity as the hypothesis that all the coefficients are equal to zero cannot be rejected. Stripval and Medref were significant on the two small producer amendments. Democratic membership was strongly opposed to relaxation of controls, as was a high ADA rating, Petcost and Urban.

TABLE 6.7 Probit Runs For S 1570–Senate Fuel Allocation Votes

Variable	S 1570A	S 1570B	S 1570C	HR 5777
Constant	2.08*	4.78**	20.80*	5.33**
Stripval	0.0095	0.031*		0.033*
Nstrip			0.0034	
Medref		0.026*		0.036**
Larref	0.023**	0.016		0.014
South				1.30**
Urban	-2.56*	-2.03	-11.69*	-2.39*
ADA	-1.93**	-2.96**	-13.55**	
Party	-1.17**	-1.51**	-4.30**	-2.24**
Petcost		-0.0049	-0.026*	-0.0092**
Sign.	0.0001	0.0002	0.138	0.000
Pseudo R^2	.435	.563	.796	.471
Observations	91	92	93	93
Oil State Dem.	6-5-2	7-4-2	1-10-2	6-4-3
Oil State Rep.	3-7-1	9-1-1	8-2-1	11-0-0

S 1570A: McIntyre amendment to Jackson substitute amendment to urge the President to take further action as necessary to effectively stabilize prices on crude oil and petroleum products (6/5/73). N=1.
S 1570B: Bartlett amendment to exempt from allocation or price restraints an oil producer with a well producing not more than 10 bpd or wells producing not more than 1,500 barrels a year (6/5/73). Y=1.
S 1570C: Bartlett amendment to provide that compensation paid for fuel allocated shall not be less than the competitive market price (the amendment would have the effect of removing price controls) (6/5/73). Y=1.
HR 5777: Bartlett amendment to exempt from price controls and allocation programs oil produced from any lease whose average daily production does not exceed 10 barrels per well (7/31/73). Y=1.

The lack of oil production influence on voting can be seen in the votes by oil state Senators. While Republicans were solidly against any controls, Democrats stayed with the party on general price controls, bolting only when the issue concerned decontrolling small producer prices, and even then a significant number voted with the party.

Analysis of the House votes (Table 6.8) also demonstrate a lack of consistent opposition to controls by oil interests. Larval is significant in opposing HR 9130 and HR 9861B, but Nonlar is significant only on the latter vote. Medref is a factor in opposition to the first two votes, and its lack of

TABLE 6.8 Probit Runs For House Oil Allocation Votes–1973

Variable	HR 9130	HR 9681A	HR 9681B
Constant	2.02**	0.92**	1.99*
Larval	0.0045*		0.0012*
Nonlar	-0.0026	0.0059	0.0010*
Medref	0.0063*	0.0044*	
South	1.52**		0.46*
Urban			0.59
McGov	2.69**	-1.87*	
Nwhite		1.29*	
ADA	-3.18**	-2.34**	-2.75**
Party			-0.40
Petcost	-0.0049**		-0.0062**
Coalcost	-0.0021	-0.0025*	-0.0089**
Sign.	0.000	0.000	0.000
Pseudo R2	.413	.321	.369
Observations	412	380	409
Oil Democrats	3-27-4	23-7-4	17-15-2
Oil Republicans	2-16-1	12-6-1	7-12-0

HR 9130: MacDonald amendment to require the President, within six months after enactment, to develop a National Petroleum Contingency Plan to allocate crude oil and petroleum products during periods of shortages (8/2/73). N=1.
HR 9681A: Pickle amendment to provide that allocation of crude oil should not apply to the production stage (but at the pipeline or refinery stage), unless the President found it necessary (10/16/73). Y=1.
HR 9681B: Passage of the bill to direct the President to establish a mandatory program to allocate crude oil and refined petroleum products (10/17/73). A "nay" was a vote supporting the President's position. N=1.

importance on the third vote is probably due to the decision of many representatives to go along on the final vote after stating their opposition on amendment votes. South is opposed to allocation but does not support an exemption for producers (HR 9681A). ADA is the only variable to be consistently in favor of allocation, while some of the other results seem a bit bizarre, such as Nwhite and McGov opposing allocation measures (the correlation between Nwhite and South may have distorted the analysis of the second vote). Party is a nonfactor in these votes but energy consumption is related to support for controls. Democratic representatives from oil states voted consistently with constituent interests until the final vote on the allocation bill.

VOTE #7: VOTES ON THE NATIONAL ENERGY EMERGENCY ACT – 1973 AND 1974

The National Energy Emergency Act (NEEA) was introduced October 18, the day after the initiation of the Arab Oil embargo, by Senate Interior Committee Chairman Henry Jackson. S 2589 required the president to issue regulations for rationing and conserving fuel supplies within 15 days of enactment, which would result in a 10 percent reduction in consumption within 10 days of being implemented, and a 25 percent reduction within four weeks. The bill passed the Senate by a 78-6 vote on November 19. The House Commerce Committee reported its version of the emergency bill on December 7 (HR 11450). The House bill set up a Federal Energy Administration (FEA) to administer the fuel allocation program and to develop energy conservation plans. The bill also eased clear air standards during the emergency, mandated that large fuel burning installations switch to coal if possible, and provided for hearings by the Renegotiation board on suspected excess profits.[27]

Numerous amendments were submitted to HR 11450 during the three days of debate. Tim Carter (R-KY) proposed exempting the coal industry from provisions of the bill restricting windfall profits, adopted by teller vote 256-155 (D 120-110; R 136-45). A number of amendments revolved around an attempt by Staggers to restrict the use of fuel for school busing. James Broyhill (R-NC) attempted to tack on the Staggers amendment an amendment to eliminate the Renegotiation Board's role in determining the existence of windfall profits, rejected by teller vote 188-213 (D 51-172; R 137-41). William Roy (D-KS) attempted to exempt from windfall taxes leases producing under 25,000 bpd, rejected 189-194 (D 101-111; R 88-83). Another issue which provoked heated debate was the attempt to ease emission standards for automobiles, with an attempt to suspend controls voted down but postponement of the effective date for compliance accepted. The bill passed the House on December 15 and went to conference, where agreement was reached on December 20. In conference, the House windfall profits provisions were retained, provisions directing the president to set prices to avoid windfall profits were added, and Congress maintained a veto over energy conservation plans.[28]

In the Senate Russell Long and Paul Fannin led a filibuster against the conference report on S 2589. Long's objection was that the Finance Committee had not held hearings on the windfall profits tax provision, which he claimed was a revenue measure. Fannin argued that the provision would inhibit the oil industry's exploration effort (Fannin is referred to as an oil state senator, which is a bit odd since Arizona's production was only 740 thousand bbls in 1974, which meant its value to the state ranked far below such industries as golf course maintenance). A compromise was reached in the cloakroom, attached to the wild scenic rivers bill (S 921) which passed 52-8 on December 21, deleting the windfall profits provisions and changing the legislation's expiration data to

April 1974 from May 1975. The House then rejected the Senate compromise due to deletion of the windfall profits section.[29]

An attempt to pass the conference committee report in January died when the Senate voted by 57-37 (D 25-30; R 32-7) on January 29 to recommit the bill, due to a combination of conservationist Senators opposed to the modifications in the Clean Air Standards and oil state Senators opposed to windfall profit provisions. The bill returned to conference and was reported February 6 to the Senate with windfall provisions replaced by a price rollback to $5.25 per barrel for uncontrolled oil (which consisted of stripper oil, exempted from price controls in the Alaskan pipeline bill, and new oil). Debate centered around a Fannin motion to recommit the bill. James Buckley (R-NY) submitted an amendment to substitute administrative procedures to adjust oil prices for provisions lowering oil prices, rejected 37-62. James Abourezk offered a motion to instruct conferees to delete provisions which would reduce the price of crude oil exempted from controls which was rejected 37-62 (D 9-49; R 28-13). The Fannin amendment to recommit was rejected 38-60 (D 10-47; R 28-13) and the conference report was adopted 67-32 (D 51-7; R 16-25) despite the President's threat to veto the bill.[30]

In the House the same scenario was played out as representatives attempted to eliminate various provisions in the bill. John Anderson (R-IL) offered a motion to remove the provision to control the price of exempt crude oil which was defeated 173-238 (D 59-173; R 114-65). An Eckhardt attempt to reinstate a congressional veto over presidential energy conservation plans was rejected 66-343. A Delbert Latta (R-OH) motion to strike the provision authorizing rationing was rejected 199-211 (D 61-169; R 138-42). Finally the conference report was adopted 258-151 (D 179-52; R 79-99) on February 27. This was followed by a veto by President Nixon on March 6 and a subsequent roll call vote a few hours later in the Senate. The President's veto, motivated primarily by his opposition to the price ceiling provisions in the bill, was upheld by a 58-40 (D 46-11; R 12-29) vote. Six senators switched on the veto override vote, John Sparkman (D-AL), Bennett Johnston, Russell Long, John McClellan (D-AR), Hugh Scott (R-PA) and James Pearson (R-KS).[31]

John Dingell attempted to retain the price control provisions of S 2589 in an amendment to HR 11793, the bill creating the Federal Energy Administration. His amendment rolling back the price of domestic crude oil was adopted 218-175 (D 157-60; R 61-115) on March 6, after addition of a clause offered by Bob Eckhardt, exempting from price controls producers of less than 30,000 bbls/day. A second vote on the amendment, requested by Chet Holifield (D-CA), resulted in its rejection 163-216 (D 129-80; R 34-136).

The oil votes are generally price control votes though HR 11450C and S 2589A were more important to small producers. Oil production is consistently correlated with increased probability of voting against oil price controls, but there is a shift from Nonlar to Larval as the significant production variable. Party, ADA rating and energy consumption are consistent in their importance in support of price and allocation controls. Oil state Democrats continued to

TABLE 6.9 Probit Runs For the House NEEA Votes

Variable	HR 11450B	HR 11450C	S 2589A	S 2589B	S 2589C
Constant	3.46**	1.68**	3.36**	2.84**	2.02**
Larval			0.0028**	0.0011*	0.0013**
Nonlar	0.0014*	0.002*	0.0010**	0.0005	0.0006
Coalval	0.0005	0.0006		0.0013**	
Medref	0.006*				0.0034
South		0.65**	0.28		
McGov	1.52				
ADA	-5.03**	-2.62**	-1.53**	-2.38**	-1.74**
Party	-0.78**	-0.82**	-0.83**	-0.90**	-0.36*
Petcost	-0.005**	-0.004*	-0.0068**	-0.0038**	-0.0044**
Coalcost	-0.007**	-0.002	-0.0033**	-0.0028*	
Signif.	0.000	0.000	0.000	0.000	0.000
Pseudo R^2	.560	.289	.313	.385	.226
Observ.	413	403	416	421	421
Oil Dem.	23-10-1	28-2-4	29-4-1	24-9-1	7-25-1
Oil Rep.	17-1-1	15-4-0	16-2-0	17-1-0	3-15-0

HR 11450B: Broyhill amendment, to the pending Staggers substitute amendment, to delete provisions permitting the President to restrict unreasonable profits in the oil and coal industries. (12/13/73). Y=1.
HR 11450C: Roy amendment to the Staggers amendment to exempt from restrictions on windfall profits owners of leases which produced less than 25,000 bpd (12/14/73). Y=1.
S 2589A: Anderson motion to strike from the conference report the provision to control the price of crude oil exempted from price controls (2/27/74). A "yea" was a vote supporting the President's position. Y=1.
S 2589B: Latta motion to strike from the conference report the provision to authorize the President to impose rationing of crude oil and oil products (2/27/74). A "yea" was a vote supporting the President's position. Y=1.
S 2589C: Adoption of the conference report on the bill to grant the President authority to impose gasoline rationing, ease clean air standards and to direct him to reduce and control the price of oil that had been exempted from controls (2/27/74). A "nay" was a vote supporting the President's position. N=1.

vote with their constituents and against the party, especially on the two votes which directly impacted small producers.

The Senate votes tend to reflect a combination of ideology, partisan sparring and constituent interest interacting on a volatile issue. The weak performance of the model on the vote on S 2589A was due to the combination

TABLE 6.10 Probit Runs For the Senate NEEA Votes–1974

Variable	S 2589A	S 2589B	S 2589C	S 2589D	S 2589E
Constant	2.12**	1.36**	1.05**	0.88**	0.99**
Stripval		0.024*	0.018*	0.012*	0.020
Nonstrip	0.0006	0.0020*	0.0018*	0.0021*	0.0017
Medref		-0.032	-0.025	-0.035*	-0.026
Larref	0.022	0.022	0.018		0.013
South	0.58	-0.65			
Urban	-1.66				
ADA	-1.03	-3.03**	-2.37**	-2.07**	-2.09**
Party	-0.94*	-1.26**	-1.40**	-1.28**	-1.05*
Signif.	0.0002	0.0006	0.000	0.000	0.000
Pseudo R^2	.291	.589	.551	.456	.474
Observ.	99	100	100	100	99
Oil Dem.	9-4-0	8-5-0	8-5-0	7-6-0	5-8-0
Oil Rep.	10-1-0	9-2-0	9-2-0	3-8-0	2-9-0

S 2589A: Nelson motion to recommit to conference the bill to grant the President authority to ration fuels, ease clear air standards, and limit windfall profits. (Supporters of the motion wanted conferees to delete the windfall profits section or change the Clean Air Act provisions) (1/29/74). A "yea" was a vote supporting the President's position. Y=1.
S 2589B: Abourezk motion to instruct conferees to delete provisions to reduce and control the price of oil exempted from controls (2/19/74). A "yea" was a vote supporting the President's position. Y=1.
S 2589C: Fannin motion to recommit to conference the bill. (2/19/74). A "yea" was a vote supporting the President's position. Y=1.
S 2589D: Adoption of the conference report on the bill to grant the President authority to ration fuels, ease clean air standards and reduce and control the price of domestic crude oil exempted from controls (2/19/74). A "nay" was a vote supporting the President's position. N=1.
S 2589E: Passage, over the President's March 6 veto, of the bill. (3/6/74) A "nay" was a vote supporting the President's position. N=1.

of liberal opposition to weakening environmental laws combined with oil producer and conservative opposition to allocation and price controls. In a sense this result lends support to the model since it is able to detect the breakdown of the normal pattern of support and opposition on oil votes. On the other votes Stripval and Nonstrip were opposed to controls, with Stripval dominating, which reflects the fact that the votes were to extend controls to oil that had been uncontrolled, including stripper well production. As usual, Party

TABLE 6.11 Probit Runs For House Oil Price Votes–1974

Variable	HR 11793A	HR 11793B	HR 13834
Constant	2.13**	3.07**	2.41**
Larval	0.0020*		
Nonlar	0.0015	0.014*	0.0092*
Medref	-0.0039		
Urban	0.47		
South	0.44*	1.24**	0.35*
Nwhite	-1.48*	-2.35**	
ADA	-2.51**	-2.05**	-2.56**
Party	-0.38*	-0.74**	-1.10**
Petcost	-0.0032*	-0.0039**	-0.0023*
Coalcost	-0.0060**	-0.0061**	-0.0021
Significance	0.000	0.000	0.000
Pseudo R2	.351	.438	.484
Observations	417	425	425
Oil Democrats	3-29-2	2-31-1	5-28-1
Oil Republicans	3-15-0	1-16-1	2-17-0

HR 11793A: Dingell amendment to roll back the price of crude oil to $5.25 a barrel and to allow prices to rise no more than 35 percent above the $5.25 ceiling (3/6/74). A "nay" was a vote supporting the President's position. N=1.
HR 11793B: Second vote on Dingell amendment requested by Holifield (3/7/74). N=1.
HR 13834: Staggers motion to suspend the rules and pass the bill to give the President authority to ration fuels and to roll back and control oil prices. A two-thirds majority vote was required for passage under suspension of the rules (5/21/74). A "nay" was a vote supporting the President's position. N=1.

and ADA rating are the most important variables supporting price controls, but Petcost was insignificant on all votes. Medref and Larref seem to be more important in the Senate than the House, with Medref supporting controls and Larref in opposition. Since medium sized refineries were more likely to be owned by non-integrated firms, while larger refineries were usually operated by the major oil companies, this result is reasonable.

Oil state Senators opposed controls, though not overwhelmingly as one would expect. There were two reasons for defections, loyalty to party, especially for Democrats, and the sense of national emergency that made this a high profile issue. Senators from Northern oil producing states (Colorado, Montana, North Dakota, and Utah) were forced to choose between oil producers and other interest groups that were adamant in their opposition to higher prices.

TABLE 6.12 Logit Runs For Oil Allocation and Oil Price Votes

Variable	*House*	*Senate*	*Senate*
Constant	2.58**	0.055	0.299
Larval (Nonstrip)	0.00049*	-0.00013	0.00057*
Nonlar (Stripval)	0.00129**	0.00708*	0.00769*
Smallref	0.0090	0.0308	
Medref	0.00225	-0.00715	-0.00657
Larref	0.00127*	0.0148**	0.0133**
South	0.308*	-0.367	-0.461
McGov	0.877	-0.259	-0.783
Urban	0.500	-0.0758	-0.053
ADA	-4.464**	-0.2.367*	-2.362**
Party	-0.816**	-0.945**	-0.939**
Petcost	-0.00394**	0.00008	0.00005
Adjusted R^2	.661	.738	.738
Observations	430	100	100

House Votes: HR 9681A, HR 11450B, S 2589A, S 2589B, S2589C, HR 11793A, HR 11793B, HR 13834. Senate Votes: S 1570A, S 1570B, S 2589B, S 2589C, S 2589D, S 2589E.

The House votes on the price control amendments are similar to the earlier votes on allocation. Party, ADA rating and energy consumption are consistent in their importance in support of price and allocation controls. Coalcost tends to dominate Petcost because coal consumption is concentrated in states that have little oil production. Oil production is consistently in support of decontrol, as oil state Democrats vote with their constituents and against the party. The support of South for decontrol and Nwhite for price controls may be partially an artifact of their interaction, though both variables have the expected sign. Nwhite may also reduce the size of the coefficients for Party and ADA due to collinearity.

I decided to aggregate a number of votes to see if this would improve model performance, as it would increase the number of observations, especially for Senate votes. I was careful to choose votes which were quite similar in their impact on various interest groups, since aggregating votes which involve differing coalitions would defeat the reason for collecting detailed data on production and refining. The eight house votes and six Senate votes were all allocation and price control votes. The only votes which involved specific small producer interests were those reimposing controls on stripper oil, but since these votes included control of the price of new oil they should be opposed by all producer interests. The dependent variable Lvote reflected the frequency, $f_i = r_i / n_i$, with which a Congressmen voted against oil allocation and price control amendments. I assumed that a nonvote meant that the Congressman had an equal probability of voting for or against controls. The method used to analyze the votes was a weighted logit technique.

A simplified model was used because of the problems with collinearity which plagued the probit runs. The House logit model was based on eight votes while the Senate used six votes. The Senate was run a second time with Smallref omitted, as this variable was highly correlated with Nonstrip. The results from the aggregated run are in line with what one would expect from both theory and observation. Oil production is related to opposition to controls, though production from marginal fields seems to be more important than that of larger fields. The presence of large refineries also influences opposition, which is reasonable given their probable ownership by integrated firms, while smaller refineries have little impact. ADA and party are the most important variables in support of controls, while the cost of petroleum products is important only in the House voting.

VOTE #8: OIL DEPLETION ALLOWANCE–1975

The next Congressional battle in the energy wars occurred in 1975 over the continuation of the tax breaks which the oil industry had historically enjoyed. Changes in the tax laws had been mentioned during the 1974 skirmishes over controls but the issue had taken a backseat in favor of attempts to control profits through reducing prices. The House Ways and Means Committee had adopted proposals by Al Ullman which would have eliminated the 22 percent depletion allowance in stages by 1980. In April 1974 the Committee approved a new plan backed by Chairman Wilbur Mills to eliminate depletion over three years, but the House Democratic Caucus voted for a full House vote on amendments that would eliminate the foreign tax credit and depletion. Speaker Carl Albert and Mills then killed the Energy Tax Bill with the help of the Rules committee.

While Nixon had committed himself to the defense of the depletion allowance, President Ford suggested that it might be necessary to sacrifice the allowance in order to achieve a free market in oil, a view shared by executives of major oil companies, though vigorously opposed by independent producers. Ford made his remarks in his October 8 message supporting the Tax Reform Act of 1974. The issue disappeared for the rest of the year, but resurfaced during consideration of the Emergency Tax Cut bill (HR 2166) in early 1975. The Ways and Means Committee twice rejected efforts to add an amendment to eliminate depletion to avoid delaying enactment of the bill. When the Rules Committee rejected a proposal to allow floor action on an amendment to repeal the allowance, while allowing action on a Republican substitute bill, Sam Gibbons and William Green took the issue to the Democratic Caucus. Green's request was approved by the Caucus February 25 by a 153-98 vote, and the

House voted 242-162 (D 227-43; R 15-119) for floor consideration of the amendment eliminating depletion on February 27. Charles Wilson then offered an amendment to keep depletion for the first 3,000 bpd of production, and Bill Archer (R-TX) offered an amendment to the Wilson motion to retain the allowance for natural gas subject to federal price regulations. The Archer amendment was rejected 157-254 (D 56-217; R 101-37) and the Wilson amendment was defeated 197-216 (D 102-172; R 95-44). The Green amendment was then adopted 248-163 (D 204-69; R 44-94).[32]

On the Senate floor Lloyd Bentsen suggested a compromise which would retain depletion for the first 3,000 bpd of production by independent producers. The Bentsen amendment, along with a provision by Vance Hartke to tighten tax treatment of foreign oil income, were added to HR 2166, the former by a vote of 47-41 (D 26-30; R 21-11), and the latter by a vote of 73-24 (D 56-4; R 17-20), after the defeat of a Carl Curtis (R-NE) to kill the Hartke amendment 44-52 (D 13-46; R 31-6). Long then succeeded in tabling a Ernest Hollings (D-SC) amendment to repeal the depletion allowance for the 12 major companies and phase out the allowance for smaller companies by a vote of 51-46 (D 26-34; R 25-12). The Finance Committee, chaired by Long, moved to offer a substitute for the House bill. After an attempt to invoke cloture on HR 2166 failed by one vote, Majority Leader Mike Mansfield maneuvered to have the bill recommitted to the Finance Committee with compromise provisions he had drawn up, thus nullifying the Bentsen and Hartke amendments.[33]

Two proposals were then offered; a Cranston (D-CA) amendment restricting use of the foreign tax credit and limiting depletion to the equivalent to 3,000 bpd and a Hollings proposal, cosponsored by Kennedy, which provided for a four-year phase-out for independent producers. At Long's request a roll-call vote was taken on the Hartke amendment, included in the Hollings proposal, which was adopted 75-20 (D 54-5; R 21-15). Bentsen's attempt to include his amendment failed 44-51, a Tower motion to table the Hollings amendment was rejected 29-67 (D 4-56; R 25-11), and an amendment from Cranston to retain depletion for independents producing less than 3,000 bpd or its equivalent was rejected 44-50 (D 23-37; R 21-13). Finally a proposal from Dewey Bartlett to raise the limit for independents to 2,000 bpd was accepted 47-46. The final version of HR 2166, hammered out in conference, retained the foreign tax credit though limiting its application. Depletion was eliminated retroactive to the first of the year except for small producers.[34]

The model seems to work very well with regard to these House votes (see Table 6.13) in distinguishing the differences between producer interests. Larval seemed to have little impact on support for retaining the depletion allowance, while Nonlar was significant in all the oil allowance votes, disappearing only on the natural gas depletion vote. Ngval was also consistent in influencing votes on depletion amendments, especially the two votes involving gas production. Since Nonlar is more likely to be associated with small producers, and Larval with large producers, this result supports the opinion at the time that the large companies had accepted loss of depletion and were more

TABLE 6.13 Probit Runs For House Votes on Depletion Allowance

Variable	HR 2166A	HR 2166B	HR 2166C	HR 2166D
Constant	3.93**	1.76**	1.31**	2.49**
Larval		-0.0005		
Nonlar	0.0027*	0.0014	0.021**	0.0052*
Ngval	0.0027	0.0035*	0.0051	0.0083**
Medref		0.012**		
Larref				-0.0027*
South				0.44*
McGov			-1.77*	
ADA	-3.86**	-5.89**	-2.15**	-2.73**
Party	-1.86**	-0.37*		-0.52*
Petcost	-0.0025			-.0034*
Coalcost	-0.014**			-.0029
Signif.	0.000	0.000	0.000	0.000
Pseudo R^2	.694	.661	.389	.476
Observations	405	412	422	417
Oil District Dem.	10-25-1	25-10-1	34-1-1	6-29-1
Oil District Rep.	0-21-2	21-1-1	21-0-2	2-19-2

HR 2166A: Adoption of the rule (H Res 259) allowing floor amendments to the bill to cut federal taxes, including one to repeal the oil depletion allowance (2/27/75). N=1.
HR 2166B: Archer amendment, to the Wilson amendment, to retain the depletion allowance for price controlled natural gas (2/27/75). Y=1.
HR 2166C: Wilson amendment, to the Green amendment, to retain the depletion allowance for the first 3,000 bpd (or 18-million cubic feet of natural gas) produced by an independent firm (2/27/75). Y=1.
HR 2166D: Green amendment to repeal the 22 percent depletion allowance retroactive to January 1, 1975; the allowance would remain for price controlled natural gas until July 1, 1976 or until the price ceiling was increased, and for natural gas sold under contract until the price was increased (2/27/75). N=1.

interested in preserving the foreign tax credit and support of price decontrol. ADA is a strong opponent of depletion as expected, but the Democrats were uninvolved in the small producer vote, and their opposition was muted in the natural gas vote, suggesting ambivalence in their attitude toward small producers. This is supported by the oil district votes which show significant Democratic opposition to depletion except on the small producer vote, while Republican votes were overwhelmingly pro-depletion.

TABLE 6.14 Probit Runs For Senate Votes on Oil Depletion

Variable	HR 2166A	HR 2166B	HR 2166C	HR 2166D	HR 2166E
Constant	2.93**	2.57**	5.01**	2.72**	1.04**
Stripval	0.040	-0.0083	0.041*		0.057**
Nonstrip	0.0051	0.0016	0.0047*	0.0046**	0.0024*
Medref	0.029*	0.025			
Larref	-0.018		-0.019*	-0.033**	
South			-1.28*	-1.23	
McGov			-8.55		
Urban	-3.21**				
ADA	-2.83**	-3.63**	-4.09**	-4.61**	-3.91**
Party		-2.00**		-2.19**	
Petcost					
Signif.	0.0004	0.000	0.0001	0.000	0.0003
Pseudo R^2	.508	.607	.557	.724	.523
Observ.	91	97	98	97	95
Oil Dem.	12-2-0	7-7-0	11-3-0	3-11-0	10-4-0
Oil Rep.	10-0-0	10-0-0	10-0-0	9-1-0	10-0-0

HR 2166A: Bentsen amendment to repeal the oil and gas depletion allowance except for the first 3,000 barrels of oil and 18 MCF of natural gas produced daily by an independent company. Price controlled natural gas or gas sold under fixed contracts was exempted until its price was raised (3/18/75). Y=1.
HR 2166B: Curtis motion to table and thus kill, the Hartke amendment to eliminate deferral of US taxation on foreign income until the income was repatriated to the United States (3/19/75). Y=1.
HR 2166C: Long motion to table, and thus kill, Hollings amendment to repeal the oil and gas depletion allowance for the 12 major oil companies, and to phase out the allowance over a four-year period for the smaller, independent companies (3/19/75). Y=1.
HR 2166D: Tower motion to table, and thus kill, Hollings amendment to repeal the foreign tax credit for taxes paid on foreign oil income, allowing such taxes to be deducted as an ordinary business expense (3/20/75). Y=1.
HR 2166E: Cranston amendment, to the Hollings amendment, to retain depletion allowance for independent companies (3/20/75). Y=1.

The Senate also demonstrated a pattern of large producer indifference or even opposition to the depletion allowance (Table 6.14). Nonstrip and Larref were either unimportant or weakly opposed, especially on the small producer votes (HR 2166A, HR 2166C and HR 2166E), while Stripval was supportive of the allowance on these votes. The unusual results were the opposition of

McGov and South to tax credits for small producers but not to continuation of the foreign tax credit. McGov could be explained by the symbolic importance of depletion, but the opposition of South to a small business benefit is inexplicable. An expected result is the consistent opposition by ADA to all tax breaks, while Party was only significant in votes concerning foreign tax credits, with Democrats opposed and Republicans strongly in support of foreign tax credits. This split can be seen on the vote on the Tower amendment, where oil state Democrats refused support for an attempt to continue the status quo, instead focusing their efforts on retaining tax breaks for small producers.

VOTE #9: HOUSE VOTES ON OIL IMPORTS AND ENERGY TAXES–1975

President Ford began 1975 by submitting a comprehensive energy program to Congress including proposals to delay auto emissions standards, decontrol oil prices, deregulate natural gas and speed licensing of nuclear power. Ford's State of the Union address included proposals for a strategic petroleum reserve (SPR) to be filled with production from the Elk Hills Naval Reserve and to lower consumption through energy taxes. The first blow to the Ford plan came when the Ways and Means Committee reported HR 1767, which suspended the president's authority to impose an oil import fee for ninety days, by a vote of 19-15 on January 24. The House passed the bill 309-114 (D 267-16; R 42-98). The Senate passed the bill on February 19 by a vote of 66-28, which was then vetoed by President Ford on March 4. Ford attempted to block a Senate overriding his veto by stressing that the impact of the import fee would fall on gasoline prices, hoping to sway Northeast Republican Senators. The President agreed to defer the fee until May 1 while the Senate indefinitely delayed a vote to override his veto.

The difficulty of passing such a program through the post-Watergate Congress can be seen by the composition of the Democratic Senate task force on energy led by John Pastore, which included Philip Hart and Edmund Muskie, ardent supporters of price controls, while excluding oil state senators. The Senate Democratic plan included the establishment of tax credits and loan and grant programs to encourage residential and commercial energy conservation, a 5¢ gasoline tax and a SPR. Liberal Democratic Congressmen dissented from the leadership, calling for lowering energy prices via a price rollback, mandatory fuel-efficiency standards and no increases in consumer prices and taxes. The Ways and Means Democrats, under the leadership of Chairman Al Ullman, opted for a plan using import quotas, gasoline taxes, a strategic stockpile, and extension of the mandatory oil allocation program.[35]

The Ways and Means Committee began consideration of an energy conservation tax bill, HR 5005, March 18, in an attempt to put together the bill by May 1, the date on which Ford threatened to impose an oil import fee along with decontrol of oil prices. The Committee approved a provision setting quotas reducing petroleum imports to 25 percent of domestic oil consumption by 1985, after it voted 11-18 to defeat a Republican sponsored motion to strike a provision allocating 2 million bpd of the 1975-1977 quotas (5.5 to 6.5 million bpd of imports) for imports of distillate and residual oil. The only two Democrats which supported the motion were Omar Burleson and Jake Pickle of Texas. Oil state Democrats joined Republicans in opposing a prohibition on increases in import duties on distillate and residual fuel oil, and a provision giving the President authority to have the government purchase and sell all oil imports. The Committee also retreated from Ullman's initial proposal for a 37¢ gasoline tax, accepting only a 3¢ increase, with a 20¢ per gallon increase triggered by gasoline consumption rising above the 1973 level. On May 8 the Committee voted 20-16 to introduce a new bill incorporating the compromises reached during the markup process, with all 12 Republicans and four liberal Democrats in opposition.[36]

On the floor of the House most of the provisions of the final Ways and Means Committee bill, HR 6860, were gutted as the battles fought within the committee were repeated on the floor. An Abner Mikva (D-IL) amendment to make federal purchase and sale of imported oil mandatory was rejected 152-265 (D 150-123; R 2-142) on June 10. This was followed by a vote on a Ways and Means Committee amendment deleting the provision allowing federal purchases of imports, adopted 216-196 (D 81-189; R 135-7). An amendment from Gibbons deleting the quota provisions was rejected 185-224 (D 59-209; R 126-15). Joe Waggonner's amendment raising quotas by a half million bpd was adopted by a narrow margin of 211-200 (D 27-163; R 133-7) while a Conte amendment to increase quotas sufficient to offset reductions in natural gas supplies was defeated 195-213 (D 71-198; R 124-15). Pete Stark (D-CA) introduced an amendment deleting the 20¢ standby gas tax, adopted 345-72 (D 211-67; R 134-5), while a Bill Alexander (D-AR) amendment eliminating the 3¢ gas tax was adopted 209-187 (D 95-170; R 114-17).[37] In effect, the House rejected provisions which would have reduced consumption by raising prices.

The analysis of the votes on HR 6860 reflect the complex politics of the bill. The vote against the import fee reflected general opposition to the measure, with Party reflecting the loyalty of Republicans to their president. The votes on HR 6860 provide some contradictory results due to maneuvering by both parties which placed them in the position of voting against their normal tendencies. The oil production variables were significant with regard to federal purchase of imports (HR 6860A & B), which they opposed, but Larval was also significant in support of the 3¢ gasoline tax (which might be due to party loyalty by oil Democrats). Both Medref and Larref are in favor of government import purchases, which is an inexplicable result, but have little impact on other issues. McGov also fails to demonstrate a consistent pattern, significant only in opposition to increasing the size of quotas and for the gasoline tax.

TABLE 6.15 Probit Runs For the House Energy Tax Votes

Variable	HR 1767	HR 6860A	HR 6860B	HR 6860C	HR 6860D
Constant	-0.78**	5.26**	3.16**	-1.20*	1.66**
Larval		0.021*	0.0014	0.0003	-0.0005*
Nonlar			0.0059*		
Coalval		0.0006*			
Medref		-0.014*			-0.0059*
Larref		-0.0039	-0.0027*	-0.0017	
South		-0.71*			
McGov					-2.79**
ADA	1.87**	-5.35**	-4.16**	-0.95**	-1.32**
Party	1.37**	-1.66**	-1.59**	-1.83**	-0.83**
Petcost				0.0060**	0.0018
Coalcost		-0.0086**			
Signif.	0.000	0.000	0.000	0.000	0.000
Pseudo R^2	.408	.644	.652	.388	.211
Obser.	428	423	418	415	339
Oil Dem.	30-5-1	4-29-3	26-7-3	14-19-3	14-20-2
Oil Rep.	3-20-0	0-20-0	20-0-0	20-0-0	18-2-3

HR 1767: Passage of the bill to suspend for 90 days the President's authority to adjust imports of petroleum and petroleum products and to negate any such action taken by the President after 1/15/75 (2/5/75). A "nay" was a vote supporting the President's position. Y=1.
HR 6860A: Mikva amendment to make federal purchase and sale of imported oil mandatory in 1976 (6/10/75). N=1.
HR 6860B: Ways and Means Committee amendment which deleted the provision to allow federal purchase and sale of imported oil (6/12/75). N=1.
HR 6860C: Gibbons amendment to delete language providing for imposition of quotas on imported oil (6/10/75). Y=1.
HR 6860D: Alexander amendment to delete a 3 cent gasoline tax (6/11/75). A "nay" was a vote supporting the President's position. Y=1.

ADA and Party demonstrate a similar pattern of influence, though ADA support is strongest for government purchases of imports and weakest with regard to maintaining quotas, while Party has its lowest value for the gasoline tax vote. Both ADA and Party reflect the Democratic dilemma–in order to justify price controls some symbol of import restraint was required without actually restricting supplies or raising prices. This explains the votes against President Ford's import fee along with votes for quotas which were more expressions of intent than a blueprint for action. Petcost was solidly against quotas and taxes as representatives from high cost states put constituent interests before partisan politics.

Oil state representatives were unified only on the issue of government involvement in import sales, probably because government involvement in any part of the industry was anathema to them. While oil state Republicans toed the party line on all votes, Democrats strongly supported the party in opposing an import fee, but only gave weak support for import quotas and the gasoline tax, with a majority deserting to vote for the Waggonner amendment.

VOTE #10: OIL PRICE CONTROLS–1975

President Ford in February 1975, made his intention clear to decontrol the price of old oil. In response the Senate Interior Committee approved a bill, S 622, on February 28, to extend the veto period for Congress to reverse decontrol from five days to a month. A second bill, S 621, which would require the President to set a price ceiling for new oil to keep it from rising as a result of the imposition of an import fee was reported March 4. The bill also required congressional review of any change in price regulations. The next day the House Commerce Committee reported a measure, HR 4035, requiring congressional review of any price increase of old oil of more than 50 cents over the average price of such oil as of January 1.[38]

The Standby Energy Powers Bill, S 622, reached the Senate floor on April 9. Bartlett lost three attempts to limit the president's power to order production at rates above the maximum efficient rate. Mike Gravel (D-AK) then submitted an amendment to end the allocation authority of the EPAA on August 31, and end price controls concurrent with the imposition of an excess profits tax, which was rejected 21-69 (D 3-54; R 18-15), and a second attempt without the clause eliminating price controls was defeated 23-68 (D 3-54; R 20-14). A similar attempt by Henry Bellmon to phase out price controls on "old oil" and impose a 100 percent tax on the extra income which was not reinvested in energy production was rejected 28-61 (D 6-49; R 22-12). A cosmetic clause submitted by Floyd Haskell (D-CO) to set a four percent reduction as the goal of conservation programs passed easily. The toughest fight came over an amendment by Johnston, to allow oil produced by enhanced recovery techniques to sell at "new" oil prices, rejected 42-47 (D 17-38; R 25-9). Two votes to reconsider the amendment barely passed, and Jackson then added an amendment to limit to $7.50 the price which could be charge per barrel of enhanced recovery oil, which passed by a 46-38 vote (D 38-13; R 8-25), after which the Johnston amendment was adopted by voice vote. The next day Bellmon attempted to slide a two-step phase-out of price controls through the Senate, but this attempt was rejected 23-62 (D 4-50; R 19-12). Glenn submitted an amendment to require the President to set a ceiling price for "new" oil at a level less than the prevailing price on January 31, 1975, which was adopted 54-31 (D 48-6; R 6-25). The bill the passed 60-25 (D 51-3, R 9-22).[39]

The price control votes during the Standby Energy Power Votes (Table 6.16) seem to follow a more consistent pattern than the votes on energy taxes. In the Senate Nonstrip is a consistent factor in opposition to decontrol while Stripval seems to be unimportant. I suspect this is due a general consensus in the oil industry against controls; since states with substantial values for Stripval are a subset of the states with large values for Nonstrip (Kansas is the only major oil producing state in which Stripval > Nonstrip), when there is solid support by oil states for a position on an issue Nonstrip should dominate Stripval. The only two other consistently important variables are ADA and Party, both of which are strongly opposed to decontrol. Oil state Republicans were strongly opposed to continuation of any sort of controls but oil state Democrats tended to split between party loyalty and support of constituent interests. These Democrats tended to oppose bills eliminating price controls and allocation, splitting on bills decontrolling old oil, but then deserting the party on the votes to decontrol prices for oil produced by secondary recovery techniques.

With the passage of S 622, the Senate turned its attention to S 621, the oil price decontrol bill sponsored by Henry Jackson. Johnston renewed his effort to exempt secondary and tertiary oil from price controls, succeeding this time by a vote of 49-37 (D 19-31; R 30-6), while Jackson failed this time to tack on his price ceiling of $7.50 for enhanced recovery oil 38-46 (D 32-18; R 6-28). The bill then passed 47-36 (D 41-9; R 6-27) on May 1. The bill included provisions to prohibit lifting of price controls for ninety days, Congressional review and veto of decontrol, and for price ceilings to be set for all uncontrolled oil.[40]

President Ford, frustrated by Congressional inaction, moved on May 27 to increase the import fee to $2 per barrel and to submit to Congress a formal proposal for ending price controls along with the enactment of a windfall profits tax with a plowback provision.[41] The House then passed HR 4035, its version of the decontrol bill June 5, by a vote of 230-151. The Senate adopted the conference report on July 16, 57-40 (D 51-8; R 6-32) and the House responded a day later by approving the conference report 239-172 (D 221-54; R 18-118). Ford then vetoed HR 4035 on July 21, and the measure was shelved as there were insufficient votes to override the veto. The House responded by voting 262-167 (D 25-53; R 27-114) to accept House Resolution 605, disapproving and thereby blocking the lifting of price controls. Ford proposed a decontrol plan on July 25 to phase out oil price controls over 39 months; the House countered with House Resolution 641 blocking his new plan, adopted 228-189 (D 207-68; R 21-121). A measure extending price controls until March 1, 1976, S 1849, passed the Senate 62-29 on July 15 and was approved by the House 303-117 (D 248-30; R 55-87) on July 31. Ford then vetoed S 1849 on September 9, but agreed to extend controls for 45 days if his veto was upheld, which it was by the Senate on September 10, 61-39. Congress sent, and

TABLE 6.16 Probit Runs For S 622–Standby Energy Power Votes

Variable	S 622A	S 622B	S 622C	S 622D	S 622E
Constant	0.57*	0.86**	1.34	7.35*	1.92**
Stripval		0.0071		0.030	
Nonstrip	0.0012**	0.0022*	0.013*	0.017*	0.010*
Medref		-0.028		-0.12	-0.029
Larref				-0.049*	
McGov				-8.09	
ADA	-2.57**	-2.77**	-4.43**	-3.98**	-3.59**
Party	-1.50**	-1.85**	-3.66**	-2.06**	-1.63**
Signif	0.0001	0.0001	0.0037	0.040	0.0001
Pseudo R^2	.477	.555	.789	.676	.647
Observ	94	94	91	90	85
Oil Dem.	3-10-1	3-10-1	6-6-2	7-5-2	4-8-2
Oil Rep.	7-3-0	8-2-0	10-0-0	10-0-0	1-9-0

S 622A: Gravel amendment to allow the allocation authority of the EPAA to expire August 31, 1975, to allow authority to control oil prices to terminate upon enactment of S 622, and to tax excessive fuel profits (4/9/75). Y=1.
S 622B: Gravel amendment to allow the allocation and price control authority of the EPAA to expire on August 31, and to tax excessive profits (4/9/75). Y=1.
S 622C: Bellmon amendment to phase out price controls on "old" oil by 1977 and impose a 100 percent tax on income produced by the resulting price increase, unless reinvested in energy (4/9/75). Y=1.
S 622D: Johnston amendment to allow oil produced from existing fields by enhanced recovery techniques to sell at the "new" oil price rather than the "old" oil price (4/9/75). Y=1.
S 622E: Jackson amendment, to the Johnston amendment, to limit to $7.50 the price which could be charged per barrel of oil produced by enhanced recovery methods (4/9/75). N=1.

Ford accepted, the extension of price controls to December 15 to provide time to develop an comprehensive energy bill.[42]

While the battle over decontrol raged between Congress and the President the House was struggling in its attempt to give birth to an omnibus energy bill. The Energy Policy and Conservation Act, HR 7014, was originally drafted by the Commerce Subcommittee on Energy and Power, and was to include a gradual phasing out of price controls. However, the full committee rejected decontrol by a one-vote margin and proceeded to extend controls, including a rollback in the price of new oil to $7.50 a barrel, before reporting the bill on

TABLE 6.16 Probit Runs For S 622–Standby Energy Power Votes

Variable	*S 622F*	*S 622G*	*S 622H*	*S 622I*
Constant	6.82*	3.08	1.09**	5.18**
Stripval	-0.011			
Nstrip	0.0069**	0.015	0.0010*	0.0086**
Larref	-0.033*			-0.017
McGov	-14.82*	13.37		
ADA	-2.97*	-6.75**	-1.95**	-2.03**
Party	-3.76**	-3.52**	-2.05**	-2.47**
Petcost		-0.010*		-0.0080*
Significance	0.016	0.012	0.000	0.000
Pseudo R^2	.759	.828	.539	.634
Observations	89	90	92	98
Oil State Dem.	3-9-2	6-6-2	11-2-1	6-8-0
Oil State Rep.	9-1-0	0-10-0	1-9-0	0-10-0

S 622F: Bellmon amendment to provide for a two-step phase-out of price controls on "old" domestic oil (4/10/75). Y=1.
S 622G: Glenn amendment to require the President to set a ceiling price for "new" domestic oil at a level no higher than the price generally prevailing for such oil as of Jan. 31, 1975 (4/10/75). N=1.
S 622H: Passage of the bill to provide the President with standby emergency powers, to provide a congressional role in any decision to lift price controls on domestic oil, to mandate a national energy conservation program and require the President to establish a ceiling price for domestic oil (4/10/75). N=1.
S 622I: Jackson motion to concur in the House amendments to the conference version of the bill including standby emergency powers for the President, creation of a national strategic oil reserve, mandatory fuel efficiency standards for automobiles, and continuation of oil price controls (12/17/75). N=1.

July 9. A number of attempts to restore decontrol were made on the House floor including a Krueger amendment providing for gradual decontrol by 1980 along with a windfall profits tax, rejected 202-220 (D 80-202; R 122-18) and a Heinz substitute for the Krueger amendment providing for gradual decontrol along with a $11.28 ceiling on new oil, rejected 94-326. Finally a Wilson amendment to delete the oil pricing provisions was adopted 215-199 (D 90-184; R 125-15) on July 23.[43]

The decontrol fight resumed the next week during consideration of a rule to allow consideration of a Krueger amendment containing the President's oil decontrol plan. The amendment would have gradually phased out controls on old oil, eliminating the category by late 1978, established a $11.50 per barrel

TABLE 6.17 Probit Runs For S 621–Decontrol of Oil Prices

Variable	S 621A	S 621B	S 621C	HR 4035
Constant	2.84*	7.79**	5.51**	14.20**
Stripval	0.100			
Nonstrip	0.017*	0.0049*	0.0056*	0.019**
Medref				-0.052
Larref	-0.026	0.0031*		-0.028
South		-1.44		-2.31
Urban	2.66	-1.81		
McGov		-11.34		
ADA	-7.00**	-3.08**	-3.08**	-8.69*
Party	-3.24**	-1.23**	-2.38**	-5.44**
Petcost			-0.0071*	-0.017*
Significance	0.067	0.000	0.000	0.221
Pseudo R^2	.748	.569	.683	.854
Observations	88	86	86	97
Oil State Dem.	10-2-2	3-9-2	5-6-3	6-7-1
Oil State Rep.	10-0-0	0-9-1	0-10-0	0-10-0

S 621A: Johnston amendment to exempt from price regulation oil produced through the more expensive enhanced methods of recovery (5/1/75). Y=1.
S 621B: Jackson amendment to place a ceiling of $7.50 on the price of oil produced by enhanced recovery methods (5/1/75). N=1.
S 621C: Passage of the bill to prohibit for 90 days the lifting of price controls on oil and, thereafter, to give Congress the right to review any presidential proposal to deregulate prices and to disapprove them within 30 days, and to require the President to set prices for all domestic oil not currently under price controls (5/1/75). A "nay" was a vote supporting the President's position. N=1.
HR 4035: Adoption of the conference report on the bill to extend authority for oil price controls to December 31, 1975, and extend for 20 days the period for congressional review and the opportunity to veto changes in oil price controls (7/16/75). A "nay" was a vote supporting of the President's position. N=1.

price ceiling on the cost of all oil, and delayed decontrol until enactment of a windfall profits tax. The rule was adopted, 310-109 on July 30, setting the stage for a series of votes on decontrol. Staggers offered substitute language to the amendment which would continue the $5.25 price ceiling for old oil and roll back other prices to an average of $7.50, which was adopted 218-207 (D 203-79; R 15-128), and the modified Krueger amendment passed 231-193 (D 214-68; R 17-125). As mentioned H Res 641 then passed the House to prevent the President from unilaterally deciding to instate decontrol.[44]

TABLE 6.18 Probit Runs For House Oil Price Control Votes

Variable	HR 4035	H Res 605	H Res 641	S 1849
Constant	5.71**	4.01**	3.32**	3.00**
Larval	0.0012			0.0006
Nonlar	0.0041**	0.0025**	0.0080**	0.0009
South	0.42			
McGov			-2.16*	
Urban			1.75**	
ADA	-4.09**	-3.94**	-3.96**	-3.41**
Party	-1.92**	-1.24**	-1.08**	-0.95**
Petcost	-0.0059**	-0.0036*	-0.0033*	-0.0033*
Coalcost	-0.017**	-0.012**	-0.0051*	-0.013**
Significance	0.000	0.000	0.000	0.000
Pseudo R^2	.700	.606	.620	.493
Observations	430	430	428	425
Oil Dem.	7-27-2	11-24-1	3-30-3	18-16-2
Oil Rep.	1-22-0	2-21-0	1-22-0	5-17-1

HR 4035: Adoption of the conference report on the bill to extend authority for oil price controls to December 31, 1975 (7/17/75). A "nay" was a vote supporting the President's position. N=1.
H RES 605: Adoption of the resolution disapproving, and thereby blocking, the proposed plan of President Ford to gradually lift price controls on domestic oil (7/22/75). A "nay" was a vote supporting the President's position. N=1.
H RES 641: Adoption of the resolution to disapprove President Ford's plan for the gradual decontrol of oil prices (7/30/75). A "nay" was a vote supporting the President's position. N=1.
S 1849: Passage of the bill to extend until March 1, 1976, from August 31, 1975, the price control and allocation authorities of the Emergency Petroleum Allocation Act of 1973 (7/31/75). N=1.

With decontrol settled the next move was to attempt to punish the large oil companies by restricting their activities. Christopher Dodd (D-CT) submitted an amendment to bar joint ventures by major oil companies on federal lands, rejected by one vote, 206-207 (D 196-76; R 10-131). Jerome Ambro (D-NY) offered an amendment to prohibit the leasing of federal non-oil bearing mineral lands to vertically integrated oil companies, rejected 158-254 (D 147-125; R 11-129). When the House returned from recess on September 17 another attempt to eliminate controls, offered by Clarence Brown (R-OH), was rejected 151-242. Moss offered an amendment to authorize the GAO to conduct verification audits of oil and gas producers, adopted 233-162 (D 203-58; R 30-104). An attempt to

TABLE 6.19 Probit Runs For HR 7014–Energy Policy and Conservation Act

Variable	HR 7014A	HR 7014B	HR 7014C	HR 7014D
Constant	4.04**	6.45**	5.46**	4.27**
Larval	0.0013	0.012**		
Nonlar	0.0059*	0.0048	0.0050**	0.0044**
Smallref			0.038*	
Medref		-0.012*		
South	0.63*		0.52*	
Urban		-0.94*		0.83*
Nwhite	-1.10	-0.89		
ADA	-3.30**	-2.81**	-3.17**	-4.80**
Party	-1.13**	-1.16**	-1.57**	-1.22**
Petcost	-0.0040*	-0.0079**	-0.0066**	-0.0048**
Coalcost	-0.0068**	-0.0043**	-0.0087**	-0.016**
Significance	0.000	0.000	0.000	0.000
Pseudo R^2	.606	.571	.632	.648
Observations	425	418	426	417
Oil Democrats	30-5-1	32-1-2	3-32-1	12-22-2
Oil Republicans	22-1-0	22-1-0	1-22-0	2-20-1

HR 7014A: Krueger amendment to provide for gradual decontrol of domestic oil prices by 1980, contingent upon enactment of a windfall profits tax on resulting oil company profits (7/23/75). Y=1.
HR 7014B: Wilson amendment to delete oil pricing provisions providing for a gradual increase in the cost of "old" controlled-price oil and a rollback in the price of "new" oil to an average level of $7.50 per barrel (7/23/75). Y=1.
HR 7014C: Staggers amendment, in the form of a substitute for the Krueger amendment, to continue the $5.25 per barrel price ceiling for "old" domestic oil, to roll back the price of most other domestic oil to an average of $7.50 per barrel (7/30/75). A "nay" was a vote supporting the President's position. N=1.
HR 7014D: Passage of the bill to provide the President with emergency energy standby authority, to authorize creation of a strategic oil reserve, to extend price controls on domestic oil, to establish a mandatory gasoline allocation program and to set fuel economy standards for domestic automobiles (9/23/75). N=1.

delete language authorizing the President to act as the purchasing agent for foreign oil was rejected 146-254. After an attempt by Brown to recommit the bill was rejected 171-232, and after an agreement by voice vote to substitute the provisions of HR 7014 for those of S 622, the bill passed the House September 23, 1975 by a vote of 255-148 (D 224-40; R 31-108).[45]

TABLE 6.20 Logit Runs For Oil Price Decontrol Votes

Variable	House	Senate (13)	Senate (8)
Constant	3.02**	3.253*	2.04
Larval (Nonstrip)	0.00040*	0.00056	0.00098*
Nonlar (Stripval)	0.00178**	0.00217	0.00233
Smallref	0.0105		
Medref	0.0038*	0.00955	0.00696
Larref	0.00001	0.00303	-0.00588
South	0.498**	-0.796*	-0.404
McGov	-0.292	-2.77	-2.79
Urban	0.294	-0.444	0.644
ADA	-3.652**	-4.629**	-3.057**
Party	-1.886**	-1.669**	-1.925**
Petcost	-0.00207*	0.00087	0.00023
Observations	432	100	100
Adjusted R2	.761	.773	.738

House Votes: H Res 605, H Res 641, S 1849, HR 7014A, HR 7014B, HR 7014C, HR 4035.
Senate Votes (8): S 622A, S 622B, S622C, S 622F, S 622H, S 622I, S 621C, HR 4035.
Senate Votes (13): Votes (8) plus S 622D, S 622E, S 622G, S 621A, S621B.

There are some significant differences between the House and Senate votes on price decontrol. On the votes blocking Ford's attempts to impose decontrol, in the Senate Nonstrip is the chief factor in opposition to continuing controls while ADA, Party and Petcost are all significant in favor of controls. On the House side Nonlar is the more important production variable in opposition to controls, while ADA, Party, Petcost and Coalcost are the chief supporters of price controls. Representatives from oil producing districts are much more likely to desert the Democratic party than Senators from oil producing states, but on the vote to maintain allocation of oil along with price controls they returned to the fold. ADA, Party, Petcost and Coalcost were consistent in support for controls, while South is significant as a factor for decontrol. Oil district Democrats voted solidly with Republicans on these control votes.

Due to the availability of numerous votes on decontrol of oil prices I was able to perform a logit analysis of the aggregated votes, similar to the one performed on the votes for the NEEA. Smallref was excluded from the Senate analysis due to its strong correlation with Nonstrip. In addition, the Senate votes included five votes that would have resulted in only partial decontrol of oil prices, involving only enhanced recovery oil or new oil. Since the voting patterns were different on these votes the logit analysis was run for all thirteen decontrol votes and for just the eight votes that dealt with general decontrol. Since these votes were performed after the probit analysis of the individual votes the distribution of the estimates is unknown, but they still provide useful information concerning both the impact of interests on votes, and in the case of the Senate votes, the effect of aggregation on the analysis of voting patterns.

The House run showed little change from the pattern exhibited during the price control votes in the previous Congress. Both Larval and Nonlar were anti-controls, with Nonlar the more important factor of the two, and its dominance had increased over the two year period. South increased its support for decontrol while McGov and Urban stayed insignificant. ADA and Petcost declined in the intensity of their opposition while Party became a more important factor, though all three remained significant factors for continued price controls.

In the Senate votes the effect of including the five partial votes is to increase the importance of the constant term while making Nonstrip insignificant, changing the sign on Larref to support from opposition to controls, making South insignificant as a supporter of price controls and increasing the importance of ADA. I conclude from examination of the individual votes that this is due to the shift in votes by Southern and oil state Democrats, who were willing to back the party on the issue of controlling prices to protect consumers, but who broke from the party on the issue of granting incentives for marginal production, both on the grounds that it helped constituents and that it would increase supply, a concern that the more conservative branch of the party considered legitimate. This points out the danger of blithely aggregating votes together without careful examination of the subtle differences that occur between various amendments, many of which are designed as compromises to allow divergent groups to support an agreed upon set of goals.

The eight Senate votes on decontrol diverge from the votes during the previous Congress. Stripval becomes insignificant as a factor for decontrol, while Nonstrip increases in importance. ADA and Party increase their opposition to decontrol while South is a weaker influence in support of price controls. The changes in the Senate and its differences with patterns in the House reflect the differences in the two branches of the legislature. In the House, specific constituent interests are more important since most people are unaware of their representative's votes while special interests have an incentive to be informed. Senators, both due to greater publicity and national ambitions, are more likely to take broader stands on highly publicized issues. Thus, Representatives felt free to vote either with constituent interests or ideology, while Senators, aware of the general antipathy toward oil companies and higher prices, were less willing to bolt the party to vote for oil interests. This explains the stronger performance of oil production variables in the House, and the change in sign of South from House to Senate, since Southern Democrats tend to vote pro-business but were constrained from doing so in the Senate.

VOTE #11: NATURAL GAS DEREGULATION–1976

Along with decontrol of oil prices President Ford also called for complete deregulation of natural gas, both to encourage increased exploration and to channel natural gas from the uncontrolled intrastate market to the interstate market. The Senate Commerce Committee responded by drafting a bill, S 692, which deregulated new gas while maintaining controls on old gas. The bill exempted from price ceilings new gas found by independent producers, required the FPC to set a ceiling price for new gas between 40¢ and 75¢ per MCF, with higher ceilings for high cost areas, and maintained the existing system of controls for old gas, with a ceiling price of 51¢ per MCF. The bill was ordered reported on May 6 by a 10-8 vote and sent to the Senate June 12. The bill fell into legislative limbo, and was supplanted by S 2310, the Emergency Natural Gas Bill, in October. The motivation for the emergency bill was the expectation of a 1.3 trillion cubic foot shortage in the winter of 1975-76, 30 percent greater than the previous winter. To prevent this the bill allowed interstate pipelines to pay intrastate prices (as of August) until June 30, 1976. Deregulation supporters felt that S 2310 was an attempt to prevent discussion of permanent deregulation and therefore attempted to add amendments providing for permanent deregulation to the bill.[46]

Paul Fannin offered an amendment to immediately decontrol natural gas which was killed by a Hollings motion to table agreed to by a 57-31 (D 47-9; R 10-22) vote on October 1. James Pearson and Lloyd Bentsen then offered an amendment allowing emergency measures for dealing with the winter shortage and providing for the gradual phase out of federal price controls on interstate gas, lifting controls on onshore gas immediately and for offshore gas by 1980 (most offshore gas is produced by major companies). Hollings also attempted to kill this amendment with a motion to table but failed by a vote of 45-50 (D 40-17; R 5-33). Deregulation opponents countered with an amendment offered by Adlai Stevenson III, drafted with Moss and Hollings, which set a ceiling price of $1.30 per MCF on all new gas, and a ceiling price of $9 per barrel on uncontrolled oil, which would also apply to controlled oil after five years. The Stevenson amendment was rejected 45-55 (D 41-21; R 4-34) on October 8.[47]

When the Senate reconvened after its October recess, deregulation opponents, realizing that the Pearson-Bentsen version of the emergency bill had sufficient votes to win adoption, concentrated on delaying decontrol and tightening the definition of new natural gas to be decontrolled. Hollings offered an amendment to make the permanent deregulation sections effective in April, 1976, after the expiration of the emergency natural gas provisions, instead of concurrent with them, adopted 37-32 (D 32-10; R 5-22). Hollings then succeeded in tabling a Bartlett attempt to reconsider this vote. John Glenn submitted an amendment to provide a price ceiling for gas sold during the emergency period set at the highest price paid for gas in that state between June and August, 1975, (in effect setting a price ceiling in the intrastate market) adopted 42-37 (D 40-11; R 2-26). John Tunney (D-CA) offered an amendment excluding from the definition of new gas all gas flowing under contract in the interstate market, adopted 56-30 (D 46-7; R 10-23). Hollings followed with an amendment excluding gas from new wells drilled in existing fields, adopted 52-37 (D 43-11; R 9-26). An attempt by William Roth (R-DE) to authorize the FPC to reallocate natural gas supplies failed twice, one by a standing vote and then by roll-call vote, 41-52 (D 36-20; R 5-32). Finally, the Pearson-Bentsen substitute amendment for the text of S 2310 was adopted 50-41 (D 17-37; R 33-4), and the bill passed 58-32 (D 25-28; R 33-4) on October 22.[48]

The House adjourned for the first session before passing its version of the Emergency Natural Gas Bill, but HR 9464, the bill reported by the House Commerce Committee, was one of the first bills considered in the second session. An attempt by Robert Krueger to add an amendment in committee to deregulate natural gas prices was blocked by Interstate and Foreign Commerce Committee Chairman Harley Staggers, who asked the House leaders to bring the bill up for floor action under suspension of the rules, prohibiting amendments. Speaker Carl Albert instead sent the bill to the Rules Committee which granted a rule allowing Krueger to offer his amendment. Staggers and Energy and Power Committee Chairman John Dingell then mounted an effort to defeat the rule allowing consideration of the amendment. Their tactics, including attacks on Speaker Albert, backfired as senior Democratic leaders rallied to support of the Speaker despite their opposition to deregulation and the resolution passed on Feb. 3 by a vote of 230-184 (D 102-175; R 128-9).[49]

Defeated on the rules vote, deregulation opponents began a delaying action to buy time to prepare a compromise measure. Bob Eckhardt offered a series of amendments, including a proposal which deleted a provision from the Krueger measure that would have ended the FPC's power to reduce prices for old natural gas, adopted 232-184 (D 207-72; R 25-112), and an amendment to allow the FPC to set varying price ceilings for offshore gas, rejected 182-236. Neal Smith (D-IA) submitted a compromise amendment to deregulate natural gas sold by small producers, but extend FPC controls to intrastate markets, which was adopted 205-201 (D 192-84; R 13-117). The amended Krueger amendment was then adopted 219-184 (D 202-72; R 17-112) on February 5. Brown (R-OH) attempted to recommit the bill with instructions to report it back with the original Krueger amendment, but his motion was rejected 198-204 (D 85-189; R 113-15). The bill passed 205-194 (D 172-101; R 33-93), effectively deregulating 5,000 to 7,000 independent producers while increasing controls on 25 to 30 major gas producers. The resulting bill was so different from the Senate-passed measure that Senate supporters of deregulation worked up a new bill, retaining controls but allowing prices to rise above the 52 cents per MCF set by the FPC. Opposition by consumer and labor groups on the one hand, and industry lobbyists on the other, kept the measure from reaching the Senate floor.[50]

TABLE 6.21 Probit Runs For S 2310–Senate Natural Gas Votes

Variable	S 2310A	S 2310B	S 2310C	S 2310D	S 2310E
Constant	1.04**	4.03*	2.87**	1.49**	-0.49
Ngval	0.0043**	0.128**	0.036*	0.0024	0.051**
McGov					7.01
ADA	-1.97**	-8.47**	-3.58**	-2.80**	-3.62**
Party	-1.41**	-3.87**	-1.29**	-0.85*	-1.55**
Ngcost		-0.011	-0.0057		
Signif.	0.000	0.0373	0.000	0.000	0.000
Pseudo R^2	.448	.788	.589	.456	.571
Observ.	93	99	100	75	80
Oil Dem.	0-8-0	0-8-0	0-8-0	0-4-4	0-4-4
Oil Rep.	0-7-1	0-8-0	0-8-0	1-7-0	0-8-0

S 2310A: Hollings motion to table, and thus kill, Fannin amendment to remove federal price controls on new natural gas, effective July 1, 1975 (10/1/75). N=1.
S 2310B: Hollings motion to table, and thus kill, the Pearson-Bentsen substitute amendment phasing out price controls on natural gas sold interstate. A "nay" was a vote supporting the President's position (10/2/75). N=1.
S 2310C: Stevenson amendment to the Bentsen-Pearson substitute to set a five-year ceiling price of $1.30 per thousand cubic feet for all natural gas, and $9 per barrel for oil (10/8/75). N=1.
S 2310D: Hollings-Glenn amendment to the Pearson substitute making the permanent natural gas deregulation sections of the Pearson-Bentsen amendment effective April 4, 1976, instead of concurrent with the emergency sections (10/20/75). N=1.
S 2310E: Glenn amendment to Pearson-Bentsen substitute to provide a price ceiling for natural gas sold during the emergency period set at the highest wellhead price paid for natural gas in that state between June 1 and August 1, 1975 (10/21/75). N=1.

The Senate votes on natural gas price decontrol demonstrate a consistency similar to that of the oil price decontrol votes, with Ngval (all the runs are with the absence of petroleum production variables because of their correlation with Ngval) a consistent factor in support of decontrol and ADA and Party, as usual, the most important factors in opposition. Ngval declines in importance on the vote delaying implementation of deregulation (S 2310D) and the one excluding gas under existing contracts (S 2310F). The Senators from major gas producing states (for both Senate and House votes the cutoff value used was $100 per capita in 1980) were consistent in their support of deregulation, with the only defections being the failure by four Democrats to vote on two minor bills (the

TABLE 6.21 Probit Runs For S 2310–Senate Natural Gas Votes

Variable	S 2310F	S 2310G	S 2310H	S 2310I	S 2310J
Constant	1.05*	-1.67	1.37**	2.50**	0.71
Ngval	0.024	0.056*	0.058**	0.043*	0.047**
Urban		4.43*			2.06
South	-0.56		1.77**		
ADA	-2.63**	-3.63**	-1.16	-3.35**	-2.79**
Party	-1.28**	-1.39**	-2.35**	-1.75**	-1.03**
Significance	0.000	0.0007	0.000	0.000	0.000
Pseudo R^2	.593	.615	.596	.610	.476
Observations	88	91	94	95	94
Oil Dem.	0-7-1	0-7-1	0-7-1	7-0-1	7-0-1
Oil Rep.	0-8-0	0-8-0	0-8-0	8-0-0	8-0-0

S 2310F: Tunney amendment to the Pearson-Bentsen substitute to narrow the definition of "new" natural gas by excluding all interstate gas (10/22/75). N=1.
S 2310G: Hollings amendment to narrow the definition of "new" gas by excluding gas from new wells in existing fields (10/22/75). N=1.
S 2310H: Roth amendment to Pearson-Bentsen substitute to authorize the FPC to redistribute natural gas supplies from pipelines in areas with sufficient supplies to those in areas with severe shortages. A "nay" was a vote supporting the President's position (10/22/75). N=1.
S 2310I: Pearson-Bentsen substitute, for S 2310, providing an 180-day exemption from price regulations for curtailed high-priority customers and providing for eventual deregulation of new natural gas prices. A "yea" was a vote supporting the President's position (10/22/75). Y=1.
S 2310J: Passage of the bill providing emergency 180-day exemptions from price regulation and providing for eventual deregulation of new natural gas prices. A "yea" was a vote supporting the President's position (10/22/75). Y=1.

four states that were considered major oil producers but not gas producers were Montana, Utah, Colorado, and North Dakota–note that the two Democrats from Montana, Mansfield and Metcalf, and Colorado, Hart and Haskell, as well as Moss of Utah and Burdick of North Dakota all voted for continued regulation of natural gas prices). The one vote in which ADA was insignificant was the Roth amendment to authorize redistribution of gas from regions of adequate supplies to areas experiencing shortages, as an ideological issue, the transfer of income from consumers to producers, was supplanted by a regional conflict. This was the only vote for which South was significant, and in opposition to the amendment.

TABLE 6.22 Probit Runs For HR 9464–House Natural Gas Votes

Variable	HR 9464A	HR 9464B	HR 9464C	HR 9464
Constant	3.62**	1.28**	2.11**	0.39
Nonlar	0.0090*	0.0048*		0.0029
Ngval	0.0090*	0.012**	0.0073*	0.0054*
South	0.79**	0.60*	1.06**	
Urban		0.90*		1.07*
Nwhite	-0.93			0.99*
ADA	-3.20**	-4.03**	-3.46**	-2.86**
Party	-1.34**	-1.05**	-1.38**	-0.25
Petcost	-0.0022			
Coalcost	-0.0054**	-0.0031		
Signif.	0.000	0.000	0.000	0.000
Pseudo R^2	.607	.629	.663	.365
Observ.	426	415	419	416
Gas Dem.	33-1-2	4-30-2	3-31-2	5-29-2
Gas Rep.	16-0-0	0-15-1	0-15-1	2-14-0

HR 9464A: Adoption of the resolution (H Res 937) for floor consideration of the bill to permit interstate gas pipelines to avert emergency shortages by buying unregulated intrastate gas. The rule permitted consideration of a Krueger amendment to end federal regulation of interstate gas prices. (2/3/76). A "yea" was a vote supporting the President's position. Y=1.
HR 9464B: Eckhardt amendment, to the Krueger substitute amendment to deregulate natural gas prices, to delete provisions that would end the FPC's authority to reduce prices on gas still under FPC regulation (2/4/76). N=1.
HR 9464C: Smith substitute amendment, to the Krueger amendment, to deregulate natural gas sold by small producers, but continue FPC regulation of major gas producers, extending controls to intrastate markets (2/5/76). A "nay" was a vote supporting the President's position. N=1.
HR 9464D: Passage of the bill to deregulate the price of natural gas sold by small producers, but continue regulation of major gas producers, and extend federal regulation of major gas producers to cover gas sales in intrastate markets (2/5/76). A "nay" was a vote supporting the President's position. N=1.

The probit runs for HR 9464 demonstrate a pattern similar to that on oil price control votes, except that there is weaker Democratic support for regulation, due to strong support by Southern Democrats for decontrol. Ngval is consistently in favor of deregulation as expected, while Nonlar is significant on the first two votes. Urban support for decontrol is inexplicable, unless one believes that urban legislators are more willing to accept higher prices to ensure supplies than rural legislators. ADA and Party continued to oppose deregulation, though Party and South become insignificant on the vote on the final bill, suggesting a return to the fold of Southern Democrats and vote switching by moderate Republicans.

A logit analysis was performed with the four votes used in Table 9.22, the votes on the amended Krueger amendment and the Brown motion to recommit for the House votes, and ten Senate gas votes. Comparing the House run and Senate results, one finds a similarity with oil decontrol votes in that the coefficient for South is significant and positive in the House but not the Senate. This suggests that while southern Democrats were more willing to stick with the party in the Senate, representatives from oil and gas producing states were more likely to defend gas producers, possibly because the issue was less volatile than decontrol of oil prices.

TABLE 6.23 Logit Runs For Natural Gas Price Decontrol Votes

Variable	HOUSE	SENATE
Constant	2.30**	2.668**
Ngval	0.0013**	0.0047**
South	0.710**	0.312
McGov	0.373	-1.411
Urban	0.0483	0.147
ADA	-4.405**	-3.071**
Party	-1.054**	-1.972**
Adjusted R^2	.760	.715

VOTE #12: MISCELLANEOUS SENATE VOTES–1976

During the votes on the natural gas deregulation bill Philip Hart and James Abourezk attempted to slip a vertical divestiture amendment through the Senate, and almost succeeded, with the bill rejected by a 45-54 vote (D 39-23; R 6-31). A second vertical divestiture amendment, submitted by Mansfield and Hart, was rejected 40-49 (D 34-19; R 6-30), while a Kennedy bill, requiring divestiture of alternative energy interests was rejected 39-53 (D 34-33; R 5-31).[51]

Oil industry supporters also used the nongermane amendment approach to forward their agenda in the Senate. Bartlett offered an amendment to the Federal Energy Administration bill to end price controls on stripper wells, which was adopted 61-29 (D 28-27; R 33-2) on June 16, 1976. Joseph Montoya

TABLE 6.24 Probit Runs For Miscellaneous Senate Votes–1976

Variable	2310AA	2310BB	2872A	2872B	RS 449
Constant	6.42**	7.63**	9.70**	4.17**	3.92*
Stripval				0.058*	
Nonstrip	0.0075*	0.0055*	0.016**	0.012*	0.003**
South					-1.26**
McGov					-7.32*
Medref			-0.043*		
Larref				-0.034*	
Urban		-1.79	-3.37*	5.62	
ADA	-5.30**	4.93**	-5.37**	-5.73**	-1.06
Party	-1.39**	-1.31**	-2.05**	-1.82**	1.17**
Petcost	-0.0061	-0.0069*	-0.0059		-0.0045
Sign.	0.0008	0.0006	0.0026	0.0036	0.0004
Pseudo R^2	.661	.631	.668	.681	.277
Observ.	99	97	92	94	93
Oil Dem.	5-9-0	5-8-1	11-2-1	12-2-0	8-3-3
Oil Rep.	0-10-0	0-10-0	10-0-0	10-0-0	5-4-1

S 2310AA: Abourezk amendment, to the Bentsen-Pearson substitute amendment, to require major oil companies to divest themselves of vertically integrated holdings within five years (8/8/75). N=1.
S 2310BB: Mansfield-Hart amendment to require major oil producers within five years to divest themselves of their petroleum refining, transportation and marketing interests (8/22/75). N=1.
S 2872A: Bartlett amendment to end federal price controls on oil from stripper wells producing an average of 10 bpd or less (6/16/76). Y=1.
S 2872B: Montoya amendment to end federal price controls on enhanced recovery oil after February 1, 1976 (6/16/76). Y=1.
S RES 449: Johnston motion to discharge from the Interior and Insular Affairs Committee the resolution to overrule the FEA revocation of a small refinery exemption from cost-equalizing entitlement payments (5/27/76). Y=1.

(D-NM) offered an amendment to end price controls on increased oil production from enhanced recovery methods, adopted 58-35 (D 28-30; R 30-5). Johnston attempted to overturn a FEA action of May 12, 1976, which revoked a small refinery exemption that allowed small refiners with access to low-cost oil to avoid making entitlement payments, subsidizing the small refiners. A Johnston motion to discharge the resolution from the Interior Committee was rejected 28-57 (D 21-31; R 7-26).[52]

The divestiture votes involved only a few factors, Nonstrip in opposition and ADA and Party, and to a lesser extent, Petcost, for divestiture. Since the bill didn't actually threaten constituents (except those whose jobs might be at stake during reshuffling in the industry) it was basically an ideological battle. The decontrol votes provide more confusing results, as Nonstrip, ADA and Party have their usual alignments, but other variable coefficients had unexpected values. Stripval was insignificant in the decontrol of stripper prices but significant in enhanced recovery decontrol (many of these wells are stripper wells) which is a bit strange. Medref opposition to stripper price decontrol and Larref opposition to enhanced oil decontrol is inexplicable, as is the flip-flop by Urban on the two votes. The entitlement vote also provides strange results, as the normal pattern on oil votes broke down completely. The small refiner issue confused the normal alignment, as oil senators wanted to support local refiners, liberals were uninterested in opposing a small business subsidy, but most senators felt free to eliminate a subsidy which had been granted without serious consideration. This suggests that support of small producers and refiners was due to hostility to the large oil companies, and not the influence of the small companies. Unable to cast themselves as the victims of monopoly, the small refiners failed to engender liberal support for continuation of this subsidy.

VOTE #13: CARTER ENERGY PLAN VOTES–1977

The Carter administration made development of a coherent and comprehensive energy policy a cornerstone of the new administration, attempting to reconcile the supply side and conservation approaches which dominated the energy debate. However, the choice of a small insular group of energy "experts" to develop the plan, while allowing rapid formulation of a blueprint for dealing with the situation also guaranteed that neither interest groups nor political leaders would be committed to support of the Carter energy program. Carter's attempts to develop a public consensus in favor of his program as a basis of support was doomed to failure simply because he misunderstood the nature of the American people. An appeal to sacrifice, no matter how reasonable the goal to be achieved, is unlikely to appeal to the public except in time of crisis. In 1977, despite the growing dependence on imported oil, there was no public perception of the need to reduce energy consumption and thus there was little support for sacrifice to achieve this goal. The objectives of the Carter energy program included reducing gasoline consumption, limiting the growth of energy consumption and lowering oil imports to 6 million bpd, to be achieved via a stiff gasoline tax, the elimination of price controls on oil and gas combined with a wellhead tax on old oil, conversion of industries and utilities to coal and development of alternative energy.

The Carter administration learned little from the struggles of the Ford administration with Congress. Transferring income from the oil companies to the government with a crude equalization tax did not deal with the fundamental objection to higher prices which was the basis of opposition to decontrol. Maintaining controls on natural gas while allowing higher prices would elicit opposition from control proponents while reducing support from the oil and industry. The gasoline tax was certain to be attacked by both consumer groups and producers. Regulation of automobile mileage and taxes on gas guzzlers was a little safer since Congress preferred hidden taxes to overt measures. Taxes on industrial use of oil and gas to encourage conversion was sure to generate general business opposition since the business community's patriotic tendencies evaporate at the bottom line. Support for conservation incentives as an alternative to relaxation of environmental standards was one of the few Carter proposals that engendered more support than opposition.

In the House Speaker Tip O'Neill attempted to facilitate movement of various portions of the program by imposing deadlines on legislative committees to complete their hearings and markups on the bill by the middle of July. The bill then went to the Ad Hoc Committee on Energy, which after three days of deliberation readied the package for submission to the House. The most important change made by the Ad Hoc Committee was the addition of an amendment, sponsored by deregulation opponent Bob Eckhardt and supporter Charles Wilson, which greatly expanded the definition of new gas. Since the price ceiling for new gas under the bill was tied to the price of old oil, starting at $1.75/MCF and expected to increase, the amendment would effectively raise the price of a substantial fraction of natural gas production. The purpose of the amendment was to garner support among deregulation advocates, among them Majority Leader Jim Wright. HR 8444, the National Energy Act, through the machinations of O'Neill and his ally, Richard Bolling, was sent to the floor from the Rules Committee with a modified "closed" rule, minimizing possible attacks by opponents.[53]

On the floor debate began with votes on a number of minor Ad Hoc Committee amendments for studies of energy conservation. The House then overwhelmingly accepted a Bolling motion to resolve itself into the Committee of the Whole for further consideration of the bill on August 3. At this point the most serious challenge to the bill was raised by a Brown amendment to end federal controls on new onshore natural gas, and on offshore gas in five years. Passage would have endangered the fragile coalition in support of the bill, but the amendment was rejected 199-227 (D 72-210; R 127-17) after an impassioned speech by Speaker O'Neill.[54]

Attempts to pass a gasoline tax were shot down by a bi-partisan coalition. An amendment submitted by the Ad Hoc Committee would have increased the tax by 4¢ and James Howard (D-NJ) submitted a proposal to raise the tax by 5¢, with funds going toward highways and conservation programs. The Howard proposal was shot down by a 82-339 vote, and then the Ad Hoc proposal was defeated 52-370. James Jones (D-OK) submitted an amendment to return to producers the money raised by the crude oil equalization tax if they plowed it back into production, rejected 198-223 (D 74-208; R 124-15). Proposals to use crude oil equalization funds to supplement the Social Security trust fund or to cut general income taxes instead of rebates to individual taxpayers were then defeated. A number of other amendments were defeated, including an Anderson (R-IL) attempt to substitute a Republican package of energy proposals for the bill. The final attempt to block passage of the bill was a William Steiger (R-WI) motion to recommit the bill with instructions to delete the crude oil equalization tax, rejected 203-219 (D 66-216; R 137-3). The bill then passed on August 5 by a vote of 244-177 (D 231-50; R 13-127).[55]

The vote on the Brown amendment to deregulate natural gas follows the expected pattern with Ngval supporting deregulation and Party and ADA in opposition. The pattern on the two oil tax equalization amendments and the final vote on the bill also follow the expected pattern with a couple of exceptions. On the Steiger amendment the Urban coefficient is probably an artifact due to correlation with other variables, though it is possible that urban representatives were opposed to any measure raising prices. The strong performance of South in opposition to the oil equalization tax suggests an alignment of conservative southern Democrats with producer interests. The importance of this shift can be seen by the defection of oil Democrats from the president's position, with a majority joining their party's leader only on the final vote on the bill.

The Senate, instead of constructing an omnibus energy bill, produced a number of different bills to cover the various issues in Carter's energy program. S 2507 was the least controversial, containing mostly energy conservation measures that had broad support. The crude oil equalization tax never survived the Senate Finance Committee as the committee killed the tax September 26 by a 10-6 vote. Chairman Long pledged to restore the tax, but only with a plowback provision. The tax was doomed by the opposition of labor, consumer and industry lobbyists. The Finance Committee also rejected user taxes on utility and industrial consumers of oil and gas.[56]

S 2104, the Natural Gas Pricing bill, reignited the conflict that had been resolved by a compromise in the House. Kennedy attempted to impose a ceiling price of $1.45/MCF on sales of new gas in interstate and intrastate markets, extending controls to the intrastate market. This proposal was killed by a Pearson motion to table the amendment, supported by the President, which carried 73-21 (D 39-19; R 34-2) on September 21. Pearson and Bentsen then offered an amendment to end federal price controls for new onshore gas immediately, and offshore gas after five years. Jackson tried to table the amendment but was defeated 46-52 (D 43-17; R 3-35). Maneuvering became fast and furious at this point as Jackson came back with an amendment to allow an increase in the price of new gas tied to the upper tier price of crude oil (estimated at $2.03/MCF in 1978) and end controls on high cost gas, tabled by a Byrd motion which carried 88-0. An Abourezk motion to kill the Bentsen amendment to permit the price of gas used for boiler fuel to rise above the cost

TABLE 6.25 Probit Runs For 1977 House Energy Votes

Variable	HR 8444A	HR 8444B	HR 8444C	HR 8444D
Constant	2.10**	2.40**	2.40**	4.28**
Nonlar		0.013**	0.0014**	0.0020**
Ngval	0.0023**			
South		0.74**	0.84**	0.64**
Nwhite	0.79			0.81
McGov			-1.73	
Urban			1.15**	
ADA	-2.79**	-2.55**	-2.47**	-2.714**
Party	-1.02**	-1.82**	-2.24**	-2.95**
Petcost	-0.0013	-0.0013		-0.0024*
Coalcost			-0.0058**	
Significance	0.000	0.000	0.000	0.000
Pseudo R^2	.404	.576	.573	.641
Observations	429	423	423	428
Oil Dem.	27-6-1	31-9-1	27-13-1	18-22-1
Oil Rep.	15-0-0	19-0-1	17-0-1	0-17-1

HR 8444A: Brown amendment to end controls on the price of new onshore natural gas, and to extend the emergency Natural Gas Act of 1977 for three years (8/3/77). A "nay" was a vote supporting the President's position. Y=1.
HR 8444B: Jones amendment to return to oil producers money from the crude oil equalization tax if used for increased energy production (8/4/77). A "nay" was a vote supporting the President's position. Y=1.
HR 8444C: Steiger motion to recommit the bill to the Ad Hoc Committee on Energy with instructions to delete the proposed crude oil equalization tax (8/5/77). A "nay" was a vote supporting the President's position. Y=1.
HR 8444D: Passage of the bill to establish a comprehensive national energy policy (8/5/77). A "yea" was a vote supporting the President's position. N=1.

of residual oil was soundly defeated 17-59 (D 16-32; R 1-27), while a Dennis DeConcini (D-AZ) amendment to require priority for sales of lower price old gas to residential and small agricultural users was adopted 61-17. A Wendell Ford (D-KY) amendment to end price controls on wells producing less than 100 MCF (17 bbl/day equivalent) survived three votes before being tabled. Finally, on October 4 the modified Pearson-Bentsen substitute amendment which deregulated gas, subject to a two year price limit equal to that of imported No. 2 oil, including provisions allocating old gas for homeowners, schools and hospitals, was adopted 50-46 (D 16-43; R 34-3) despite Presidential opposition.[57]

The Senate Republicans then moved to bring Carter's energy taxes to the floor before they could be discussed in conference. The Finance Committee had already rejected the tax on gas guzzling automobiles, the tax on crude oil and the tax on utility and industrial use of oil and gas. Russell Long had announced that his intent was to take a non-controversial tax bill through the Senate to conference with the House, there to engineer a compromise which could pass the Senate. The Republican intent was to force votes on the controversial measures in order to put the Senate on record as opposed to the measures.[58]

Ironically, it was liberal Democrats who pushed tax measures through the Senate over the objections of Long. Metzenbaum submitted an amendment to HR 5263, the Energy Tax Incentives bill, to impose a tax on utility and industrial use of oil as boiler fuel in new or coal capable facilities adopted 51-37 (D 39-16; R 12-21) on October 27. The Metzenbaum measure differed from the Carter proposal in that the latter applied to all industrial and utility use of oil and gas. Long managed to include a "section 1054" which provided that no taxes would go into effect unless they were combined with tax incentives for increased production, conservation and conversion to alternative fuels, designed to improve his negotiating position on the crude oil equalization tax. Jackson submitted an amendment to delete the section, a statement of no confidence in Long as negotiator for the Senate, rejected by a 38-56 (D 32-27; R 6-29) vote.[59]

The next important challenge was a McIntyre amendment to delete a provision allowing the President to impose tariffs or quotas on imported petroleum products, rejected 23-58 (D 14-37; R 9-21). Long then blocked a move by Dole to recommit the bill for redrafting by a vote of 51-30. Long lost on a motion to kill the Kennedy amendment to reduce from 40 to 10 percent the tax credit for the costs of conversion from oil and gas to coal, one of the chairman's few defeats on the bill. An amendment by Roth to put the Senate on record as opposing any tax on crude oil was rejected 30-47 (D 11-39; R 19-8). Roth was also thwarted by a Long motion to table his amendment to transfer to the Social Security trust fund excess revenues from energy taxes which carried 57-18. The antagonism to the major oil companies made itself felt in the Kennedy amendment to deny tax benefits to the 17 major companies, defeated by a motion to table submitted by Spark Matsunaga (D-HI). Bartell obtained approval for an amendment expanding the number of oil wells exempted from price controls, but discontinued his efforts in response to Long's request. The amended bill passed the Senate on October 31 by 52-35.[60]

The analysis of the votes on natural gas deregulation are given in Table 9.26. The results for S 2104A and S 2104C differ from the other three votes, which follow the familiar pattern of Ngval in favor of deregulation, while ADA and Party are opposed (the hypothesis that the coefficients for S 2104B are jointly equal to zero cannot be rejected). Moderate price control advocates found the provisions of the Kennedy amendment too severe to support, while the two southern votes against the tabling motion were sufficient to make South appear in opposition to gas producers. This may seem odd until one realizes that given the ADA ratings of southern congressmen, any votes for the

TABLE 6.26 Probit Runs For S 2104–Senate Natural Gas Pricing

Variable	*S 2104A* *Run 3*	*S 2104B* *Run 3*	*S 2104C* *Run 3*	*S 2104D* *Run 3*	*S 2104E* *Run 3*
Constant	8.28**	8.71*	3.10	4.23**	6.47*
Ngval	0.022	0.065**		0.033*	0.036*
McGov	-8.87*	-12.04			
Urban		3.26*			
South	-2.01*	-1.63			-1.32*
ADA	-4.55**	-7.27**	-2.65**	-6.18**	-8.03**
Party	-0.91	-3.55**	-1.02*	-1.62**	-2.86**
Ngcost	-0.0061	-0.0087			-0.0065**
Sign.	0.0014	0.088	0.0001	0.0010	0.011
Pseudo R^2	.489	.739	.260	.627	.727
Observ.	97	100	81	82	100
Gas Dem.	6-0-0	0-6-0	0-2-4	0-2-4	6-0-0
Gas Rep.	10-0-0	0-10-0	0-6-4	0-9-1	10-0-0

S 2104A: Pearson motion to table, and thus kill, the Kennedy amendment to impose a ceiling price of $1.45 per MCF for new natural gas in all markets (9/21/77). A "yea" was a vote supporting the President's position. Y=1.
S 2104B: Jackson motion to table, and thus kill, the Pearson-Bentsen amendment to end federal price controls for new onshore gas, and cease regulation of new offshore gas after five years (9/22/77). A "yea" was a vote supporting the President's position. N=1.
S 2104C: Abourezk motion to table, and thus kill, the Bentsen amendment to permit increases in gas prices for boiler fuel users until the cost reached 120 percent of the cost of residual oil (9/24/77). N=1.
S 2104D: Metzenbaum motion to table, and thus kill, the Ford amendment to end price controls on natural gas produced from wells that produce an average of less than 100 MCF a day (9/24/77). N=1.
S 2104E: Adoption of the modified Pearson-Bentsen substitute amendment to end price controls for natural gas, and cease regulation of new offshore gas after 1982, imposing a price limit equal to that of imported heating oil for two years and an incremental pricing provision to protect homeowners, schools and hospitals (9/4/77). A "nay" was a vote supporting the President's position. Y=1.

Kennedy amendment would be unexpected. The vote on S 2104C is similar in that only hard core Democratic liberals felt it necessarily to control the price of gas used in boilers. The other three votes involved a closely fought conflict between producer interests and liberal opposition to higher prices, in which party loyalty was an important influence on the marginal Democratic Senators.

TABLE 6.27 Probit Runs For HR 5263–Senate Energy Tax Votes

Variable	HR 5263A	HR 5263B	HR 5263C	HR 5263D	HR 5263E
Constant	1.28**	4.12**	9.38**	0.26	8.84**
Stripval		0.014			0.042*
Nonstrip	0.0068**		0.013	0.0027*	
McGov		-5.55	-15.46**		-22.14**
Urban		2.10		-1.86	3.78**
South			-0.76		-0.89
ADA	-4.46**	-2.75**			
Party		-0.85*		-1.51**	-2.21**
Petcost			-0.006**	0.0027**	-0.0023
Signif.	0.000	0.000	0.0043	0.005	0.0022
Pseudo R^2	.581	.420	.380	.305	.487
Observ.	94	96	82	80	84
Oil Dem.	3-9-0	5-7-0	0-12-0	4-6-2	7-4-1
Oil Rep.	0-12-2	0-12-2	0-10-4	8-1-5	12-0-2

HR 5263A: Metzenbaum amendment to impose a tax on utility and industrial use of oil and natural gas as boiler fuel in new facilities or in existing facilities designed to use coal instead (10/27/77). A 'yea" was a vote supporting the President's position. N=1.
HR 5263B: Jackson amendment to delete from the bill a section requiring that energy taxes had to be tied to tax incentives for energy production, conservation and conversion (10/27/77). N=1.
HR 5263C: McIntyre amendment to delete from the bill a provision allowing the president to impose tariffs, fees or quotas on imports of refined petroleum products (10/28/77). N=1.
HR 5263D: Roth amendment to state that it was the sense of the Senate that any tax on crude oil was unacceptable (10/29/77). Y=1.
HR 5263E: Matsunaga motion to table, and thus kill, the Kennedy amendment to deny tax benefits in the bill to the 17 major oil and gas producing companies in the United States (10/31/77). Y=1.

South mattered on the final vote, when Southern Senators who had voted against the President on earlier votes returned to the fold on the final vote.

The energy tax votes reflect the limited role of oil production as a influence on the tax issues. Stripval is important only on the votes on tying taxes to production incentives and the major company tax benefit denial, while Nonstrip matters on the vote on the industrial user tax, the crude oil tax vote, both of which threatened oil markets, and the quota vote, where it is a factor in favor of quotas. In the Senate when Stripval is the only oil production variable that is significant this usually indicates that oil state support is limited to a core consisting of Texas, Oklahoma, Wyoming, New Mexico and Kansas. There is a collapse of the usual division of oil producers, Republicans and conservative southern Democrats against liberals and the rest of the Democratic party. ADA disappears on the votes on quotas, crude oil taxes and the vote denying tax benefits to major oil companies. Party distinctions fade on the industrial use votes and the quota amendment. With a Democratic President supporting oil taxes, liberals opposed the President along with producers and Republicans. Therefore ADA becomes unimportant and Party depends on whether the amendment is so onerous that Democrats defect from the Administration's position. The importance of McGov reflects this shift, especially on the quota and tax benefit bill as Senators from more liberal states voted with their constituents. Under Ford this rarely occurred since his stress on production incentives strengthened the partisan/ideological divide.

The energy bill then went to conference where it was stalled for most of 1978. The key issue upon which the two chambers of Congress could not agree was the deregulation of natural gas prices. A tentative compromise was formulated in March, under which the Carter administration would accept a deregulation compromise in exchange for Russell Long's active support of the industrial oil and gas use tax. The energy tax conference had been blocked by Long's insistence that energy tax decisions would be delayed until the natural gas pricing issue had been resolved. Long also made it clear that the crude oil equalization tax was a dead horse during an election year. The natural gas compromise would end price controls for new gas in 1985, and controlled prices would start at $1.75/MCF, with the annual rate of increase equal to the consumer price index plus 3.5 percent through April 1981, and 4 percent after that time. High cost gas would be freed from price controls, while gas from "stripper wells" (producing less than 60 MCF per day) would have an initial price of $2.09/MCF.[61]

The tentative compromise fell apart after the Senate conferees voted 10-7 to offer the delicately balanced compromise to the House, but House conferees voted 13-12 to accept the offer but with an amendment, in effect substituting their own version, though going on record in support of ending price controls for the first time. The House plan was similar to the Senate version except that it established a much tighter definition of new gas. Conservative Republicans joined with consumer-oriented Democrats in sabotaging an agreement. Voting for the Senate compromise were five deregulation supporters, as well as five supporters of Carter's plan, while three liberal Democrats, along with Weicker and three anti-regulation Republicans opposed the Senate compromise. All eight Republican House conferees plus Waggonner, along with three liberal Democrats were against the House compromise. A second attempt to reach a compromise was made on April 21, but pressure on two liberal Democratic House conferees who voted for the initial House version resulted in their opposing the new compromise on the grounds that it was too generous to the oil and gas industry. Attempts during May to obtain the one additional vote needed from a House conferee were unsuccessful, leading O'Neill to decide to split the energy package, dooming the chances of passage for the crude oil equalization tax.[62]

In September a compromise was reached in conference, resulting in a gas deregulation bill, HR 5289. The battle then moved to the Senate where the administration conducted an intensive lobbying effort to ensure passage. The administration's strategy was to seek support from special interest groups which in turn would put pressure on legislators. The key was to split interest groups to allow Senators to vote with the administration without facing unified opposition. Opponents included oil companies such as Amoco, consumer groups such as Energy Action and the Consumer Federation of America, the Chamber of Commerce, the AFL-CIO and the UAW. Bankers supported the administration while farm groups, the steel and automobile industries and the oil industry were split between support and opposition. Republicans were motivated by the desire to embarrass the President during an election year while Democrats were divided between party loyalty and political self-interest. The two key attempts to kill the bill was a motion by Metzenbaum to recommit the conference report, rejected 39-59 on September 19 and a Dole motion to recommit, rejected 36-55 on September 26. After these attempts failed the conference report was adopted 57-42.[63]

VOTE #14: HOUSE ENERGY DEPARTMENT AUTHORIZATION VOTES –1979

President Carter, in a televised address on April 5, 1979, announced his decision to end price controls on domestic oil, beginning on June 1, in a gradual process stretching until October 1981. Oil discovered before the Arab oil embargo was given the lowest price, about $6 a barrel, and was labeled "lower tier." "Upper tier" oil, discovered after that date, was allowed a higher price of around $13 a barrel. Marginal and enhanced recovery oil would be immediately decontrolled, lower tier oil would be gradually redefined as upper tier after June 1, and the price of upper tier oil would rise by monthly increments until reaching world levels by October 1, 1981, at which time all controls would end. These dates coincided with the expiration of the price controls provided by the Energy Policy and Conservation Act of 1975, under which controls were at the President's discretion as of June 1979, and automatically ended at the end of September 1981. One factor encouraging Carter's action was the passage of the natural gas deregulation bill during 1978, evidence that political support for price controls was diminishing.[64]

Liberal Democrats in the House attempted to reverse the decontrol decision, but were unable to obtain a floor vote on a bill extending price controls. In the Senate, Energy Chairman Jackson didn't even try to get a bill extending controls out of his own committee, and though he threatened to submit the bill as an amendment on the floor, a fellow Senator estimated that there were at best 25 votes in that chamber for continuing controls. In the House the best bet for the pro-control forces was to add amendments to the DOE authorization bill since the ban on non-germane amendments limited opportunities to submit amendments on the floor. An attempt to cut funding for decontrol regulations was defeated 11-10 in the Energy and Power Subcommittee of the House Commerce Committee on April 3.[65]

The next test occurred in the full committee, but the liberals lost this challenge on May 2 by a 21-21 vote, although committee Democrats voted against the president 18-9, including chairman Harley Staggers. One of the strongest supporters of decontrol, John Dingell, saw the effort as a lost cause, as the votes were not available in the Senate, an view shared by Speaker O'Neill. Momentum to reverse the President's decision seemed to be growing when on May 24 the House Democratic Caucus voted 138-69 against decontrol, but this was more of a last gasp than a rising tide. The actual vote on decontrol was delayed until the Energy Department Authorization bill, HR 3000, reached the House floor in October. Toby Moffett (D-CT) began by offering an amendment prohibiting the use of funds for expenditures related to lifting price controls, rejected 135-257 (D 128-120; R 7-137) on October 11. Peter Kostmayer (D-PA) followed with an amendment restoring price controls on heating oil and diesel fuel by restricting refinery profit margins, rejected 124-243 (D 116-120; R 8-123). A James Courter (R-NJ) measure to prohibit gasoline price control was adopted 191-188 (D 61-176; R 130-12). Gramm, aware that Courter's amendment had eliminated the government's authority to ensure that priority users had access to supplies, and that Dingell was planning to use that argument to reverse the Courter vote, offered an amendment to resolve this problem, adopted 257-119 (D 141-95; R 116-24). Eckhardt attempted to restore authority for the President to impose gasoline rationing and price controls, but his amendment was rejected 182-191 (D 169-67; R 13-124).[66]

The changes in alignments that had occurred during the 1970s can be seen in a subtle fashion in Table 6.28. On the crude oil price control only Larval is significant, and neither production variable is important on the oil product vote. It is only on the much closer allocation votes that Nonlar becomes significant, while Larval remains unimportant. This is due to the effect of a general consensus in favor of decontrol obscuring the influence of oil production. Party is significant, but far less important on the price control votes than on the allocation votes. With the President in favor of decontrol Southern Democrats are free to vote with the oil industry. It is only on the Gramm bill, a compromise designed to block support for more extensive allocation measures, that the normal divisions disappear.

TABLE 6.28 Probit Runs For HR 3000–Energy Department Votes

Variable	HR 3000A	HR 3000B	HR 3000C	HR 3000D	HR 3000E
Constant	2.64**	2.36**	2.20**	1.81	-2.73**
Larval	0.0042*				
Nonlar		0.0007	0.0019**	0.0018**	
McGov	-2.35**		-1.51	-1.91*	
Urban				0.87*	2.05**
Nonwhite					0.92
South	0.48**	0.63**			-0.46*
ADA	-2.04**	-1.44**	-1.78**	-2.00**	1.14**
Party	-0.72**	-0.84**	-1.53**	-1.38**	
Signif.	0.000	0.000	0.000	0.000	0.000
Pseudo R^2	.430	.352	.486	.456	.228
Observ.	404	377	379	373	371
Oil Dem.	2-31-2	2-31-2	21-8-6	7-22-9	29-0-6
Oil Rep.	0-30-0	0-27-3	27-1-2	1-26-3	24-3-3

HR 3000A: Moffett amendment to prohibit use of funds for expenditures related to lifting price controls on crude oil (10/11/79). A "nay" was a vote supporting the president's position. N=1.
HR 3000B: Kostmayer amendment to restore price controls on heating oil and diesel fuel by restricting refinery profit margins (10/11/79). A "nay" was a vote supporting the president's position. N=1.
HR 3000C: Courter amendment to prohibit the use of funds to allocate or control the price of gasoline (10/12/79). A "nay" was a vote supporting the president's position. Y=1.
HR 3000D: Eckhardt amendment, to the Gramm amendment to authorize the president in an emergency to reimpose the gasoline allocation system and price controls (10/12/79). N=1.
HR 3000E: Gramm amendment to authorize the president to allocate gasoline to agriculture and other priority users (10/12/79). Y=1.

VOTE #15: WINDFALL PROFIT TAX VOTES

The windfall profits tax proposal was initiated by President Carter in order to tax the profits resulting from his decision to allow the price of oil to rise to the world market level over a two year period. Carter learned from his attempt to establish a crude oil equalization tax in 1977, which required legislators to vote for higher oil prices, a move which was not greatly appreciated in Congress. Decontrol allowed the administration to accept the blame for higher prices as legislators merely needed to refuse to actively support attempts to block the president's action At the same time the windfall profits tax gave Congressmen the opportunity to present themselves as opponents of the detested oil companies, as they voted for taxes on the profits the companies received from decontrol. Thus, though oil price decontrol and a windfall tax would leave more money in the hands of oil companies than the crude equalization tax, it was a more politically palatable combination from a Congressional vantage point.

The initial administration proposal established a three part tax: a lower-tier tax on oil which is shifted to upper-tier at a rate greater than 2 percent per month; an upper-tier tax on the difference between the current upper-tier price (about $13) and the uncontrolled price; and a world price tax on the difference between the world price and a base price of $16 per barrel, with an exemption for Alaskan oil. Given a 50 percent tax rate the administration estimated that decontrol would produce $8.4 billion through 1981 in increased Federal revenue. The Federal share of the windfall income was slated to go into an energy Security Trust Fund to help low income families, improve mass transit, and support programs to limit domestic oil consumption.[67] The proposal faced two separate threats; from liberals in both chambers whc would try to increase the level of the tax and from Russell Long who wanted a plowback provision.

The Ways and Means Committee split the administration proposal into two parts, a tax bill (HR 3919) and a spending bill, to be developed at a latter date. Within the committee chairman Al Ullman had to fight off challenges by members attempting to substitute a weaker bill and by liberal members who wanted a stiffer tax. The Democratic liberals attempted to eliminate the overseas tax credit for the oil companies as well as pushing for a higher tax rate. The committee settled on an amendment by Dan Rostenkowski for a 70 percent rate by a vote of 22-14, and also raised the base price for calculating the tax to $17, to be escalated by the inflation rate plus 2 percent. One of the proposed substitutes was offered by James Jones and Henson Moore, which would increase the tax rate to close to 60 percent, tax oil from marginal wells using a base price of $13 instead of $6, and end the tax after 1990. The Jones-Moore substitute was supported by the committee Republicans, oil state Democrats and Ken Holland (D-SC) and Ed Jenkins (D-GA). The committee voted 20-16 on June 19 to send HR 3919 to the floor but then fought over proposed rules. A proposal to allow floor votes on two substitute amendments lost by two votes before the panel voted 18-17 to seek a rule preventing changes on the floor. The Rules Committee then voted to allow floor votes on four amendments; the restrictive Ways and Means rule, the Jones-Moore substitute, a Shannon-Gephardt substitute to tighten the tax on old oil and a Republican recommittal motion to add a plowback provision. Rules then refused a floor vote on an amendment exempting producers of less than a 1,000 bpd.[68]

The House overwhelmingly adopted the rule providing floor consideration of the bill by a 292-118 (D 213-42: R 79-76) vote. The Ways and Means Committee amendments were adopted 230-185 (D 215-45; R 15-140), followed by adoption of the Jones-Moore substitute by a 236-183 (D 90-173; R 146-10) vote on June 28. An attempt by Shannon to extend the tax on old oil for an additional year was rejected 236-183 (D 165-94; R 7-147). A Barber Conable (R-NY) motion to recommit and add a plowback provision giving a 75 percent credit for investment above historic levels was rejected 172-241 (D 46-215; R 140-14).[69] The final House bill was tougher than the administration's proposal over the short-run, but more lenient over the long-run, given expectations of escalating oil prices.

The Senate Finance Committee took its time considering its version of the windfall profits tax bill. One question was the amount of money which would be raised by different versions of the tax, and how the money would be spent. Another issue was the effect on production of various versions of the tax bill. The proliferation of estimates, and the uncertainty surrounding them, made the choice of a scenario, and thus the impact of various options, a matter of convenience dependent upon the legislator's personal inclination. The final version of the committee bill established a 75 percent tax on old oil, a 60 percent rate on tier two oil, and provided exemptions for oil discovered after 1978, high cost oil, stripper wells owned by independent producers, and oil owned by governments and non-profit institutions. The committee also placed a ceiling on revenues collected by the tax, phasing out the tax when revenues reached 90 percent of $138 billion.[70]

On a high profile bill such as HR 3919 the version leaving a Senate committee is of little importance, since unlike the House there are few restrictions on offering amendments on the Senate floor. David Boren (D-OK) began by attempting to reduce the tax to 50 percent, a move which was rejected 32-53 (D 8-40; R 24-13) on November 26. A Dole amendment to reduce the tax on old oil to 60 percent, already rejected in committee, failed on the floor by a 32-58 vote (D 9-45; R 23-13). This was then followed by a Long motion to table the Bumpers amendment to substitute the House version for the Finance Committee bill, agreed to 50-32 (D 18-27; R 32-5). A Bentsen amendment to exempt the first 1,000 bpd of independent producers carried 53-41 (D 27-31; R 26-10), while a Moynihan motion to kill the amendment was rejected 32-52 (D 22-25; R 10-27).[71]

Bill Bradley and John Chaffee offered an amendment to raise the tax rate for oil discovered between 1973 and 1978 from 60-75 percent, and the failure of a tabling motion by Long, by a 39-58 (D 13-44; R 26-14) vote on November 28, brought on a filibuster threat. An attempt to reach a compromise failed but oil-state senators agreed to allow a vote on the Bradley-Chaffee amendment in return for a chance to offer several amendments weakening the tax. The Bradley amendment was adopted 58-35 (D 44-10; R 14-25) on December 4. Metzenbaum followed with an amendment to delete the requirement that the tax be phased out, rejected 39-54 (D 38-16; R 1-38), but Moynihan responded by raising the level at which phase-out began to 90 percent of $210 billion instead of $138 billion, adopted 68-26 (D 48-7; R 20-19). The next day producer supporters found themselves fighting a holding action against attempts by liberals to extend their gains. Dole was able to table the Pat Leahy (D-VT) amendment to eliminate the depletion allowance for independent producers for income in excess of $15.30/bbl, 57-37 (D 26-30; R 31-7). On the other hand a Bellmon attempt to grant producers a credit against the windfall tax for increased production was stopped by a Ribicoff motion to table, agreed to 50-35 (D 44-6; R 6-29).[72]

The conflict became heated during the next week as an amendment to stiffen the tax by setting a 20 percent tax on newly discovered and high cost oil, offered by Ribicoff, survived a Long motion to table, 44-53 (D 10-47; R 34-6). The loss by Long was followed by a filibuster supported by oil state Democrats and most Republicans. A motion by Majority Leader Byrd to invoke cloture failed 53-46 (D 48-10; R 5-36; 60 votes are required for cloture), but Byrd kept the Senate in session to 3:45 am on Dec. 13, hoping to encourage a compromise. A second attempt to invoke cloture was rejected when the Senate reconvened that morning, but the vote was a closer 56-40 (D 51-6; R 5-34) as four conservative Democrats returned to the fold. A third attempt the next day was rejected 56-39, but Byrd offered a compromise substitute amendment lowering the tax rate to 10 percent on new oil, adopted 78-13. The Ribicoff amendment was then adopted 52-38 (D 41-10; R 11-28). John Danforth (R-MO) offered an amendment to remove an exemption for state owned oil, but opposition by Long and Bentsen resulted in acceptance of Long's motion to table, 65-28. The bill passed 74-24 on December 17.[73]

The wide differences between the two versions of the windfall profit tax bill raised problems reaching an agreement in conference. To facilitate the process the Carter administration finagled the conferees into splitting the difference in the revenues raised by the two versions, with both sides agreeing on a final tax that would raise $227 billion. Despite opposition by three Senators, the concurrence of Long and Bentsen insured the compliance of the Senate conferees. Once the revenue figure had been decided the key decision became the choice of Senate provisions which would be modified to collect additional income. The choice came down to increasing taxes on major companies or on independents exempted under the Senate version. The solution was to maintain the House tax rate of 30 percent on new discoveries for all producers, using a base price of $16.55, while providing tax breaks for independents on old oil. Old oil would have a tax rate of 70 percent of income

TABLE 6.29 Probit Runs For HR 3919–House Windfall Profits Tax

Variable	HR 3919A	HR 3919B	HR 3919C	HR 3919D	HR 3919E
Constant	1.93**	2.98**	3.13**	2.51**	-0.53*
Larval	0.0008*	0.0011	0.0050**		0.0006
Nonlar	0.0027**	0.0033**		0.010**	0.0026**
South	0.41	0.47*	0.39*		
McGov		-1.63		-4.12**	
Urban		-0.50	-0.87*		1.05**
ADA	-3.06**	-2.46**	-3.40**	-1.70**	-3.04**
Party	-1.99**	-1.42**	-1.29**	-1.32**	-0.31
Sign.	0.000	0.000	0.000	0.000	0.000
Pseudo R^2	.677	.585	.610	.549	.412
Observ.	415	419	421	417	415
Oil Dem.	23-12-0	10-24-1	32-3-0	26-10-1	31-3-1
Oil Rep.	13-17-0	0-29-1	28-1-1	28-1-1	28-1-1

HR 3919A: Adoption of the rule providing for floor consideration of the windfall profits tax bill (6/28/79). N=1.
HR 3919B: Ways and Means Committee amendments providing for a tax rate of 70 percent and a permanent tax on revenue above $17 a barrel, adjusted for inflation plus 2 percent annually (6/28/79). N=1.
HR 3919C: Jones substitute providing for a tax rate of 60 percent and discontinuation of the windfall tax at the end of 1990 (6/28/79). Y=1.
HR 3919D: Conable motion to recommit with instructions to add a "plowback," a credit of 75% of incremental investment in energy (6/28/79). A "nay" was a vote supporting the president's position. Y=1.
HR 3919E: Archer motion to recommit the conference report with instructions to exempt a 1,000 bpd produced by independent firms (3/13/80). A "nay" was a vote supporting the president's position. Y=1.

above the base price of $12.81 a barrel, but independents would be exempt from taxes on their first 1,000 bpd of production. Stripper wells would be taxed at 60 percent for major companies, but independents would have a tax rate of 30 percent on their first 1000 bpd, calculated from a base price of $15.20. High cost oil was given a base price of $16.55 and a tax rate of 30 percent. After some discussions on tax credits and spending agreement was finally reached on February 27, 1980.[74]

In the House an attempt to recommit the conference report was offered by Bill Archer, who asked for recommittal with instructions to exempt the first 1,000 barrels of independent production. The Archer motion was rejected

TABLE 6.30 Probit Runs For HR 3919–Senate Windfall Profits Tax

Variable	HR 3919A*	HR 3919B	HR 3919C	HR 3919D	HR 3919E
Constant	2.49**	3.08**	2.02**	1.92**	1.74**
Stripval	0.0092	0.0061		0.048**	0.014
Nonstrip		0.0045*	0.011*	0.050**	0.024**
South					-0.71
ADA	-10.23**	-14.27**	-3.23**	-6.15**	-4.88**
Party			-1.29**		
Signif.	0.0001	0.015	0.0001	0.0006	0.0003
Pseudo R^2	.723	.805	.540	.693	.547
Observ.	88	92	85	85	97
Oil Dem.	8-3-2	5-7-1	10-2-1	2-9-2	13-0-0
Oil Rep.	15-0-0	13-0-2	14-0-1	0-13-2	14-0-1

*Failed to converge on the first run.

HR 3919A: Boren amendment to reduce to 50 percent, instead of 60 percent in the House-passed bill, the tax rate on the difference between the price of controlled and decontrolled oil (11/16/79). A "nay" was a vote supporting the President's position. Y=1.
HR 3919B: Dole amendment, to the Finance Committee substitute, to reduce the tax on "old" oil to 60 percent from 75 percent (11/19/79). A "nay" was a vote supporting the president's position. Y=1.
HR 3919C: Long motion to table the Bumpers amendment substituting the House version of the tax bill (11/26/79). A "nay" was a vote supporting the president's position. Y=1.
HR 3919D: Moynihan motion to table the Boren amendment to exempt all stripper wells from the windfall profits tax (11/26/79). A "yea" was a vote supporting the president's position. N=1.
HR 3919E: Bentsen substitute amendment to exempt from the tax the first 1,000 bpd of oil production by independent producers (11/27/79). A "nay" was a vote supporting the president's position. N=1.

185-227 (D 61-198; R 124-29), and the bill passed the House on March 13, 302-107. A similar attempt in the Senate by Dole to refer the conference report to the Finance Committee to hold hearings on the effect of the tax on independent producers was rejected 35-61 (D 7-49; R 28-12), and the conference report was adopted 66-31 (D 49-8; R 17-23) on March 27.

The House votes followed the usual pattern for oil votes; oil production was opposed to the tax and a liberal ADA rating and membership in the Democratic party predisposed a representative to vote for the tax. Oil interests were

TABLE 6.30 Probit Runs For HR 3919–Senate Windfall Tax Votes

Variable	*HR 3919F*	*HR 3919G*	*HR 3919H*	*HR 3919I*	*HR 3919J*
Constant	3.77**	0.85*	0.88	3.70	2.38**
Stripval		0.014			0.0037
Nstrip	0.0045**	0.0015	0.0020**	0.0021**	
Urban					
McGov			3.81		
ADA	-4.93**	-3.64**	-2.41**	-4.93**	-4.83**
Party	-0.62	-0.70	-1.84*	0.65	-1.63**
Petcost	-0.0028			-0.0019	
Signif.	0.000	0.000	0.000	0.0003	0.000
Pseudo R^2	.604	.532	.498	.567	.612
Obser.	97	94	93	94	87
Oil Dem.	8-5-0	7-5-1	8-4-1	6-6-1	7-4-2
Oil Rep.	15-0-0	0-15-0	0-15-0	0-12-3	0-14-1

HR 3919F: Long motion to table the Bradley amendment to increase to 75 percent, from 60 percent, the tax rate on oil discovered between 1973 and 1979 (11/28/79). A "nay" vote was a vote supporting the president's position. Y=1.
HR 3919G: Bradley amendment to increase from 60 to 75 percent the tax rate on oil discovered between 1973 and 1978, thus increasing expected windfall tax revenues by an estimated $22.5 billion by 1990 (12/4/70). A "yea" was a vote supporting the president's position. N=1.
HR 3919H: Metzenbaum amendment to delete the requirement that the tax be phased out when 90 percent of the anticipated revenues had been collected (12/4/79). N=1.
HR 3919I: Moynihan amendment to phase out the tax when 90 percent of $210 billion had been collected (12/4/79). N=1.
HR 3919J: Ribicoff motion to table the Bellmon amendment allowing oil producers a credit against the windfall profits tax for 3 percent of increased production compared to the previous year (12/6/79). A "yea" was a vote supporting the president's position. N=1.

dominated by Nonlar, with Larval important only on the Jones amendment vote. Party became unimportant while Urban actually supported oil interests on HR 3919E, an amendment which discriminated in favor of independents.South was a weak influence in opposition to the tax, suggesting that southern Democrats were continuing to oppose the party's stance on oil issues, though less forcefully when it meant opposing a Democratic president as well the majority opinion of the party membership. Oil state Republicans voted

TABLE 6.30 Probit Runs For HR 3919–Senate Windfall Tax Votes

Variable	HR 3919L	HR 3919M	HR 3919N	HR 3919O	HR 3919P
Constant	2.09**	1.37**	1.59**	1.87	-.22
Stripval				0.022**	0.0094*
Nstrip	0.0025**	0.0031**	0.0043**		
Urban					1.62
ADA	-3.69**	-2.04*	-4.31**	-2.26**	-4.70**
Party	-1.93**	-2.53**	-1.67**	-1.13**	
Petcost				-.0014	
Signif.	0.000	0.000	0.000	0.000	0.000
Pseudo R^2	.631	.668	.659	.433	.459
Observ.	91	94	92	96	98
Oil Dem.	7-6-0	7-6-0	5-8-0	7-5-1	6-6-1
Oil Rep.	15-0-0	0-15-0	1-14-0	15-0-0	1-14-0

HR 3919L: Long motion to table the Ribicoff amendment for a 20 percent tax on new, tertiary and heavy oil (12/12/79). Y=1.
HR 3919M: Byrd motion to invoke cloture on the bill to tax excess profits of oil companies (12/13/79). N=1.
HR 3919N: Ribicoff amendment as amended to set a 10 percent tax on new oil and a 20 percent tax on tertiary and heavy oil (12/14/79). A "yea" was a vote supporting the president's position. N=1.
HR 3919O: Dole motion to refer the conference report to the Finance Committee for hearings on the effect of the tax on independent producers (3/27/80). A "nay" was a vote for the president's position. Y=1.
HR 3919P: Adoption of the conference report on the windfall profits tax bill, with expected revenues of $227 billion by 1990 (3/27/80). A "yea" was a vote for the president's position. N=1.

consistently against the tax, but a sizable number of oil state Democrats stuck with the party on the Ways and Means vote and against the plowback provision.

The Senate vote also reflected the normal pattern of voting on oil issues with oil production in opposition and Party and ADA the most important factors in favor of the tax. Party was less important in the Senate than the House. Stripval replaced Nonstrip on four votes, the original Boren amendment, the amendment establishing a plowback tax credit, and the two conference report votes. Although Stripval was significant on the three small producer votes, HR 3919D, HR 3919E and HR 3919O, there did not seem to be a substantial difference in voting patterns for these three votes.

TABLE 6.31 Logit Analysis of Senate Windfall Tax Votes

Variable	17 Votes	14 Votes
Constant	3.00**	2.66**
Stripval	0.00365**	0.00374**
Nonstrip	0.00010**	0.00011**
South	0.163	0.149
McGov	-0.700	-0.377
Urban	-0.221	-0.218
ADA	-6.04**	-5.86**
Party	-1.37**	-1.27**
Petcost	0.00021	0.00021
Adjusted R^2	.782	.785

This was confirmed by a logit analysis (Table 6.31), performed with 17 votes and then with the 14 votes which did not involve small producer interests. The analysis gave no evidence that there was any difference when the small producer votes were included. The logit runs also demonstrated the dominance of Stripval over Nonstrip, while rejecting a role for any other variables except for ADA and Party. The importance of Stripval is due to the Senators from the core oil states Senators voting more consistently against the windfall tax than the other oil state Senators.

VOTE #16: MISCELLANEOUS REAGAN ERA VOTES 1981-1982

The Reagan administration entered office ideologically committed to a free market approach. At the same time Reagan was aided by two factors which allowed him to pursue this approach to energy policy; the hard fought victories won by Carter which had made deregulation inevitable and the weakening world oil market, resulting in lower oil prices and diminished importance of energy as a political issue. Thus most of the votes on energy during the 1980s have been symbolic in nature, either by self-appointed consumer representatives endeavoring to present themselves as defenders of the "public interest" or oil state legislators attempting to salvage the economy of oil states under the guise of protecting the nation from dependence on foreign oil.

Reagan ended price controls on oil January 28 by executive order, eight months before they were scheduled to expire, removing the requirement, as of March 31, that large oil companies sell crude oil to small refiners. A number of Democratic Senators attacked the president for the move, and Howard

Metzenbaum moved to initiate efforts to block Reagan's action. Small refiners acted quickly in requesting aid from Congress with suggestions ranging from exemptions from the windfall profits tax, an import fee on imported products or tax credits for sales by large companies to small refiners. Given the prevailing mood in Congress, blocking decontrol or providing subsidies for smaller refiners was a losing proposition. Senator Metzenbaum was aware of this when he offered an amendment to nullify the order terminating controls, but he wanted to put the Republicans on record as voting for decontrol. The amendment, offered to a bill extending anti-trust exemptions to firms participating in programs of the International Energy Agency, was easily defeated on March 10 by a vote of 24-68 (D 21-21; R 3-47).[75]

There were a few other attempts by liberals to push their energy agenda through Congress against administration opposition. Dale Bumpers (D-AR) tried to provide a $450 million increase for conservation and solar in an attempt to reverse the pro-nuclear bias of the Reagan energy program, but his attempt was rejected 35-63 (D 33-13; R 2-50). In the House a motion to prevent a vote on HR 5274 was defeated 73-249 (D 20-151; R 53-98) and a bill to prohibit acquisitions of other oil companies by the nine largest companies or foreign governments for six months passed in December (D 148-33; R 75-74).[76] The administration did not take a stand on the issue, and it was more a symbolic gesture of concern with regard to merger activity than a serious attempt to prevent consolidation within the industry.

The bill to authorize the president to allocate petroleum during an oil shortage was also a symbolic gesture, but one which the administration felt required active opposition. Both the House and Senate passed legislation directing the president to draw up contingency plans for petroleum allocation in case of a new shortage. Since the president would have to declare that there was a "severe shortage" before he would implement allocation regulations, the legislation left the actual allocation power in the hands of the Executive. In March of 1982 the adoption of the conference report on the bill, S 1503, by the House by a 246-144 vote (D 190-23; R 56-121) set the stage for a confrontation with the President. Reagan vetoed the bill on March 20 and the Senate upheld his veto by a 58-36 (D 38-4; R 20-32) vote four days later.[77] While the administration was able to resist pressure for allocation (even though implementation was a hypothetical issue) the desertion of twenty Republican senators on the veto override vote suggests that a lack of confidence in the free market (at least in the case of oil) persisted on both sides of the Senate aisle.

The anti-merger votes in the House are characterized foremost by a large number of absentees and secondly by the weak performance of the model. Party and oil production seemed to be unimportant in the vote, evidenced by both the parameter coefficient values and voting by oil state representatives. ADA is the strongest factor in favor of a merger moratorium, while Urban is the most important factor in opposition, possibly due to the presence of corporate headquarters or merely a lower degree of hostility to big business. The

TABLE 6.32 Probit Runs For Miscellaneous House Votes

Variable	HR 5274A	HR 5274B	S 1503
Constant	-0.21	-1.55**	0.87
Larval		0.0003	
Nonlar	0.0001*	0.0003	
McGov	-1.79		
Urban	1.30**	1.50**	1.65**
ADA	-2.60**	-3.30**	-4.35**
Party			-0.77**
Petcost		0.0012	-0.0008
Signif.	0.000	0.000	0.000
Pseudo R^2	.243	.357	.489
Observations	322	332	398
Oil District Dem.	12-17-5	11-17-6	23-8-3
Oil District Rep.	12-15-3	10-15-5	7-21-2

HR 5274A: Livingston motion that the House adjourn, thus blocking consideration of the bill to place a moratorium on the acquisition of US oil companies by larger oil companies (12/16/81). Y=1.
HR 5274B: Dingell motion to suspend the rules and pass the bill to prohibit the nine largest US oil companies or foreign oil producing governments or companies from acquiring any of the top 40 US oil-producing companies for six months (12/16/81). N=1.
S 1503: Adoption of the conference report on the bill to authorize the president to allocate petroleum supplies during a severe oil shortage (3/3/82). A "nay" was a vote supporting the president's position. N=1.

allocation vote is basically a partisan split, with oil production irrelevant to the issue but Urban in favor of the administration's position.

The oil control vote in the Senate demonstrated the traditional alignment of Democrats and liberals against oil producers, with South in support of the pro-control forces. The Bumpers vote results, which should be viewed skeptically since the chi-square score was so low, suggests a Democrat and liberal core confronted by a general climate of opposition. While Republican oil senators were unanimous in their opposition a slim majority of Democratic oil senators actually supported the bill. The vote on allocation was also interesting as ADA played no part, but oil production (against) and party (Democrats for) were significant.

TABLE 6.33 Probit Runs For Miscellaneous Senate Votes

Variable	S 573	S 1377	S 1503
Constant	5.12**	-6.56*	1.33
Stripval	0.027**		0.0037*
Nonstrip			0.0005
McGov	-7.35	-12.28	
Urban			1.53
South	-1.31*	-1.40	1.08**
ADA	-2.58*	12.81**	
Party	-1.20*	1.26	-2.01**
Petcost		0.0037*	-0.0029*
Signif.	0.000	0.041	0.000
Pseudo R^2	.517	.826	.441
Observations	97	99	96
Oil State Dem.	1-9-2	7-5-0	9-2-1
Oil State Rep.	0-16-0	0-16-0	3-12-1

S 573: Metzenbaum amendment to nullify President Reagan's Jan. 28 order terminating controls on oil and gasoline (3/10/81). A "nay" was a vote supporting the president's position. N=1.
S 1377: Bumpers amendment to provide a $450 million increase in fiscal 1982 funding for energy conservation and solar energy (6/24/81). Y=1.
S 1503: Passage, over President Reagan's March 20 veto, of the bill to authorize allocation of petroleum supplies in a severe shortage (3/24/82). A "nay" was a vote supporting the president's position. N=1.

VOTE #17: MISCELLANOUS REAGAN ERA VOTES 1983-1987

During the rest of the Reagan administration action on energy, as little as there was relative to other issues, was basically confined to the Senate. The two major issues were natural gas price control and oil import restrictions, but a number of minor issues managed to reach the Senate floor. Metzenbaum, playing his well practiced role as oil company gadfly, attempted to add an amendment to the Export Administration Act (S 979) to halt oil company mergers for five years. A Heinz motion to table the amendment, submitted February 29, 1984, was agreed to 61-34 (D 19-21; R 42-13). This was followed by a Johnston amendment to deny federal leases to companies involved in mergers, also killed by a Heinz motion to table, 52-42 (D 9-31; R 43-11).

TABLE 6.34 Probit Runs For Miscellaneous Senate Votes–1983-87

Variable	S 979A	S979B	S 979C	S 1420B	S 1420C
Constant	1.35*	0.61**	-3.33**	1.52**	1.43**
Stripval	0.0055*	0.0032*			
Nonstrip			-0.0010	0.0084*	0.0096*
Urban			1.90		
South				0.74*	0.87*
ADA			4.64**	-3.05**	-3.16**
Party	-0.73**	-1.45**			
Petcost	-0.0010				
Signif.	0.016	0.000	0.000	0.000	0.000
Pseudo R^2	.124	.251	.574	.483	.514
Observ.	94	94	99	93	100
Oil Dem.	6-5-2	4-7-2	8-4-1	0-13-1	13-1-0
Oil Rep.	15-0-0	14-1-0	0-15-0	0-14-0	14-0-0

S 979A: Heinz motion to table the Metzenbaum amendment to halt major oil company mergers for five years (2/29/84). Y=1.
S 979B: Heinz motion to table the Johnston amendment to the Mineral Lands Leasing Act to deny federal leases to major oil companies involved in a merger for six months (2/29/84). Y=1.
S 979C: Bradley amendment to establish a standby block grant program to states to aid low-income citizens in the event of an oil shortage (6/18/85). A "nay" was a vote supporting the president's position. Y=1.
S 1420B: Metzenbaum motion to table Gramm motion to recommit the bill to the Finance Committee with instructions to report it with an amendment repealing the windfall profits tax (7/14/87). N=1.
S 1420C: Boren amendment to repeal the windfall profits tax on domestic oil (7/15/87). A "yea" was a vote supporting the President's position. Y=1.

The following year Bill Bradley, one of the new Democratic senators who supported decontrol in 1980, tried to add an amendment to the Energy Policy and Conservation bill (S 979), which would have established a standby block grant program to aid low-income citizens in the event of an oil shortage. His attempt, opposed by the Reagan administration, failed on June 18 by a 44-55 (D 35-11; R 9-44) vote. In 1987 the windfall profits tax came under attack as oil prices plummeted to the extent that the tax was merely a formality imposing a paperwork burden but raising no revenue. A Gramm motion to recommit the Omnibus Trade bill with instructions to add an amendment repealing the windfall profits tax was challenged by Metzenbaum. The motion to table S 1420 failed 36-57 (D 26-21; R 10-36) on July 14. This was followed the next day by a vote on the Boren amendment to repeal the tax, adopted 58-40 (D 23-29; R 35-11).[78]

The merger votes were split mostly along partisan lines with some support from Stripval for the anti-restriction forces. It is interesting to note that ADA was not a factor, unlike the House vote a few years earlier. Unlike many issues the split on merger activity is not a simple liberal/ conservative dichotomy but more a measure of comfort with concentration in the industry. The vote on the Bradley amendment showed only liberal support, with a suggestion of Urban support, for energy aid for low-income recipients. While the general mood against spending doomed the amendment, if legislators were rational actors one would expect that oil interests would have favored this piece of legislation since it would have reduced political pressures for price controls during a future shortage, while the cost of the program would have been borne by the Treasury and not oil producers. The votes on the windfall tax were as expected except for the absence of partisan influence, with ADA the only significant factor in opposition to repeal, and South along with Nonstrip the major supporters of repeal.

VOTE #18: NATURAL GAS VOTES DURING THE REAGAN ERA

While decontrol of oil was a simple matter of an executive order, natural gas decontrol was a far more complex affair, involving legislation which was sure to generate tremendous controversy, as the issue had been contested for forty years. A wide range of interest groups were actively involved on various sides of the issue. Gas producers supported decontrol, but pipeline companies opposed it, since higher price would mean lower demand and a decreased volume of gas to be transported. Heating oil marketers supported decontrol to reduce competition. Business organizations wanted a repeal of the incremental pricing provisions of the 1978 act which put the burden of gradual decontrol on large industrial users. Coal interests opposed any change in the prohibition of gas use in large boilers. Consumer groups were adamant in their opposition to further decontrol.[79]

The new regulatory environment that the pipeline companies were operating under the Natural Gas Act of 1978 encouraged rapid increases in prices. Since pipeline companies are regulated utilities they had no incentive to keep the cost of purchasing gas at a minimum, and the existence of excess demand encouraged the companies to buy deregulated gas at prices far exceeding the market level as they could roll in the price of the expensive gas with the cheap price controlled gas they were already purchasing. The companies therefore agreed to long-term contracts to purchase gas at above market prices, and the price of gas to consumers rose more rapidly than expected, reaching the cost of heating oil and exceeding the cost of residual oil. The result was that a shortage of natural gas began to turn into a glut, and high prices created political pressure for action.

A two-part amendment to the Transportation Assistance Act of 1982 (HR 6211) was offered by Nancy Kassebaum, the first part which would allow natural gas pipelines to circumvent contractual obligations to buy higher cost gas and the second part which froze the wellhead price of natural gas for a two-year period. Baker (R-TN) then offered motions to table each division of the amendment separately, with the first motion agreed to 56-38 (D 25-18; R 31-20) and the second agreed to 62-33 (D 25-19; R 37-14) on December 14, 1982.[80]

While it was surprising for a Senator from a major gas producing state to propose a freeze on wellhead prices, Kassebaum was influenced by the reaction of her constituents to a 20 percent increase in their gas rates. President Reagan accepted the verdict of consumers by submitting a proposal that while providing for eventual decontrol, ended automatic pass-through of costs by pipeline companies but allowed pipeline companies to break commitments to buy expensive gas. The proposal also repealed the authority of the president to reimpose controls and eliminated restrictions imposed by the 1978 Fuel Use Act. The first test for the administration came in the Senate Energy Committee in April when an attempt to eliminate deregulation of old gas was defeated by an 8-10 vote. However progress stalled in committee after four weeks as decontrol opponents were able to block a move to extend market out provisions (which allow contracts to be broken) to producers which would have resulted in de facto decontrol. The key vote was that of Don Rickles (R-OK), who opposed market out provisions because he represented high cost producers.[81]

A natural gas bill finally reached the Senate floor on November 1, though differing substantially from the original administration proposal. S 1715, the Natural Gas Policy Act, was reported July 29 without a recommendation in order to obtain a majority vote of the Energy Committee. The bill decontrolled old gas over a 36 month period while lowering the price of expensive new gas over 12 months, limited take and pay liability to 50 percent the first year, and allowed gas users to buy directly from producers and requiring pipelines to act as common carriers. The major alternative bill, S 996, developed by Kassebaum, would keep controls on old gas, roll other gas prices back to August 1982 levels, delay scheduled decontrol of some new gas, and limit take and pay liability to 50 percent. A number of attempts were made to develop a compromise proposal as it was obvious that neither bill could pass the Senate. The predictions of failure proved correct when the Senate voted November 15. The Kassebaum bill was offered as an amendment to S 1715, but was rejected 26-71 (D 17-26; R 9-45). The original bill was rejected 28-67 (D 2-39; R 26-28).[82]

TABLE 6.35 Probit Runs For Senate Natural Gas Votes

Variable	HR 6211A	HR 6211B	S 1715A	S 1715B	HR 4515
Constant	3.77**	1.59**	2.96**	1.94	0.071
Ngval	0.0038**	0.0028*	0.0014	0.0007*	0.0077**
McGov	-7.76**			-9.90**	
Urban			-1.95*	3.80*	2.19
South					0.50
ADA		-1.86**	-3.13**	-1.89	-1.45*
Party		0.50	0.70	-1.19*	-0.96*
Ngcost	-0.0044**	-0.0036**		-0.0043*	-0.0040*
Signif.	0.0001	0.0010	0.000	0.0002	0.0004
Pseudo R^2	.299	.205	.323	.466	.411
Observ.	94	95	96	94	96
Gas Dem.	5-0-0	5-0-0	0-6-0	1-4-1	6-0-0
Gas Rep.	9-2-0	9-2-0	1-9-0	7-3-0	9-1-0

HR 6211A: Baker motion to table division I of the Kassebaum amendment to authorize FERC to allow gas pipelines to modify their contractual obligations to buy higher cost gas when lower cost gas is available (12/14/82). Y=1.
HR 6211B: Baker motion to table division II of the Kassebaum amendment to freeze the wellhead price of natural gas at October 1 levels for a two-year period (12/14/82). Y=1.
S 1715A: McClure amendment (offered on behalf of Kassebaum) to the McClure amendment, that was the text of S 996, to retain price controls on gas from wells in production prior to April 1977 and roll other prices back to 1982 levels (11/15/83). N=1.
S 1715B: McClure amendment, to the McClure amendment, that was the text of S 1715, which after 44 months would decontrol prices of natural gas and allow current gas purchase contracts to be adjusted (11/15/83). Y=1.
HR 4515: Hatfield motion to table the Exon amendment to prevent FERC from deregulating the price of old natural gas (6/6/86). N=1.

The House Energy and Commerce Committee approved a natural gas bill in April, 1984, which retained existing price controls on old gas and capped the price escalation of newer gas, by a narrow vote of 22-20. The bill retained the NGPA freeze on wellhead price of old gas, froze prices of new gas, allowed pipelines to force gas producers to renegotiate contracts for high priced gas, and limited take or pay requirements to 50 percent. The bill was allowed to lie dormant for several months until Speaker O'Neill, after receiving a letter urging action signed by 218 House members, referred it to the Rules

Committee. However complaints by members that forcing a vote on a no-win issue would hurt them in the impending elections caused the bill to quietly disappear from the Rules Committee agenda. O'Neill then announced on September 26 that there would be no natural gas bill during the session.[83]

A final attempt to stop decontrol of natural gas occurred when Exon (D-NE) attempted to add an amendment to the Urgent Supplemental Appropriations bill in 1986 to stop the Federal Energy Regulatory Commission from deregulating the price of old natural gas. Hatfield offered a motion to table the Exon bill which was agreed to 48-47 (D 12-33; R 36-14) on June 6.[84]

Natural Gas votes exhibit the usual distortions associated with lopsided votes. Ngcost is better behaved in the 1980 votes, suggesting that the increased use of gas by nonproducing states decreased the correlation between production and consumption. Ngval supports decontrol, and Ngcost is associated with opposition for every vote but the one on the Kassebaum amendment, and in this case the only Senators voting for the alternative bill were liberals from Northern urbanized states. Party was only important on two votes. The only vote which tends to match the historical pattern for energy voting is the one on the Hatfield motion, which is also the closest vote.

VOTE #19: OIL IMPORT FEE / QUOTA VOTES

Oil import fees were a controversial issue for Ford (see Vote #9) and remained so for the Carter administration. When Congress rejected Carter's crude oil tax in 1977 Energy Secretary James Schlesinger threatened to impose quotas or a $5 import fee on imported oil, which the President was authorized to order under the 1962 Trade Expansion Act. The Senate responded by adopting an amendment to the Finance Committee version of the energy tax bill, HR 5263, which prohibited the imposition of import fees except in time of war or national emergency (see Vote #13). Since the energy tax bill had little change of passage, Bob Dole and 30 cosponsors introduced S Con Res 73 in April 1978, expressing the sense of the Senate that the president should not impose oil import fees. Although the Finance Committee ordered the resolution reported on May 9 by a 11-6 vote, it did not reach the floor, which Dole charged was due to the action of Finance Chairman Russell Long in concert with the Senate leadership. Dole proceeded to take a third step to prevent an oil import fee by offering an amendment to the Treasury appropriations bill, HR 12930, which would prohibit the use of funds appropriated in the bill for enforcement of import fees. Opposition to the Dole move was led by Herman Talmadge, the ranking Democrat on the Finance Committee and Henry Jackson. Talmadge submitted a motion to table the Dole amendment on June 27, which was defeated 39-49 (D 34-19; R 5-30), after which the Dole amendment was adopted by voice vote. On the House side Clarence Brown offered a motion to instruct the House conferees on HR 12930 to agree to the Senate amendment to prohibit oil import fees or quotas. The motion was rejected 194-201 (D 70-195; R 124-6) on Sept. 7.[85]

The rise in world oil prices in 1979 brought on new threats by the administration to impose quotas or import fees if conservation measures did not succeed in constraining consumption. Carter's pledge of July 15 to limit 1979 oil imports to 8.2 million bpd, going beyond his pledge at the Tokyo summit in June to hold imports to 8.5 million bpd, gave warning to Congress that the threat of new fees or quotas was real. Charles Vanik, Chairman of the Ways and Means Trade Subcommittee, declared on October 10 that imposing quotas would boost inflation and disrupt the economy. Two weeks later the Senate Energy Committee voted 10-8 in favor of requiring congressional approval of import quotas, attaching it to an omnibus energy bill (S 932). The Carter administration strongly disapproved of the amendment, and worked out a compromise with its author, Bennent Johnston, which substituted an amendment requiring both houses to approve a resolution blocking a plan to impose quotas. This actually meant a two-thirds majority would be required in both houses to overcome a presidential veto. The amendment was attached to S 1871, the Oil Company Antitrust Exemption, when Johnston's motion carried 70-23 (D 37-17; R 33-6) October 30.[86]

When Carter imposed a $4.62 per barrel import fee on May 15, 1980, as part of his March 15 anti-inflation package Congress reacted quickly to block his action. The fee was attacked for being too little to stimulate conservation, designed to raise revenue to finance a tax cut, and badly timed given the large increases in gasoline prices. The Senate Finance Committee voted for S J Res 159, a resolution disapproving the fee, by a 14-4 vote, and the House Ways and Means Trade Subcommittee voted 17-4 to approve a similar measure the same week. Ways and Means then approved the resolution 27-7 on May 22. Concurrently a suit was filed by three consumer groups, a gasoline dealers association and Marathon Oil, asking for temporary and permanent injunctions on the grounds that it was a tax because the fee was designed to be levied on gasoline and not crude oil.[87]

In order to circumvent the opposition of the House leadership, opponents of the fee offered the resolution as an amendment to the debt limit bill. Although the bill had been sent to the floor under a closed rule, the House rejected the rule by a 312-74 procedural vote, only to have the amendment barred on a point of order. The bill was then returned to the Rules Committee, which refused to budge under pressure, but Rules Committee Chairman Richard Bolling did promise that the committee would take up the fee resolution. When the resolution finally reached the floor on June 4 it passed by an overwhelming 376-30 (D 222-30; R 154-0) vote. Most of the thirty Democrats were part of the party leadership including Rostenkowski, Bolling, Wright, Vanik, Robert Giaimo and Sam Gibbons. The Senate vote went smoother since Majority Leader Robert Byrd was in favor of the resolution. Dole offered the resolution as an amendment to HR 7428, the debt limit bill, and it was adopted 73-16 (D

TABLE 6.36 Probit Runs For Oil Import Votes

Variable	HR 12930	HR 12390	HR 7428	H J Res 668	S 1420A
Constant	-0.32	0.92	4.92**	-2.59**	1.58**
Nonstrip	0.0005				
Stripval				0.0087**	0.049**
South	-0.74			0.81*	0.64
Urban	-1.17		-2.91*		
ADA	-1.81*	-1.50**	-3.24**	1.95**	-3.27**
Party	-0.97*	-2.04**			
Petcost	0.0054		0.0023	-0.0020	
Signif.	0.0005	0.000	0.000	0.000	0.000
Pseudo R^2	.284	.405	.311	.281	.467
Obser.	91	395	77	97	93
Oil Dem.	8-2-2	15-18-6	8-1-4	5-7-1	1-13-0
Oil Rep.	1-12-1	16-0-4	10-0-5	12-2-1	6-7-1

HR 12930: Talmadge motion to table Dole amendment to prohibit use of funds to enforce the imposition of import fees on foreign oil (6/27/78). Y=1. A "yea" was a vote supporting the President's position.
HR 12930: Brown motion to instruct House conferees on the fiscal appropriation bill to agree to the Senate amendment to prohibit the President from imposing fees or quotas on oil imports (9/7/78). A "nay" was a vote supporting the President's position. N=1.
HR 7428: Passage, over the President's veto, of the bill to extend the existing debt limit and to disapprove the $4.62 fee per barrel of imported oil that President Carter imposed effective March 15 (6/6/80). A "nay" was a vote supporting the president's position. N=1.
H J RES 668: Heinz motion to table the Hart amendment to the Finance Committee amendment to increase the existing tariff on crude oil and refined petroleum products by $10 a barrel (7/31/85). N=1.
S 1420A: Bradley and Packwood amendment to strike provisions granting the President power to limit dependence on foreign oil, requiring a plan for keeping imports below 50 percent of domestic consumption (7/1/87). A "yea" was a vote supporting the president's position. N=1.

38-13; R 35-3), followed by passage of the bill on June 4. Carter then vetoed the bill on June 5, but the House passed the bill by a 335-34 vote and the Senate passed the debt bill by a 68-10 vote the next day.[88]

With the advent of the Reagan administration the threat of an oil import fee or quota subsided along with OPEC dominance of the world oil market.

However, the Reagan deficit generated pressure to find additional revenue that would be politically palatable, and with declining energy prices an oil import fee seemed a possible candidate. In 1982 Senate Republican leaders were toying with the idea of a $5 per barrel import fee, but the idea drew little support from Ways and Means, as Northeastern representatives were in opposition. The precipitous drop in oil prices that occurred in late 1985 generated new attempts to institute an oil fee, though this time the goal was not to reduce dependence on foreign oil but to avoid bankruptcy for domesticoilmen and their bankers. Gary Hart offered an amendment to S J Res 77 to impose a $10 fee but a McClure motion to table was agreed to 78-18 (D 30-16; R 48-2) on November 15. Hart tried again in July 1986, adding an import fee amendment to H J Res 668, but a Heinz motion to table eliminated that attempt by a 82-15 (D 35-11; R 47-4) vote. The closest supporters of an oil fee came to passage was when Bentsen slipped provisions into the Omnibus Trade Bill, S 1420, granting the president broad power to limit dependence on foreign petroleum and requiring him to file a plan to limit imports to less than 50 percent of domestic consumption. A Bradley and Packwood amendment striking those provisions was adopted 55-41 (D 22-29; R 33-12) on July 1, 1987.

On the Carter oil fee votes the general consensus was that nobody liked the fees. Republican Senators had no qualms about voting against the fee, and on subsequent votes everyone deserted the President. On the House side party loyalty evaporated when legislators had to actually vote for permitting a fee. The analysis of the vote on HR 7248 and the Hart amendment vote should be taken with a grain of salt because there were so few Senators supporting a fee (this causes both a statistical problem and a question of inference; when a vote is lopsided a legislator can vote frivolously, secure in the knowledge that his vote cannot influence events). The Bentsen amendment vote shows the emergence of domestic oil producer support for import fees, opposed by a liberal consensus against raising costs to consumers.

CONCLUSIONS

While it is hard to generalize over a series of votes that span a thirty year period and cover a diverse range of energy issues a number of observations can be made. Going back as far as the fifties there seems to have been a conflict between oil industry supporters and congressmen who vote in concurrence with the judgment of those who analysis votes from what is called a "liberal" perspective (i.e. ADA ratings of votes). While Peltzman considers ADA ratings merely a reflection of the economic interests of a Congressman's constituents I feel this is a simplistic dismissal of a complex phenomenon. Since Senators from the same state have the same set of constituents we should expect them to have similar ADA ratings under this reasoning, but a quick perusal of unlikely pairs[89] provides empirical evidence that the Peltzman explanation lacks predictive power. In the case of oil voting my analysis included the major economic interests which should influence legislators; energy production and energy consumption. Although collinearity problems obscure relationships between variables it is obvious that ADA ratings dominated voting when included along with both production and consumption variables.

My suspicion is that the ADA ratings reflect the legislator's ideological leanings as they are expressed over a wide variety of votes tempered by his constituent's interests. Measures of constituent ideology, expressed through votes for polarizing presidential candidates (Goldwater, McGovern) were not strongly correlated with Congressmen's ADA ratings, nor did they play a significant role in many energy votes. The chain of causality flows from the selection of individuals, or at least the tolerance of individuals with a divergent (from the constituency) ideological bent, as long as they take care to satisfy the important economic concerns of the constituency.

Unfortunately the measures of ideological content of voting behavior are collected by interest groups with a specific agenda, only partially related to a measure of economic ideology . Therefore a measure such as the ADA rating is actually a proxy for a number of ideological attitudes, only some of which are relevant with regard to political economy. In the case of oil politics I suspect that ADA ratings are a proxy for antagonism to big business, a concept of "public interest" which includes a type of consumerism (to be distinguished from the economic concept of welfare maximization) that equates low prices with maximum general welfare, and a faith in the ability of government to micromanage economic markets.

Given this concept of ADA as a measure it is not surprising that it is associated with opposition to every bill favorable to the oil industry, from depletion to price decontrol. However it should be noted that on most issues the majority of legislators are neither from oil states (or districts) or do they have a strong ideological bias. Their votes are determined by a combination of factors including the stand taken or avoided by the president and the leadership of their party, the rewards and punishments for divergence from the party leadership (especially in the 1950s and 1960s), the specifics of the bill, and even their sense of the public interest.

A key factor influencing oil policy during the 1970s was an innate prejudice in favor of small business which resulted in subsidies to small refiners and oil producers which are difficult to explain via the constituent interest hypothesis. I will expand upon this theme further in the next chapter. However, it should be noted that this was primarily an expression of hostility toward large companies. Toward the end of the decade, when the value of small oil producers and refiners as constraining factors on the "oil monopoly" came into question, support for these subsidies faded. While the Reagan-Bush administrations favored small business, they preferred the policies of tax preferences and deregulation to those of subsidies and economic controls.

NOTES

[1] Herbert F. Weisberg, "Evaluating Theories of Congressional Roll-Call Voting," *American Journal of Political Science* 22 (1978): 554-77; John Kingdon, *Congressmen's Voting Decisions,* 2nd ed. (New York: Harper and Row, 1981); and Aage R. Clausen, *How Congressmen Decide: A Policy Focus* (New York: St. Martins Press, 1973).

[2] Morris P. Fiorina, *Representatives, Roll Calls, and Constituencies* (Lexington, Mass.: Lexington Books, 1974).

[3] Sam Peltzman, "Constituent Interest and Congressional Voting," *Journal of Law and Economics*, 27 (Apr. 1984): 181-210; James Kau and Paul Rubin, "Self-Interest, Ideology, and Logrolling in Congressional Voting," *Journal of Law and Economics* 22 (Oct. 1979): 365-84; Joseph Kalt and Mark Zupan, "Capture and Ideology in the Economic Theory of Politics," *American Economic Review* 74 (June 1984): 279-300.

[4] Keith T. Poole, "Dimensions of Interest Group Evaluation of the US Senate, 1969-1978," *American Journal of Political Science* 25 (1981): 49-67; Richard Carson and Joel Oppenheimer, "A Method of Estimating the Personal Ideology of Political Representatives," *American Political Science Review* 78 (1984): 163-78.

[5] Edward J. Mitchell, "The Basis of Congressional Energy Policy," *Texas Law Review* 57 (1979): 591-613; Sally Lopreato and Fred Smoller, "Explaining Energy Votes in the Ninety-Fourth Congress," (Unpublished) Austin: Center for Energy Studies, University of Texas at Austin, June 1978; James Riddlesperger and James King, "Energy Votes in the US Senate, 1973-1980," *Journal of Politics* 44 (Aug. 1982): 838-47; Robert Bernstein and Stephen Horn, "Explaining House Voting on Energy Policy: Ideology and the Conditional Effects of Party and District Economic Interests," *Western Political Quarterly* 34 (June 1981): 235-45.

[6] Joseph Kalt, *The Economics and Politics of Oil Price Regulation* (Cambridge: MIT Press, 1981), pp. 253-67.

[7] Kau and Rubin, "Self-Interest, Ideology, and Logrolling," p. 378; James Kau, Donald Keenan and Paul Rubin, "A General Equilibrium Model of Congressional Voting," *Quarterly Journal of Economics* 97 (May 1982): 271-94; and *Congressional Quarterly Almanac*, 1974, p. 12-H.

[8] George Judge, Carter Hill, William Griffiths, Helmut Lutkepohl and Tsoung-chao Lee, *Introduction to the Theory and Practice of Econometrics* (New York: John Wiley & Sons, 1982), pp. 522-26; Robert Pindyck and Daniel Rubinfeld, *Econometric Models and Economic Forecasts* (New York: McGraw-Hill, 1981), pp. 280-87; G.S. Maddala, *Limited-Dependent and Qualitative Variables in Econometrics* (Cambridge: Cambridge University Press, 1983), pp. 22-26.

[9] George Judge, Carter Hill, William Griffiths, Helmut Lutkepohl and Tsoung-chao Lee, *The Theory and Practice of Econometrics*, 2nd ed. (New York: John Wiley & Sons, 1985), pp. 896-930.

[10] David Belsley, Edwin Kuh and Roy Welch, *Regression Diagnostics* (New York: John Wiley & Sons, 1980).

[11] Judge et al., *Theory and Practice of Econometrics*, p. 869.

[12] *Congressional Quarterly Almanac, 1949*, p. 362-69.

[13] Ira Sharkansky, "The Utility of Elazar's Political Culture: A Research Note," in *The Ecology of American Political Culture: Readings*, ed. Daniel Elazar and Joseph Zikmund (New York: Thomas Crowell, 1975), p. 252.

[14] Congressional Quarterly, *Congress and the Nation: 1945-1964*, (Washington D.C.: CQ Press, 1965), p. 983-84.

[15] *Congressional Quarterly Almanac, 1953*, p. 215, 252-53.

[16] *Congress and the Nation: 1945-1964*, p. 983-84.

[17] Nash, *United States Oil Policy*, p. 198-99; Bruce Oppenheimer, *Oil And The Congressional Process* (Lexington, Mass.: D.C. Heath, 1974), pp. 34-35; *CQ Weekly Report*, Oct. 5, 1951, p. 1484.

[18] *Congressional Quarterly Almanac*, 1958, p. 454; *CQ Weekly Report*, July 3, 1959, p. 927, June 24, 1960, p. 1112, Aug. 26, 1960, p. 1499, Sept. 7, 1962, p. 1514.

[19] Manley, *The Politics of Finance*, pp. 25-27.

[20] Oppenheimer, *Congressional Process*, pp. 72-73, 104-111.

[21] *CQ Weekly Report*, July 25, 1968, pp. 1315-16, Aug. 8, 1969, p. 1430.

[22] *CQ Weekly Report*, Aug. 8, p. 1431, Oct. 31, 1969, p. 2132.

[23] *CQ Weekly Report*, Nov. 28, p. 2430, Dec. 5, p. 2417, Dec. 12, 1969, p. 2591.

[24] *CQ Weekly Report*, Jan. 9, 1970, p. 105.

[25] De Marchi, "Energy Policy Under Nixon," p. 432; *CQ Weekly Report*, June 9, p. 1467, June 16, 1973, p. 1536.

[26] *CQ Weekly Report*, June 9, p. 1467, Aug. 4, 1973, pp. 2172-74, Oct. 20, 1973, p. 2818; *Energy Policy*, p. 136.

[27] *Energy Policy*, p. 138.

[28] *Energy Policy*, p. 138.

[29] *CQ Weekly Report*, Dec. 29, 1973, p. 3458.

[30] *CQ Weekly Report*, Feb. 23, 1974, p. 522.

[31] *CQ Weekly Report*, Mar. 9, 1974, p. 632.

[32] *Oil and Gas Journal*, Oct. 21, 1974, p. 90-91; *CQ Weekly Report*, Feb. 22, p. 351-352 and Mar. 1, 1975, p. 458.

[33] *CQ Weekly Report*, Mar. 22, 1975, p. 571-573.

[34] *CQ Weekly Report*, Mar. 29, 1975, p. 620-622.

[35] *CQ Weekly Report*, Jan. 18, pp. 131-137, Feb. 22, p. 351-352, March 8, 1975, p. 474-77; and US Congress, House, Committee on Interstate and Foreign Commerce, *Energy Conservation And Oil Policy, Hearings before the*

Subcommittee on Energy and Power, 94th Cong., 1st Sess., March 1975, pp. 79-107, 155-61.

[36] *CQ Weekly Report*, Apr. 26, pp. 858-59, May 3, pp. 952-54, May 10, 1975, p. 959-61.

[37] *CQ Weekly Report*, June 4, 1975, p. 1260-62.

[38] *CQ Weekly Report*, Mar. 8, 1975, p. 475-77.

[39] *CQ Weekly Report*, Apr. 12, pp. 764-65, Apr. 19, 1975, pp. 765, 825.

[40] *CQ Weekly Report*, May 3, 1975, p. 949.

[41] *CQ Weekly Report*, May 31, 1975, p. 1123.

[42] *CQ Weekly Report*, Aug. 2, 1975, pp. 1726, 1734, *CQ Almanac*, 1975, pp. 43-S, 92-H, 94-H; *Energy Policy*, p. 154.

[43] *CQ Weekly Report*, Aug. 2, 1975, p. 1726; *CQ Almanac, 1975*, p. 96-H; and *Energy Policy*, p. 152.

[44] *CQ Weekly Report*, Aug. 2, 1975, p. 1657, 1726.

[45] *CQ Weekly Report*, Aug. 2, p. 1734; Sept. 20, p. 1999, Sept. 27, 1975, pp. 2043, 2082.

[46] *CQ Weekly Report*, June 21, pp. 1278-79, Oct. 4, 1975, pp. 2144-46.

[47] *CQ Weekly Report*, Oct. 4, p. 2141; Oct. 11, 1975, pp. 2200, 2205.

[48] *CQ Weekly Report*, Oct. 25, 1975, pp. 2288-89, 2292-94.

[49] *CQ Weekly Report*, Feb. 7, 1976, pp. 260-61, 293; and *Energy Policy*, p. 174.

[50] *CQ Weekly Report*, Feb. 7, 1976, pp. 260-61, 293-95; and *Energy Policy*, pp. 174-75.

[51] *CQ Weekly Report*, Oct. 11, p. 2200, Oct. 25, 1975, p. 2288; and *Energy Policy*, p. 175.

[52] *CQ Weekly Report*, June 16, 1976, p. 1549; *CQ Almanac*, 1976, p. 29-S.

[53] *CQ Weekly Report*, Aug. 6, 1977, pp. 1623-28.

[54] *CQ Weekly Report*, Aug. 6, p. 1627; Aug. 13, 1977, p. 1732.

[55] *CQ Weekly Report*, Aug. 6, p. 1628, Aug. 13, 1977, p. 1734.

[56] *CQ Weekly Report*, Oct. 22, 1977, pp. 2234-36.

[57] *CQ Weekly Report*, Oct. 8, 1977, p. 2172; and *CQ Almanac*, 1977, pp. 50-S, 60-S.

[58] *CQ Weekly Report*, Oct. 22, 1977, p. 2236.

[59] *CQ Weekly Report*, Oct. 29, 1977, pp. 2288, 2337.

[60] *CQ Weekly Report*, Oct. 29, pp. 2337, 2390, Nov. 5, 1977, pp. 2355-58.

[61] *CQ Weekly Report*, Mar. 11, 1978, pp. 631-32.

[62] *CQ Weekly Report*, Mar. 25, pp. 743-44, Apr. 29, p. 1039, May 13, 1978, p. 1163.

[63] *CQ Weekly Report*, Sept. 2, pp. 2395-96, Sept. 16, pp. 2451-54, Sept. 30, 1978, pp. 2615-16.

[64] *CQ Weekly Report*, Apr. 7, 1979, pp. 619-20; and *Energy Policy*, pp. 207, 211.

[65] *CQ Weekly Report*, Apr. 14, p. 712, 1979.

[66] *CQ Weekly Report*, May 5, pp. 822-23, May 26, pp. 997-98, Oct. 13, pp. 2262, 2302, Oct. 20, 1979, pp. 2312, 2370.

[67] *CQ Weekly Report*, Apr. 28, p. 772, June 2, 1979, p. 1040.

[68] *CQ Weekly Report*, June 23, p. 1277, June 30, 1979, p. 1284.

[69] *CQ Weekly Report*, June 30, 1979, p. 1328.

[70] *CQ Weekly Report*, Oct. 27, 1979, pp. 2414-16.

[71] *CQ Weekly Report*, Dec. 1, 1979, pp. 2743-45.

[72] *CQ Weekly Report*, Dec. 8, 1979, p. 2759-60, 2804-08.

[73] *CQ Weekly Report*, Dec. 15, pp. 2818-19, 2860-61, Dec. 22, 1979, pp. 2875-76, 2921, 2929.

[74] *CQ Weekly Report*, Dec. 22, 1979, p. 2875, Jan. 26, pp. 183-84, Mar. 8, 1980, p. 668-69.

[75] *CQ Weekly Report*, Jan. 31, p. 231-32, Mar. 14, 1981, p. 465, 492.

[76] *CQ Almanac*, 1981, p. 116-H.

[77] *CQ Almanac*, 1982, p. 4-H, 13-S.

[78] *CQ Weekly Report*, June 22, 1985, p. 1231, July 18, 1987, p. 1613; *CQ Almanac*, 1984, p. 6-S.

[79] *CQ Weekly Report*, Feb. 14, 1981, pp. 313-14.

[80] *CQ Weekly Report*, Dec. 18, 1982, p. 3068.

[81] *CQ Weekly Report*, Apr. 23, p. 779-80, May 14, p. 941, June 18, 1983, p. 1246.

[82] *CQ Weekly Report*, Nov. 5, p. 2335-38, Nov. 19, 1983, pp. 2453, 2460.

[83] *CQ Weekly Report*, Apr. 14, p. 888-89, Sept. 29, 1984, p. 2409.

[84] *CQ Almanac*, 1986, p. 22-S.

[85] *CQ Weekly Report*, July 1, pp. 1659-60, 1707, Sept. 9, 1978, pp. 2440, 2442.

[86] *CQ Weekly Report*, July 21, pp. 1438-39, Oct. 13, p. 2265, Oct. 27, p. 2417, Nov. 3, 1979, pp. 2480, 2500.

[87] *CQ Weekly Report*, May 17, pp. 1307-8, May 24, 1980, p. 1389.

[88] *CQ Weekly Report*, June 7, 1980, pp. 1539-40, 1599, 1604.

[89] In 1980: California had Cranston, ADA rating of 85 in 1978 and Hayakawa (15); Colorado, Hart (65), Armstrong (5); Delaware, Roth (15), Biden (50); Idaho, Church (70), McClure (0); Indiana, Bayh (85), Lugar (10); Maine, Muskie (75), Cohen (30); Maryland, Mathias (50), Sarbanes (90); North Dakota, Young (5), Burdick (55); South Dakota, McGovern (75), Pressler (25). Michael Barone, Grant Ujifusa and Douglas Matthews, *The Almanac of American Politics, 1980* (New York: E.P. Dutton, 1979).

VII

Interest Groups, Rent Seeking, and Oil Politics

Explanations of economic regulation have ranged from a naive belief that government endeavored to further the public interest to an overly mechanical and cynical belief that self-interested behavior controls politics and policy development. The evolution of the theory of economic politics has been convoluted, involving a number of strands of thinking in two interlinked disciplines. In political science, interest group theory and pluralism can be traced back to the Federalist. The emergence of Public Choice theory with the application of microeconomic theory to the analysis of politics raised questions about the assumptions underlying traditional interest group theories. Parallel with this development, economists began to model political behavior as merely another form of economic behavior, challenging the traditional view of economic regulation as motivated by a desire to further the public interest. Finally, a backlash against the simplistic formulas of economic politics has emerged, creating a far richer and more complex picture of political behavior, but one which is analytically far more intractable, providing new legitimacy for the economic and political historian. I place myself firmly in this last camp, contending that social phenomena are too complex to be fully explained by simple models of rational self-Interested behavior.

INTEREST GROUP THEORY

Concern with the power of interest groups, especially economic interest groups, to determine government policy can be traced back to essay number 10 of the *Federalist*, written by James Madison. Madison's concern was to "control the violence of faction," which he considered that most dangerous vice of popular government. What Madison called a faction, "a number of citizens, whether amounting to a majority or minority of the whole, who are united and actuated by some common impulse of passion, or of interest" is similar to the modern definition of an interest group in its broadest sense. For Madison, causes of faction included both ideology ("a zeal for different opinions concerning religion, concerning government"), and economic interests. It was the latter, however, which he most feared:

> But the most common and durable source of factions has been the various and unequal distribution of property. Those who hold and those who are without property have ever formed distinct interests in society. Those who are creditors, and those who are debtors, fall under a like discrimination. A landed interest, a manufacturing interest, a mercantile interest, a moneyed interest, with many lesser interests, grow up of necessity in civilized nations, and divide them into different classes, actuated by different sentiments and views. The regulation of these various and interfering interests forms the principal task of modern legislation and involves the spirit of party and faction in the necessary and ordinary operations of government.[1]

Madison felt that establishment of the Federal government, by expanding the sphere of interaction, would limit the ability of factious combinations to do mischief to society. "Extend the sphere and you take in a greater variety of parties and interests; you make it less probable that a majority of the whole will have a common motive to invade the rights of other citizens; . . . where there is a consciousness of unjust or dishonorable purposes, communication is always checked by distrust in proportion to the number whose concurrence is necessary." Hamilton, in essay 35, is more explicit about the limited ability of various economic interest groups to be represented in government. Hamilton surmised that the actual representatives would consist of merchants representing manufactures and artisans as well as their own interests, learned professionals, chosen by reputation, and landowners of various sizes. He concluded that even this limited division of interests was sufficient to check abuses of the taxing power of the new government.[2]

It could be said though the modern concern with interest group theory began with Arthur F. Bentley, in his seminal work, *The Process of Government*, published in 1908, it was not until David Truman returned to Bentley's work in his *The Governmental Process*, first published in 1951, that group theory reemerges as a central issue in American political science. During the period between these two books the most influential work, especially with regard to economic interest groups, was E.E. Schattsneider's *Politics, Pressures and the Tariff.*

Schattsneider examined the interest group pressure involved with the determination of tariffs with the Smoot-Hartley tariff bill. He saw public policy as the product of "effective demands" upon the government. In many ways he anticipated Mancur Olson's classic work on group action:

> It seems probable that the spontaneous political response to economic interest is far less general than is commonly supposed. . . . In general, it may be said that the capacity of a group for organization has a great influence on its activity. In this respect great numbers are usually a handicap, though large

> groups organized for nonpolitical purposes may sometimes become effective in politics.
> . . . Selectivity is the basis of pressure politics. If all interests and all groups were equally active, pressure politics would be futile. The assumption is that a few can exert great influence on the process of government because they are organized and because they are alert and have access to information, know what they want, and have power in the economic regime outside the formal organization of government, while the mass remains inert. . . .[3]

Schattsneider's pessimistic view of interest group politics was supplanted by a more optimistic view, shared by certain political scientists in the 1950s, of the positive aspects of political power shared among a wide array of interest groups. Exponents of pluralism argued that the dispersal of governmental power yielded numerous benefits for society, including political stability, widely distributed political satisfaction, and a process of negotiation and compromise between interests that promoted political moderation and acceptance of others' reasonable claims.[4]

The Governmental Process (1951) was the pivotal work in the group theory tradition, reintroducing Bentley's approach to the study of group politics along with the recognition of political and social constraints on the actions of groups. Truman saw group politics as a dynamic process in which constantly changing patterns of relationships produce continual shifts in relative influence. The increased importance of group politics in the American system was attributed to the growth in complexity of society and the accelerated pace of change, resulting in disturbances affecting the interests of various groups. As organized political interest groups appeared, their success in influencing government motivated other groups to protect their own interests.[5]

Truman noted the importance of influencing public opinion, access to points of power, and control of the scope of debate in the conflict between political interest groups. An important group strategy was to control the size of the public on a given measure, with propaganda as a main tool for extending the public, especially when the group can appeal to the "national interest." Truman was aware of the superior resources available to business groups in influencing public opinion, with the caveat that this power was subject to a discount when applied to the political arena. Along with mobilization of publics, the other key element in influence in Truman's schema was access to key points of decision in the government. Politicians facilitate access in order to obtain both technical information defining the content of a policy issue and political knowledge of the relative strength of competing claims.[6]

The structure of the government gave an advantage to groups defending the status quo. Most issues involve only limited publics, as there are few proposals in Congress that receive extended debate and attention in the mass media, and little attention is paid to detailed provisions. This gives an advantage to the well-organized, cohesive group with skillful leadership. The factors limiting group conflict were overlapping membership, the existence of potential interest groups, and social expectations. Overlapping membership limited the intensity of conflict as accommodation of the divided loyalties of significant segments of groups potentially reduced the intensity with which groups pursued their interest. Potential groups were postulated as a check on interest group conflict, as they could be prompted to organize if these submerged potential interests were sufficiently threatened. Truman considered the underlying expectations of society, the rules of the game, as the most important check on interest group activity. Since members of groups also possess these attitudes, interest groups face both internal and external limits to behavior.[7]

Grant McConnell expressed the more pessimistic view of political interest group influence on American politics which emerged in the 1960s. He traced the strength of interest group influence to the deep seated faith in the virtue of small units of social and political organization, in contrast to the prevailing view of government as a threat to liberty. McConnell explicitly rejected this ingrained attitude, fnding that "organization of political life by small constituencies tends to enforce conformity, to discriminate in favor of elites, and to eliminate public values from effective political consideration. . . ." The representatives chosen in small local units are accountable to narrow constituencies and more likely to take specific and extreme positions favoring particular interests. Minority interests and especially the diffuse interest most commonly called "the public interest" will be discriminated against in a system of many small constituents. Instead of the mutual checking and resultant compromise associated with large constituencies, logrolling becomes the means of accommodating the various narrow interests associated with small units. The importance of centralization is the support of public values, public in the sense that they are shared by broad constituencies and therefore must be achieved through mobilization of large constituencies. The major difficulty with Congress is that it embodies the localist aspect of federalism, constrained only insofar as the party system and the presidency are influential.[8]

In recent years many political scientists have become skeptical of both the pluralist view of interest group influence as benign, and the pessimistic view which sees corporate interests and the wealthy as dominate in policy formation. This skepticism arises principally from empirical studies of interest groups participation in the legislative process. Pluralist optimism about the balance and universality of group representation is contradicted by the evidence. Registered interest groups disproportionately represent corporate business interests despite recent growth among citizen, civil rights, and social welfare groups. The extent to which one's interests are represented also tends to vary positively with wealth. Schattsneider, as early as 1960, pointed out that party politics played a far more important role than interest group theorists had surmised. While the "flaw in the pluralist heaven is that the heavenly chorus sings with a strong upper-class accent," the success of pressure groups was also exaggerated. Public conflict was the last resort of weaker interests who wanted to enlarge the scope of conflict. Powerful business groups tended to avoid political conflict, preferring to subsume their differences in search of a consensus. While "potential groups" cannot redress the balance of power, neither can powerful groups dominate policymaking.[9]

A case study of tariff politics, by Raymond Bauer, Ithiel de Sola Pool, and Lewis Dexter, showed a very different pattern for the period from 1953 to 1962. The authors rejected the pluralist belief that legislative policy is a product of the balance of interest group power, finding that interest groups were weak and under-financed, tended to offset each other, and thus left legislators substantial freedom to follow their own inclinations. "When we look at a typical lobby, we find that its opportunities for maneuver are sharply limited, its staff mediocre, and its major problem not the influencing of congressional votes but the finding of clients and contributors to enable it to survive at all."
Lobbying proved unimportant compared to other functions of pressure groups. Their effect in buttonholing, cajoling, and persuading congressmen was far less than their effect in organizing and channeling communications.[10]

Lowi examined the implications of the Bauer, Pool and Dexter study for existing theory in his classic 1964 *World Politics* article. He rejected the pluralist model, substituting a general interpretative scheme, based upon the premise that political relationships are determined by the type of policies at stake, with a distinctive type of political relationship for each type of policy. Lowi splits the possible arenas of power, or areas of policy, into three types, distributive, regulatory and redistributive. Distributive policies are those made without regard to limited resources, such as public land and resource policies; pork barrel type programs, defense, clientele services and the tariff. They "are characterized by the ease with which they can be disaggregated and dispensed unit by small unit, each unit more or less in isolation from other units and from any general rule." Regulatory policies involve a direct choice between individuals by application of a general rule, with decisions cumulative largely along sectoral lines. Redistributive policies are similar to regulatory policies except the categories of impact are broader, involving the transfer of property between aggregates. Lowi considers these short-run distinctions, as in the long run "all government policies may be considered redistributive, because in the long run some people pay in taxes more than they receive in services. Or, all may be thought regulatory because, in the long run, a governmental decision on the use of resources can only displace a private decision about the same resources."[11]

The distributive arena is pluralistic in the sense that a large number of small, highly organized interests are operating, since benefits can be disaggregated so that actors pursue their own self-interest while refraining from opposition toward similar behavior on the part of other actors. The log roll is the predominate form of distributive politics, and the Congressional committees which control the distribution of favors take the role of "broker" leadership. The regulatory arena is composed of groups organized around tangential relationships, with policy a residue of the interplay of group conflict. The power structure in regulatory politics will be less stable than in the distributive arena as coalitions shift when interests change or conflicts of interest merge. Decision-making will be centered in Congress as a whole, as committees and administrative agencies are unable to contain the scope of conflict. In effect, regulatory politics are pluralistic politics. Redistributive politics are the closest one comes to class oriented politics in America. The only type of negotiation which can occur is to determine the extent of the transfer between classes, and bargaining is dominated by elites. Decision-making is removed from Congress and shifted to the national party leadership and the administration.[12]

The primary weakness in the group theory literature is the absence of a model of group formation and action. Pluralist theory, and critiques of pluralism implicitly accept the assumption that political behavior is economic in character; economic in the sense of goal oriented, and that both groups and the individuals within the groups are rational in pursuit of their desired outcomes. The study of economics as the study of the rational pursuit of goals by actors constrained by the scarcity of resources certainly applies to the competition between groups with regard to the formation of economic policy (and to a lesser extent with regard to other types of politics). So it is not surprising that an extensive literature has developed applying microeconomic theory to political behavior.

PUBLIC CHOICE THEORY

There are three works which established the basis for an "economic" theory of politics; Anthony Downs, *An Economic Theory of Democracy*, Mancur Olson, *Logic of Collective Action*, and James Buchanan and Gordon Tullock's *Calculus of Consent*. Downs work, published in 1957, applied the concept of the rational economic man to political behavior.

Economists had traditionally dealt with the economic policy form a normative standpoint in which some notion of public interest was defined to allow identification of optimal policies to achieve this goal. This tradition can be traced as least as far back as the Mercantilists, who wanted to maximize the power of the sovereign, and their opponents, lead by Adam Smith, who wanted to maximize the economic wealth of the nation. Modern welfare economics defined the public interest through some form of social welfare function, usually by aggregating individual preferences to satisfy a set of normative criteria. This literature is closely related to the procedures for aggregating preferences using voting rules in social choice theory. The public finance literature also influenced thinking on policy analysis, as for example, Musgrave's separation of government activity into allocation and redistribution decisions.[13] However, despite these various approaches to public decision-making until Downs no systematic treatment of the implications of rational behavior on politics had been constructed.

Downs began by assuming rationality, which implies that actors have preferences which are complete and transitive, that is, they can rank alternatives and are consistent in doing so, always choosing the most preferred position to others. Downs also incorporated a weak version of rational expectations by assuming that "A rational man who is systematically making some mistake will cease to do so if (1) he discovers what the mistake is and (2) the cost of eliminating it is smaller than the benefits therefrom. . . ." Downs was aware of both the importance of information and the problem of uncertainty in making reasonably accurate forecasts of costs and benefits required for rational behavior. "Some ambiguity is inevitable, but whenever uncertainty increases greatly, rationality becomes difficult." The core assumptions of the model were that every government seeks to maximize political support, that its primary goal is reelection (and election for parties out of power), and that the elected party controls the entire government until the next election.[14]

A political party is defined as a coalition whose members agree on all their goals, which seeks to control the government by gaining office in an election. Party members seek power in order to attain income, prestige, and power, and never as a means of carrying out particular policies. Voters attempt to maximize their utility by choosing the party which provides the most individual and social benefits. The government maximizes political support by equating the marginal vote-gain from spending with the marginal vote-loss from revenue collection.[15] Downs accepts that this is an approximate explanation:

> In practice, no government actually carries out such elaborate calculations. Not only does it lack information about the shapes of individual utility functions, but also it cannot possibly make such staggering calculations for each decision. Nevertheless, the rudiments of this kind of thinking appear in the government's keeping an eye on various groups in society to see how they are doing and to discover what actions should be taken to appease them or ensnare their votes. By simplifying the millions of votes into a small number of blocs, and merging the thousands of acts into a few major policy groups, the government can actually make the kind of recalculations discussed. . . .[16]

Downs observes that under conditions of certainty, a government's best strategy is to adopt choices which are favored by a majority of voters, planning the budget by taking a hypothetical poll on each decision. However, when there is a lack of a strong consensus the opposition can form a coalition of dissenters and win by upholding the minority view on key issues. Therefore majority rule will prevail on specific issues in a two-party system when the majority strongly favors a certain policy but not necessarily on other issues.[17]

The source of uncertainty is lack of knowledge, which provides an opportunity for persuaders to provide facts which are favorable to the group which they support. Downs then defines leadership "as the ability to influence voters to adopt certain views as expressing their own will," requiring uncertainty to be effective. Political parties as leaders develop policies to gain votes and then attempt to convince the voters of the value of those policies. Interest groups as leaders follow a dual strategy; they claim to represent voters so as to convince the government to adopt policies beneficial to themselves while attempting to persuade voters to support those policies. Favor buyers trade resources needed to finance persuasion for specific acts.[18]

Ideologies are useful to voters because they save them the cost of becoming informed, instead simply supporting the party with ideologies most like his own. Reliability requires that policy statements can be used to make accurate predictions of behavior while responsibility implies that policies are consistent over time. When neither reliability nor responsibility exists, no political predictions can be made and rational behavior is impossible. A party which is both responsible and reliable will probably have an ideology which is relatively coherent and immobile. Therefore, while ideologies develop out of the desire to gain office the maintenance of ideologies may become a subsidiary end, especially if a change in ideology is regarded by the public as loss of integrity or responsibility.[19]

Downs applies marginal cost analysis to the decision to purchase information, with the decision-maker acquiring information until the marginal value (the expected benefit) equals the marginal opportunity cost of acquisition. There is a constant flow of free (except for time required for perception and assimilation) and partially subsidized (by purchasers of advertised products or government) information. The major drawback of using these data is that the selection principles embodied in the data may differ from those which would be optimal for the decision-maker. Citizens acquire political information to help them decide how to vote, and to form opinions with which they can influence government policy formation during the period between elections. It seems probable that for many citizens rational behavior excludes any investment in political information since his vote has almost no chance of influencing the outcome. Most citizens will rely on free information received in the course of nonpolitical pursuits, and the cost of acquiring information and communicating opinions to government will determine the structure of political influence.[20]

While Downs focused on the economics of government, voting and the cost of information, Mancur Olson was concerned with the economics of group activity. Olson's most important contribution was to show how the internal dynamics of groups, under the assumption of rational individual behavior, determined group behavior. Olson challenged the traditional theory of group behavior, derived from the social psychology literature, which assumed that private organizations and groups are ubiquitous, due to a fundamental human propensity to form and join associations, or were an aspect of the evolution of modern, industrial societies. The key element in his argument revolves around the indivisibility (and therefore difficulty of exclusion, at least to group members) of a collective good. If members are rational, they would only be willing to pay for a collective good up to the point that their marginal cost of the good equals the marginal benefit to themselves. This will lead to a suboptimal result since ideally the group should equate total marginal cost with total marginal benefit. Instead, most individuals will stop contributing well before the optimum supply of the good is obtained. In large groups if there are lumpy costs (quasi-fixed costs) the cost may exceed the benefits of the collective good for each individual and none is provided. In small groups, the most extreme situation would have one individual providing the good, and the other members would "free ride."[21]

Olson then develops a logic of group structure from the need to provide incentives or coercion to overcome the "free rider" problem. In a small group where one member would be willing to provide the collective good there is little need for organization. The intermediate case is the group in which no member is willing to pay the whole cost of providing the good, but each member is important enough that their defection would have a noticeable effect on the costs to other members. At this point some sort of organization or coordination is required, sufficient to motivate a subset of the group into providing the good. In a large group in which no single individual's contribution has a noticeable impact on the organization, some form of inducement and coercion is required to lead the members in acting in their common interest. As the size of the group increases, one must include the cost of organization along with the cost of obtaining the collective good. This includes costs of communication among group members, the costs of bargaining among them, and the costs of the formal group organization. When the group has already been organized for other purposes significant savings in organization costs are obtained. However for large groups the initial cost of organization adds an additional barrier to group action.[22]

The extreme situation is the very large group, which Olson refers to as the "latent" group. Latent groups require separate and selective incentives for group actions such as social sanctions and rewards. One method of generating social pressures is to employ propaganda transmitted through the mass media, but this requires a pre-existing organization and resources. The primary flaw of pluralism was that it took for granted that economic interest groups will act to defend or advance their group interests, ignoring the problem of organizing large, latent groups. Those large and powerful economic lobbies which exist are usually the by-products of organizations that perform functions in addition to lobbying, such as trade associations. Olson accepted that his theory was limited in applicability to ideologically motivated groups, such as the occasional band of committed people who work for admittedly lost causes. He observed the relatively minor ideological sacrifices made for political parties in the United States, and the benefits provided by parties for people with personal political ambitions in the form of public office, with 700,000 officials elected.[23]

Terry Moe dropped the assumption that individuals are perfectly informed about marginal costs and marginal benefits, and he assumed that behavioral expectations were contingent upon perceptions. An individual is more likely to contribute the higher his estimate of marginal benefits, the lower his estimate of marginal costs, and the lower his estimate of the total level of supply of the collective good, that is, if he has a perception of personal efficacy. Group leaders then have the potential to manipulate information in order to attract members and contributions. Socialization which leads individuals to believe in political efficacy enhances the likelihood that they will think that their contributions make a difference. If estimates of efficacy are randomly distributed, those who tend to overestimate will join groups for political reasons even though a person with an average perception of efficacy would refuse to contribute. This implies the existence of a sizable resource base for organization by a leader able to locate individuals who overestimate their political efficacy. The ability of public interest groups to organize and obtain funding in the 1970s and 1980s attests to the empirical basis for this conclusion.[24]

The third major work in establishing the economic perspective of political behavior was Buchanan and Tullock's theoretical treatment of political philosophy. They made little distinction between individual and group interests: "Throughout our analysis the word 'group' could be substituted for the word 'individual' without significantly affecting the results." The authors reject any concept of the public interest apart from, or independent of, the separate interests of the individual participants in social choice. Collective action is then the product of individual action in choosing to accomplish purposes collectively, and government is only the set of processes which allow collective action. Group decisions are the product of individual decisions combined through a specific rule of decision-making. The decision-making rule adds an element of uncertainty to individual choices, as the final outcome is unknown when the individual choice is made, unless the individual vote can be exchanged for votes in later choices. The reason for collective action is the reduction of external costs imposed on the individual by private action. The other costs which individuals take in account are the costs of participation in organized activity, which they refer to as decision-making costs.[25]

The individual balances the costs that the actions of others impose on him with the costs that he can expect to incur through the operation of a rule to prevent those actions. The individual, aware that collective decisions which modify individual human or property rights may impose very severe costs on him, will tend to place a high value on the attainment of his consent. With regard to other potential collective activities, including those most characteristically undertaken by governments, the individual will recognize that private organization will impose some interdependence costs on him, and he will support a shift of such activities to the collective sector. Since time and effort will be required to participate in these decisions, and these costs increase with the size of the group required for decisive action, he will try to choose a decision-making rule that will minimize both the costs imposed on him and those which he will incur as a decision-maker. In the political process, opportunities for bargaining will result in waste as each individual, attempting to minimize his costs, invests resources in bargaining which result in a higher total cost of decision-making. Therefore specific decision-making rules are required to reduce the inefficiencies associated with bargaining.[26]

Since a vote has an economic value the individual will be interested in opportunities to sell or trade votes, which is usually accomplished through the process of logrolling. A system of voting provides each individual with an equal "political" income, but given a minority that feels more intensely about an issue than a majority, logrolling will take place as the minority will trade votes on other issues to detach marginal votes from the majority. Explicit logrolling occurs when a small number of individuals vote on a continuing sequence of measures, a common occurrence in representative government. Implicit logrolling can occur when large bodies of voters must decide on a complex issue. Political entrepreneurs develop a mix of policies to attract support, taking advantage of the fact that certain voters feel strongly enough on one issue to vote for a party despite the fact he opposes the party stand on other issues.[27]

A majority rule allows members of the decisive coalition to impose external costs on other individuals. The members of the majority coalition will receive larger shares of expected benefits and/or smaller shares of the costs. Therefore collective decisions reached by a less-than-unanimity rule will always include a redistributive element. Vote-trading, by producing something akin to side payments, will reduce the resource-wasteful aspect of majority voting. If full side payments are allowed to take place, any decision-making rule for collective action will lead to positions that may properly be classified as Pareto-optimal. Strong and effective ethical and moral restraints which prevent vote-trading make it essential to place constitutional curbs on the political decisions of the majority.[28]

Public choice theory has evolved into a formal theory of political decision making, especially the literature on collective choice.[29] Like microeconomic theory, the increase in formal rigor and intellectual elegance has been associated with decreasing relevance to the real world policy analyst. There is a more explicitly economic approach to politics, reflected in the literature on economic rent seeking and the political economy of trade and regulation. This literature basically perceives the politics of economic policy as the pursuit of profit by means of investing resources to persuade governmental actors to permit them to obtain transfers or economic rents. It is an extension of the bargaining process of Buchanan and Tullock.

THE POLITICAL ECONOMY OF RENT SEEKING

The economic explanation for economic regulation and the rent-seeking literature, which are closely related, evolved as a response to traditionally theories of regulation in the public interest. Public interest theory, under which the government attempts to improve the general welfare by financing public goods and correcting instances of market failure, can be traced back to at least the progressive era. Antitrust, health regulation and railroad regulation are among the types of activity which can be explained by some form of public interest motive.[30] The public interest view emphasizes the importance of ideology and the desire to make good policy, which are seen as motivating legislators to seek to improve society.

Economists generally defined a public interest policy as one which maximized social welfare, whether through explicit maximization of some social welfare function or correction of perceived distortions in the economic system. Theoretically, any pareto-optimal decision (one which made one person better off without making anyone worse off) would be in furtherance of the public interest. Economists worked out an approximation where if the beneficiaries of a policy could potentially compensate the losers, the policy was a potential improvement over the status quo, the basis for cost-benefit analysis. There are two problems with this methodology. One is that a policy could change income and prices such that reversal of the policy ex post would also be socially efficient. A more important objection is that the compensation principle requires the assumption that winners and losers are randomly distributed over policies and there are no distributional impacts. In the real world winners and losers tend to be segregated by class, occupation or region, and we must ask whether the value of the gain in utility to one group fairly balances the lost in utility. Since there is no method for comparing the satisfaction of various people, and, even if we could determine the preferences of each person for social policies, there is no unique set of policies which would be chosen under majority rule, no unique optimal set of policies can be shown to exist.

This is only a problem if we wish to determine an optimal set of policies. Politicians do not do sophisticated calculations of social preference aggregation to determine whether a policy seems to be in the public interest. They have a general sense of what constitutes the public interest, shaped by their own experiences as members of the society and culture, sharpened through self-selection as monitors of public sentiment via elections. It is this perception of the public interest that constitutes the political ideology of the policymaker. Economic theory matters only to the extent that it influences the policymaker's perception. However, the fact that a policymaker has some concept of the public interest is independent of whether pursuit of the public interest is one of the policymaker's goals.

The public interest approach came under attack by Marver Bernstein in 1955, when he redefined "independence" as lack of accountability, resulting in "capture" by those presumably to be regulated. Regulation was initiated against the will of those regulated, but once the issue lost salience regulated groups could apply political pressure to counter the original intent of regulation.[31] It was a small step from eventual capture of regulation to postulating that regulation was initiated for the benefit of the regulated.

George Stigler applied economic theory to economic regulation, rejecting the views of regulation that saw it as primarily for the protection and benefit of the public at large or as the product of an "irrational" political process, which allowed for acts of great moral virtue concurrent with those of extreme venality. Instead Stigler proposed that "as a rule, regulation is acquired by the industry and is designed and operated primarily for its benefit." He noted that an industry does not usually attempt to obtain money, since unless beneficiaries can be limited subsidies will soon be dissipated among new entrants. Three limits on political benefits restrict their pursuit. Distribution of control of the industry will shift to reflect the political strength of firms, "so small firms have a larger influence than they would possess in an unregulated industry." Government controls add additional costs to firms due to procedural safeguards. "Finally, the political process automatically admits powerful outsiders to the industry's councils."[32]

Limitations of the political process, such as the cost of voting on each issue and the absence of accurate signals as to the true preferences of voters, are dealt with by employing representatives organized in firms called political parties. "The representative and his party are rewarded for their discovery and fulfillment of the political desires of their constituency by success in election and the perquisites of office." Political parties have operating costs, and the industry seeking regulation must be prepared to pay with votes and resources. Resources are provided directly or indirectly through various types of contributions, including programs to generate votes. Stigler suggested that one reason so many politicians are lawyers is that this facilitates extra-political payments, as one simply employs the congressman's law firm. He argued that political competition is similar to economic competition, and that voters attempt to achieve their preferences in political markets.[33]

Richard Posner rejected the contention that economic regulation was motivated by economic distortions due to research that regulation is not positively correlated with the presence of externalities or monopolistic market structure. Posner faults public interest theory for the lack of an mechanism for turning a perception of the public interest into legislative action, and offered as a possible solution Coase's suggestion of the moral distinction between public and private action, where individuals feel little guilt concerning their market decisions but are morally constrained from voting for policies which exploit fellow citizens. A second explanation is the willingness of dominate political leaders to impose their own vision of the public interest on the legislative agenda.[34]

Posner then turned to the economic theory of regulation as postulated by Stigler. He pointed out that the one characteristic which discourages cartelization, a large number of firms in an industry, encourages regulation. The industries with the lowest cost of obtaining protective legislation, a small number of firms with a large number of employees, also have the lowest cost of private collusion, and therefore rarely seek regulation. Posner pointed out that the theory is so spongy that virtually any observations can be rationalized.

Some actions such as oil import quotas could be the product of a combination of public interest and interest group pressures. The difficulty of determining the actual effect of economic regulations raises the question whether participants or observers can identify winners and losers. The rhetoric of public interest policy is also unexplained, along with its role concerning the costs of information in the determination of the quality of legislation and the value of fraud.[35]

Peltzman developed a general model of economic rent seeking in which politicians choose to grant transfers of wealth to maximize their political benefit. Peltzman concludes that both natural monopolies and competitive markets are more politically attractive to regulate than oligopolies because there are more opportunities for regulators to manipulate profits and prices to obtain maximum political benefit. Gary Becker also developed an economic theory of competition among interest groups for political influence, modeling the political game as zero-sum in influence and negative-sum in taxes and subsidies. Pressure groups can purchase votes with lobbying and other political activities and influence depends only on pressures exerted by groups. He agrees with Peltzman that policies reducing deadweight losses are favored since they allow interest groups to obtain more resources.[36]

The development of the concept of economic rent seeking put a sharper focus on the economics of policy, concentrating on economically unproductive activities which presumably were prima facie evidence of policies opposed to the public interest. The initial article in the rent seeking literature was written by Gordon Tullock for the *Western Economic Journal* in 1967. Tullock challenged the conventional measures of the cost of distortions in markets, arguing that using the loss of consumer surplus (deadweight loss) as the measure of loss grossly underestimated the real cost of tariffs and monopolies. He suggested that domestic producers would invest resources in lobbying until marginal cost equaled marginal return on the transfer payments obtainable from the tariff, and there could be additional expenditures by opponents to the tariff. Applying this reasoning to other situations, though theft involves only a transfer payment and therefore should have no welfare effects, the large investments in capital and labor to commit and protect against theft are socially unproductive and thus constitute waste. In the same fashion, the true cost of monopoly is not only the loss of consumer surplus but includes the investment by potential monopolists to establish monopolies, and the cost of government policies to discourage such behavior.[37]

An application of the rent seeking paradigm to international trade was undertaken by Anne Kreuger. Competition for import licenses can take various forms, depending on the decision-rule used by government. Allocation in proportion to firm capacity encourages overinvestment, licenses allocated pro rata in proportion to applications will cause a larger than optimal number of firms, and licenses allocated by government decision will cause resources to be invested in influencing officials, and, if the rewards are sufficient then there will be resources invested in becoming government officials. She estimated that competitive rent seeking in some countries consumed up to 7-15% of GNP. One factor determining the level of rent seeking was the perceived legitimacy of the economic system. If the market mechanism was suspect, due to advantages provided by wealth and connections in rent seeking, there will be political pressure for additional intervention, resulting in additional rent seeking opportunities, resulting in a vicious political circle.[38]

Browning offered a critique of the Tullock concern with transfer costs, resorting to the free rider problem in explaining why there would be limited investment in lobbying for income redistribution or even in forming collusive agreements among firms. When lobbying does occur Olson's observations on group investment in collective goods suggest that expenditures will be less than the potential benefits. Also, some lobbying may provide information and thus cannot be considered a cost. Given the existence of a large conventional excess burden from redistribution policies (administrative, compliance and tax distortion costs), Browning felt it was unlikely that "Tullock welfare losses" were of the same magnitude. Tullock suggested that while there might be less resources wasted by outside lobbying, internally generated lobbying suggested a large amount of waste by legislators and bureaucrats competing for various programs.[39]

McChesney expanded the model by making politicians active participants, offering regulation to collect a share of the resulting rents or threatening regulation to extort payments. Flowers noted that rent creation should be distinguished from rent allocation. Rent creation requires an investment in initial political action, permitting free riding from those who plan to invest in obtaining a share of these rents after creation. Investment in rent creation will depend on the expectation of the quantity of rents received after creation, and uncertainty concerning allocation will reduce initial investment.[40]

Attempts have been made to add further complexity to the rent seeking model of policy formation. Applebaum and Katz extend the rent seeking model by making the regulator's utility function include consumer support. Congleton developed a model in which ideology affected voters' responsiveness to persuasion by rent seeking groups.[41] However, there is a paucity of empirical studies on rent seeking behavior, and associated costs and benefits and their allocation among actors. The assumptions required for these models are too simple to provide a meaningful explanation of actual policy formation. Politicians are modeled as simple economic actors with no personal agenda, no institutions and no diversity. Economic actors know the level of influence purchased with resources, and the economic return from this influence. Agency problems are ignored, as if lobbyists and lawyers were closely monitored employees of the firms seeking favors.

Douglas North contrasted the rent-seeking paradigm with his transactions cost approach to institutions. Transactions costs are defined as the costs of negotiating and enforcing contracts, and institutional arrangements are contractual agreements. The three assumptions of the transactions cost framework are that individuals maximize their self interest, rule making and enforcement is costly, and ideology modifies maximizing behavior.

Measurement costs lead to imperfect enforcement, permitting poor performance or even contract abrogation, but behavior can be influenced by convictions concerning fairness or justice of contracts. Ideology acts as a device to economize on information gathering and processing but also involves judgments concerning the legitimacy of institutional arrangements. In a diverse society there will exist conflicting ideologies, requiring precise definition and enforcement of rules. Agency theory is a major component of both rent-seeking and transactions costs, though the former tends to ignore measurement and enforcement factors that are an important component of the transactions approach. Transaction cost analysis implies that no meaningful standard of Pareto efficiency exists since one cannot specify a least-cost government structure for any economic output.[42]

CONSTITUENTS, INTEREST GROUPS AND LOBBYS

Whether the goal of the congressman is to get reelected, as public choice and economic regulation advocates would assert, or to make policy, access to positions which allow control of the levers of power is of prime importance. The public choice approach to political behavior would assume in the case of congressional politics, the congressman carefully calculates his election probabilities and then "sells" his votes in order to obtain the necessary resources to maximize his probability of election. The motivation of the congressman will be important in determining the kind of policies which he champions. If reelection is the primary goal then constituent concerns will dominate his agenda, with interest groups important only insofar as they either reflect constituent desires or provide resources which allow him to address those concerns. On the other hand, if reelection is merely a constraint in the pursuit of policy goals dictated by either ideology or crass self-interest, then constituents and resources will be important only to the point of satisfying the constraint. While this distinction may seem trivial, there are numerous decisions that a congressman makes which involve subtle trade-offs between election and policy goals. When the consequences for reelection are minor, a policy oriented congressman can choose to follow his own judgment. The congressman, as the voters' agent, has substantial slack under these circumstances.

A key question is just how often a congressman must concern himself with the electoral consequences of his actions. Given the complexity of legislative procedures and processes, the subtle nuances of many legislative provisions, and the lack of attention by constituents this may be a weak constraint on behavior. Miller and Stokes found that constituents had limited awareness of the identity of their representatives, much less the positions that they take on major issues. In fact constituents were not well informed on the general positions of the parties on major issues. While it has been shown that voters recognize candidates' names more readily than they can recall them, there is no evidence that voters' congressional information has increased over the last couple of decades, and even when issues seem to affect voting, it is difficult to separate perceptions of issues from attitudes toward parties and the incumbent.[43]

Fenno found that congressmen perceive a complex web of constituencies, reacting to expectations of the demands of the wider groups but receiving input mostly from the more tightly defined constituencies. A basic problem is that uncertainty concerning the composition of constituents cannot be avoided; weak opponents mean the reelection constituency includes weak supporters, new challengers mean new issues, and population shifts obscure the size of the reelection constituency. However, the representative doesn't passively respond to his perception of his constituency, since the description, interpretation and justification of his actions helps determine the response to his Washington behavior. Most of the members Fenno interviewed believed that no single vote could defeat them but their overall performance could. "There are at most only a very few policy issues on which representatives are constrained in their voting by the views of their reelection constituencies. On the vast majority of votes, however, representatives can do as they wish--provided only that they can, when they need to, explain their votes to the satisfaction of interested constituents."[44]

This is reflected in the difficulty of defeating a congressman over his voting record. General ideological or partisan attacks are common but rarely effective, especially during recent years as incumbents have learned to avoid ideological categorization and responsibility for party decisions when it would hurt them politically back home. Fiorina suggests that representatives have learned to use their offices in soliciting support as individuals by providing constituency services, separating reelection from both party identification and issues. Party identification has declined in importance, as growth in defections has been entirely at the expense of challengers, reflecting the growing advantage of incumbency. House members have been better positioned to take advantage of incumbency, since while Senators are better known, the voter recall gap between incumbents and challengers for the Senate is smaller. Senators receive far more news media attention over which they have limited control, unlike House members who produce most of their own press. The result is that House incumbents are substantially better regarded than Senate incumbents and so have an additional advantage over challengers. In addition there is more money available for Senate campaigns and challengers use paid television more extensively and more efficiently, increasing their ability to catch the attention of voters, an essential ingredient of electoral success.[45]

When incumbents are defeated it is usually due to a well-financed challenger who is able to focus voter attention on items which evoke a negative reaction to the incumbent. In this case voter assessments on ideological and policy dimensions, in the general sense of who would do the better job of dealing with important problems, is strongly related to voting behavior. Campaigning is subject to diminishing returns as congressional incumbents have usually exploited their official resources for reaching constituents, providing limited gains from additional publicity. Spending by the challenger, both to gain name recognition and to generate a negative impression of the incumbent, is far more effective. The incumbent's strategy is to discourage serious opposition by avoiding the appearance of electoral vulnerability, which means raising sufficient funds to intimidate potential challengers. The 1980 elections demonstrated that cultivation of the district is not sufficient to guarantee reelection in the face of a vigorous challenge by a well-financed opponent, since this sort of challenge allows voters to make ideological and policy distinctions between House candidates, producing contests that are more like Senate elections. Jacobson suggests that when conditions appear to favor one party, the favored party will field a larger proportion of formidable challengers with well-funded campaigns, implying that partisan issues can still have a substantial indirect influence on Congressional elections.[46]

The importance of money in elections raises the question of the impact of contributions on the Congressman's behavior. While public choice theorists see this as the driving force behind political behavior the evidence gathered by political scientists is certainly more ambiguous in its conclusions. Reliable figures on expenditures are not available for the era prior to the 1970s, but $25,000 would probably be sufficient for a House campaign around 1960. During this period the lack of available information concerning the sources of campaign funds meant that a Congressman could be easily tempted by a significant contribution, as well as various types of indirect aid. The Corrupt Practices Act required disclosure by candidates and political committees operating in two or more states, and imposed ceilings on expenditures by candidates for federal office, while individual contributions were limited to $5,000 by the Hatch act, but establishment of multiple committees working for the same candidate allowed both limits to be bypassed.[47]

Though direct contributions by corporations were banned, donations to independent campaign committees and private contributions by executives were permitted. This led to the practice of executive "bonuses" which would then be contributed to candidates. Corporations often extended free use of company facilities, including employees loaned to campaigns while kept on the corporate payroll. A notorious method used for contributions was the "double envelope" technique, where each employee's contribution was sealed in a small envelope, and these envelopes were collected together in a larger envelope with the company's name prominently displayed on the outside.[48] There is a sizable body of anecdotal evidence which suggests that political money was widely distributed during this period with generally beneficial results for contributors.

One of the most consistent sources of funds was the oil and gas industry. A $2,500 contribution offered to Senator Francis Case by a Superior oil lobbyist led to the veto of natural gas deregulation in 1956. A $100 a plate dinner for House Minority Leader Joseph W. Martin raised $100,000 from oilmen for the

GOP in 1958, although only 500 persons attended. Senator Robert Kerr used his personal wealth and his contacts in the oil industry in consolidating his strength in the Senate by helping other Senators with their financial problems. Lyndon Johnson began his march to power by arranging for $100,000 in oil contributions to Democratic candidates in 1938. Bobby Baker hinted to Senator Thomas McIntyre (D-NH) that support for the oil depletion allowance might result in the elimination of a $10,000 campaign debt. Senator Frank Moss (D-UT) was offered a $5,000 oil donation to his campaign to be funneled through the Democratic Senatorial Campaign Committee.[49]

People who had worked for or were familiar with oil industry practices in that period are quite explicit about the fact that the amounts of oil money flowing through Washington during this period far exceed contributions during the 1970s. The most blatant use of oil money was the $5 million contributed to President Nixon's reelection campaign in 1972, including illegal $100,000 corporate campaign donations from Gulf, Phillips and Ashland, as well as legal payments of $1,176,500 from Gulf officials ($1,003,000 of this from Richard Scaife, heir to the Gulf Mellon fortune), $261,956 by Amerada Hess, $179,292 for Getty Oil and $166,000 from Socal, and over $100,000 from Sun, Pan Ocean, Phillips, Exxon, and Ashland Oil.[50]

Along with the under the table payments the oil money was funneled into lobbying and other activities designed to extend its influence. The industry reportedly spent $50 million for publicity and to subsidize its lobbying effort, although reported expenditures on lobbying were much smaller than many industries. The API reported total lobby expenses of $39,119 for 1968, though it was paying Frank Ikard, former representative from Texas, a reported $100,000 a year, and had an organizational budget of about $10 million, with 47 out of 255 employees working in Washington. An example of an unreported expenditure was the $500,000 advertising campaign in 1969 in support of the depletion allowance by the Petroleum Industry Information Committee, unconnected with the API despite their overlapping membership. The API continued to report minor sums spent on lobbying, increasing only to $121,000 in 1973, although it spent $4 million on advertising in 1971, $3.6 million in 1972 and $2.5 million on magazine advertising alone in 1973.[51]

The changes in the campaign finance laws in the 1970s, combined with the much greater sums needed to run for office, has greatly reduced the influence of money on politics. The reason is twofold; the requirement for disclosure makes it embarrassing for a legislator to receive large amounts of money from a specific interest group while the limits on individual and political action committee (PAC) contributions forces politicians to raise money from a large number of sources. The changes were initiated with the passage of the Federal Election Campaign Act of 1971 (FECA), replacing the Federal Corrupt Practices Act of 1925, and the Federal Revenue Act of 1971. FECA required each political committee and candidate to report total contributions and identity of contributors for each person giving $100 per year, raised to $200 by the 1979 Amendments. It also explicitly granted the right to corporations and labor unions to create, administer and raise funds for PACs. The 1974 Amendments limited to $1,000 the amount contributed to federal candidates in each separate election, the same limit for independent expenditures to influence elections, and a cumulative cap of $25,000 in contributions by an individual in one year to all federal candidates. Multicandidate PACs were limited to $5,000 per candidate per election with no overall limit on giving.[52]

While oil companies took advantage of PACs they are not in the same league with many interest groups involved in campaign financing. In 1975-76 the Union Oil PAC was third among corporate PACs, contributing $86,190, and Texaco ranked fourth, at $83,300. Of the corporate PACs in 1981-82, Tenneco was #1 with $500,000 contributed, and Amoco ranked at #8 by contributing $200,000, but no other oil company cracked the top ten. In the 1975-76 period the AMA contributed $1.79 million, Dairy Committees $1.36 million, and the AFL-CIO and the Maritime unions about $1 million each, compared to a total contribution of all oil, coal and gas interests of $810,000. In 1982 the Realtors PAC contributed $2.1 million, the AMA $1.7 million, Home Builders $1 million, Bankers and Auto Dealers $0.9 million each, while total oil and gas contributions approached $3 million.[53] On the other hand, when there is a consensus in the industry on some issue, a sizable quantity of money can be concentrated on important congressmen due to both the number of oil PACs and individuals with sizable incomes.

Though significant sums are contributed by PACs there is some question as to importance of PAC money, given the total amount of money raised by politicians. The combination of contribution limits and the growth in the cost of campaigning has forced politicians to access a wide range of contributors, limiting the importance of any single group. The average cost of a House campaign was $51,000 in 1972, rising to $222,600 in 1982, and Senate costs increased from $353,900 to $1,771,200 over the same period. The share contributed by PACs rose to 30 percent of House expenditures but peaked at only 21 percent of Senate costs, and corporate PACs accounted for only one-third of all PAC contributions by 1982.[54]

Studies that account for the simultaneity of votes and money support the conclusion that compared to other influences on congressional voting such as ideology, party, and constituency interests, the effect of PAC spending is small. One reason is the unsophisticated strategies followed by most PACs in supporting incumbents, especially those with seniority and relevant committee members. The oil industry tends to be more willing to support Republicans and challengers than most industries. 72 percent of the $2.1 million contributed in 1977-78 went to Republicans, rising to 81 percent of the $4.5 million contributed in 1979-1980. Incumbents received only one-third of the oil money, compared to a 58 percent average for all corporate PACs. An explanation may lie in the group of oil industry PACs associated with people who made it big since 1973, such as the Dallas Energy PAC, the third largest nonconnected (noncorporate) PAC in contributions to candidates in 1982. These PACs were more ideological than the older oil groups, interested in promoting laissez-faire economics as well as supporting measures like decontrolling energy prices and ending the windfall-profits tax.[55]

Their influence may be limited by the same factors which weaken the impact of some trade PACs. Wright found that the organizational arrangements that allow these PACs to raise large sums of money, by soliciting contributions at the grassroots level, gives local officials discretion over allocation decisions. This encourages congressmen to cultivate favorable relationships with the local officials, undercutting the influence of professional lobbyists. Supporting this conclusion is the pattern of giving by the oil industry which tends to make its influence less than its ability to raise money would indicate. The $3 million contributed by 183 oil and gas PACs during the 1982 election cycle was the most of any single industry group. However the pattern of contributions indicates that substantial amounts of money went to congressmen who were guaranteed to vote with oil interests for either constituency or ideological reasons irregardless of the source of their funding. In 1982, Senator Orrin Hatch (R-UT) received $77,615 from oil interests; Senator Richard Lugar (R-IN), $65,133; Senator Malcolm Wallop (R-WY), $59,400; Senator Harrison Schmitt (R-NM), $52,750, Lloyd Bentsen, $83,408, and Phil Gramm, $66,750.[56]

Although PAC contributions became an important sources of funds during the 1970s, individual giving was still the primary source of political money. While executives of large oil companies may have accounted for some funds via this method, it worked primarily to increase the importance of the "small" oilmen. In the period from 1973 to 1977 most oil related contributions were from independent producers, drilling contractors, and oil and gas attorneys. Individual contributions are often made for ideological reasons but obtaining access to politicians is an important motive. Candidates may find themselves with a difficult trade-off between attachment to emotional symbols and causes, such as environmentalism or new right issues, which increase their ability to raise funds but may have a negative impact on their ability to win votes.[57]

Whatever the source of political money the empirical evidence linking money to congressional voting during the 1970s is both weak and issue specific. Studies examining floor votes found that members' ideological outlooks were better predictors of votes than campaign contributions. Examination of narrowly focused economic regulation bills found a weak relationship between contributions and votes. During the 1981-1982 period, despite the large increases in business contributions most attempts to pass pro-business regulatory changes failed. Even when there is a vote where campaign contributions seem to have a significant influence, it is necessary to separate the impact of contributions from the effect of lobbying activities. The resources devoted to lobbying are far greater than election contributions, as single votes can motivate the expenditure of millions of dollars by interested parties. Many lobbyists feel extorted by the demands of politicians, as their increased needs for funds has them raising money almost continuously. Some of the major oil companies are among the PACs tired of the pressure, as they are aware of the minor impact of their contributions in million dollar campaigns.[58]

One could describe campaign contributions as merely another method of lobbying, since the ultimate aim is to influence a congressman's decisions. Whether this is accomplished by expenditures on advertising, initiating, organizing and stimulating constituent responses, provision of information, or by direct contributions to either the congressman or potential opponents, for the interest group the major consideration is to use its resources efficiently. As competition for influence increased, new strategies have been developed.

There has been a proliferation of interest groups operating in Washington, as new groups have formed and older groups have established offices in the last twenty years. Most of these groups have increased their activity over this period, but they attribute increased governmental intrusions, and not the efforts of their competitors, as the reason for their political involvement. The other reason for increased activity were the reforms in Congress which altered the environment of legislation formation by increasing the number of access points. Classic lobbying techniques such as contacting officials, testifying at hearings, providing technical information and helping to draft legislation are still the dominant means used by interest groups, but they also employ constituent based strategies such as grassroots lobbying and working with the mass media to publicize issues. It is generally accepted that "old boy" networking is no longer sufficient, and a more professional approach is necessary for persuasion. Interest groups are also more aware of the advantages of coalition formation with other groups.[59]

The advantage of coalitions is to increase resources available for persuasion and to minimize conflict before issues reach the legislative arena through a process of interest aggregation. By combining demands into a smaller number of policy alternatives followed by bargaining to determine common interests, interest groups can reach a consensus allowing formation of a wide coalition in support of a policy. These ad hoc coalitions tend to be transitory, disappearing once the issue in dispute is resolved. More permanent coalitions can form, referred to as "peak associations" in the case where there is an attempt to promote a united front in the pursuit of a set of policies, such as labor/business conflicts.[60]

The oil industry has had difficulty in maintaining this sort of united front as there are so many actors with conflicting interests. Even before the energy crisis there were a number of major oil lobbies who spent much of their time at odds. The API was the industry's umbrella organization, which limited its activities to issues such as the depletion allowance and intangible drilling expensing, since the only group hurt were the taxpayers. The IPAA was the major group representing independents, though many of its members are large enough to have world-wide operations. The regional organizations such as TIPRO, the Mid-Continent Oil and Gas Association and the Western Oil and Gas Associations tended to be more homogeneous. Gas pipelines were represented by the Independent Natural Gas Association of American (renamed the Interstate Natural Gas Association in 1974) while gas producers had the American Gas Association (AGA). Municipally owned gas companies organized the American Public Gas Association (APGA), which while not registered as a lobbyist in the 1960s could not be ignored as its constituents were very sensitive to the price of gas. Refiners were organized in the National Petroleum Refiners Association (NPRA) and independent marketers protected their interests via the National Oil Jobbers Council. The larger oil companies had their own lobbyists at work in the capital on company specific issues. The industry also worked with groups representing general business interests such as the National Association of Manufacturers or the Chamber of Commerce.[61]

CONGRESSIONAL POLITICS

The literature on the impacts of constituents and contributions on congressional voting casts doubt on the explanatory value provided by economic models of politics. Before judging the extent of their applicability to energy politics it is necessary to discuss the environment in which economic policy is formulated in the American political system. Microeconomic policy, while influenced by the President (as the most important potential political entrepreneur), is determined through Congressional action. In turn, Congressional action is molded by the institutional framework of the two chambers. The changes which have occurred over the past few decades in Congress have had significant repercussions on the shape of energy politics. The politics of energy in the 1950s and 1960s was shaped by the operation of Congressional institutions, and the Congressional reforms of the 1970s had a sizable influence on energy policy during this latter period.

One of the major influences on Congress has been the tremendous workload that is the product of the volume of legislation and the need for continual campaigning. Between the 80th and 90th Congresses the number of bills introduced in the House increased from 7,163 to 20,587, while in the Senate the increase was from 2,945 to 4,199.[62] The increased volume of legislation created pressures to streamline the process, as well as limiting the ability of legislators to monitor many of the issues on which they are expected to vote.

In the House the large number of legislators creates a need for formal rules to control debate and the introduction of legislative measures. In the Senate the means used to make the workload manageable are more informal in nature. Traditionally, Senators were expected to serve a proper apprenticeship, to keep their mouths shut as freshmen and follow the lead of their elders. Among the mores of the Senate were the need for specialization on a few topics, willingness to reciprocate and bargain on bills (i.e., logrolling as a social more), and to choose diligent service over publicity seeking. Even in the 1950s these rules were bent by those with political ambitions or a large complex constituency which required more attention to reelection and therefore a need to seek publicity and engage in activity on a wider range of issues. The larger size of the House requires a more hierarchical structure and organization than the Senate. Leadership is more powerful, with the Rules Committee controlling the flow of legislation. House members serve on less committees, chair less committees and subcommittees, and influence fewer issues than Senators. Traditionally House members served a longer apprenticeship period and have a lower probability of rising to a position of leadership. House members also represent a narrower constituency relative to that of most Senators.[63]

While party leaders possess limited formal powers in the House, their knowledge of procedures, control of rewards, supply of information and influence with members provide resources for shaping the legislative program. The Speaker, as leader of the majority party, is the presiding officer of the House, and exercises a series of procedural controls over House activity. Since 1910 the Speaker's formal authority has been modest, and successful Speakers such as Sam Rayburn have been men who were skillful at dealing with their fellow House members. The mere fact of being the leader gives the Majority Leader superior access to information about both legislation and procedure. Strong leaders will reinforce their power by helping members make legislative records that will aid them in reelection and granting or withholding favors that make a legislator's role easier and more effective. A limit on leadership power is the independence of committees which can act too slowly or not at all, altering legislative proposals and changing the whole tenor of floor action.[64]

A fundamental limitation on majority party influence in the House until the 1970s was the inability of its leaders to select committee chairmen, usually determined by seniority. On the Democratic side, committee assignments were made by the Democratic members of the Ways and Means Committee, along with the Speaker and Majority (or minority) Floor Leader. Members seeking assignments channeled their requests through the senior member of their state party delegation. Manipulation of the selection of Democrats for Ways and Means by the Democratic caucus was one of the keys to a Democratic Speaker's powers. The Republican Committee-on-Committees was composed of one member from each state with a republican representative, with assignments handled by a subcommittee composed of senior members, mostly from large states. Personal judgments of party leaders were the key to major committee assignments, providing rewards for loyal members and placating those too influential to disregard. Membership was exclusive on the three most important and prestigious committees, Appropriations, Rules, and Ways and Means. To be selected for these committees a legislator needed to have the respect of the party leaders, which required moderation in his approach toward legislation.[65]

Policy leadership for the parties comes most importantly from the President. To the members of his party in the House, his program provides a unifying, centralizing influence, and targets to shoot at for the members of the other party. He is therefore cast in the role of a negotiator with the elements of his party coalition and, when necessary, with elements of the other party coalition. The same holds true for the Senate, with leaders in both parties concerned with maintaining party unity on discreet pieces of legislation, conceding to the President the role of developing a program of legislative action. The Senate Democratic party leadership in the Senate was traditionally highly personal and informal. Senators Barkley (1947-50) and Johnson (1953-1960) were basically the Democratic party organization, while other leaders had less power, and depended on formal party agencies for generation of support. The Republican leadership was more formalized and decentralized, with the Floor Leader, Conference chairman, chairman of the Policy Committee, and chairman of the Committee on Committees held by different men.[66]

Seniority in the Senate was an important factor in committee assignments with Senators ranked according to the length of their uninterrupted service. In 1953, Democratic Leader Lyndon Johnson departed from rigid application of the seniority rule with the so-called Johnson Rule providing that no member of the party, regardless of seniority, would receive a second top committee seat until every Democratic senator had been given at least one such assignment. Republicans, less inclined to centralize power in one party leader than the Democrats, did not adopt a similar procedure until 1959.[67]

The system of standing committees is the mechanism for screening most of the bills and resolutions introduced in the House. Hearings are held on a small fraction of these bills, and only a portion of those will then be sent as a modified bill out to the floor of the House for final action. The full House usually accepts the version of the bill produced by the committee, especially when the issues are technical and complicated, the subject is noncontroversial, or committee members are united. Committee claims to expertise stem from the talents of their leaders as well as the abilities of their professional staffs. Committee chairmen have a formidable set of prerogatives over procedure, agenda, staff, hearings, subcommittee creation, membership and jurisdiction, giving them a potential for influence, though reduced somewhat by the reforms of the 1970s. A chairman's actual influence in the House depends upon both the prestige of his committee and the extent of his control over the committee. A chairman who is the acknowledged expert in his field, skillful at political maneuver and who exploits his prerogatives can dominate his committee or subcommittee. Many important committee leaders do not hold the position of chairman but are subcommittee chairmen, ranking minority members of committees, and occasionally members who hold no formal committee position.[68]

Subcommittee and committee leaders are involved in the initial stages of formation of legislation; committee leaders dominate review, reformulation, and recommendation of legislation; the Committee on Rules and a few party leaders control the agenda stage; committee leaders, party leaders, and a cluster of other interested members dominate the floor debate and amending stage; committee leaders speak for the House in the conference committee. Therefore supporters of a given bill must build a series of majorities, in the substantive committee, in the Rules Committee, on the floor, and in conference to be successful while opponents of a bill need only a single majority at any stage in the process to achieve their ends.[69]

Senate party leaders make more use of policy committees in scheduling than do House leaders. The majority policy committee of the Senate could be considered a distant parallel to the House Rules Committee, performing some scheduling functions though with far less control over sequence and timing than the House committee. Bills reported from committees are placed on the Calendar in the order in which they are received, but more important legislation is taken up out of order by special action of the Senate, either by majority vote or unanimous consent. A committee may be by-passed by means of suspension of the rules, for example when a House bill is laid before the Senate for reference, though such a motion is debatable and requires a two-thirds vote. Another means of by-passing a committee is to bring up a measure in the form of an amendment to a bill that is up for consideration. General appropriations bills and bills considered under the cloture rule are subject to points of order as to germaneness, while unanimous consent agreements frequently provide that amendments must be germane. Beyond this there are no Senate restrictions as to the relevancy of amendments, even on House bills, as precedent denies members of the House the right to challenge the germaneness of Senate amendments.[70]

Once the bill has passed both houses, if differences exist in the two versions it is necessary to go to conference to iron out a compromise. Prior to 1965 a Senate-amended measure could be sent to conference in the House only by unanimous consent or a special rule from the Committee on Rules. As a special rule was usually needed for controversial measures, the Rules committee could instruct the conference committee how to handle Senate amendments or even refuse a rule. In 1965 the 89th Congress gave the Speaker the power to recognize a member of the relevant committee for the purpose of moving that a measure be sent to conference by a majority floor vote. In the Senate, the majority leader makes a motion to send a House-amended measure to conference which is confirmed by a majority vote. Conference committee members are appointed by the presiding officers of their respective bodies, though usually accepting the nominations of committee chairmen. On important matters, the senior members of both parties, including the chairman of the subcommittee that handled the bill, are designated.[71]

Conference committees are supposed to deal only with matters in disagreement, but these rules are often violated to facilitate the process of compromise. When one chamber has passed an essentially different piece of legislation and all provisions are considered to be in disagreement, the conference need only come up with a measure germane to the original versions. The final conference report must be agreed to by a majority of the members of each delegation. Pressures from constituents, interest groups, and political parties are kept at a minimum, giving greater freedom to the expertise or prejudices of the senior committee specialists. A conference delegation may receive instructions from the parent body, and it pays close attention to the support its chamber gave various amendments, rarely ignoring a lopsided roll-call vote. Reports must be accepted or rejected without change, and if rejected, may be returned to either the same or a newly constituted conference.[72]

The Legislative Reorganization Act of 1946 determined the post-war structure of the congressional committee system through consolidation of committees and changes in committee organization and procedure. The number of standing committees was reduced from 33 to 15 in the Senate, and from 48 to 19 in the House, committee jurisdiction was more clearly defined, and provisions were made for professional and clerical staff for all committees. Subcommittees in the House decreased from 97 to 60, but in the Senate increased from 34 to about 60. During the next twenty years the number of subcommittees grew to more than 100 in the House and to more than 80 in the Senate. Chairmen in charge of important control committees of Congress, such as the Senate Finance and the House Ways and Means Committees, tended to block formation of subcommittees in order to retain personal control of all legislation at the full committee level. In other committees, the political advantages of subcommittees encouraged their formation, since the chair determined the subcommittee structure and membership, providing rewards for loyal colleagues. This helped the chairman maintain the support of senior members in struggles over control over the committee agenda. Subcommittee members were usually designated by the chairman alone, or in consultation with the ranking minority member.[73]

During the two decades preceding the emergence of the "energy crisis" the two committees which controlled almost all legislation dealing with oil and gas were the Senate Finance Committee and the House Ways and Means (Interior was limited to issues such as leasing of federal oil land and Interstate Commerce had jurisdiction over natural gas regulation). These two committees had jurisdiction over tax and trade issues, which meant they dominated discussion of the oil depletion allowance and other tax breaks as well as oil import quotas and tariffs. Oil issues were not high on the agenda of either committee, since tax and trade bills engendered active participation by practically every economic interest group in the country.

The Finance Committee was the most centralized committee in the Senate due to the consistent opposition by its chairmen to expansion. Both the budget and staff were limited, as Chairman Harry Byrd (1955-64) hired only one professional during his tenure, supplementing his services with the staffs of the Legislative Reference Service and the Joint Committee on Internal Revenue Taxation. When Russell Long (1965-80) succeeded as chairman, he slowly increased staffing and committee proceedings became more open and informal, but he continued to hold tight reins on the committee. Ways and Means, the opposite number of the Senate Finance Committee, was the most centralized committee in the House. Wilbur Mills (1961-74) abolished all subcommittees a short time after he succeeded to the chairmanship of the full committee. When the Democrats on the Ways and Means Committee acted as the Committee on Committees the chairman, especially with a Democratic majority in the House, possessed unique added powers.[74]

The importance of Ways and Means to the Democratic leaders explained their active participation in filling vacancies on the Committee. Approximately one-quarter of Ways and Means members were given the more prestigious assignment without requesting it because none of the applicants were deemed acceptable. Geographical representativeness was an important characteristic of potential members along with a criterion described as a responsible legislative style, consisting of good interpersonal relations and institutional loyalty, and conformity to expectations with regard to free trade, Medicare and the oil depletion allowance, especially during the Rayburn period. On other issues membership tended to be more liberal than the party, as Rayburn favored members who shared his southern populist attitudes, men such as Wilbur Mills, Hale Boggs, Frank Ikard and Clark Thompson. Republicans were consistently chosen from the conservative wing of the party. Ways and Means members were pragmatic in their outlook on politics and inclined to compromise on most issues, usually possessing safe seats which allowed the time required to master the complex subject matter of the Committee.[75]

The make up of the Committee supported the norm of restrained partisanship. Members of Ways and Means, despite serious disagreements over policy, agreed on the importance of maintaining the status of the committee. The care taken to develop a general perception of the Committee's expertise permitted virtually all major bills reported to the House by the Committee to be debated on the floor under a closed rule, precluding floor amendments. The House acquiesced in rules and customs allowing the Committee a special degree of procedural autonomy in order to strengthen it with respect to the Senate Finance Committee and the Executive Branch. The key element in the development of a consensus was the skill of Wilbur Mills as chairman to compromise and cajole the membership until he obtained maximum support for a measure, and acceptance by the whole committee of the strategy to develop a bill which would pass the House. The desire for consensus allowed members to bargain for the addition or deletion of minor provisions which would protect their constituents or aid their friends. Partisan demands from the executive branch and party leaders were contained until the final stages of decision making. At this point full committee executive sessions were held to facilitate development of a bipartisan consensus before sending a bill to the floor. The strategy was highly successful as the Ways and Means Committee won 103 of 109 roll calls in the period from 1953 to 1964.[76]

Ways and Means members saw themselves as responsible legislators in contrast to a Senate characterized by irresponsible logrolling and capitulation to politically popular demands. Interest groups and lobbyists would probably concur with that opinion as the Senate provides an opportunity to appeal House decisions and add new provisions to House-numbered bills, due to the legislative sequence which has the Senate acting in response to a House initiative. Finance Committee Senators are aware of the value of membership for pursuit of both policy and re-election goals. Senators can score political points by offering to aid interest groups in their attempts to change decisions made in the House committee, and their wider constituency increases the probability that some segment will be affected by a bill. Finance Committee members face the concerted efforts of clientele groups, party leaders and the executive branch. The influence of interest groups extended to membership on the committee, an example being the two Democrats who claimed to have been blocked by "the oil interests primarily," which explains why the Committee tended to be more conservative than the Senate as a whole.[77]

CONGRESSIONAL REFORM

The reform efforts of the 1970s resulted in significant changes in the operation of Congress as a whole, the committee system in general and the development of energy policy in particular. The first step was the formation of the Joint Committee on the Organization of Congress to study organizational reform in 1965. Despite quick action by the Senate, resistance to several proposed reforms delayed passage of the Legislative Reorganization Act until 1970, and provisions regarding seniority, electronic voting, reform of the Congressional Record, campaign spending and lobbying reform were deleted by the Rules Committee. The most dramatic change involved the recorded teller vote in the House. Traditionally, most major bills were considered in the Committee of the Whole, and most amendment votes were teller votes in which the number of votes but not names were recorded. This allowed members to avoid accountability, both in reference to how they voted as well as whether they voted. The other major provision was the reduction in size of Senate committees and the limitation of members to two major and one minor committee assignment, with membership on the four major committees mutually exclusive.[78]

The Democratic Study Group (DSG), formed in 1959, was the source of reform efforts in the House. While the organization fought for changes in House procedures almost from its inception, a change in strategy in 1968 involving a push for monthly party caucuses led to the March 1970 caucus resolution calling for creation of a seniority-study committee. Julia Butler Hansen (D-WA) was picked to head the Committee on Organization, Study, and Review. The committee's first set of recommendations were adopted by the Caucus in January, 1971. These included limiting members to one subcommittee chair (stopping the practice of committee chairmen chairing several subcommittees), allowing subcommittee chairs selection of a professional staff member, and allowing a vote on committee membership and chairmen.[79]

The second set of Hansen committee recommendations were the product of a review of a set of proposals drawn up by the DSG and Common Cause. The Hansen Committee recommended a group of reforms in 1973 which were ultimately ratified by the Caucus. Among them were automatic votes on committee chairs, including secret ballots when requested by 20 percent of the caucus, and expansion of the Committee on Committees to include the majority leader, the caucus chairman, and the Speaker as chair. A new Steering and Policy Committee was created to act as a leadership forum for developing party policy positions. A subcommittee "Bill of Rights" was approved which granted subcommittees fixed jurisdictions, guaranteed powers, referral of appropriate legislation and adequate budget and staffing. Each Democratic committee member was guaranteed choice of one subcommittee membership according to seniority.[80]

The adoption of the Hansen Committee proposals was followed by a bipartisan study of the committee system in the House, embodied in the House Select Committee on Committees, led by Richard Bolling. Its report, released in March 1974, proposed to reorganize the House into 15 exclusive (including an Energy and Environment committee) and 7 nonexclusive committees with most members serving on a single exclusive committee. Intense opposition from committee leaders who opposed jurisdictional shifts and interest groups which feared loss of access to sympathetic committees doomed the Bolling proposal. The Bolling plan was referred to the Hansen Committee, where it was replaced with a scheme that limited changes to a series of minor adjustments, and this alternative was adopted in October 1974. The Speaker was granted the power to refer a bill jointly, in parts, or sequentially, to two or more committees. Committees with 15 or more members were directed to establish at least four subcommittees and committee staff size was increased with a third of staffs dedicated to the minority party. However, energy and environmental issues remained scattered and uncoordinated.[81]

In early 1975 reform Democrats moved to strip Ways and Means Democrats of their role as the party's Committee on Committees, transferring the responsibility to the Steering and Policy Committee. The Ways and Means Committee was expanded to 37 members, giving junior and liberal members access to the committee. The Speaker was also allocated the authority to nominate Democratic members of the Rules Committee.[82]

A new attempt to consolidate energy jurisdiction in 1979 was attempted by the Patterson Committee, with an unspoken goal of resolving conflicts with John Dingell's Commerce Subcommittee on Energy and Power, which had come to dominate energy policy. Dingell, Interior Chairman Morris Udall, Science and Technology Chairman Don Fuqua, and Tom Ashley, chair of the Ad Hoc Energy panel, all competed to head the hypothetical energy committee. Affected committee chairmen appeared before the Rules Committee to denounce the plan, and a substitute which changed little other than to rename the Interstate and Foreign Commerce Committee the Energy and Commerce Committee, was adopted.[83]

Senate reform was initiated with the creation of a bipartisan Temporary Select Committee, headed by Adlai Stevenson in March 1976. The Stevenson Committee proposed reducing the number of committees from 31 to 15, including consolidating energy matters from several committees into one with Interior becoming the Energy and Resources Committee. A proposal to combine energy and environmental issues failed, not because of environmentalist opposition (as in the House) but to preserve the domains of Jackson (energy) and Randolph (environment). In February 1977 an amendment was adopted prohibiting full committee chairmen from heading more than two subcommittees on major panels, and one other chair on a nonmajor committee. Committees were reduced from 31 to 24, the number of subcommittees declined by one third, from 174-117, and average assignments declined from 18 to 11 posts.[84]

Impacts on energy policy were centered in the House as changes in the Rules Committee and the Ways and Means Committee affected the development of legislation and the balance of power between the two chambers. A focus of the reformers was to open up the Ways and Means Committee. The mandate to hold open hearings unless closed by a roll call vote, and the secret ballot vote for committee chairmen put pressure on Wilbur Mills. The change which allowed a majority of the Democratic caucus to instruct the Democratic members of Rules to report a rule to allow a vote on an amendment offered a way around Mills's obstruction. The requirement for four subcommittees, removal of the power to make Democratic committee appointments, enlargement of the committee and the change of the party ratio from 3:2 to 2:1 all challenged his authority. However the results were unexpected. Open hearings led not to public attention but increased efforts by interest groups to insert special provisions, countered by public interest group efforts to modify tax bills. The use of open rules was a disaster, as the energy tax bill was gutted on the floor of the House with over 200 amendments proposed. Al Ullman, who succeeded Mills, found himself challenged in the committee, on the floor, and in conference, especially as Russell Long was well aware of Ullman's lack of authority in both the committee and the House.[85]

There was a large turnover in House membership during the 1970s because of the electoral consequences of Watergate, responsible for seventy-eight freshmen Democrats after the 1974 election, and an increase in voluntary retirements attributable to the heavier demands of entrepreneurial electoral politics. The new members were more adept with the new styles of electoral politics and most of them had come of political age during a decade of assassination, violent protest, the Vietnam War, the emergence of consumer and environmental issues, energy "shortages," and a decline in respect for authority. They shared the conviction that they were elected to lead, and these impatient new House Democrats provided the votes that effectively undermined the seniority system, disaggregated legislative authority from the committee to the subcommittee level, solidified the power of subcommittee chairmen, and distributed subcommittee chairmanships and seats on desirable subcommittees more widely.[86]

The result of the reforms was the collapse of the discipline of the House. The destruction of the traditional basis for power, cooperation between a persuasive Speaker and the chairmen of the important committees, resulted in a less cohesive policy process. The augmentation of subcommittee independence and the elevation of numerous members to positions of influence meant that legislative development would require negotiating trades and compromises with additional actors. The increased recording of votes and open hearings meant that interest groups were able to monitor representatives' activity and hold them accountable, weakening party discipline. The increased importance of the media in campaigning encouraged members to latch onto issues which could attract its attention. In the case of energy, the high salience of the issue combined with the overlapping jurisdiction of various committees and subcommittees guaranteed that numerous members would be tempted to become involved with some aspect of energy legislation.

This was illustrated by the events which took place in the Commerce Committee in 1975 as John Dingell and John Moss (D-CA) took advantage of the new rules to stage a revolt against the chairman, Harley Staggers. They removed Staggers from the chairmanship on Oversight and Investigations, transferring the chair to Moss, and gave budgeting and staffing control to subcommittee chairmen. John Dingell's Energy and Power Subcommittee became a major player in House energy policy deliberations, while the Oversight and Investigations Subcommittee devoted itself almost exclusively to energy for the next several years, especially to investigating alleged industry withholding of natural gas. Moss, by using his subcommittee as a platform for opposing energy deregulation, clashed with Dingell's legislative responsibilities.[87]

The separation of authority over energy policy increased the opportunities for input to and obstruction of legislation. The 1975 attempt to develop an energy policy was crippled by the separation of authority. The Ways and Means Committee was putting together an energy tax bill while the Energy and Power Subcommittee was writing a conservation bill, but there was little coordination between the two. Although Dingell's subcommittee approved the Kreuger decontrol and windfall tax amendment, Ullman objected to the windfall profits tax provision as an encroachment on Ways and Means' jurisdiction, while Staggers mobilized the Commerce Committee against the amendment. Decontrol opponents took advantage of these conflicts to prevent the merger of the tax and conservation bills, and to prevent a windfall tax and decontrol bill from being considered together.[88]

Lobbyists were quick to take advantage of these changes. The increased number of subcommittees, the requirement for open meetings and the rapid turnover in congressional membership provided new opportunities for interest groups to exert influence on policy formation. At the same time the increase in overlapping jurisdictions and the elimination of legislative roadblocks made it necessary to deal with far more actors than in the past. The wave of new congressmen also meant that a significant part of the institutional memory of Congress had departed office, while the issues became more technical and complex, making information more valuable.[89] Many of these new lobbyists were associated with public interest groups which benefited from the shifting balance of power in Congress. The new politics, based around gaining media exposure and other forms of positive advertising to constituents, encouraged the rise of political entrepreneurship, both by politicians and by individuals who desire to become involve in setting the political agenda.

POLITICAL ENTREPRENEURS AND AGENDA SETTING

The increased importance of political entrepreneurship is due to the rise of the mass media and its impact on the cost of communication, allowing mobilization of Truman's "latent interests." Down's and Olson's observations on the cost of organizing opposition to private interests are dependent on the cost of organization and the cost of disseminating information. The mass media, by providing a means to reach a large number of people at low cost, and thereby encouraging the organization of Truman's potential interests and transmitting their preferences to politicians, changed the economics of politics. Since the key to profitability in the media business is to obtain a large audience, the issues which will garner publicity are those which are either highly relevant to a great number of people, or those which lend themselves to dramatization. Obviously, a leader who is able to turn an issue into entertainment will have a competitive advantage in the struggle for influence.

Richard Wagner first pointed out that the existence of the political entrepreneur solved the dilemma of large unorganized groups obtaining political benefits, as Olson's system of group activity could not explain why latent interests like the elderly, consumers and migrant farm workers received beneficial public action. By investing resources into reaching these potential members the political entrepreneur hopes to solicit sufficient payments to make a "profit" (usually some other goal than monetary compensation) in return for providing services. The difference that pressure groups make is to accelerate the flow of information, providing political entrepreneurs with the knowledge of "profit" opportunities more rapidly.[90]

One problem in the literature on political entrepreneurship has been the lack of a well formulated definition of entrepreneurship. The most useful description was provided by Joseph Schumpeter. Schumpeter saw capitalism as an organic process, operating within a perennial gale of creative destruction, in which the conventional concepts of competition and equilibrium are meaningless because of their essentially static nature. The role of the entrepreneur "is to reform or revolutionize the pattern of production by exploiting an invention or, more generally, an untried technological possibility . . . To act with confidence beyond the range of familiar beacons and to overcome that resistance . . . define the entrepreneurial type as well as the entrepreneurial function. This function does not essentially consist in either inventing anything or otherwise creating conditions which the enterprise exploits. It consists in getting the things done."[91]

Frohlich and Oppenheimer employed Schumpeter's concept in developing their definition of the political entrepreneur, as the person who bears the initial costs of supplying incentives to collect resources from group members necessary to meet the group's needs. "A political entrepreneur is an individual who invests his own time or other resources to coordinate and combine other factors of production to supply collective goods." Unfortunately they confuse the management function with the entrepreneurial function. "The role of a political leader can be viewed as a job, much like any other . . . The political entrepreneur must balance a complex set of incentives in his business. He must tax. He must reward. He must manipulate information to co-ordinate expectations. And in doing so he must raise enough to cover his cost and adequately reward his efforts."[92] That is a description of management, which while important to a political leader, is not the role of the political entrepreneur. The entrepreneur creates new organizations, raises new issues, employs new technologies and redefines the political environment through his actions.

Robert Salisbury's "entrepreneur/organizer" tends to mix both roles, investing capital to initiate the group but also acting as organizers of offshoots of existing organizations, involved in market expansion more than innovation (for example, forming a farm group organized around a different crop or region than existing farm groups). In either case the entrepreneur or leader must provide benefits to members to encourage their participation. However the distinction is important when discussing the types of policy the organizer will lobby for when the group has a political mission. For Salisbury the organizer has a number of options, either spending his "profit" to further his own policy positions or lobbying to obtain benefits for the group. In the case where the two options are similar there are no decision costs accrued to the leader.[93]

I distinguish the entrepreneur from the leader, with the latter faced with Salisbury's set of choices. The political entrepreneur creates the group in order to pursue some personal political agenda for which the group is instrumental but not determinate; it will not decide her behavior as the reason for its creation was to provide political capital to achieve a pre-existing goal. In fact, the political entrepreneur can operate without a group affiliation if she can convince other actors that she represents a latent constituency which she is capable of mobilizing. Taken to the limit, this could even include the occasional intellectual whose ideas are sufficiently powerful to motivate action by other individuals (Karl Marx). The group leader must take in account the desires of the membership as his primary goal is maintenance of the group and therefore his "profit," and political action is but one means for providing benefits to his constituents. While he may be able to shirk to a certain extent, the group's interests, not his own policy preferences, will be the primary concern of the group leader.

This distinction explains the rise in the number and activity of interest groups over the last few decades. Political entrepreneurs establish new markets (perceived needs or areas for government action) which creates an atmosphere which makes funding available (consumer revenue). An inevitable conflict between entrepreneurs and group leaders occurs when entrepreneur's goals conflict with the desire of existing group leaders to obtain some of these resources (market share) for their constituents. The extent of the conflict and the ability to form coalitions depends on the ability and willingness of group leaders to compromise constituent interests in order to facilitate achievement of the entrepreneur's policy goals.

A problem arises when amorphous policy goals, "justice," "fairness," "protecting the environment," "equal opportunity," "consumer protection," which become popularized by entrepreneurs must be translated in concrete policy decisions. At this point clearly defined winners and losers become apparent, and interest group leaders are able to make cost-benefit calculations with regard to their constituents, and calculate their reaction to alternative actions. The political entrepreneur often finds herself in the unpleasant position of choosing between the lesser of two evils, compromising her political goals or losing the political battle. This is where the professional politician as entrepreneur has a tactical advantage, as the requirements of office teach him to temper personal principle in order to achieve some of his policy goals, while amateurs often self-destruct politically rather than accept the necessity for compromise.

Political entrepreneurs can include legislators, presidential candidates and presidents, party and group leaders, revolutionary figures, and interest group leaders. The entrepreneur often lacks information about who and where potential members are, making communication essential, but he has an incentive to communicate as cheaply as possible by using the most efficient means available for obtaining and exchanging information with clients. Among the methods employed the mass media will reach the most people, but media appeals may not be persuasive, personal, intensive, or specialized enough, though it can often be obtained for free if the entrepreneur can supply "entertainment" value in return for exposure. Direct mail techniques allow specialized communication but at a high cost (Richard Viguerie's use of direct mail established him as both a political and economic entrepreneur). Careful description of political goals, the effectiveness of his political methods, and the need for funds (none of which is necessarily close to the truth), will increase the levels of contributions and political support especially with low-information members who depend heavily upon him for the "facts." Group publications allow creation of markets for selective incentives by advertising new services, especially provision of specialized information likely to appeal to a select clientele, which probably cannot easily be obtained elsewhere.[94]

The two most prominent ways in which political entrepreneurs have operated in the energy arena have been either as a congressman (or the President) employing his power to engage media attention to publicize an issue, or as an outsider who uses the media to organize a latent interest by focusing public attention on an issue. A third method has been the development of intellectual or technical arguments which shift perceptions on a subject, making a non-issue an issue or changing the way an existing issue is perceived. These courses of action are interlinked, as an intellectual argument can motivate a politician to exploit an opportunity to take credit for recognizing a "problem," while outsiders are sufficiently inspired by the revelation to initiate the formation of a group or activation of a latent interest. Both the consumer movement and the environmental movement combined these approaches. In fact, most of the "public interest" groups have emerged from these combinations of intellectual, political and personal initiatives. The three key public interest movements which impacted energy politics were consumerism, the environmental movement, and the "good government" interest groups.

The emergence of the consumer movement began with traffic safety, which also serves as an example of technical initiation of an issue. Although there was an existing policy community which concentrated on improving the performance of the driver, a small group of traffic engineers began to advocate reform of the automobile. The key to converting their technical arguments into political action was the intervention of Ralph Nader, who by publishing the book, *Unsafe at Any Speed* in 1966, generated public pressure. Robert Kennedy was the insider who aligned himself with Nader, both obtaining credit for himself and providing additional media access and credibility to Nader.[95] The emergence of auto safety as an issue was both a catalyst and a product of the consumer movement.

Previous attempts to motivate consumerism as a public issue include the hearings on the prescription drug industry held by Senator Estes Kefauver's Antitrust and Monopoly Subcommittee from 1959 through 1962, a truth-in-lending bill sponsored by Senator Paul Douglas in 1960 and a consumer message by President Kennedy in 1962. Although a drug safety law passed in 1962 due to the thalidomide scare there was no further attention paid to the issue until 1966, and only 19 articles on consumer protection appeared in popular magazines from 1963-65. In 1966 consumer affairs became a hot issue, with Ralph Nader receiving a *Newsweek* cover story as the "Consumer Crusader," and 128 articles on consumerism appeared during the next four years. The chief inside political entrepreneur was President Johnson, eager to seize upon a popular but inexpensive issue. Consumerism was the classic presidential issue in the sense of representing a wide diffuse constituency.[96]

On the outside Ralph Nader was poised to become the spokesman for the movement, providing the hero for the type of morality play (good guys, i.e., consumer activists vs. bad guys, i.e. big business) that the mass media loves to focus attention upon as it makes for good entertainment. Journalists were also interested in promoting the efforts of congressional activists. Congressional activists may have felt that consumer legislation was necessary and in the public interest, but they were well aware that it was good politics, especially health and safety issues, and played well as a campaign theme. Hearings provided an opportunity for committee and subcommittee chairmen to play the role of defender of the public interest in confrontations with business. Congressional activists are convinced that this sort of positive media exposure provides a far greater economic benefit to a politician than the support provided to their electoral opposition by affected industries.[97] The pattern of congressional activity on consumer issues was repeated with energy, as many energy issues could be considered an extension of consumer-industry conflicts.

Consumers Union was one of only two consumer groups in existence before 1968, a classic case of a large group which exists because of the benefits provided members independent of the collective goods which it seeks. The economic value of the organization's publication far exceeds its cost, explaining its large membership (1.8 million subscribers). The group was limited in its ability to lobby on energy issues because of its diverse membership, which raised the specter of internal division if it took strong stands on controversial issues. Given that the organization sees its primary mission as disseminating information about products there was little motivation to adopt the role of a major actor in consumer politics (Ralph Nader resigned from the group's board of directors in 1975 because he couldn't convince the board to take a more activist position).[98]

During the next four years 11 new consumer groups would be organized, but the most important development in terms of energy politics was the Consumer Federation of America (CFA). The group originated with a nominal coalition of groups called the Consumer Clearing House which set up a consumer assembly in 1967. A small group of consumer activists primarily from labor unions, the President's Consumer Advisory Council, and Consumers Union met and decided to establish the CFA in November, 1967, with the intent to serve as an information and not a lobbying service.[99]

The CFA took an active role in energy politics, both through its two lobbyists assigned to energy issues, Lee C. White (FPC chairman from 1966-69) and Ellen Berman, and the efforts of the CFA Energy Policy Task Force, formed in March 1973. Member organizations of the Task Force as of September 1975 included fourteen public power or rural electric cooperatives and organizations, twelve AFL-CIO unions including the Oil, Chemical and Atomic Workers, the UAW and the Steelworkers, five farm organizations, the Conference of Mayors and Consumers Union. The public power membership was a constraint on the CFA with regard to coal and nuclear environmental issues. The group focused on opposition to deregulation and appointments of energy company executives to governmental positions, and support for a federal oil and gas corporation.[100]

The CFA exemplifies the shift from policy entrepreneurship to a leadership focus in an organization. The initial concept of organizing a diverse group of organizations to support consumer issues was an act of political entrepreneurship. However, once this coalition organization was established the conflict between advocacy and organizational survival limited the intensity with which ideological goals could be pursued. The group was able to actively support only those issues for which there was consensus agreement. This meant that consumer issues which adversely affected member organizations would be ignored. In effect the CFA was unable to follow an ideologically consistent political agenda because of the requirement for group maintenance.

Ralph Nader continued to act as an political entrepreneur through the 1970s, creating groups to further ideological concerns, and forming coalitions with other interest groups when it served to further his agenda for action. His various organizations, developed as instruments for implementing his opposition to corporate interests, had no inhibitions toward taking strong positions on energy issues. Nader organized Public Citizen as a shell to collect money for a number of satellite groups, along with financing obtained through speaking fees, book royalties, foundation grants and private contributions. His talent at garnering publicity encouraged congressmen to work with him. Nader's appearance at a hearing ensured publicity for the chairman. As a media star he brought a new level of impact to his efforts at political entrepreneurship, becoming a one-man issue facilitator. Nader in 1973 joined with the Sierra Club and Friends of the Earth in opposing nuclear power. Critical Mass was then formed to further antinuclear efforts. While the need to maintain his basic consumer constituency limited Nader's actions on energy (due to resource limits, not ideological conflicts), anti-nuclear, anti-oil industry actions and opposition to deregulation were all issues well suited for his participation.[101]

The environmental movement had a core of existing groups which only needed to be activated. The Sierra Club, begun in 1892 by John Muir to protect wilderness areas against encroachment, had a membership of around 150,000, mostly in the western states, and influence with the California congressional delegation due to the 78,000 members in the state. Other long established conservationist organizations include the Izaak Walton League, devoted to the interests of fishermen, the National Wildlife Federation, basically a hunters' group, and the Audubon Society. The onslaught of the environmental decade saw the birth of groups such as the League of Conservation Voters, Friends of the Earth, and Environmental Action (which originated out of the ad hoc organization responsible for Earth Day in 1970).[102] These new groups tend to be more activist, but like the Sierra Club their involvement in the energy debate was basically as the opposition to coal, nuclear and offshore drilling. Environmentalists tended to be ambivalent toward the issue of energy prices, as sympathy with ideologically convergent groups in supporting regulation conflicted with the desire for high prices to encourage conservation.

The most important citizens group was Common Cause, organized by former HEW Secretary John Gardner in the summer of 1970, who managed to sign up 230,000 members in the space of one year, and reached 300,000 in early 1974. The organization's original priorities were support for government reforms such as campaign contribution disclosures and the passage of FECA amendments in 1974, and opposition to the Vietnam war. The group lobbied against the oil depletion allowance in 1975 but was unable to take a position on other issues because of its membership divided over energy issues. Since it uses a consensual form of decision-making there was no position on conservation or deregulation which could harness an overwhelming majority that would have permitted action.[103] The reluctance to take a strong position on energy issues was due to a realization by the group members that energy issues were unrelated to the group's perceived mission of promoting "good government."

Amory Lovins was an example of an political entrepreneur whose influence was due to his intellectual skills. His *Foreign Affairs* article, "The Road Not Taken?" which introduced the concepts of "hard path" and "soft path" energy technologies gave respectability to alternative energy advocates, as it was the most requested article in the journal since George Kennan's famous 1947 analysis of the Cold War. Following the article Lovins was asked to testify before a number of congressional hearings on energy issues, further cementing his credibility. Lovins challenged the assumptions used by pro-production spokesman to justify lowering environmental standards and increasing production incentives, thus providing intellectual ammunition for opponents of the energy industries.[104]

The Lovins example points out one of the most important influences on agenda setting during the last two decades, the resources to create persuasive arguments to support a policy position. Knowledge becomes power when decisions on highly complex, technical issues must be made, because of its persuasive effect on politicians in convincing them of the merits of a policy, its importance in defending a vote by a congressman to his constituents, and its power to elicit media support by adding a patina of credibility to a policy. Committee and personal staff are both a source of knowledge and the usual channel for transmitting information from other sources to congressmen, as well as the interpreter of information received from other sources. While staff doesn't always dominate, its opposition can often be fatal to a proposal.[105]

The executive branch, bureaucratic agencies, congressional research arms, private interest groups and research foundations are all active sources of information. The expansion of information has made both the President and congressmen dependent on staff. President Nixon estimated that he personally saw less than one percent of the 42 million pieces of presidential materials generated during his administration, while the average Representative has less than a half-hour a day available to work on legislation.[106]

The importance of knowledge in shaping and setting the legislative agenda can been seen in the rapid growth in committee staffs over the last forty years. In the House, committee staff increased from 193 aides in 1947 to 1,548 in 1976 while the Senate committee staff members rose from 290 to over 1,500 over the same period. At the same time there has been a dramatic increase in both the number and size of the congressional support agencies; the Congressional Research Service of the Library of Congress (CRS), the General Accounting Office (GAO), the Office of Technology Assessment (OTA), and the Congressional Budget Office (CBO). Some agencies such as CBO and OTA undertake major studies only at the request of a committee while the GAO and the CRS may detail staff to congressional committees for assistance. The staff of the CRS increased from 323 in 1970 to 778 in 1975. The GAO grew from 4,471 in 1970 to 5,126 in 1976. The OTA, which began operations in 1973-1974, was established to assist Congress in evaluating the technological implications of proposed legislation and to study and assess the impact of scientific and technological developments. The CBO was established by the Congressional Budget and Impoundment Control Act of 1974 to provide independent economic and fiscal analysis to the Congress.[107]

The search for solutions and sometimes for problems has increased the influence of academics and other experts, though more as a resource to be selectively tapped than as a force shaping the policy agenda. The proliferation of policy programs and social scientists with research programs which have policy implications has provided a background of knowledge (or to be more accurate, informed speculation). Some of this research is the direct product of governmental financing of attempts to develop solutions to perceived problems, but a large portion is due to the curiosity of researchers and their need to produce a product to be successful in their profession. The policy entrepreneur can co-opt an idea volunteered by an academic or use the results of research as ammunition to support an issue in which resources have already been invested. In either case the ideas developed by policy experts will be molded by the need to conform to the established professional consensus in order to obtain professional esteem and recognition. Therefore the policy entrepreneur will have available to him at any time a limited set of professionally credible alternatives, though the set will change over time.

Heclo suggests that issue networks are beginning to dominate agenda formation. It is impossible to identify the dominant actors in the webs of influence that surround many issues. The issue network is defined as a shared knowledge group dealing with some aspect of public policy, to be distinguished from shared-action groups (coalitions) and shared-belief groups (interest groups). The Carter administration energy proposals were a typical example, evolving from an initial proposal worked out among experts tied to conservation oriented networks, to one in which macroeconomic concerns were addressed, then in congress expanding to include tax issues, nuclear power, employment impacts, etc.[108] Each node of the issue network was dependent on the accepted state of knowledge in the area of specialization; for example the effectiveness of energy conservation depended upon econometric estimates of energy elasticities, macroeconomic impacts were anyone's guess as the profession was split between Keynesians, Monetarists, Rational Expectation Classicalists, Post-Keynesians, and various eclectics; tax issues depend on conclusions drawn from public finance theory, accounting theory and tax law; the availability of nuclear power on the economic estimates of costs and engineering estimates of reliability and safety; employment effects depended on your choice of macroeconomic and labor market theories and analyses of regional impacts.

The diffusion of authority in Congress and along issue networks makes the President's role as the pre-eminent policy entrepreneur even more important than in the past. His ability to set the policy agenda focuses attention on certain issues. His set of institutional resources, including the veto and the power to fill key policy making positions, along with the hierarchical structure of the executive branch provide the means to promote a chosen agenda. The President's command of public attention, i.e., Roosevelt's "bully pulpit," provides him with the essential resource required for policy entrepreneurship. He also has tremendous information resources at his command, including agencies such as OMB, the Council of Economic Advisors (CEA), and the control of bureaucratic resources by his political appointees. The bureaucracy may hinder implementation, and even sabotage parts of the president's agenda, but its power is essentially limited. However, setting the agenda is not the same as controlling the alternatives or determining the final outcome.[109]

Congress is the second actor in the process of agenda setting, but given its diverse and ambitious membership, congressional agenda-setting can easily become an exercise in competition for a limited supply of media attention and institutional resources. On the other hand, the president's agenda must be implemented through congressional action, where submission of alternatives allows congressmen to substitute their own perception of the issues for that presented by the executive branch. Therefore presidential entrepreneurship to be effective must take in account the restrictions imposed by congressional politics.

Policy entrepreneurs in Congress tend to be egocentric in wanting to be the leading spokesmen on a particular, even if relatively narrow, policy area, willing to cooperate only to the extent that they can retain the personal satisfaction and recognition that comes with effective control of a policy agenda. The reforms of the 1970s increased the opportunities open to entrepreneurs, as committees, such as Interior and Insular Affairs, moved from the realm of the "pork barrel" to concern for issues of energy and the environment. The policy entrepreneur's interest in policy does not seem predicated on the re-election goal, as the worlds of policy formation and seeking reelection seem generally distinct. Policy entrepreneurship seems to be a common orientation of House members possibly because they have become such experts in the science of constituency service that they have considerably more freedom to take whatever policy positions they want in committee or on the House floor.[110]

In turn, congressmen are dependent on their staff, who in effect determine the alternatives available to the members. Committee staff can also serve as a source of policy proposals, developing ideas, framing alternatives, encouraging input from potential allies, and generally determining the form in which issues come to a committee for decision. The ability of staff members to engage in policy entrepreneurship is determined by the willingness of committee or subcommittee chairman to allow them this latitude, which in turn depends on how well they serve the chairman's interests. The staffer who is successful at developing issues which enhances the congressman's reputation will increase his own power and influence. The staff member can also be used as a middleman, to search for ideas in academic journals or through contacts with experts which the congressman can then employ to develop legislation. The staff entrepreneur will look for issues requiring governmental action which can attract publicity, package the idea to stimulate interest by the press, and in turn media attention will increase the probability that the issue will become part of the policy agenda.[111]

The ultimate key to the shape of the issue agenda is the tendency of the body politic to focus on a small set of issues, raising them to prominence. There is a small group of politically sophisticated activists, perhaps a few percent of the population, with a level of knowledge and cognitive style similar to professional politicians, journalists, and political analysts. They are articulate and active, and their views and concerns make up much of what is heard as "public opinion." The bulk of the population can be mobilized to political action, as they half-attentively monitor the flow of political news, but they run for the most part on a psychological automatic pilot. Most of the time public opinion is passive and unresponsive, but when aroused, it has effects on the polity that are significant and immediate. In the public at large there is awareness or concern about only a few issues, perhaps a half-dozen or so that receive prominent attention in the media. "The key to the democratic process is the fluidity of the public agenda, the possibility that at any minute what was once the concern of a tiny group of activists may suddenly crystallize the attention of the mass electorate and become a matter about which they do indeed have real opinions and real knowledge."[112]

OIL POLITICS AND ECONOMICS: 1945-1968

The evolution of oil politics reflects a shift from the politics of special interests to a more pluralist conflict in which previously unorganized interests obtained a voice with which to express their policy preferences. While the potential for activating these interests had been demonstrated during the debate over natural gas deregulation in the 1950's, it would not be until the 1970s that the combination of an amenable public and the increased sophistication of political entrepreneurs would provide a mechanism to active these latent interests into taking a dominant role in energy debates.

The emergence of consumer and environmental groups possessing strong ideological agendas resulted in the presentation of new political issues, while providing a focus for public resentments. The influence of these groups was increased by the formation of coalitions with existing organizations around specific issues. Finally, the increasingly receptive atmosphere for opposition to entrenched interests combined with the perception of an "energy crisis" led to the development of groups which focused on energy issues. The actions of groups like Common Cause, by initiating action on campaign contribution restrictions and congressional reforms, favorably changed the legislative environment in which these "public interest" groups operated.

During the 1950s, the three issues which made up the oil and gas political agenda were tax breaks, natural gas deregulation, and oil import restrictions. In the 1960s oil and gas policy was basically concerned with competition for the economic benefits of oil quota regulation. In the 1970s the emergence of "public interest" groups changed the conflicts over oil and gas policy to an ideological battle over the proper relationship of government and the economic system under American capitalism, the values of efficiency vs. equity, and the desire to establish countervailing power to contain the influence of large, powerful corporations.

The oil and gas issues in the 1950s and 1960s can be separated into import restrictions, a significant part of the larger trade debate, tax breaks such as the oil depletion allowance and intangible drilling expensing, included among the many decisions comprising tax politics, and natural gas regulation, one of the major regulatory issues. In the first two cases oil politics were a subset of a larger group of issues decided in isolated policy arenas where the scope of conflict could be restricted to interested parties. Natural gas deregulation began in this fashion but the conflict quickly spread in scope as consumers objected to proposed price increases.

The naiveté of the conventional political economy approach to free trade can be seen in the convolutions required to explain support for protectionism. Free trade is theoretically superior to protectionism, since the increase in efficiency allows complete side payments to compensate the losers from the removal of tariffs. The failure to support free trade is attributed to the cost of voting, and imperfect information, allowing narrow economic interests to obtain protection by investing in economic rent seeking activity. However, this framework of analysis ignores two problems: (1) no enforcement mechanism to insure the delivery of side-payments exists; (2) there are real transaction costs for labor separate from the direct economic losses due to increased competition, as workers who lose their jobs must bear the loss of the human capital accrued as well as the psychic costs of loss of identity, status, and possible dislocation. The lack of information and liquidity constraints faced by workers makes losses and gains nonsymmetric (the cost of losing a job is greater than the social benefit of adding one) and politicians faced with discontented voters and increased demands on social services may weight the costs of free trade differently than economists concerned only with efficiency considerations.

Robert Pastor explains much of the protectionist activity in Congress as symbolic posturing to injured groups that the legislator is listening and aware of their plight and to signal to the executive branch and foreign governments that relief is in order. A bill is sponsored, hearings are held, speeches are made and a log roll is mobilized. In response economists and free trade groups mobilize and cry protectionism, as in 1955 when Sam Rayburn warned that without a closed rule to debate the trade bill, the result could be another Smoot-Hawley. The Administration shows, in various ways, that it is listening to the Congress by satisfying some of the interest groups, and the others take satisfaction that the issue they raised was addressed. The march toward freer trade continued unabated until the 1970s, with average duties on imports declining from 25 percent to 11 percent between 1946 to 1967. "Perhaps of greatest importance, however, in determining US trade policy and in explaining the cry-and-sigh paradox, is the ideological underpinning of the debate: that is, the almost universal belief in the importance and the benefits of freer trade to the domestic and global political economy. . . . Economists are the reigning priests of this freer trade ideology, and at the slightest threat that the gates of their hallowed temple might be closed, they respond with great vigor and not a little exaggeration."[113]

Oil trade policy was an example of the use of minor concessions to defuse conflict which characterized the struggle over trade policy. Interest groups attempted to influence policy against a backdrop of an ideological belief in the social value of free trade. Attempts to legislate an oil import quota were blocked by the strong consensus among Democrats in favor of free trade. Republican support for protectionism was based as much on ideology as self-interest, as can be seen by the 15 Republicans from states with neither oil or coal production who voted for the Thomas amendment in 1949. Trade was a partisan issue, and a large portion of the Republican party would vote for any protectionist bill, while Democrats would support free trade unless an important local industry was threatened.

The eventual success of the supporters of oil import restrictions can be traced to the institutional framework of Congress. As long as the Democratic leadership supported the president in opposing import restrictions the most that coal and oil interests could achieve would be symbolic concessions. The key to action was the distribution of economic interests among states and nations, and congressional power among individuals. Coal production was concentrated in two states, West Virginia and Kentucky, with six other eastern states with some production, Pennsylvania, Virginia, Alabama, Illinois, Ohio and Indiana. No important political leaders came from the two major coal states, while only Harry Byrd (D-VA) on the Senate Finance Committee, and Minority Leader Everett Dirksen (R-IL) hailed from the other coal states. Small oil producers and New England residual oil consumers had access to power brokers like Sam Rayburn, John McCormack (D-MA) and Joseph Martin (R-MA) in the House, and Lyndon Johnson and Robert Kerr in the Senate.

From the point of view of trade policy and foreign policy Venezuela and Canada were only oil exporting countries for which an oil quota would cause problems. The Middle Eastern countries had not begun major exports to the US, they were unimportant as markets, and it could be left to the major international oil companies to adjust production and trade patterns if a quota was imposed. It is not surprising then when oil import restrictions were imposed they protected small, inefficient oil producers in the Southwest, provided numerous loopholes for New England's residual oil import requirements and Venezuelan and Canadian exports, and provided limited aid for the coal industry. It was good politics for the executive branch to grant concessions on minor trade issues to maintain good relations with the major supporters of free trade in Congress, and the use of an administrative program prevented the kind of logroll which might have occurred had a quota bill been passed in Congress.

Once quotas were established the political arena shifted from Congress and trade policy to the Department of the Interior and regulatory policy. The ideological backdrop was removed from the issue, and it became a simple question of allocating costs and benefits. It is not surprising that most consumers failed to perceive the cost of the program, other than residual oil consumers, as the impact was not to increase product prices but to stop the gradual decline in real prices that had been occurring. There were actually two separate conflicts, a redistributive battle between East Coast industrial residual oil consumers and the coal industry, and a regulatory battle over the allocation of entitlements, transferring benefits to inland refiners from oil importers. The coal industry had an advantage through its long relationship with the Interior Department, but the East Coast consumers could mobilize greater political pressure on the department. The entitlements battle was the classic example of rent seeking behavior in action as various actors attempted to obtain a larger share of the entitlements pie, and new firms attempted to enter the market. It should be noted that there is no evidence that expenditures by oil companies to obtain entitlements approached in magnitude the value of these regulatory rents. However, the sliding scale for allocating entitlements can only be seen as an ideologically inspired policy, since small refiners did not possess the resources to obtain such a lucrative subsidy. Given the size, wealth and political influence of the competition there is no other explanation for the subsidy to small, mostly inefficient refiners.

The inefficient structure of the entitlements programs, both in the 1950s and 1970s, casts doubt upon the validity of the Peltzman/Becker model of economic regulation.

The ability of the industry to preserve its tax breaks depended on internal unity and the invisibility of tax policy. Most tax breaks to business can usually be justified by some sort of pseudo-economic argument which allows a congressman to explain his behavior to constituents. Since tax deductions are the most difficult of all microeconomic policies to monitor, in terms of cost and economic impact, it requires a well financed and dedicated effort to expose and explain to the public the effect of decisions on tax policy. The oil depletion allowance seemed reasonable on the surface; since oil is a depleting resource the owner should be allowed to receive a deduction for the decline in his capital in the same way a factory owner is permitted to deduct the deterioration of his physical plant. The fallacy here is that the oil in the ground is not the oil producer's capital, it is the cost of drilling the well which constitutes his investment, including the intangible drilling costs. The second argument used to justify the depletion allowance was the national security rationale for the encouragement of domestic production. The observation that a strategic reserve and targeted subsidies would be more effective in achieving this goal was absent from the policy debate.

The key to maintaining the depletion allowance had little to do with intellectual arguments and a lot to do with oil money--everyone in the oil and gas industry, as well as members of other extractive industries, had a vested interest in investing resources in politicians in return for protection of the status quo. Unlike oil quotas, no one in the industry from producers to marketers had an incentive to fight the allowance, and opposition was restricted to a few lonely liberal crusaders. The necessity for a united front, for the majors were aware that their tax breaks were dependent upon the cooperation of the small producers, provided a bargaining tool for the latter in the conflict over import restrictions. The only group which had reason to oppose the allowance was the taxpayers, and they were silent during this period. Even oil consumers profited from tax breaks, which by encouraging excess drilling and production caused downward pressure on prices.

Natural gas deregulation was an exception to the normal case where consumer interests were ignored; despite tremendous efforts (or perhaps because of those efforts) the industry was unable to eliminate natural gas regulation. The natural gas industry had developed slowly up to WWII as most of the gas produced with oil was flared off or reinjected. During the war construction of long-distance pipelines was subsidized and the industry was firmly established on an interstate basis. The industry was segmented into three parts, producers (initially oil companies but as the market grew gas producers emerged), pipelines and local distribution companies (LDCs). Unlike oil pipelines or railroads, gas pipelines were not common carriers, although they were natural monopolies. Pipelines had partially integrated backward into gas production (producing 23 percent of their sales volume in 1946), but most gas of their supply was purchased from producers and sold to the LDCs, who were prohibited from integrating backward by the Public Utility Holding Company Act of 1935. The LDCs were regulated by the states, and the Federal Power Commission (FPC), created to regulate electric utilities, was granted authority to regulate gas pipelines by the Natural Gas Act of 1938.[114]

The FPC had stated that it had no authority over sales of natural gas at the wellhead in 1940, but ruled in a 1945 case that the price of gas produced by a pipeline company fell under its jurisdiction, confirmed by the Supreme Court in 1947. Consumerist spokesmen such as FPC Commissioner Leland Olds proposed constraints on gas consumption for uses such as electricity generation, a position supported by the National Coal Association. The gas industry, alarmed by these developments, began a campaign to amend the Natural Gas Act. A bill was introduced by Representative Ross Rizley (R-OK) and Senator E.H. Moore (R-OK) to exempt non-pipeline gas producers from wellhead price regulation, and prohibit end-use restrictions. The Rizley-Moore bill was supported by the oil and gas industry and passed the House, but the administration blocked the bill in committee in the Senate.[115]

A natural gas investigation instigated by the FPC resulted in two reports in 1948, each of which was supported by two commissioners. The Smith and Wimberly report recommended legislation to clarify the FPC's jurisdiction, including a regulatory exemption for independent producers. The Olds-Draper report concluded that since the industry was oligopolistic it was necessary for the FPC to monitor field prices and to intervene if it felt that competition was insufficient to protect consumers. Robert Kerr and Representative Oren Harris (D-AK) introduced bills to exempt producers from FPC jurisdiction, but they were opposed by the new majority on the FPC, led by Olds, who argued that the bill was a $500 million raid on midwestern farmers and northeastern homeowners. The press suggested the bill would result in higher prices and Truman decided to oppose the bill. Kerr and Johnson organized a vicious attack on Olds during the confirmation hearings for his fourth term, and his reappointment was soundly rejected despite the active support of the President. After the Kerr-Harris bill passed it was vetoed by Truman in 1950.[116]

The 1954 Supreme Court decision in the Phillips case established federal price controls over the entire industry, and a 1955 ruling by an appellate court overturning a FPC decision determined that wellhead prices must be based on costs instead of market value. With prices rising to 20¢ per thousand cubic feet the FPC declared a price freeze, which brought on a storm of protests, especially from small producers and royalty owners, who considered the action an attack on states' rights and the free enterprise system. Harris and Senator William Fulbright (D-AK) introduced a bill to exempt producers from direct regulation, though it would permit the FPC to reject contracts or rate increases by pipelines which exceed the reasonable market price.[117] The bill passed both Houses, but disclosure of the attempted bribe of Senator Case forced Eisenhower to veto the bill in February 1956.

The natural gas battle illustrated the limits of the influence of the oil and gas industry. While it was able to push its bills through Congress the industry could not convince Presidents from both parties that they allow themselves to be identified as pro-industry and anti-consumer. While industry money probably had some influence in Congress, a stronger factor in favor of deregulation was a conservative pro-business stance. Only four senators in 1950 who voted against the industry failed to vote for the Fair Employment Practice Act. Southern Democratic Senators (from non-gas producing states) voted 10-3 for deregulation in 1950 but changed to a 5-9 vote against in 1956. Southern Democratic Representatives were also overwhelmingly pro-industry in 1950, by a vote of 56-4 (excluding producing districts), declining to 37-29 six years later. Given the unity of Southern Democrats during this period, the most plausible explanation for the change was the negative publicity surrounding the bill in 1956.

Comparing the three issues highlights the importance of perception of the issue in determining whether it is placed on the public agenda or if decisions will be made quietly, dominated by interest groups. An increase in natural gas costs was directly transmittable to residential customers, which made mobilizing opposition far easier, while casting the issue in the mold of farmers and consumers against big business made it an attractive issue. The ability of natural gas consumers to prevent deregulation is even more surprising since of the three issues, import quotas were more costly to consumers and oil depletion was a larger burden for taxpayers. While there is no data concerning expenditures, it is likely that far more money was spent to obtain deregulation of natural gas than on oil import quotas, and probably similar to the expenditures to protect the depletion allowance. The power of consumer interests were still fairly limited, as a majority of Congress voted for producers while the defeat of the Leland Olds reappointment was evidence of liberal weakness. Given the ideological tinge to the natural gas votes, antagonism to government interference with industry may have been a more important factor than the resources invested by gas producers.

OIL POLITICS AND ECONOMICS: 1969-1976

During the 1960s the cumulative impact of economic criticism and the lonely protests of liberal politicians began to hurt the image of the oil industry. Oil prices peaked after Suez, and there was a substantial decline in product prices until 1965. The end of the trend of falling prices came just when the resurgence of inflation due to Johnson's refusal to raise taxes to finance the Vietnam War increased sensitivity to price increases. With gasoline prices rising a cent a year from 1965 to 1967, and heating oil increasing by a half a cent a year, consumers and politicians were becoming aware of oil prices. The potential for making energy prices a political issue began to be recognized during this period.

Economists writing on oil were indirectly responsible for the change in attitude toward the industry from passivity to anger. McKie and McDonald's 1962 article and the two articles published in 1964 by Kahn and Adelman, as well as the Petroleum Study Committee report made it clear that waste, tax breaks and oil import controls were costing consumers billions of dollars. The 1965 article by Gilbert Burck in Fortune and Milton Friedman's 1967 column in Newsweek brought the economic critiques to the attention of the public. The CRA study of the oil import program, Ronnie Dugger's Atlantic article and finally the Oil Import Task Force report kept the focus on the industry. At the same time a flurry of articles on oil pollution incidents during the 1967-70 period reinforced the negative impression of the industry.

The combination of the exposure of academic criticism and the burgeoning consumer movement culminated in the March 1969 hearings before the Subcommittee on Antitrust and Monopoly, chaired by Philip Hart, with technical support from the Subcommittee's chief economist, John Blair, who was highly critical of oil policy. The Task Force report was the subject of hearings before the House Interior Subcommittee on Mines and Mining and the Ways and Means Committee hearings on trade legislation. East Coast congressmen argued for eliminating restraints on heating oil imports before the Senate Banking and Currency Subcommittee on Small Business. The key event was the attack on the depletion allowance in 1969, which signified that the industry was becoming politically vulnerable.

Wilbur Mills was able to prevent the House from voting on the allowance because of the power of Ways and Means Committee to obtain a closed rule on tax bills, though he made minor concessions to tax reform. In the Senate voting the most interesting development was the split into three groups, the 30 member pro-industry group which voted to restore the allowance to 27.5 percent, the 40 member anti-industry group which voted to reduce it to 20 percent, and the 28 Senators who rejected both changes to keep the allowance at 23 percent. While there had been some increase in the liberal group (during the 1960s anti-depletion votes ranged between 30 to 34) the change was concentrated in non-oil state Democrats, with only three voting for the industry, of which two were from Arkansas, a marginal producing state. Congressmen were beginning to weigh the value of oil money against the negative impact of association with the industry.

By 1971 energy policy had become one of the issues on the congressional agenda. A number of investigations were held in response to both rising prices and occasional shortages of fuel, as curtailments of natural gas occurred for the first time during the winter of 1970-71, increasing demand for heating oil, in short supply due to distortions in refining patterns as a result of price controls. The emergence of an awareness of an "energy crisis" in 1973 triggered the movement of energy policy to the forefront of the policy agenda, and the general perception that the large oil companies were responsible, though there was little confidence in the ability of the Federal government to manage the issue. An FTC report alleging cartel behavior and attacks by numerous congressmen did little to improve the industry's image.

Along with escalating public pressure on Congress to do something, the independent refiners, struggling to find crude supplies, and hurting due to the loss of subsidies, found an ally in the Energy Task Force of the CFA in their demand for an allocation program. Thus, the basic alignment of interests and the precedent that the government should intervene in the market was established three months before the Arab Oil Embargo. The political alignments which would repeat themselves over the next seven years were already in place; Democrats and liberals for allocation and controls, oil state congressmen and Republicans against, occasional exemptions for stripper well owners and small producers, but a large consensus in favor of controlling and even punishing the larger companies.

The effect of the embargo was to increase visibility of the issue and the size of the economic stakes, without changing the attitudes of most of the actors. Negative attitudes toward the oil industry intensified to the point that oil lobbyists in late 1973 refrained from lobbying on the emergency energy legislation, as according to an API representative: "Our credibility is so bad right now that we don't get anywhere by being visible."[118] While the oil companies were keeping a low profile just about every other group with a conceivable interest in energy policy converged on Washington. Numerous groups either formed or added energy staffs after the embargo. They generally fell into two categories, economic interest groups which wanted to ensure the protection of their specific interest and public interest groups with a pro-consumer orientation.

The three important public interest groups involved in energy lobbying, the Energy Policy Task Force of the CFA, Energy Action, and the Citizen-Labor Energy Coalition, all opposed deregulation of oil prices on the grounds that current prices provided generous returns on investment, a free market did not exist in petroleum, demand for petroleum products was price-inelastic and therefore higher prices would be inflationary without reducing demand or increasing supply. Organized labor lent nearly unanimous support to this view, feeling that their members would suffer harsher consequences from higher oil prices than jobs lost due to curtailments. Other public interest groups were sympathetic to these arguments, but were more doubtful concerning the assumption that demand was not responsive to price. Environmental groups felt that higher oil prices might promote conservation and renewable energy, but were reluctant to advocate the industry's position. Their reticence on the issue was influenced by the fear of losing consumer groups as future political allies. The Sierra Club and the Natural Resources Defense Council were the only environmental groups which openly favored deregulation, though insisting that profits be taxed or otherwise limited.[119]

Other groups supporting regulation included public transportation associations, the Petrochemical Energy Group, composed of twenty-one major petrochemical firms that use petroleum gases as feedstocks, the American Petroleum Refiners Association (APRA), representing small refiners, the Independent Refiners Association of America, representing a somewhat wider range of refiners, the Society of Independent Gasoline Marketers of America (SIGMA), which represents independent marketers of gasoline. APRA and SIGMA were divided on crude pricing but favored oil allocation, and APRA also supported the small refiner entitlement bias.[120]

The major integrated oil firms and the independent producers, represented through the API, IPAA and TIPRO, were united against regulation, arguing that higher prices would result in increased production, greater conservation, elimination of shortages, prevent the waste of resources on administrative costs and onerous reporting requirements, and increased oil investment due to the elimination of uncertainty caused by regulation. The Council of Active Independent Oil and Gas Producers, the National Oil Jobbers Council (NOIC), and the partially integrated, large refiners such as Ashland Oil were in general agreement with this position. The National Petroleum Refiners Association (NRPA) was neutral with regard to crude price regulation but advocated product decontrol. A number of commercial users and business groups, including the Business Roundtable, the Chamber of Commerce, the American Trucking Associations (ATA), and the Motor Vehicle Manufacturers Association, favored deregulation because they were more concerned about ensuring adequate supplies or due to a general antagonism toward government intervention in the marketplace.[121]

Once the price and allocation regulatory apparatus was in place the same scramble for advantage occurred as with oil import quota regulation. Although the FEA conducted numerous hearings and encouraged input into rulemaking, the consensus among corporate lobbyists, trade association representatives, and public interest advocates was that hearing participation had a negligible impact on regulatory activities, and formal channels were seen only as opportunities to go on the record. The two methods considered effective were direct pleading, using technical expertise to deluge the administrator with arguments in favor of the group's position, and indirect political persuasion, usually through Congress. This worked to the detriment to the major oil companies, as consumer groups, small refiners, marketers, and widely distributed oil consuming interest groups were treated more sympathetically.[122]

The influence of groups supporting regulation was due not only to hostility to the oil industry but also the awareness of politicians that the public wanted low-cost energy and supported price controls and allocation. Even without the participation of interest groups it was unlikely that Congress would accept any measure that raised the price of energy, either through taxes, tariffs or price decontrol. Among public interest groups only Common Cause was willing to take a stand in favor of higher energy taxes, as most groups in favor of conservation preferred mandated efficiency standards. Once the energy debates became arguments over the details of tax provisions both the consumer coalition and the oil industry experienced fraternal warfare, as public interest groups tried to dump the cost of conservation on industry, industries attempted to transfer costs to each other, independent oilmen fought for tax breaks, and the majors accepted decontrol even at the price of a windfall profits tax.[123]

The attempt by Ford to decontrol oil in 1975 and the move to deregulate natural gas in 1976 revived the basic cleavage in energy policy between producer and consumer interests. Decontrol united the coalition of consumers because it threatened to raise prices in general while eliminating the channels by which organized interests could lobby for an economically valuable favor. Industrial customers such as airlines and petrochemical companies possessed sizable resources which could be used for persuasion. Consumer groups had large potential constituencies, access to the media, and expressed the general attitude toward price controls. The support in Congress for small refiners is harder to explain, since they employed few people, were widely dispersed, and did not contribute a great deal of money to politicians. The bias had been institutionalized under the oil import program, and a mythology had sprung up that justified supporting inefficient small refiners to increase competition with monopolistic oil companies, and it had become a legitimate claim over the years. Since this subsidy was ostensibly paid by large companies (by forcing them to share low-cost oil), to politicians it seemed a costless method of supporting small business.

The result was a strange combination of redistributive and distributive policies. The effect of price controls was to redistribute the potential profits of producers to subsidize lower costs of energy for consumers. A secondary redistributive program was the refinery entitlement program, which subsidized refiners of imported oil at the expense of the refiners of domestic oil, and in addition included a subsidy for smaller refiners. Stripper well exceptions to price controls provided a subsidy to inefficient producers. This allowed Congress to have the best of both worlds; they could redistribute from the hated oil companies to consumers, then intervene with the bureaucracy to provide subsidies and transfers for interest groups, an economic benefit unlikely to generate resentment since it was rarely recognized as such by the general public. However, instead of a simple relationship between benefits received as a function of the resources invested in rent seeking, success in obtaining rents was strongly influenced by ideological considerations with small business (producers, refiners and marketers) receiving rents far in excess to their investment in rent seeking.

THE ECONOMICS OF OIL POLITICS: 1977-1980

The Carter administration began its tenure with significant advantages in making energy policy over Ford. Instead of facing a Congress dominated by the opposition party, with a sizable cadre of legislators elected in the wake of Watergate, who had no desire to cooperate with Nixon's appointed president, Carter would try to implement a policy as the leader of the majority party in Congress. The problem Carter faced was that most of his proposed solutions were unpopular with the public, as people opposed gasoline taxes and higher energy prices. Spending for new energy sources, creating a Department of Energy and imposing conservation regulations were acceptable, probably because they seemed to be relatively costless. But any attempt to use the market system to encourage conservation and allocate resources was sure to trigger intense opposition.

Carter also demonstrated the limits of the Presidency as a "bully pulpit." He was able to significantly increase the proportion of people who saw energy as a serious problem, but only for about a year, after which the issue dropped from the public agenda until the Iran-Iraq war and the resulting price escalation. The increased attention to the issue did not translate into willingness to make economic sacrifices as the one issue on which public interest groups and the oil industry agreed was opposition to the crude oil equalization tax. Most of the elements of the Carter's national energy plan (NEP) which might have reduced demand were removed or watered down to the point of ineffectiveness. Lobbyists fell on his plan like vultures on a cow's carcass in West Texas. The only group that practiced restraint were the large integrated oil companies, aware that they were still persona non grata in Washington. The API's most visible lobbyist was a former aide to Edward Kennedy and John Glenn. The industry's position wasn't helped when oil money became a touchy issue as Congress Watch released a report showing that nine members on the Senate Finance Committee received a total of $547,000 from oil and gas interests.[124]

One of the myths of the energy debate is the illusion that it was a regional conflict. While northern states had higher energy costs, and there was a Southwest bloc of energy producing states, support for deregulation was more ideological than regional. The analysis of votes in Chapter 9 show a tendency of southern Representative to support oil and gas interests but a reverse trend by Southern senators, with the effect much weaker than ideology or party variables. Western state congressmen demonstrated even less unity, as their economic interests in energy were highly divergent, and public attitudes ranged from fiercely pro-environment to intense hatred of governmental interference. Even in the Northeast and Midwest, there were significant differences between the energy poor New England and East Coast states and the coal states. The type of energy produced and consumed in a state were also important in determining policy attitudes, especially the conflict between natural gas and coal.

When movement came on the tough issues it was because the market forced Congress into reluctant acceptance of reality, whether due to growing curtailments of natural gas during the winters of 1975-76 and 1976-77 or the dramatic increase in oil prices in 1979 which made it self-evident that decontrol could not isolate domestic consumers from the world market. Natural gas deregulation had been defeated in 1976 but it reemerged as part of the NEP, as the unwillingness of the FPC to allow sufficient increases in prices accelerated the diversion of gas from interstate to intrastate markets. The two year battle over the Natural Gas Policy Act (NGPA) reflected the reluctance of pro-control forces to allow price increases and the intransigence of deregulation supporters. The final agreement created a complex set of rules leading to eventual price decontrol, which resulted in a chaotic natural gas market when the market fundamentals changed in the 1980s.

The strongest supporters of gas decontrol, and the most adamant critics of the NGPA were the small producers, while the larger companies generally were quiet on the issue or mildly supportive. This reflected the political sophistication and poor bargaining position of the major oil and gas producers, who focused their lobbying activities through a coordinating organization, the Natural Gas Supply Committee, making use of third parties at the state or district level instead of direct contacts with legislators. An important factor in their favor was the shift by major groups in support of decontrol, as continual curtailments had changed attitudes toward favoring reliable supplies at higher prices (especially industrial users, first to be cutoff). Farm groups, trade associations and even the NAACP, along with the Governors' Conference (by a 34-5 vote) and the National Coal Association came out in favor of decontrol.[125]

The NGPA was in serious trouble before the final Senate vote as it managed to alienate almost everyone; the anti-control people led by Russell Long, John Tower, Howard Baker and Robert Dole, and the pro-control group featuring George McGovern, James Abourezk, Edward Kennedy and William Proxmire. Instead of two groups conflicting, providing a majority for one side of the issue, there were two extremes with a core willing to compromise. Provisions to shift the cost of decontrol to industry through incremental prices, while protecting congressmen from the wrath of residential consumers, generated intense business opposition. Passage of the bill was a product of a series of White House conferences with business leaders, which generated reports of heavy handed pressure to go along with the administration or deal with a hostile bureaucracy when it came time to ask for emergency allocations and other favors. An administration official, commenting on the steel's industry switch from opposition to silence, noted that "If you're going to suck the federal teat, I guess you're going to have to pay for it once in a while." The administration also applied intense pressure on House Rules Committee members to go along with Tip O'Neill's strategy to combine the Senate compromise with the more popular parts of the president's program to guarantee passage in the House.[126]

With the impending end of oil price controls in 1979 and elections on the horizon Carter found himself between the rock and a hard place. Oil decontrol merely required the president to do nothing, since passage of a new control act was unlikely, but this would make him responsible for higher prices. Maintaining controls, and blaming possible shortages on the oil companies would be politically more palatable but bad policy. His choice of a gradual decontrol policy, even with the inclusion of a windfall profits tax, was not a viable strategy to increase the president's popularity, as an overwhelming majority opposed raising prices to encourage production. This was proven by an October 1979 poll that found twice as many people had confidence in Edward Kennedy's handling of energy as President Carter's approach. A President can only be successful as a policy entrepreneur by proposing solutions which at least acquiesce to public sentiment.

The politics of oil decontrol were the same as for the NGPA; producers hated the windfall tax, consumers detested higher prices, and both groups were willing to sabotage the bill in order to present themselves as defenders of the faith. Once passage became inevitable the battle shifted to the structure of the windfall tax and determining recipients of the windfall revenues. As usual, the independent producers were able to hustle favorable treatment relative to the major companies. With both natural gas and oil, decontrol began as a redistributive issue, but was converted into a distributive issue once the size of transfer of income from consumers was decided. First Congress baked a pie, then everyone scrambled for a piece.

POLITICS OF THE POST-CRISES YEARS

The real impact of deregulation was to diminish the arena for pursuit of economic rents in energy regulation for interest groups, and credit for price containment by Congressmen. At the same time the decline in importance of energy as part of the public agenda contributed to "issue fatigue," as politicians simply don't want to deal with controversial issues with limited potential political profits. One group which felt the sea change were the small independent refiners who wanted to retain the system of allocation. With the administration opposed to regulation, public interest groups on the defensive and with more important issues on their agenda, the refiners had no leverage to use in maintaining their benefits.

The important issues for the oil and gas industry in the 1980s were tax benefits and regulation of natural gas. During the race to grab credit for tax cuts in the 1981 tax reduction plan the oil industry was able to obtain a number of favorable provisions, but many of these came in the form of exemptions to the Windfall Profits Tax. A $2,500 exemption for royalty owners was accepted in Ways and Means, along with a stripper oil exemption starting in 1983, but an attempt to freeze the depletion allowance for independents was rejected. Over in the Senate Lloyd Bentsen tried to exempt independent producers from windfall taxes, but Robert Dole led the administration position against the measure, and the Bentsen bill was defeated. Dole tried to phase out the tax on new oil, but Kennedy and Howard Metzenbaum threatened a filibuster, and a vote to table the measure demonstrated that Dole lacked the votes to obtain closure. A compromise to gradually phase out the tax on new oil was accepted.[127] The only logic to the success and failure of various provisions is that depletion had such emotional connotations that it could not survive the conference, lowering the tax on new oil made economic sense, while raising the royalty limit was a "widows and children" clause.

The success of Reagan's tax cut lead to pressure for increased revenues as budget deficits began to mount. Oil and gas companies found themselves aligned with consumer groups in 1983 in opposing gasoline taxes and a possible import fee and taxes on natural gas, with independent producers on the opposing side of the import fee issue. However two years later consumer and environmental groups were concerned about possible tax breaks for oil companies and removal of tax breaks for alternative energy industries.[128] In effect the tax arena had returned to the good old days of fighting for tax breaks by interest groups, but with the difference that the public interest groups were permanent, if not dominant, actors on the scene.

Natural gas deregulation was the one area where consumer groups were able to win some battles. The reason was the effect of higher prices on heating bills for residential consumers, a volatile political issue, and the lack of a national security argument which could justify the resultant hardships. Since gas was gradually rising in price, and new supplies were available, there was no feeling of dire necessity motivating Congress into action on the issue, especially since the issue was a low priority matter for the Reagan administration. The collapse of oil prices completely changed the issue, as the problem became adjusting to a low price energy regime. The conflict turned into a struggle between pipelines and producers over who would bear the cost of the lower prices required to maintain market share.

The inability of the oil and gas industry to obtain relief from the impact of falling prices is evidence of both the reluctance in the industry to encourage government interference, and the underlying hostility to the industry which makes any sort of subsidy politically unpalatable. The independent oilmen who were able to wrangle tax breaks and exemptions from regulations find it much more difficult to arrange for subsidies in a time of income constraints, or an import fee which would impose higher prices on consumers. Thus while the consumer groups do not have the power over energy issues that they wielded in the 1970s, there is still a residue of the stigma of identification with oil interests which makes Congress reluctant to grant visible aid to the industry.

The inability to pass a meaningful energy bill in the aftermath of the Gulf War demonstrated that higher energy prices are not a viable energy policy option. It has become obvious that domestic production will inevitably decline over the next few decades, even if coastal waters and the Alaskan Wildlife Refuge are opened to drilling. Despite the threat of increasing oil imports, and huge budget deficits stretching indefinitely into the future, the Clinton Administration could only obtain a 4.3 cent per gallon increase in the gasoline tax at a time when real gasoline prices were approaching historical lows for the post-WWII era.

NOTES

[1] Alexander Hamilton, James Madison and John Jay, *The Federalist Papers* (New York: New American Library, 1961), p. 81.

[2] ibid., p. 83, 214.

[3] E.E. Schattschneider, *Politics, Pressures and the Tariff* (Hamden, Conn.: Archon Books, 1963), pp. 283, 287.

[4] Robert Dahl, *A Preface to Democratic Theory* (Chicago: University of Chicago Press, 1956); Earl Latham, *The Group Basis of Politics* (Ithaca: Cornell University Press, 1952); David B, Truman, *The Governmental Process*, 2nd ed. (New York: Alfred Knopf, 1971).

[5] Truman, *Governmental Process*, pp. 37, 65, 106-7. Truman's work was soon followed by numerous studies of pressure groups, many which focused on economic interest groups.

[6] ibid., pp. 213-14, 260-69, 333-36.

[7] ibid., pp. 353-58, 507-12.

[8] Grant McConnell, *Private Power and American Democracy* (New York: Random House, 1966), pp. 3-9, 108-11, 554-65.

[9] E.E. Schattschneider, *The Semi-Sovereign People* (Hinsdale, Ill.: Dryden Press, 1975), pp. 30-43; Michael T. Hayes, *Lobbyists and Legislators* (New Brunswick: Rutgers University Press, 1981), pp. 15-17.

[10] Raymond A. Bauer, Ithiel De Sola Pool, and Lewis Dexter, *American Business and Public Policy,* 2nd ed. (Chicago: Aldine Atherton, Inc. 1972), pp. 324-25, 396-99.

[11] Theodore J. Lowi, "American Business, Public Policy, Case-Studies, and Political Theory," *World Politics* 16 (1964): 678-91.

[12] ibid., p. 692-99, 707-15.

[13] Dennis C. Mueller, *Public Choice*, (Cambridge: Cambridge University Press, 1979), pp. 2-5.

[14] Anthony Downs, *An Economic Theory of Democracy* (New York: Harper & Row, 1957), pp. 9-11.

[15] ibid., pp. 23-25, 37-38, 45-46, 52.

[16] ibid., p. 70.

[17] ibid., pp. 73-74.

[18] ibid., pp. 79, 83, 87.

[19] ibid., pp. 98-110.

[20] ibid., pp. 219-55.

[21] Mancur Olson, *The Logic of Collective Action* (Cambridge: Harvard University, 1965), pp. 16-17, 26-36.

[22] ibid., pp. 44-46.

[23] ibid., pp. 59-63, 147-64. George Stigler tested the "cheap rider" proposition with regard to trade associations and found a positive relationship between number of members and resources available to the association but a

weak negative correlation of association resources and concentration ratios. He suggested that an explanation might be that large companies undertake many of the activities allocated to associations in less concentrated industries. George Stigler, "Free Riders and Collective Action: An Appendix to Theories of Economic Regulation," *Bell Journal of Economics* 5 (Autumn 1974): 364-65.

[24] Terry M. Moe, *The Organization of Interests* (Chicago: University of Chicago Press, 1980), pp. 31-34; Michael McCann, "Public Interest Liberalism and the Modern Regulatory State," *Polity* (Winter 1988): 384-85.

[25] James M. Buchanan and Gordon Tullock, *The Calculus of Consent* (Ann Arbor: University of Michigan Press, 1962), pp. 9-13, 35-36, 44-48. Note that Buchanan and Tullock, while claiming to be developing a positive theory, rigged their argument in favor of their libertarian values (a normative approach) by restricting costs to only these two categories. Inclusion of other categories of costs for which government could provide economies of scale would significantly modify their conclusions.

[26] ibid., pp. 70-80, 111.

[27] ibid., pp. 123, 133-135.

[28] ibid., pp. 164-169, 190-196.

[29] Mueller provides a general review of this area, and other works which also provide an introduction to this literature include Robert Abrams, *Foundations of Political Analysis* (New York: Columbia University Press, 1980); Peter Ordeshook, *Game Theory and Political Theory* (Cambridge: Cambridge University Press, 1986); Thomas Schwartz, *The Logic of Collective Choice* (New York: Columbia University Press 1986); Richard Cornes and Todd Sandler, *The Theory of Externalities, Public Goods, and Club Goods* (Cambridge: Cambridge University Press, 1986).

[30] McCraw, *Prophets of Regulation*, Michael Reagan, *Regulation: The Politics of Policy* (Boston: Little, Brown & Co., 1987), pp. 20-25. Gabriel Kolko's explanation of railroad regulation as an example of capture has been refuted, see Albro Martin, *Railroads Triumphant* (New York: Oxford University Press), p. 350-58.

[31] Reagan, *Politics of Policy*, pp. 28-29, 56-59.

[32] George J. Stigler, "The Theory of Economic Regulation," *Bell Journal of Economics* 2 (Spring 1971): 3-7.

[33] ibid., pp. 10-13; idem, "Economic Competition and Political Competition," *Public Choice* 13 (Fall 1972): 91-106. Note that Stigler's hypothesis concerning lawyers has some empirical support. A survey of 50 law firms with partners serving in Congress found that: "40 represent banks, 31 represent insurance companies, 11 represent gas and oil companies, 10 represent real-estate firms. Some of the biggest corporate names in America are listed as clients of Congressmen's law firms in such out-of-the-way places, say, as Nicholasville, Kentucky, and Pascagoula, Mississippi. . . ." Drew Pearson and Jack Anderson, *The Case Against Congress* (New York: Simon and Schuster, 1968), p. 102.

[34] Richard Posner, "Theories of Economic Regulation," *Bell Journal of Economics* 5 (Autumn 1974): 335-41.

[35] ibid., pp. 352-55.

[36] Sam Peltzman, "Toward A More General Theory of Regulation," *Journal of Law and Economics* 19 (August 1976): 222-30; Gary S. Becker, "A Theory of Competition Among Pressure Groups For Political Influence," *Quarterly Journal of Economics* (August 1983): 371-400.

[37] Gordon Tullock, "The Welfare Costs of Tariffs, Monopolies, and Theft," reprinted in *Toward A Theory of the Rent-Seeking Society*, eds. James Buchanan, Robert Tollison, and Gordon Tullock (College Station: Texas A&M University Press, 1980), pp. 39-50.

[38] Anne O. Krueger, "The Political Economy of the Rent Seeking Society," *American Economic Review* 64 (June 1974): 291-94.

[39] Edgar Browning, "On the Welfare Cost of Transfers," reprinted in *Toward A Theory of the Rent-Seeking Society*, pp. 283-88; Gordon Tullock, "More on the Welfare Cost of Transfers," reprinted in *Toward A Theory of the Rent-Seeking Society*, pp. 288-92.

[40] Fred McChesney, "Rent Extraction and Rent Creation in the Economic Theory of Regulation" *Journal of Legal Studies* 16 (Jan. 1987): 101-9; Marilyn Flowers, "Rent Seeking and Rent Dissipation: A Critical View," *Cato Journal* 7 (Fall 1987): 431-33.

[41] Elie Appelbaum and Eliakim Katz, "Seeking Rents by Setting Rents: The Political Economy of Rent Seeking," *Economic Journal*, 97 (Sept. 1987): 686-91; Roger Congleton "Ideological Conviction and Persuasion in the Rent-Seeking Society," *Journal of Public Economics* 44 (Feb. 1991): 69.

[42] North, Douglas, "Three Approaches to the Study of Institutions," in *Neoclassical Political Economy*, ed., David C. Colander (Cambridge, Mass.: Ballinger Publishing, 1984), p. 33-37.

[43] Warren Miller, and Donald Stokes, "Constituency Influence in Congress," in *New Perspectives on the House of Representatives*, 2nd. ed., eds. Robert L. Peabody and Nelson W. Polsby (Chicago: Rand McNally & Co., 1969), pp. 31-54; Donald Stokes and Warren Miller, "Party Government and the Saliency of Congress," *Public Opinion Quarterly* 26 (1962): 531-46; Thomas E. Mann, "Public Awareness of Congressional Candidates," in Richard Niemi and Herbert Weisberg, *Controversies in Voting Behavior,* 2nd ed. (Wash., D.C.: CQ Press, 1984), pp. 252-67; Gary Jacobson, *The Politics of Congressional Elections* (Boston: Brown, Little and Company, 1983), p. 88; and Barbara Hinkley, *Congressional Elections* (Wash., D.C.: CQ Press, 1981), p. 109.

[44] Richard F. Fenno, *Home Style: House Members in Their Districts* (Boston: Little, Brown & Co., 1978), pp. 8-17, 136-51.

[45] Jacobson, *Politics*, pp. 66, 89-95, Morris P. Fiorina, "The Case of the Vanishing Marginals: The Bureaucracy Did It," in *Congress: Structure and Policy*, ed. Mathew McCubbins and Terry Sullivan, (Cambridge: Cambridge

University Press, 1987), pp. 30-38; and Thomas Mann and Raymond Wolfinger, "Candidates and Parties in Congressional Elections," in Niemi and Weisberg, *Controversies in Voting Behavior*, pp. 269-86.

[46] Jacobson, *Politics*, pp. 43-44; idem, *Money in Congressional Elections* (New Haven: Yale University Press, 1980), pp. 117-19, 139-40.

[47] Herbert E. Alexander, *Financing Politics*, 3rd ed. (Wash., D.C.: CQ Press, 1984), p. 131.

[48] Larry Sabato, *PAC Power* (New York: W.W. Norton, 1984), pp. 5-7.

[49] *Congress and the Nation* (Wash. D.C.: CQ Press, 1965), p. 983; and Pearson and Anderson, *Case Against Congress*, pp. 134-41.

[50] Sabato, *PAC Power*, p. 5; *CQ Weekly Report*, Jan. 19, 1974, p. 113-15.

[51] *CQ Weekly Report*, Sept. 26, 1969, pp. 1811-12, Feb. 16, 1974, p. 361; Pearson and Anderson, *Case Against Congress*, p. 139.

[52] Alexander, *Financing Politics*, pp. 31-35; Sabato, *PAC Power*, pp. 8-9.

[53] Sabato, *PAC Power*, pp. 19-23; Dennis Ippolito and Thomas Walker, *Political Parties, Interest Groups, and Public Policy* (Englewood Cliffs, NJ: Prentice-Hall, 1980), pp. 352-53.

[54] Gary Jacobson, "Money and Votes Reconsidered: Congressional Elections, 1972-1982," *Public Choice* 47 (1985): 11-12.

[55] Theodore Eismeier and Philip Pollock, "Political Action Committees: Varieties of Organization and Strategy," in *Money and Politics in the United States*, ed. Michael J. Malbin (Chatham, NJ: Chatham House Publishers, 1984), p. 123; M. Margaret Conway, "PACs, the New Politics, and Congressional Campaigns," in *Interest Group Politics*, ed. Allan Cigler and Burdett Loomis (Wash. D.C.: CQ Press, 1983), pp. 135-37; Amitai Etzioni, *Capital Corruption* (New York: Harcourt Brace Jovanovich, 1984), pp. 36-37; and Margaret Ann Latus, "Assessing Ideological PACs: From Outrage to Understanding," in *Money and Politics in the United States*, p. 143.

[56] John Wright, "PACs, Contributions and Roll Calls: An Organizational Perspective," *American Political Science Review* 79 (1985): 411; Etzioni, *Capital Corruption*, pp. 53-54.

[57] *National Journal*, Dec. 3, 1977, p. 1894; Jacobson, *Money in Congressional Elections*, pp. 66-67.

[58] Michael Malbin, "Looking Back at the Future of Campaign Finance Reform," in *Money and Politics in the United States*, pp. 248-51; *National Journal*, Feb. 21, 1987, p. 418-22.

[59] Kay L. Schlozman and John T. Tierney, "More of the Same: Washington Pressure Group Activity in a Decade of Change," *Journal of Politics* (May 1983): 351-68.

[60] W. Douglas Costain and Anne N. Costain, "Interest Groups as Policy Aggregators in the Legislative Process," *Polity* (Winter 1981): 260-61.

[61] *CQ Weekly Report*, Nov. 21, 1968, p. 3211-14.

[62] George Goodwin, *The Little Legislatures: Committees of Congress* (Mass.: University of Massachusetts Press, 1970), p. 173.

[63] Lewis A. Froman, "Differences Between the House and Senate," in *Readings on Congress*, ed. Raymond Wolfinger (New Jersey: Prentice-Hall, 1971), pp. 61-66.

[64] George Goodwin, *Little Legislatures*, pp. 178-81; Richard F. Fenno, "The Internal Distribution of Influence: The House," in *Readings on Congress*, pp. 206-7; and Randall B. Ripley, *Party Leaders in the House of Representatives* (Wash. D.C.: Brookings Institute, 1967), pp. 6-8, 54-80. Speaker Sam Rayburn and Majority Leader John McCormack (Mass) served together from 1940 to 1961, with the exception of two Republican-controlled Congresses; Speaker McCormack and Majority Leader Carl Albert served together, from 1962 to 1970. Speaker Albert served from 1971-1976, with Hale Boggs as Majority Leader from 1971-1972 and Tip O'Neill from 1973-1976. O'Neill then served as Speaker from 1976-1986 with Majority Leader Jim Wright, who succeeded him as Speaker.

[65] Nicholas A. Masters, "Committee Assignments," in *New Perspectives on the House of Representatives*, 2nd. ed., ed. Robert Peabody and Nelson Polsby (Chicago: Rand McNally & Co., 1969), pp. 229-41; Fenno, "Internal Distribution of Influence," pp. 206-13.

[66] Fenno, "Internal Distribution of Influence," pp. 213-14; Donald R. Matthews, *U.S. Senators and Their World* (New York: Vintage Books, 1960), pp. 119-25.

[67] Goodwin, *Little Legislatures*, pp. 83-92.

[68] Fenno, "Internal Distribution of Influence," pp. 200-3.

[69] ibid., pp. 203-4.

[70] Goodwin, *Little Legislatures*, pp. 225-34.

[71] ibid., pp. 244-45.

[72] ibid., pp. 247-51.

[73] ibid., pp. 19-22, 26-28, 36-37; Steve Smith and Christopher Deering, *Committees in Congress* (Wash. D.C.: CQ Press), pp. 26-29.

[74] Goodwin, *Little Legislatures*, pp. 60-61. Other Finance Chairmen: Eugene Miliken (R-CO; 1947-48, 1953-54), Walter George (D-GA; 1949-52), Robert Dole (1981-86), Lloyd Bentsen (1986-1992). Other Ways and Means Chairmen: Robert Doughton (D-NC; 1949-52), Daniel Reed (R-NY; 1953-54), Jere Cooper (D-TN; 1955-56), Al Ullman (1974-80), Dan Rostenkowski (1981-1994)

[75] Manley, *The Politics of Finance*, pp. 25-31, 36-38, 49-51; and Richard F. Fenno, *Congressmen In Committees* (Boston: Little, Brown & Co., 1973), pp. 20-21.

[76] Manley, *The Politics of Finance*, pp. 64-74, 82, 106, 114; and Fenno, *Committees*, pp. 18, 54-56, 84, 106, 203.

[77] Manley, *The Politics of Finance*, pp. 250, 264-67, 277-79; and Fenno, *Committees*, pp. 144-46, 153-60.

[78] Congressional Quarterly, "1970 Legislative Reorganization Act: First Year's Record," in *Congress in Change*, ed. Norman Ornstein (New York: Praeger Publishers, 1975), pp. 190-194.

[79] Smith and Deering, *Committees in Congress*, pp. 43-44; and Norman J. Ornstein, "Causes and Consequences of Congressional Change: Subcommittee Reforms in the House of Representatives," in *Congress in Change*, pp. 91-98.

[80] Smith and Deering, *Committees in Congress*, pp. 44-45; and Ornstein, "Consequences of Congressional Change," pp. 105-6.

[81] Smith and Deering, *Committees in Congress*, pp. 45-47; and Roger H. Davidson, "Two Avenues of Change: House and Senate Committee Reorganization," in *Congress Reconsidered,* 2nd ed., ed. Lawrence Dodd and Bruce Oppenheimer (Wash. D.C.: CQ Press, 1981), p. 112-18. The two most adamant opponents of the Bolling plan were Philip Burton (D-CA), the DSG chairman, and John Dingell. Dingell and Burton enlisted the aid of Ralph Nader and various environmental groups to block the transfer of authority over energy from Commerce to a proposed Energy Committee, arguing that the new committee would favor the energy industry to the environment's detriment. James Everett, *Congress and National Energy Policy* (New Brunswick: Transactions Books, 1984), p. 207.

[82] Smith and Deering, *Committees in Congress*, p. 47

[83] Davidson, "Two Avenues of Change," pp. 119-20.

[84] ibid., pp. 121-28.

[85] Catherine E. Rudder, "Committee Reform and the Revenue Process," in *Congress Reconsidered,* 3rd ed., ed. Lawrence Dodd and Bruce Oppenheimer (New York: Praeger Publishers, 1977), pp. 118-32.

[86] Jacobson, *Politics*, pp. 161-62.

[87] Everett, *Congress and National Energy Policy*, p. 209.

[88] Bruce Oppenheimer, "Policy Effects of US House Reform: Decentralization and the Capacity to Resolve Energy Issues," *Legislative Studies Quarterly* (February 1980): 16-18.

[89] Schlozman and Tierney, "Washington Pressure Group Activity," pp. 369-71.

[90] Richard E. Wagner, "Pressure Groups and Political Entrepreneurs: A Review Article," *Papers on Non-Market Decision-Making*, ed. Gordon Tullock (Charlottesville: University of Virginia, 1966): pp. 165-66.

[91] Schumpeter, p. 82-84, 132.

[92] Norman Frohlich and Joe A. Oppenheimer, *Modern Political Economy* (Englewood Cliffs, NJ: Prentice-Hall, 1978), pp. 68-69.

[93] Robert H. Salisbury, "An Exchange Theory of Interest Groups," *Midwest Journal of Political Science* (February 1969), pp. 15-30.

[94] Larry Sabato, *The Rise of Political Consultants* (New York: Basic Books, 1981), pp. 221-23; Moe, *Organization of Interests*, pp. 36-41.

[95] Jack L. Walker, "Performance Gaps, Policy Research, and Political Entrepreneurs: Toward a Theory of Agenda Setting," *Policy Studies Journal*

(Autumn 1974): 113-15. It should be noted that Kennedy was not the first politician to involve himself with automobile safety, as Representative Kenneth Roberts (D-AL) held hearings on the issue from 1956 to 1964, and even obtained passage of three minor safety bills, with practically no publicity given to his efforts. Mark V. Nadel, *The Politics of Consumer Protection* (Indianapolis: Bobbs-Merrill, 1971), p. 31.

96 Nadel, *The Politics of Consumer Protection*, pp. 32-42.

97 ibid., p. 110-12.

98 Andrew McFarland, *Public Interest Lobbies: Decision Making on Energy* (Wash. D.C.: American Enterprise Institute, 1976), pp. 89-91.

99 Nadel, *The Politics of Consumer Protection*, pp. 159-64.

100 McFarland, *Public Interest Lobbies*, pp. 93-95.

101 ibid., pp. 68-74.

102 ibid., pp. 85-88; J. Clarence Davies and Barbara S. Davies, *The Politics of Pollution*, 2nd ed. (Indianapolis: Bobbs-Merrill Company, 1975), pp. 91-93.

103 McFarland, *Public Interest Lobbies*, pp. 64-66; and idem, *Common Cause: Lobbying in the Public Interest* (Chatham NJ: Chatham House Publishers, 1984), pp. 5-10.

104 *National Journal*, Oct. 1, 1977, pp. 1532-34.

105 John Pierson, "Aspects of Legislative Persuasion: Media," *National Tax Journal*, Sept. 1979, p. 314.

106 Hugh Heclo, "Issue Networks and the Executive Establishment," in *The New American Political System*, ed. Anthony King (Wash., D.C.: American Enterprise Institute, 1978), pp. 99-100.

107 Harrison Fox, Jr. and Susan Webb Hammond, *Congressional Staffs* (New York: The Free Press, 1977), pp. 24, 130-134.

108 Heclo, "Issue Networks," pp. 103-6.

109 John W. Kingdon, *Agendas, Alternatives and Public Policies* (Boston: Little, Brown and Company, 1984), pp. 23-36.

110 Eric M. Uslaner, "Policy Entrepreneurs and Amateur Democrats in the House of Representatives: Toward a More Party-Oriented Congress?" in *Legislative Reform*, ed. Leroy N. Rieselbach, (Lexington, Mass.: Lexington Books, 1978), pp. 106-112.

111 David E. Price, "Professionals and 'Entrepreneurs': Staff Orientations and Policy Making on Three Senate Committees," *Journal of Politics* 33 (1971): 320-25; Michael Malbin, *Unelected Representatives* (New York: Basic Books, 1979), pp. 18-19, 34-35, 44.

112 W. Russell Neuman, *The Paradox of Mass Politics*, (Cambridge: Harvard University Press, 1986), pp. 6-7.

113 Pastor, *Politics of US Foreign Economic Policy*, pp. 191-99; and idem, "The Cry-and-Sigh Syndrome: Congress and Trade Policy," in *Making Economic Policy in Congress*, ed. Allen Schick (Wash., D.C.: American Enterprise Institute, 1983), pp. 159-67.

114 Veitor, *Energy Policy in America*, pp. 60-70.

[115] ibid., pp. 71-77.

[116] ibid., pp. 77-81.

[117] ibid., pp. 85-88.

[118] *National Journal*, Jan. 5, 1974, p. 26.

[119] John E. Chubb, *Interest Groups and the Bureaucracy* (Stanford: Stanford U. Press, 1983), p. 131.

[120] ibid., p. 136.

[121] ibid., pp. 132-33.

[122] ibid., pp. 138-41, 148-53.

[123] *CQ Weekly Report*, May 3, 1975, p. 939-46.

[124] *National Journal*, Nov. 26, 1977.

[125] Pietro S. Nivola, "Energy Policy and the Congress: The Politics of the Natural Gas Policy Act of 1978," *Public Policy* (Fall 1980): 504-7, 527

[126] ibid., p. 527-29.

[127] *National Journal*, Aug. 15, 1981, p. 1454-60.

[128] *CQ Weekly Report*, June 25, 1983, p. 1279-82, Aug. 17, 1985, p. 1641.

CONCLUSION

This book began as an attempt to answer what seemed a simple question, the reasons for the change in the fortunes of advocates of oil import quotas between the 1950s and 1980s. To explain oil policy in the 1980s it was necessary to review the evolution of oil politics, as well as the political system in general, during the 1960s and 1970s. The rise of public interest groups, the increasingly negative public perception of authority and business, and the reforms of campaign financing and congressional procedures changed the rules for interest group activity. The expansion of the activity of policy entrepreneurs outside the government, and political entrepreneurs within, and the growing importance of the electronic media which facilitated their efforts to arouse latent interests, broadened the scope of political conflict. These factors operated against a background of ingrained social attitudes, including a "public economics" encompassing a set of beliefs concerning economic behavior which define the range of activities considered "fair" as well as rules for attributing responsibility for economic outcomes, a general political culture influenced by a vestigial Populism which viewed large economic institutions and government with suspicion and potential hostility, and a more elitist belief, reflecting the Progressive tradition of scientific management, in the ability of government to successfully intervene in the management of economic affairs.

Interest group, public choice, economic regulation and rent seeking theories were all found to be inadequate explanations of the structure of oil policies adopted during the last few decades. The importance of ideology and party in Congressional voting, the continued influence of the underlying American political culture in framing the political agenda, as well as the pervasive and powerful role of "public interest" groups in the policy debate contradict the predictions of the economic model of political behavior. This is not to say that self-interested behavior is absent from politics; both the scramble for economic rents and the battles over redistribution of resources attest to the importance of selfish behavior in politics. Some of the groups labeled "public interest" were interested only in their specific definition of the public interest, and had little regard for the overall social impact of the policies which they championed. But simple models in which economic policy is determined by the market for political influence (or the balance of power between interest groups) ignore the complexity of the process and the diverse motives, costs and resources of political actors.

Amartya Sen distinguishes between two concepts which are useful in understanding the roots of nonrational social behavior, sympathy and commitment. "The former corresponds to the case in which the concern for others directly affect one's own welfare. If the knowledge of torture of others makes you sick, it is a case of sympathy; if it does not make you feel personally worse off, but you think it is wrong and you are ready to do something to stop it, it is a case of commitment." In economic terms, sympathy is a case of an externality, while commitment involves counterpreferential choice.[1]

I contend that this idea of commitment when applied on a social scale corresponds to Truman's "rules of the game," the set of social relations that are generally accepted and define the limits of legitimate behavior in both the political and economic spheres. The existence of community standards for economic behavior with regard to the setting of prices and wages are an example of these rules. Fairness as an factor in firm behavior has been explained by economists as instrumental to the maximization of long-run profit, but experimental studies have shown that fair behavior can be produced in the absence of enforcement. Expectations of fairness in markets include the belief that it is unfair to raise prices in the short run when there is excess demand unaccompanied by an increase in supply costs, that prices at the margin bear some relation to average prices (social limits to price gouging), it is more acceptable for prices to respond to variations in costs than changes in demand, and that price increases should be justified by cost increases.[2]

This underlying concept of fairness certainly explains the shift in attitudes toward the oil industry. During the period from WWII to 1965 the general decline in product prices meant that the industry must be behaving in an acceptable manner, even if it was actually pursuing a number of policies which a knowledgeable observer might find offensive. The only exception would be the natural gas industry, where attempts to raise the extremely low prices (relative to the cost of substitutes) prevailing until through the 1960s created the perception of an attempt to price gouge, encouraging the imposition of price controls. The oil price increases of the 1970s, especially when accompanied by large increases in profits, implied that the industry was engaging in unfair behavior and should be constrained, if not punished. The underlying political culture, shaped by the Populist and Progressive movements, was responsible for the assumption that the major oil companies must be responsible for this unfair behavior, and that it was necessary for the federal government to step in and take control of energy markets.

It is the existence of underlying belief systems which complicate the politics of microeconomic policy. While the battle between interest groups and the investment of resources to obtain influence have explanatory power in explaining policy decisions, they operate in a system where some interest groups are considered more legitimate than others, and some outcomes are more acceptable than others. Groups that can portray themselves as representing the public interest, or who can portray themselves as victims of the traditional villains of American politics, big business, have a competitive advantage. The irony is that in the name of free enterprise and efficiency programs which are economically inefficient are passed. This is not a new phenomenon as shown when Louis Brandeis, as a leader of the campaign to protect small business against large corporations and trusts in the name of preserving competition during the Progressive era, ended up favoring cartels and retail price fixing.[3]

This is not to say that merely allowing the market to operate free of governmental interference would necessarily lead to a social optimum. Before an optimal policy can be chosen there must be agreement upon the goals of policy, whether it is restoring an efficient market, maximizing social welfare, preventing an undesirable transfer of income, finding the most efficient method of preserving national security, or some combination of the above. The choice of goals is the essence of politics, but when there is no clear articulation of goals, it is impossible to develop rational policies to achieve these ends. When politicians used terms like "national security" as a smokescreen to disguise transfer payments to economic interest groups, and not as a signpost for policy, it is not surprising that an inefficient, irrational system of oil regulation was the end result.

Even when the problem could be framed in economic terms there was no assurance that the available data was sufficient to provide a definitive answer. There had long been extensive controversy over the amount of oil and gas available, and the costs of producing these resources. Paul MacAvoy and Robert Pindyck predicted in 1972 that if the price of gas was allowed to rise to 90¢ per thousand cubic feet (MCF) at the wellhead then production should reach 26.6 trillion cubic feet (TCF), but when the FPC allowed prices to rise to $1.42 MCF production in 1977 was only 20.3 TCF. The Stanford Research Institute claimed that the increase in production due to deregulation by 1980 would be higher than the quantity which the FEA predicted was possible by 1985. It is no wonder that liberal Democrats assumed that allowing prices to increase above $1.75 MCF would elicit little or no additional production. Estimates of the elasticity of demand for gasoline varied between -0.1 and -0.4 for the short-run elasticity, and -0.4 to -1.4 for the long run value, allowing both sides to develop widely divergent estimates of the impact on demand of oil price increases. During the debate over the windfall profits tax expected revenues rose from $20 to $200 billion, as price changes in the world market made projections obsolete the day they were printed. Studies conducted in 1977 and 1978, assuming constant prices, gave the expected demand in the US for oil in 1985 in a range from 21-24 million bpd, with expected domestic supplies of little over 10 million bpd. The actual numbers were 15.7 million bpd for demand and 10.6 million bpd production. Demand for OPEC oil was expected to be in the range of 36-50 million bpd, instead it actually declined to 16 million bpd.[4] The unexpectedly high prices of 1979 increased supply less than anyone expected, and reduced demand far more than would have been acceded by either producers or consumer advocates.

Apart from the inability of quantitative economists to come up with credible estimates of the parameters of their models (in energy models, supply and demand elasticities and the length of lags are essential for any meaningful projection of the impacts of policies) there were fundamental disagreements concerning the qualitative effects of policy. One key issue was whether the United States had an appreciable effect on the world oil market. In trade theory there is a concept of an "optimum tariff," which by lowering demand for an imported good can cause a decline in its price. Pindyck criticizes Stobaugh and Yergin's Energy Future for this supposition, yet one can turn around and find numerous articles claiming that the entitlements program, by subsidizing imports, increased the world price of oil, or that an optimal tariff would be good policy by reducing the world oil price and lowering the potential cost of supply disruptions.[5] To a non-economist this would appear to be a fundamental disagreement, but the core of the argument is about the magnitude of certain parameters; the extent to which US imports affect the world market, the elasticity of supply from different regions in the world and the elasticity of substitution among oil, other forms of energy and other inputs to production. Thus we cannot escape the necessity of quantitative estimation even if past experience casts doubt upon the validity of our calculations.

The problem for economists is that the data they use, and the theories applied to that data, only permit a rough approximation of the reality which is to be modeled. One of the most discouraging discoveries I had as a graduate student was the realization that there is no "right" way to conduct an econometric investigation which will give the "correct" answer. The problem is that the choice of relevant variables and specification of the functional form are basically matters of opinion. As the number of potential variables and functional forms is practically unlimited, the researcher is guaranteed to have specification error in his work, as the available degrees of freedom forces choice of a narrow variable set, while ignorance of reality requires choice of some functional form. As Leamer points out:

> The false idol of objectivity has done great damage to economic science. Theoretical econometricians have interpreted scientific objectivity to mean that an economist must identify exactly the variables in the model, the functional form, and the distribution of the errors. Given these assumptions, and given a data set, the econometric method produces an objective inference from a data set, unencumbered by the subjective opinions of the researcher.
>
> This advice could be treated as ludicrous, except that it fills all the econometric textbooks. Fortunately, it is ignored by applied econometricians. The econometric art as it is practiced at the computer terminal involves fitting many, perhaps thousands, of statistical models. One or several that the researcher finds pleasing are selected for reporting purposes. This searching for a model is often well intentioned, but there can be no doubt that such a specification search invalidates the traditional theories of inference. The concepts of unbiasedness, consistency, efficiency, maximum-likelihood estimation, in fact, all the concepts of traditional theory, utterly lose their

> meaning by the time an applied researcher pulls from the bramble of computer output the one thorn of a model he likes best, the one he chooses to portray as a rose.[6]

Policymakers don't have the luxury of going to conferences, arguing over model specifications, then publishing another paper and obtaining tenure and promotions by participating in this sort of controversy. They are trying to make decisions with consequences for political allies, constituents and even their own sense of justice and fairness. However, if economists don't agree on the numbers the politician is free to choose the estimate which corresponds to his policy prejudices. In the natural gas debates of 1977-78 numerous studies were used as ammunition to support both sides of the issue. House supporters of the administration relied on an analysis produced by the staff of John Dingell's Subcommittee on Energy and Power. House deregulation advocates used Dave Stockman's figures. Senate administration supporters, lead by Henry Jackson, used CBO numbers. Proconsumer critics used a study produced by Lawrence Kumins of the Congressional Research Service. Senate deregulation supporters used an Edward Erickson study prepared for the Natural Gas Supply Committee. A GAO study was cited by people on all side of the issue.[7]

The supplier of expertise was the traditional concept of the mechanism by which social scientists were felt to affect policy deliberations, beginning with Frederick Taylor's gospel of "scientific management", and the Progressive Movement's support for professional administration. Oil policy reinforces Banfield's skepticism of the value of social science research for policy formation, as "tens of thousands of Ph.D. dissertations later, there are few social science theories or findings that could be of much help to a policy maker . . . one can think of no program that was terminated, or even very substantially revised, because of an evaluation by policy scientists."[8] The abundance of contradictory energy studies meant that politicians were free to choose among researchers for ammunition to support their policy preferences.

George Stigler shared Banfield's pessimistic view with regard to the value of economics in the policy debate, as only economic theories developed as a response to the demand for rationalizations of desired policies will be used. Bronfenbrenner contended that most issues required inclusion of historical, legal, sociological, philosophical and political considerations, while the training of economists limited their ability to comment on many issues, exacerbated by the lack of agreement among economists and skepticism toward the opinions of economists. Economists who worked for the government felt that their work had little significant impact in the making of policy, being largely confined to support of programs determined by others. Explanations for the lack of influence ranged from the irrelevance of academic economics to the real world to the refusal of policymakers to apply even simple price theory. The suppression of a report by DOE's Policy and Evaluation office opposing continuing subsidies for small refiners by Deputy Secretary Jack O'Leary in 1980 illustrates the difficulties faced by the bearer of bad economic news.[9]

There are some observers with more optimistic views of the economist's influence. Kaysen, while granting the failure of economists to change certain policies after decades of criticism, asserted the value of economics as a filter. The economist selects out of the range of possible alternative policies those few which are acceptable given the knowledge of the profession and relevant to the perspective of his political client. Nelson claimed a victory for economists in the deregulation movement of the 1970s, gained by their undermining of the Progressive era ideology used to justify regulation, and the demonstration of the inefficiency of existing regulatory policies. Niskanen suggests that policy advisers are more influential than is generally perceived because they are in a position to shape how those with direct political authority think about issues. He suggests economists are most effective when defending views widely shared in the profession such as free trade.[10]

A number of authors support the "molders of perceptual frameworks" concept, suggesting that the real value of research is not the direct impact on policy-making but its effect on the intellectual climate in which policy is made. Carol Weiss proposed an enlightenment model of research as social criticism, providing the background of concepts, orientations, and empirical generalizations that inform policy. James Wilson claimed the influence of intellectuals comes from providing the conceptual language and ruling paradigms that become accepted assumptions for policy-makers. He is skeptical of the value of research but grants an important role to the influence of theory.[11] Craufurd Goodwin expressed similar thoughts concerning the contributions of economists:

> The most important element of the economist's mattering is the way in which he has affected the entire perception and discourse of society. This includes the focus of national attention on the C.P.I, The GNP, the unemployment rate, the balance of payments and the basic style of cost-benefit analysis. . . . Economics has become inextricably imbedded in the American consciousness, and it will not be possible again to reject it even if economists make mistakes.[12]

If the president's (and other political actors') economic ideology is an important determinant of the economic policies which he will support in office,[13] then the intellectual environment in which that ideology is formed will have a significant impact on the policy alternatives considered. This implies that a key factor in determining policies will be the perceptions of the public and the politicians concerning the nature of the issues under discussion. The role of social scientists in general, and economists in particular, in the policy formation process shifts from being merely short-run suppliers of specific expertise to that of long term molders of perceptual frameworks.

The profusion of economic reports generated in the debates over economic issues during the last two decades substantiates the argument that economists have little direct impact on policy. When nearly every interest group and government agency involved in a policy debate can find an economist to lend support to their side, economic analysis would seem to have little effect in policy debates.[14] However, the fact that everyone engaged in shaping economic policy finds it necessary to obtain some sort of economic blessing for their policy preferences suggests that economic reasoning has acquired legitimacy as the preferred method for resolving such disputes. This supports the hypothesis that economics has shaped the policy agenda by defining the language of discourse and the set of acceptable arguments. The importance of economics in the policy debate is as a restraint on the actions of interest groups, whether pursuing economic goals or an ideological agenda. While the balance of political power will determine which policies are implemented, the ubiquitous presence of economists delineates limits to the rapacity of interest group demands.

The oil industry's awareness of the potential dangers of economic criticism can be seen by the resources invested in refuting Eugene Rostow, the active participation of industry economists such as John Boatwright and Richard Gonzales at Congressional hearings and economic seminars, and the attacks on "academic thinkers" in criticizing the Oil Import Control report. However, by the early 1960s almost all of the industry's arguments in support of its privileges had been decisively refuted. By removing the veneer of respectability which the phrase "national security" provided for oil subsidies, economists changed the perception of these policies to one of special favors lacking an underlying rationale.

An additional factor which was added to the political equation was the emergence of public interest groups. These groups are based on the organization of people sharing specific sympathies and commitments (the environmental movement is a perfect example of a group which mixes sympathy and commitment) which may conflict with those held by the general public, or at least large segments of the public, as well as those of economically motivated interest groups. Within the political arena public interest groups behave in similar fashion to other interest groups with the exception that their motivations, and therefore the rewards needed to placate them, differ considerably. An actor pursing an economic interest will attempt to estimate costs and benefits (equating marginal cost with marginal benefit is an ideal, but unrealistic description of this behavior) and act accordingly, but a group which is motivated by sympathy and commitment cannot be satisfied unless it is victorious in achieving its agenda. Public interest groups therefore are constrained only by the resources which they command and not the efficacy with which they can apply those resources to desired goals. This explains why they pursue lost causes, i.e., engagement in actions for which costs exceed probable benefits.

These three factors, the public perception of certain economic behavior as "unfair," the erosion of legitimacy of current oil policies due to criticism from economists (transmitted to the public through the media), and the emergence of the public interest movement, changed the environment in which oil policy had been made. The potential latent power of consumers was apparent in the battles over natural gas policy, but except for the involvement of New England consumers in the conflict over residual imports, it had been absent from oil policy debates. The reason that public interest theorists overlooked the potential of consumer interests is that these interests are rarely expressed directly in the policy debate. Instead they are generated by an underlying perception of "unfairness" which can then be exploited by a political entrepreneur who transmits this attitude both to his fellow policymakers and to his constituents. The initial cost is minimal, since the barrier to entry consists of passing the media's filter, which requires a minimum of resources though it does demand the ability to locate a suitable issue and package it as entertainment. This process allows the entrepreneur to increase the intensity of support for his position at low cost, while informing other politicians of the potential value of identification with the issue.

Oil depletion in 1969 was the first example of this factor entering oil policy. While liberals had tried to eliminate the allowance for two decades the growing public perception that the allowance was an unfair advantage forced industry supporters to concede to small cuts as symbolic gestures to defuse opposition. The price increases which began in 1970, followed by the shortages resulting from the Nixon price controls, granted increasingly unfavorable publicity to an industry already scarred by its association with oil pollution. Had the oil import quota not become irrelevant it would have surely become the next political target. It is not surprising that rumors of major oil companies driving gasoline operators, marketers and small refineries out of business in order to drive up prices were given credence by politicians at this time, well before the oil embargo. The media, faced with the choice between blaming government regulation and market forces or a conspiracy by big oil, tended to lean toward the latter, more salable alternative.

The oil embargo of 1973 merely intensified the hostility toward the oil industry, as the gasoline lines and sharply higher prices which followed escalated the level of public indignation, creating additional opportunities for politicians to use oil companies as scapegoats. Oil policy was influenced by emotional considerations as much as either interest group pressure or economic logic. In the name of protecting consumers a complicated system of price controls, allocation, and entitlements was established, as Congress blocked price increases and left it to the bureaucracy to make the oil market function. Somehow the FEA was supposed to preserve an economically sound and competitive petroleum industry, equitably distribute crude oil and products at equitable prices to all regions of the country and sectors of the industry, maximize economic efficiency and minimize economic distortions. Of course it was impossible to accomplish such contradictory goals, and the FEA created a system designed to remove resources from the most politically unpopular group and grant them to more deserving beneficiaries. The simple solution of taxing economic rents and transferring them to deserving groups through lower tax rates or rebates was unacceptable (there were a number of problems with this solution; people perceived the problem as higher prices so transfers wouldn't address that source of discontent, it would be difficult to design a program which would be politically acceptable across regions; redistributing income to the low-income consumers most seriously affected by the price increases was not very popular). The political solution was to transfer income from oil producers to refiners, while adding special provisions to help out small oil companies. The hope was that transfers would work as a negative tax with a proportion shifted to consumers.

The actual system which was installed was the product of institutional inertia, ideological deadlock, and interest group machination. The problem began with the Nixon's price controls, and the need to have some form of allocation system to address the distortions resulting from the foolish attempt to administer an inflation solution. When price controls were placed on old oil, refiners with access to this oil would have lower average costs than refiners dependent on imported oil, which was thought to provide a competitive advantage. As Kalt points out, access to old oil did not provide an advantage in competition since product prices depended on the marginal, and not the average cost of oil.[15] The logic behind entitlements was the fear was that integrated companies with access to supplies of old oil would use their lower average costs to drive independent refiners out of business. While this argument does not hold up, controlling the price of old oil created a large source of economic rent, which accrued to refiners with access to this oil. The entitlements program was a way of spreading these rents over the whole universe of refiners, under the guise of increasing competition in order to benefit consumers.

Kalt estimates the size of the rents transferred from producers to refiners and consumers to be about $160 billion (1980$) from 1974-1980, of which $100 billion of the rents were gathered by refiners and about $45 billion were transferred to consumers. Assuming a ten percent normal decline rate, I estimate (using back-of-the-envelope methods) that approximately $100 billion of the $160 billion in transfers were economic rents earned on old oil due to the action of the OPEC cartel. Among the negative impacts of the entitlements program were the deadweight losses, due to misallocation of resources, plus the resources used to administer the program, both by government and industry. Kalt estimates these costs at $12-15 billion for the period from 1974 to 1980. The cost of additional oil imports over the period was about $100 billion, attributable to subsidization of imports through entitlements and domestic supply lost because of the misallocation of resources due to the price control regulations.[16]

While these are crude estimates, and may vary widely depending on the values of various elasticities, they provide some idea of the size of the economic benefits which interest groups were pursuing through the political process. The motivation of the liberal politicians who were responsible for the system of controls was to protect consumers and punish the large oil companies, who they felt were partially responsible for the escalating energy prices. The large multinational oil companies, although they had practically no influence on energy policy during this period, on average felt little pain from these actions. Large integrated firms can shift investments upstream or downstream, and across borders to maximize their returns, and all price controls did was increase domestic refining profits while encouraging them to shift production investments to foreign countries. The program provided additional profits to many refining firms, especially small refiners who found their inefficient operations subsidized.

The losers under the program were the large oil companies which were concentrated in production, and that had invested heavily in domestic production as smaller oil producers were able to obtain tax breaks and exemptions from oil controls. The other group of losers under oil regulation were foreign oil consumers, who paid higher oil prices because of American subsidization of oil imports. There were also significant losses with regard to the amorphous concept referred to as the public interest. This cost includes the direct deadweight losses due to the program, the resources invested in political activity to gain a share of the economic goods produced by the program, and the failure to address other social problems due to attention paid to energy issues. Subsidizing oil consumption was socially inefficient since direct transfer of the economic rents would have allowed consumers to choose where they wished to spend the additional income.

The small refiner provisions of both the oil import quota and the entitlements program are evidence that special interests need not invest tremendous resources to achieve large amounts of economic rents. The small refiners benefited from the perception of a lack of competition in the oil industry which prompted bureaucratic support (due to the FTC and the Justice Department) for subsidies, a sliding scale for import tickets followed by a sliding scale for entitlements under price controls. Small refiners had limited financial resources compared to their competitors, and small refineries were located in a small fraction of congressional districts. One would not expect them to be able to obtain substantial favors from Congress, yet the small refiners received an economic good worth about $700 million per year, and exceptions and appeals relief was worth an additional $460 million. Efficiency was definitely not a goal of the program, as providing a $2/bbl subsidy to refineries with higher costs merely encouraged the building of inefficient refineries. Proof of their inability to compete without subsidies can be seen in the fact that of the 3 million bpd of refinery capacity shutdown between 1981 and 1990, 129 refineries totaling 1.9 million bpd refining capacity were small company refineries, including 35 small refineries built since 1973 with 345,000 bpd capacity (compared to 24 refineries belonging to large companies, most of which were over 25 years old).[17]

Natural gas regulation, though established in the 1950s, can be criticized on the same grounds as oil price control. The distortions due to control of gas prices became apparent in the 1970s as "shortages" appeared. To an economist, the existence of a shortage implies a distortion in the market, since if prices are allowed to rise market forces will eliminate the shortage. The effect of price control of natural gas was to provide subsidies to customers with access to gas, and to encourage wasteful consumption since in a free market users who value the special qualities of natural gas would pay a premium. While gas pipelines, as natural monopolies, may require regulation, there had never been any logic behind the regulation of natural gas producers. To add insult to injury, while gas consumers were using political pressure to hold prices down, natural gas producers were receiving tax breaks to encourage production.

The usual explanation for these sort of misguided policies is the power of interest groups, and more recently, the resources invested in obtaining economic rents due to regulation. However, I contend that the problem has its roots in public economic myths concerning prices, profits and markets. Liberals tend to see business as automatically engaged in collusion unless constrained by government, and feel that it is both legitimate and effective to redistribute income by government action in the marketplace. Conservatives confuse the free market with maximum profits, and assume that any action that benefits business must increase public welfare, opposing regulations which increase market efficiency by internalizing externalities. Neither group understands the difference between the market as means of allocating resources and making production and consumption decisions using prices as signals, and capitalism, the system of private ownership of the means of production. Therefore they confuse questions of efficiency, involving the optimal functioning of an economic system, and questions of equity, which concern the distribution of wealth and income within society.

During the 1960s a third group emerged which combined the Conservative dislike of government interference with the Liberal distrust of business. This public interest movement is deeply rooted in the American political tradition, combining elements of Populism and Progressivism. The primary difference between the public-interest movement or Neopopulism, and New Deal Liberalism is their attitude toward containing the power of the corporation. Liberals call for increased authority of government over business, requiring a more centralized and powerful bureaucracy. Neopopulists oppose attempts to centralize regulatory authority in the executive branch, preferring a decentralized system which allows them access. They want consumers and taxpayers and environmentalists to have the same influence over public policy as business. Public interest groups desire participation in policy decisions as one of their goals.[18]

Schneider found among congressmen a similar three-fold dichotomy between Conservatives, Liberals and Progressives (similar to my Neopopulists), with the first two groups mirroring the splits with the economic profession, but Progressives rejecting the efficacy of the market. Liberals and Progressives agreed that there is an outstanding social agenda that must be met by government policies and programs. The major difference was the belief by Progressives that much of the economy was structured by powerful economic forces that did not behave in accordance with market models of economics.

Schneider found that this attitude was far more prevalent in the Congress than one would expect, shared by politicians who would be considered to the right of senators such as Kennedy, Humphrey, and Mondale. A Senator described as moderately liberal told Schneider that: "He deeply believed that the laws of supply and demand do not operate to determine rationally what is invested in, what is produced, and what is consumed, but that an inherently asocial process, skewed by the implications of enormous concentration of wealth and of the control over capital deployment, determined economic events."[19]

The coalition of Liberals and Neopopulists which dominated Congress during the 1970s would find price control a solution which they could agree on. Production incentives, either higher prices or subsidies, generated opposition by environmentalists and pleased only Conservatives. Using governmental power to allocate oil and control prices was a traditional Liberal solution which pleased consumer groups which saw it as a victory over business interests, especially since the regulations were administered by agencies in which they could exercise influence. Mandated conservation measures were also acceptable solutions, as they reflected the expectation that the market would not respond sufficiently without governmental intervention. Neopopulists welcomed the concentration of authority in agencies open to interest group influence. The collapse of support for price controls in the late 1970s was due to the defection of Liberals such as Bill Bradley, Paul Tsongas and John Dingell, who had become disillusioned with governmental interference in the market and accepted the need for prices to provide necessary incentives.[20]

This discussion leads back to the question asked at the beginning of this thesis: why the oil industry is unable to implement an oil import fee. The interest group and public choice theorists would assume the failure to pass such a bill was due to a split in the industry, and opposition by other industries. The problem is that there was also a split in the industry in the 1950s, and the large companies owned far more foreign oil in the past than at present. While some business groups may oppose a measure which would raise raw materials costs, many business leaders would favor a tax which would reduce the deficit. Certainly the national security argument for limitation of oil imports is far more legitimate in the 1980s than in the 1950s.

The resistance to import fees in the 1980s can be explained by the changes in the structure of Congress limiting the power of oil state members such as Russell Long, Lloyd Bentsen, and Jim Wright, and the mobilization of latent interests in the 1970s. The inability of party leaders to discipline their membership makes it impossible to pass unpopular legislation even with Presidential support. The result is that the set of acceptable policy alternatives is restricted to those which will not be perceived as unpleasant or painful, or those issues which are too complicated or obscure to receive media attention. Energy policies which depend on price signals to encourage conservation or to limit dependence on foreign oil are unacceptable, since the problem is price.

The illusion behind the "Energy Crisis" mentality was that the problem was the threat to national security from foreign oil. As Eisenhower made clear in the 1950s, any threat to Middle East oil, irregardless of its impact on American oil supplies, would trigger US military involvement because of the threat to our allies and trade partners. This was emphasized by the demonstration of American military might during the Gulf War. Therefore the only energy problem is the problem of price, since in a free market, shortages cannot persist if prices are allowed to return the market to equilibrium. High energy prices have two undesirable impacts; they cause slower economic growth and increased unemployment, and have undesirable redistributional effects, similar in nature to a sales tax.

Once we accept that price is the problem, the solution becomes a tradeoff between increases in oil prices to lower demand, the cost of maintaining an oil stockpile to bring stability to spot markets in times of market disruption, and the cost of market disruptions. Natural gas, coal and electricity can be treated like other industries, allowing the market to make supply and allocation decisions, with proper allowances for the costs of externalities such as pollution.

The political immaturity of the American people presents a difficult dilemma as pursuit of good policy is often bad politics, whether formulating a rational energy policy or dealing with the budget deficit. The reluctance to court disfavor produces a Congress which prefers to dawdle when faced with controversial issues requiring public sacrifice. Even a courageous President, willing to exercise leadership, may find himself blocked by a Congress which would rather obfuscate issues than resolve them, while a weak President may act as an accomplice, choosing to engage in symbolic politics as an alternative to action. A society built on the principle of instant gratification is incapable of the vision, will and wisdom to develop long term policies to deal with national problems.

NOTES

[1] Amartya K. Sen, "Rational Fools: A Critique of the Behavioral Foundations of Economic Theory," *Philosophy & Public Affairs*, (Summer 1977): 320-321.

[2] Daniel Kahneman, Jack Knetsch, and Richard Thaler, "Fairness as a Constraint on Profit Seeking: Entitlements in the Market," *American Economic Review* (September 1986): 728-39.

[3] McGraw, *Prophets of Regulation*, p. 141.

[4] Pietro S. Nivola, "Energy Policy and the Congress: The Politics of the Natural Gas Policy Act of 1978," *Public Policy*, Fall 1980, p. 531-33; *National Journal*, Aug. 26, 1978, p. 1350; API *Basic Petroleum Data Book*, Sept. 1987; *O&GJ*, Jan. 25, 1988, pp. 48-49.

[5] Robert S. Pindyck, "The American Energy Debate," *Public Interest*, Spring 1980, p. 102; Rodney T. Smith, "In Search of the 'Just' U.S. Oil Policy: A Review of Arrow and Kalt and More," *Journal of Business* 54, 1981, pp. 91-92; Douglas R. Bohi and Milton Russell, *US Energy Policy: Alternatives For Security* (Baltimore: Johns Hopkins Press, 1975), p. 1-2; and Harry Broadman and William Hogan, *Oil Tariff Policy in an Uncertain Market*, Discussion Paper, E-86-11, Harvard University, JFK School of Government, Nov. 1986, pp. 5-6.

[6] Edward Leamer, "Let's Take the Con Out of Econometrics," *American Economic Review*, March 1983, p. 36-37.

[7] Michael J. Malbin, *Unelected Representatives* (New York: Basic Books, 1979), p. 211-12.

[8] Edward C. Banfield, "Policy Science As Metaphysical Madness," in *Bureaucrats, Policy Analysts, Statesmen: Who Leads?*, ed. Robert Goldwin, Wash., D.C.: American Enterprise Institute, 1980, p. 3-10.

[9] George Stigler, "Do Economists Matter?" *Southern Economic Journal* (Jan. 1976): 347-54; Martin Bronfenbrenner, "Mutterings About Mattering," *Southern Economic Journal* (Jan. 1976): 358-60; William R. Allen, "Economics, Economists, and Economic Policy: Modern American Experiences," *History of Political Economy* (Spring 1977): 78-87, Hank Jenkins-Smith, "Analytical Debates and Policy Learning: Analysis and Change in the Federal Bureaucracy," *Policy Sciences* 21 (1988): 179-181.

[10] Carl Kaysen, "Model-Makers and Decision-makers: Economists and the Policy Process," *Public Interest* (Summer 1968): 94, Robert Nelson, "The Economics Profession and the Making of Public Policy," *Journal of Economic Literature* (March 1987): 61; William A. Niskanen, "Economists and Politicians," *Journal of Policy Analysis and Mangement* 5 (1986): 241-43.

[11] Carol H. Weiss, "Research for Policy's Sake: The Enlightenment Function of Social Research," *Policy Analysis* (Fall 1977): 544; James Q.

Wilson, "Policy Intellectuals and Public Policy," *Public Interest* (Summer 1981): 33, 46.

[12] Letter by Craufurd Goodwin to Martin Bronfenbrenner, in Bronfrenbrenner, "Mutterings About Mattering," p. 356.

[13] M. Stephen Weatherford, "The Interplay of Ideology and Advice in Economic Policy-Making: The Case of Political Business Cycles," *Journal of Politics* (Nov. 1987), p. 926-27.

[14] *National Journal*, June 6, 1987, p. 1457-1458.

[15] Kalt, *Economics and Politics of Oil Price Regulation*, p. 35.

[16] ibid., pp. 46, 191-234.

[17] Tom Alexander, "How Little Oil Hit a Gusher on Capitol Hill," *Fortune*, Aug. 14, 1978, p. 148-153; API, *Basic Petroleum Data Book*, Jan. 1995, Sec. VIII, Table 10c.

[18] David Vogel, "The Public-Interest Movement and the American Reform Tradition," *Political Science Quarterly*, Winter 1980-81, p. 616-27.

[19] Jerrold E. Schneider, *Ideological Coalitions In Congress* (Westport Conn.: Greenwood Press, 1979), p. 160-66, 176-89.

[20] *CQ Weekly Report*, Oct. 25, 1980, p. 3208.

Bibliography

Abrams, Robert, *Foundations of Political Analysis.* New York: Columbia University Press, 1980.

Adelman, M.A. "Efficiency of Resource Use in Crude Petroleum." Southern Economic Journal 31 (1964): 101-122.

———. "Oil Prices in the Long Run (1963-1975)," *Journal of Business* 37 (1964): 143-161.

———. *The World Petroleum Market.* Baltimore: Johns Hopkins Press, 1972.

———. "Is the Oil Shortage Real? Oil Companies As OPEC Tax-Collectors." *Foreign Policy* 9 (Winter 1972-1973): 69-107.

———. "The World Oil Cartel: Scarcity, Economics, and Politics," *Quarterly Journal of Economics and Business* 16 (Summer 1976): 7-18.

———. "OPEC As a Cartel." in *OPEC Behavior and World Oil Prices.* edited by James M. Griffen and David J. Teece, 37-63. London: George Allen & Unwin, 1982.

———. "An Unstable World Oil Market," *Energy Journal* 6 (1985): 17-22

———. "The Competitive Floor to World Oil Prices," *Energy Journal* 7 (1986): 9-31.

———. "Scarcity and World Oil Prices," *Review of Economics and Statistics* (1986): 387-397.

Akerlof, George A. "The Economic Consequences of Cognitive Dissonance." In *An Economic Theorist's Book of Tales*, edited by George Akerlof and William T. Dickens, 124-140. Cambridge: Cambridge University Press.

Akins, James E. "The Oil Crisis: This Time the Wolf is Here," *Foreign Affairs* 51 (April 1973): 462-490.

Aldrich, John H., and Forrest D. Nelson. *Linear Probability, Logit, and Probit Models.* Beverly Hills: Sage Publications, 1984.

Alexander, Herbert E., *Financing Politics*, 3rd ed. Washington, D.C.: Congressional Quarterly Press, 1984.

Allen, William R. "Economics, Economists, and Economic Policy: Modern American Experiences," *History of Political Economy* 9 (Spring 1977): 48-88.

Alt, James E., and K. Alec Chrystal. *Political Economics.* Berkeley: University of California Press, 1983.

Amacher, Ryan C., Robert D. Tollison, and Thomas D. Willett. "Import Controls on Foreign Oil: Comment," *American Economic Review* 63 (December 1973): 1031-1034.

American Petroleum Institute, *Basic Petroleum Data Book*, September 1987, January 1995.

Areskoug, Kaj. "US Oil Import Quotas and National Income." *Southern Economic Journal* 37 (January 1971): 307-317.

Bain, Joe S. "Rostow's Proposals for Petroleum Policy." *Journal of Political Economy* 57 (February 1949): 55-60.

Banfield, Edward C. "Policy Science As Metaphysical Madness." In *Bureaucrats, Policy Analysts, Statesmen: Who Leads?* edited by Robert Goodwin, 1-19. Washington, D.C.: American Enterprise Institute, 1980.

Barber, William J. "Eisenhower Energy Policy." In *Energy Policy in Perspective: Today's Problems, Yesterday's Solutions.* edited by Craufurd D. Goodwin, 205-286. Washington D.C.: Brookings, 1981.

________. "Studied Inaction in the Kennedy Years," In *Energy Policy in Perspective: Today's Problems, Yesterday's Solutions.* edited by Craufurd D. Goodwin, 287-336. Washington D.C.: Brookings, 1981.

Barnett, Harold J., and Chandler Morse. *Scarcity and Growth.* Baltimore: Johns Hopkins University Press, 1963.

Barro, Robert. "The Control of Politicans: An Economic Model." *Public Choice* 14 (Spring 1973): 19-42.

Barry, Brian. *Sociologists, Economists and Democracy.* Chicago: University of Chicago Press, 1978.

Barzel, Yoram, and Christopher D. Hall. *The Political Economy of the Oil Import Quota.* Stanford: Hoover Institute Press, 1977.

Bauer, Raymond, Lewis Anthony Dexter and Ithiel De Sola Pool. *American Business and Public Policy.* 2nd ed. Chicago: Aldine Atherton, Inc., 1972.

Becker, Gary S. "A Theory of Competition Among Pressure Groups For Political Influence." *Quarterly Journal of Economics* 48 (August 1983): 371-400.

Belsley, David, Edwin Kuh and Roy Welch. *Regression Diagnostics.* NY: John Wiley & Sons, 1980.

Bennett, Lance W. *Public Opinion In American Politics.* New York: Harcourt Brace Jovanovich, 1980.

Bernstein, Robert A., and Stephen R. Horn. "Explaining House Voting on Energy Policy: Ideology and the Conditional Effects of Party and District Economic Interests." *Western Political Quarterly* 34 (June 1981): 235-245.

Berstrom, Theodore C. "On Capturing Oil Rents with a National Excise Tax." *American Economic Review* 72 (March 1982): 194-201.

Blair, John M. *The Control of Oil.* New York: Random House, 1976.

Bohi, Douglas R., and Milton Russell. *US Energy Policy: Alternatives For Security.* Baltimore: Johns Hopkins Press, 1975.

________. *Limiting Oil Imports.* Baltimore: Johns Hopkins Press, 1978.

Bond, Jon R. "Oiling the Tax Committees in Congres, 1900-1974: Subgovernment Theory, the Overrepresentation Hypothesis, and the Oil Depletion Allowance," *American Journal of Political Science* 23 (November 1979): 651-664

Bradley, Paul G. *The Economics of Crude Petroleum Production*. Amsterdam: North-Holland Publishing Company, 1967.

Brand, Donald R. "Corporatism, the NRA, and the Oil Industry." *Political Science Quarterly* 98 (Spring 1983): 99-119.

Breyer, Stephen. *Regulation and Its Reform*. Cambridge: Harvard University Press, 1982.

Broadman, Harry G. "The Social Cost of Imported Oil," *Energy Policy* 14 (June 1986): 242-252.

Broadman, Harry G., and William W. Hogan. "Oil Tariff Policy in an Uncertain Market." Discussion Paper, E-86-11, Harvard University, JFK School of Government, November 1986.

Bronfenbrenner, Martin. "Mutterings About Mattering," *Southern Economic Journal* 42 (January 1976) 355-363.

Brown, Jonathan, C. "Why Foreign Oil Companies Shifted Their Production from Mexico to Venezuela during the 1920s." *American Historical Review* 90 (April, 1985): 378-379.

Browning, Edgar, "On the Welfare Cost of Transfers," reprinted in *Toward A Theory of the Rent-Seeking Society*, edited by James Buchanan, Robert Tollison, and Gordon Tullock, 283-288. College Station: Texas A&M University Press, 1980.

Brunner, Ronald D., and Vivian E. Weston. "Citizen Viewpoints on Energy Policy." *Policy Sciences* 12 (1980): 147-174.

Buchanan, James M., and Gordon Tullock. *The Calculus of Consent*. Ann Arbor: University of Michigan Press, 1962.

Burrows, James C., and Thomas A. Domencich. *An Analysis of the United States Oil Import Quota*. Charles River Associates Research Study, Lexington, Mass.: Heath Lexington Books, 1970.

Cabinet Task Force on Oil Import Control, *The Oil Import Question*. Washington: GPO, 1970.

Caddell, Patrick H. "Why Kevin Phillips Misfired." *Public Opinion* 1 (July/August 1978): 54-58.

Campbell, Angus, Philip Converse, Warren Miller and Donald Stokes. *The American Voter*. Chicago: University of Chicago Press, 1960.

Campbell, Duncan R.G., "Public Policy Problems of the Domestic Crude Oil Industry: Comment," *American Economic Review* 54 (March 1964): 114-118.

Carlisle, Donald, "The Economics of a Fund Resource With Particular Reference to Mining," *American Economic Review* 44 (September 1954): 595-616.

Carson, Richard T., and Joe A. Oppenheimer. "A Method of Estimating the Personal Ideology of Political Representatives." *American Political Science Review* 78 (March 1984): 163-178.

Cassady, Ralph. *Price Making and Price Behavior in the Petroleum Industry.* New Haven: Yale University Press, 1954.

Chubb, John E. *Interest Groups and the Bureaucracy.* Stanford: Stanford University Press, 1983.

Cicchetti, Charles J., and William J. Gillen, "The Mandatory Oil Import Quota Program: A Consideration of Economic Efficiency and Equity." *Natural Resources Journal* 13 (July 1973): 399-430.

Ciriacy-Wantrup, S.V. *Resource Conservation: Economics and Policies.* Berkeley: University of California Press, 1952.

Clark, Colin. "Afterthoughts on Paley." *Review of Economics and Statistics* 36 (August 1954): 267-273.

Clausen, Aage R. *How Congressmen Decide: A Policy Focus.* New York: St. Martins Press, 1973.

Cochrane, James L. "Energy Policy in the Johnson Administration: Logical Order versus Economic Pluralism." In *Energy Policy in Perspective: Today's Problems, Yesterday's Solutions*, edited by Craufurd D. Goodwin, 337-394. Washington, D.C.: Brookings 1981.

________. "Carter Energy Policy and the Ninety-fifth Congress," In *Energy Policy in Perspective: Today's Problems, Yesterday's Solutions*, edited by Craufurd D. Goodwin, 547-600. Washington, D.C.: Brookings 1981.

Congressional Quarterly Almanac, 1950-1986.

Congressional Quarterly Weekly Report, various issues.

Congressional Quarterly, *Congress and the Nation.* Washington D.C.: Congressional Quarterly Press, 1965.

________. *Energy and the Environment, The Unfinished Business.* Washington D.C.: Congressional Quarterly Press, 1985.

________. *Energy Policy.* 2nd ed., Washington D.C.: Congressional Quarterly Press, 1981.

CONSAD Research Corporation, "The Economic Factors Affecting the Level of Domestic Petroleum Reserves," December 1968. In *Tax Reform Studies and Proposals*, Joint Publication Committee on Ways and Means and Committee on Finance. 91st Cong., 1st Sess., 1969.

Conway, M. Margaret, "PACs, the New Politics, and Congressional Campaigns." In *Interest Group Politics*, edited by Allan J. Cigler and Burdett A. Loomis, 126-144. Washington D.C.: Congressional Quarterly Press, 1983.

Costain, W. Douglas, and Anne N. Costain, "Interest Groups as Policy Aggregators in the Legislative Process," *Polity* 14 (Winter 1981): 249-272.

Cox, James C., and Arthur W. Wright, "A Tariff Policy For Independence From Oil Embargoes," *National Tax Journal* 28 (March 1975): 29-41.

Cremer, Jacques, and Martin L. Weitzman, "OPEC and the Monopoly Price of World Oil." *European Economic Review* 8 (1976): 155-164.

Daly, George, James M. Griffin, and Henry B. Steele. "The Future of OPEC: Price Level and Cartel Stability." *Energy Journal* 4 (1983): 65-77.

Dam, Kenneth W. "Implementation of Import Quotas: The Case of Oil." *Journal of Law and Economics* 14 (April 1971): 1-60.

Davidson, Paul. "Public Policy Problems of the Domestic Crude Oil Industry," *American Economic Review* 53 (March 1963): 85-108.

———. "Public Policy Problems of the Domestic Crude Oil Industry: A Reply," *American Economic Review* 54 (March 1964): 125-134.

———. "The Depletion Allowance Revisited," *Natural Resources Journal* 10 (January 1970): 1-9.

Davidson, Paul, Laurence H. Falk and Hoesung Lee. "Oil: Its Time Allocation and Project Independence." *Brookings Papers on Economic Activity* 2 (1974): 411-448.

Davidson, Roger H. "Two Avenues of Change: House and Senate Committee Reorganization." In *Congress Reconsidered*, 2nd ed. edited by Lawrence C. Dodd and Bruce I. Oppenheimer, 107-133. Washington D.C.: Congressional Quarterly Press, 1981.

Davies, J. Clarence, and Barbara S. Davies. *The Politics of Pollution*. 2nd ed., Indianapolis: Bobbs-Merrill Company, 1975.

Davis, David Howard. *Energy Politics*. 2nd ed. New York: St. Martin's Press, 1978.

Day, Barbara S. T. *The Oil and Gas Industry and Texas Politics, 1930-1935*, Ph.D. diss., Rice University, 1973.

De Chazeau, Melvin G., and Alfred E. Kahn. *Integration and Competition in the Petroleum Industry*. New Haven: Yale University Press, 1959.

De Marchi, Neil. "Energy Policy Under Nixon, Mainly Putting Out Fires." In *Energy Policy in Perspective: Today's Problems, Yesterday's Solutions*, edited by Craufurd D. Goodwin, 395-474. Washington D.C.: Brookings, 1981.

________. "The Ford Administration: Energy As a Political Good," In *Energy Policy in Perspective: Today's Problems, Yesterday's Solutions*, edited by Craufurd D. Goodwin, 475-546. Washington D.C.: Brookings, 1981.

Derthick, Martha, and Paul J. Quirk. *The Politics of Deregulation*. Washington D.C.: Brookings, 1985.

Devarajan, S., and A.C. Fisher. "Hotelling's 'Economics of Exhaustible Resources': Fifty Years Later." *Journal of Economic Literature* 19 (March 1981): 65-73.

Downs, Anthony. *An Economic Theory of Democracy*. New York: Harper & Row, 1957.

________. "Up and Down With Ecology--the 'Issue-Attention Cycle." *Public Interest* 28 (1972): 38-50.

Drapkin, Dennis B., and Philip K. Verleger, Jr. "The Windfall Profit Tax: Origins, Development, Implications." *Boston College Law Review* 22 (May 1981): 631-704.

Dugger, Ronnie. "Oil and Politics." *Atlantic Monthly* 224 (September 1969): 66-90.

Eaton, Jonathon, and Zvi Eckstein. "The US Strategic Petroleum Reserve: An Analytic Framework." In *The Structure and Evolution of Recent US Trade Policy*, edited by Robert E. Baldwin and Anne O. Krueger, 237-276. Chicago: University of Chicago Press, 1984.

Edelman, Murray. *The Symbolic Use of Politics.* Urbana: University of Illinois Press, 1964.

Eismeier, Theodore, and Philip Pollock. "Political Action Committees: Varieties of Organization and Strategy." In *Money and Politics in the United States*, edited by Michael J. Malbin, 122-141. Chatham, NJ: Chatham House Publishers, 1984.

Eldridge, Douglas, H. "Tax Incentives for Mineral Enterprise." *Journal of Political Economy* 58 (June 1950): 222-240.

———. "Rate of Return, Resource Allocation and Percentage Depletion," *National Tax Journal* 15 (June 1962): 209-217.

Engler, Robert. *The Politics of Oil.* Chicago: University of Chicago Press, 1961.

Epple, Dennis N. *Petroleum Discoveries And Government Policy.* Cambridge, Mass.: Ballinger Publishing, 1975.

Erickson, Edward W. "Crude Oil Prices, Drilling Incentives and the Supply of New Discoveries." *Natural Resources Journal* 10 (January 1970): 27-52.

Etzioni, Amitai. *Capital Corruption.* New York: Harcourt Brace Jovanovich, 1984.

Everett, James. *Congress and National Energy Policy.* New Brunswick: Transactions Books, 1984.

Ezzati, Ali. "Future OPEC Price and Production Strategies as Affected By Its Capacity to Absorb Oil Revenues." *European Economic Review* 8 (1976): 107-138.

Federal Trade Commission. *The International Petroleum Cartel.* 82nd Cong., 2nd sess., Staff Report, Washington: GPO, 1952.

Feis, Herbert. "Oil for Peace Or War." *Foreign Affairs* 32 (April 1954): 416-429.

Fenno, Richard F., "The Internal Distribution of Influence: The House." In *Readings on Congress*, edited by Raymond E. Wolfinger, 199-224. New Jersey: Prentice-Hall, 1971.

———. *Congressmen In Committees.* Boston: Little, Brown & Co., 1973.

———. "If, As Ralph Nader Says, Congress Is 'The Broken Branch,' How Come We Love Our Congressmen So Much?" In *Congress in Change*, edited by Norman J. Ornstein, 277-287. New York: Praeger Publishers, 1975.

———. *Home Style: House Members in Their Districts.* Boston: Little, Brown and Company, 1978.

Fiorina, Morris P., *Representatives, Roll Calls, and Constituencies.* Lexington, Mass.: Lexington Books, 1974.

——— . "The Case of the Vanishing Marginals: The Bureaucracy Did It." *American Political Science Review* 71 (March 1977): 177-181.

——— . *Retrospective Voting in American National Elections.* New Haven: Yale University Press, 1981.

——— . "The Presidency and the Contemporary Electoral System," in *The Presidency and the Political System*. Edited by Michael Nelson, 204-226. Washington, DC: Congressional Quarterly Press, 1984.

Fisher, Franklin M. *Supply And Costs In The US Petroleum Industry*. Baltimore: Johns Hopkins Press, 1964.

——— . "The Social Costs of Monopoly and Regulation: Posner Reconsidered," *Journal of Political Economy* 93 (1985): 410-415.

Fox, Harrison W. Jr., and Susan Webb Hammond. *Congressional Staffs*. New York: The Free Press, 1977.

Folkerts-Landau, Elena. "The Social Cost of Imported Oil," *Energy Journal* 5 (1984): 41-58.

Frank, Helmut J. *Crude Oil Prices in the Middle East*. New York: Frederick A. Praeger, 1966.

Frankel, P.H. *Essentials of Petroleum*. London: Chapman & Hall, 1946.

——— . "Integration in the Oil Industry," *Journal of Industrial Economics* 1 (July 1953): 202-211.

——— . "Crude Oil Prices--A Comment," *Journal of Industrial Economics* 6 (October 1957).

Frohlich, Norman, and Joe A. Oppenheimer. *Modern Political Economy*. Englewood Cliffs, NJ: Prentice-Hall, 1978.

Froman, Lewis A. "Differences Between the House and Senate." In *Readings on Congress*, edited by Raymond E. Wolfinger, 61-66. New Jersey: Prentice-Hall, 1971.

Fusso, Thomas E. "The Polls: The Energy Crisis in Perspective." *Public Opinion Quarterly* 42 (Spring 1978): 131-133.

Gallup, George Jr. *Gallup Poll*. Wilmington, Del.: Scholarly Resources, 1987.

Gans, Herbert J. *Deciding What's News*. New York: Vintage Books, 1980.

Gately, Dermot. "A Ten Year Retrospective: OPEC and the World Oil Market." *Journal of Economic Literature* 22 (September 1984): 1100-1114.

Gately, Dermot, "The Prospects For Oil Prices, Revisited," *Annual Review of Energy* 11 (1986): 513-538.

Gilbert, Richard J. "Dominant Firm Pricing Policy in a Market For an Exhaustible Resource." *Bell Journal of Economics* 9 (Autumn 1978): 385-395.

Gilbert, Richard J., and Knut Anton Mork, "Efficient Pricing During Oil Supply Disruptions," *Energy Journal* 7 (1986): 51-68.

Goodwin, Craufurd D., "Truman Administration Policies toward Particular Energy Sources," in *Energy Policy in Perspective: Today's Problems, Yesterday's Solutions*. Edited by Craufurd D. Goodwin, 63-204. Washington., D.C.: Brookings, 1981.

Goodwin, George. *The Little Legislatures: Committees of Congress*. Mass.: University of Massachusetts Press, 1970.

Gordon, Richard L. "A Reinterpretation of the Pure Theory of Exhaustion." *Journal of Political Economy* 75 (June 1967): 274-286.

Gordon, H. Scott. "The Economic Theory of a Common-Property Resource: The Fishery." *Journal of Political Economy* 62 (April 1954).

Gormley, William T., Jr. "Regulatory Issue Networks in a Federal System. *Polity* 18 (Summer 1986): 595-620.

Gray, Lewis C. "Rent Under the Assumption of Exhaustibility." *Quarterly Journal of Economics* 28 (May 1914): 66-89.

Gray, Horace M. "Percentage Depletion, Conservation, and Economic Structure." In *Federal Tax Policy for Economic Growth and Stability*. Joint Committee on the Economic Report, 84th Cong., 1st sess., 430-439. Washington: GPO, 1955.

Greenberger, Martin. *Caught Unawares: The Energy Decade in Retrospect*. Cambridge, Mass.: Ballinger Publishing Co., 1983.

Griffin, James M., and Clifton T. Jones. "Falling Oil Prices: Where Is the Floor?" *Energy Journal* 7 (1986): 37-50.

Hamilton, Alexander, James Madison and John Jay. *The Federalist Papers*. New York: The New American Library, 1961.

Hammond, Susan Webb. "Legislative Staffs." *Legislative Studies Quarterly* 9 (May 1984): 284-317.

Harberger, Arnold C., The Taxation of Mineral Industries." in *Federal Tax Policy for Economic Growth and Stability*. Joint Committee on the Economic Report, 84th Cong., 1st sess., 440-449. Washington: GPO, 1955.

Hause, John C. "The Economic Consequences of Percentage Depletion Allowances." *National Tax Journal* 16 (December 1963): 405-409.

Hay, George A. "Import Controls on Foreign Oil: Tariff or Quota?" *American Economic Review* 61 (September 1971): 688-691.

Hays, Samuel P. *Conservation and the Gospel of Efficiency*. Cambridge: Harvard University Press, 1959.

Heclo, Hugh. "Issue Networks and the Executive Establishment." In *The New American Political System*, edited by Anthony King, 87-124. Washington: American Enterprise Institute, 1978.

Heitman, George. "US Oil Import Quotas and National Income: A Comment." *Southern Economic Journal* 38 (January 1972): 430-432.

Herfindahl, Orris C. "Some Fundamentals of Mineral Economics." *Land Economics* 31 (May 1955): 131-138.

________. "Depletion and Economic Theory." In *Extractive Resources and Taxation*, edited by Mason Gaffney, 63-90. Madison: University of Wisconsin Press, 1967.

Hinkley, Barbara. *Congressional Elections*, Washington, D.C.: Congressional Quarterly Press, 1981.

Hnyilicza, Esteban, and Robert S. Pindyck. "Pricing Policies for a Two-Part Exhaustible Resource Cartel.," *European Economic Review* 8 (1976): 149-154.

Hofstadter, Richard. *The Age of Reform*. New York: Vintage Books, 1955.

Holloway, Harry, and John George. *Public Opinion: Coalitions, Elites, and Masses*. New York: St. Martin's Press, 1979.

Hoskins, Halford L. "Needed: A Strategy For Oil." *Foreign Affairs* 29 (January 1951): 229-237.

Hotelling, Harold. "The Economics of Exhaustible Resources." *Journal of Political Economy* 39 (April 1931): 137-175.

Humphrey, Don D. *American Imports*. New York: The Twentieth Century Fund, 1955.

Ise, John. *The United States Oil Policy*. New Haven: Yale University Press, 1926.

Issawi, Charles, and Mohammed Yeganeh. *The Economics of Middle East Oil*. New York: Praeger Publishers, 1962.

Iyengar, Shanto, Mark D. Peters, and Donald R. Kinder. "Experimental Demonstrations of the 'Not-So-Minimal' Consequences of Television News Programs." *American Political Science Review* 76 (1982): 848-858.

Iyengar, Shanto and Donald R. Kinder. "More Than Meets The Eye: TV News, Priming, And Public Evaluations of the President." In *Public Communication And Behavior*, Vol. 1., edited by George Comstock, 136-171. Orlando, Fla.: Academic Press, 1986.

Jacobson, Gary, *Money in Congressional Elections*, New Haven: Yale University Press, 1980.

———. *The Politics of Congressional Elections*. Boston: Brown, Little and Company, 1983.

———. "Money and Votes Reconsidered: Congressional Elections, 1972-1982." *Public Choice* 47 (1985): 1-12.

———. "Running Scared: Elections and Congressional Politics in the 1980s." In *Congress: Structure and Policy*, edited by Mathew D. McCubbins and Terry Sullivan, 39-81. New York: Cambridge University Press, 1987.

Jenkins-Smith, Hank C. "Analytical Debates and Policy Learning: Analysis and Change in the Federal Bureaucracy." *Policy Sciences* 21 (1988): 169-211.

Johanny, Ali D. "OPEC and the Price of Oil: Cartelization or Alteration of Property Rights." *Journal of Energy and Development* 5 (Autumn 1979): 72-80.

———. *Myth of the OPEC Cartel: The Role of Saudi Arabia*. Chichester: John Wiley & Sons, 1980.

Johnson, G. Griffith to the Secretary of State, Report of the United States Delegation to the United States-Canadian Discussions of Petroleum Policies and Programs, Ottawa, Canada, December 13-14, 1962, declassified 11/26/79.

Jones, Charles O. "Speculative Augmentation in Federal Air Pollution Policy-Making." *Journal of Politics* 36 (1974): 438-464.

Jones, Charles O., and Randall Strahan. "The Effect of Energy Politics on Congressional and Executive Organization in the 1970s." *Legislative Studies Quarterly* 10 (May 1985): 151-179.

Judge, George, Carter Hill, William Griffiths, Helmut Lutkepohl and Tsoung-Chao Lee. *Introduction to the Theory and Practice of Econometrics.* New York: John Wiley & Sons, 1982.

———. *The Theory and Practice of Econometrics.* 2nd ed. New York: John Wiley & Sons, 1985.

Kahn, Alfred, "The Depletion Allowance in the Context of Cartelization." *American Economic Review* 54 (June 1964): 286-314.

———. "The Combined Effects of Prorationing, the Depletion Allowance and Import Quotas on the Cost of Producing Crude Oil in the United States." *Natural Resources Journal* 10 (January 1970): 53-61.

Kahneman, Daniel, Jack Knetsch, and Richard Thaler. "Fairness as a Constraint on Profit Seeking: Entitlements in the Market." *American Economic Review* (September 1986): 728-739.

Kalt, Joseph P. *The Economics and Politics of Oil Price Regulation* Cambridge: MIT Press, 1981.

Kalt, Joseph P., and Mark A. Zupan. "Capture and Ideology in the Economic Theory of Politics." *American Economic Review* 74 (June 1984): 279-300.

Kalt, Joseph, and Robert Leone. "Regional Effects of Energy Price Decontrol: the Roles of Interregional Trade, Stockholding, and Microeconomic Incidence." *Rand Journal of Economics* 17 (Summer 1986): 201-213.

Kau, James B., and Paul H. Rubin. "Self-Interest, Ideology, and Logrolling in Congressional Voting." *Journal of Law and Economics* 22 (October 1979): 365-384.

———. "Public Interest Lobbies: Membership and Influence." *Public Choice* 34 (1979): 45-54.

Kau, James, Donald Keenan and Paul Rubin. "A General Equilibrium Model of Congressional Voting." *Quarterly Journal of Economics* 97 (May 1982): 271-294.

Kaysen, Carl. "Model-Makers and Decision-Makers: Economists and the Policy Process." *Public Interest* 12 (Summer 1968): 80-95.

Kennedy, Michael. "An Economic Model of the World Oil Market." *Bell Journal of Economics* 5 (August 1974): 540-577.

Kennedy, William J., editor. *Secret History of the Oil Companies in the Middle East.* Salisbury, North Carolina: Documentary Publications, 1979.

Keohane, Robert O. "State Power and Industry Influence: American Foreign Oil Policy in the 1940s." *International Organization* 36 (Winter 1982): 168-178.

Kernell, Samuel. *Going Public: New Strategies Of Presidential Leadership.* Washington DC: Congressonal Quarterly Press, 1986.

Key, V.O. Jr. *Public Opinion and American Democracy.* New York: Alfred A. Knopf, 1961.

Keynes, John Maynard. *The General Theory of Employment, Interest and Money*. New York: Harcourt Brace Jovanovich, 1964.

Kinder, Donald R., and Walter R. Mebane. "Politics and Economics in Everyday Life." In *The Political Process and Economic Change*, edited by Kristen R. Monroe. New York: Agathon Press, 1983.

Kingdon, John W. *Congressmen's Voting Decisions*, 2nd ed., New York: Harper and Row, 1981.

———. *Agendas, Alternatives and Public Policies*. Boston: Little, Brown and Company, 1984.

Kolb, Jeffrey Alan. *An Econometric Study of Discoveries of Natural Gas and Oil Reserves In the United States, 1948 to 1970*. New York: ARNO Press, 1979.

Kouris, George. "Energy Demand Elasticities in Industrialized Countries: A Survey." *Energy Journal* 4 (1983): 73-94.

Krasner, Stephen D. *Defending the National Interest*. New Jersey: Princeton University Press, 1978.

Krueger, Anne O., "The Political Economy of the Rent Seeking Society," *American Economic Review* 64 (June 1974): 291-303.

Landeau, Jean-Francois G. *Strategies of US Independent Oil Companies Abroad*. Ann Arbor: UMI Research Press, 1977.

Latham, Earl. *The Group Basis of Politics*. Ithaca: Cornell University Press, 1952.

Latus, Margaret Ann. "Assessing Ideological PACs: From Outrage to Understanding." In *Money and Politics in the United States*, edited by Michael J. Malbin, 142-171. Chatham, NJ: Chatham House Publishers.

Leeman, Wayne A. "Crude Oil Prices in the United States at the Gulf Coast." *Journal of Industrial Economics* 5 (July 195): 180-191.

Leamer, Edward. "Let's Take the Con out of Econometrics." *American Economic Review* 73 (March 1983): 31-43.

Levine, Michael E., and Jennifer L. Forrence. "Regulatory Capture, Public Interest, and the Public Agenda: Toward a Synthesis." *Journal of Law, Economics, and Organization* 6 (1990): 167-212.

Levy, Brian. "World Oil Marketing In Transition." *International Organization* 36 (Winter 1982): 113-133.

Levy, Walter J. "The Past, Present and Likely Future Price Structure For the International Oil Trade." In *Third World Petroleum Congress*, Volume 10, The Hague, 1951.

________. "Oil Power." *Foreign Affairs* 49 (July 1971): 652-668.

Lewis, Tracy, and Richard Schmalensee. "On Oligopolistic Markets For Nonrenewable Natural Resources." *Quarterly Journal of Economics* 95 (November 1980): 475-491.

Libecap, Gary D., and Steven N. Wiggins. "The Influence of Private Contractual Failure on Regulation: The Case of Oil Field Unitization." *Journal of Political Economy* 93 (1985): 690-713.

Lipset, Seymour Martin, and William Schneider. *The Confidence Gap*, rev. ed., Baltimore: Johns Hopkins University Press, 1987.

Logan, Leonard M. *Stabilization of the Petroleum Industry*. Norman: University of Oklahoma Press, 1930.

Lopreato, Sally C., and Fred Smoller, "Explaining Energy Votes in the Ninety-Fourth Congress." (Unpublished) Austin: Center for Energy Studies, University of Texas at Austin, June 1978.

Lovejoy, Wallace F. "Oil Import Policies As They Relate To National Security." *Quarterly of the Colorado School of Mines* 64 (October 1969): 157-180.

________. "Oil Conservation, Producing Capacity, and National Security." *Natural Resources Journal* 10 (January 1970): 62-96.

Lovejoy, Wallace F., Paul T. Homan, and Charles O. Galvin. *Problems of Cost Analysis In The Petroleum Industry*. Dallas: Southern Methodist University Press, 1964.

Lovejoy, Wallace F., and Paul T. Homan. *Economic Aspects of Oil Conservation Regulation*, Baltimore: Johns Hopkins University Press, 1967.

Lowi, Theodore J. "American Business, Public Policy, Case-Studies, and Political Theory," *World Politics* 16 (1964): 677-715.

———. *The End of Liberalism*, 2nd ed., New York: W.W. Norton, 1979.

MacDougall, A. Kent. *Ninety Seconds To Tell It All*. Homewood, Ill: Dow Jones-Irwin, 1981.

MacKuen, Michael B., and Steven L. Coombs. *More Than News: Media Power In Public Affairs*. Beverly Hills: Sage Publications, 1981.

Maddala, G. S. *Limited-Dependent and Qualitative Variables in Econometrics*. Cambridge: Cambridge University Press, 1983.

Malbin, Michael J. *Unelected Representatives*. New York: Basic Books, 1979.

———. "Looking Back at the Future of Campaign Finance Reform." In *Money and Politics in the United States*, edited by Michael J. Malbin, 232-276. Chatham, NJ: Chatham House Publishers, 1984.

Mancke, Richard. "The Long-Run Supply Curve of Crude Oil Produced in the United States." *Anti-Trust Bulletin* 15 (Winter 1970): 727-756.

Manes, Rene Pierre. "More Fuel on the Depletion Fire," *National Tax Journal* 16 (March 1963): 102-106.

———. "Import Quotas, Prices and Profits in the Oil Industry." *Southern Economic Journal* 30 (July 1963): 13-24.

Manley, John F. *The Politics of Finance*. Boston: Little, Brown and Company, 1970.

Mann, Thomas E. "Public Awareness of Congressional Candidates." In *Controversies in Voting Behavior*, 2nd ed., edited by Richard G. Niemi and Herbert F. Weisberg, 251-268. Washington D.C.: Congressional Quarterly, 1984.

Mann, Thomas E. and Raymond E. Wolfinger. "Candidates and Parties in Congressional Elections." In *Controversies in Voting Behavior*, 2nd ed., edited by Richard G. Niemi and Herbert F. Weisberg, 269-291. Washington, D.C.: Congressional Quarterly, 1984.

Manne, Alan S., and Leo Schrattenholzer, "International Energy Workshop: A Summary of the 1983 Poll Responses," *Energy Journal* 5 (1984): 45-64.

Mason, Edward S. "The Political Economy of Resource Use." In *Perspectives on Conservation: Essays on America's Natural Resources*. 157-186. Baltimore: Johns Hopkins Press, 1958.

Mason, Edward S. "Comment," *Review of Economics and Statistics* 36 (August 1954): 273-278.

Masters, Nicholas A. "Committee Assignments." In *New Perspectives on the House of Representatives*, 2nd. ed., edited by Robert L. Peabody and Nelson W. Polsby, 227-252. Chicago: Rand McNally & Company, 1969.

Matthews, Donald R. *U.S. Senators and Their World*. New York: Vintage Books, 1960.

McCaffery, David P. "Corporate Resources and Regulatory Pressures: Toward Explaining a Discrepancy." *Administrative Science Quarterly* 27 (1982): 398-419.

McCann, Michael W. "Public Interest Liberalism & the Modern Regulatory State." *Polity* (Winter 1978): 373-400.

McClosky, Herbert, and John Zaller. *The American Ethos*. Cambridge: Harvard University Press, 1984.

McConnell, Grant. *Private Power and American Democracy*. New York: Random House, 1966.

McDonald, Stephen L. "Percentage Depletion and the Allocation of Resources: The Case of Oil and Gas." *National Tax Journal* 14 (December 1961): 323-336.

———. "Percentage Depletion and Tax Neutrality: A Reply to Messrs. Musgrave and Eldridge." *National Tax Journal* 15 (September 1962): 314-326.

———. *Federal Tax Treatment of Income From Oil And Gas*. Washington D.C.: Brookings, 1963.

———. "On the Non-Neutrality of Corporate Income Taxation: A Reply to Steiner." *National Tax Journal* 17 (March 1964): 101-104.

———. "Conservation Policies and National Security." *Quarterly of the Colorado School of Mines* 64 (October 1969): 145-156.

———. "Distinctive Tax Treatment of Income From Oil and Gas Production." *Natural Resources Journal* 10 (January 1970): 97-112.

———. *Petroleum Conservation in the United States: An Economic Analysis*. Baltimore: Johns Hopkins University Press, 1971.

McFarland, Andrew S. *Public Interest Lobbies: Decision Making on Energy*. Washington D.C.: American Enterprise Institute, 1976.

———. *Common Cause: Lobbying in the Public Interest*. Chatham NJ: Chatham House Publishers, 1984.

McGee, John S. "Conservation, Integration, and Pricing in the Oil Industry of the United States: A Review Article." *Journal of Industrial Economics* 9 (November 1960): 63-81.

McGraw, Thomas K. *Prophets of Regulation*. Cambridge: Harvard University Press, 1984.

McGuire, William J. "The Myth of Massive Media Impact: Savagings And Salvagings." In *Public Communication And Behavior*, Vol. 1, edited by George Comstock, 173-257. Orlando, Fla.: Academic Press, Inc., 1986.

McKie, James, and Stephen L. McDonald. "Petroleum Conservation in Theory and Practice." *Quarterly Journal of Economics* 76 (February 1962): 98-120.

McLean, John G., and Robert W. Haigh. *Growth of the Integrated Oil Companies*. Boston: Harvard University Press, 1954.

Mead, Walter, J. "An Economic Analysis of Crude Oil Price Behavior in the 1970s." *Journal of Energy and Development* 4 (Spring 1979): 475-491.

Mead, Walter J. and Philip E. Sorensen. "A National Defense Petroleum Reserve Alternative to Oil Import Quotas." *Land Economics* 47 (August 1971): 211-224.

Menge, John A. "The Role of Taxation in Providing for Depletion of Mineral Reserves." *Tax Revision Compendium*. Submitted to the Committee on Ways and Means, Vol. 2, 967-978. Washington: GPO, 1959.

Miller, Warren E., and Donald E. Stokes. "Constituency Influence in Congress." In *New Perspectives on the House of Representatives*, 2nd. ed., edited by Robert L. Peabody and Nelson W. Polsby, 31-54. Chicago: Rand McNally & Co., 1969.

Mitchell, Edward. "The Basis of Congressional Energy Policy." *Texas Law Review* 57 (1979): 591-613.

Mitchell, Neil. "Corporate Power, Legitimacy, and Social Policy." *Western Political Quarterly* 39 (June 1986): 197-212.

Moe, Terry M. *The Organization of Interests*. Chicago: University of Chicago Press, 1980.

Monroe, Alan D. "Consistency Between Public Preferences and National Policy Decisions." *American Politics Quarterly* 7 (January 1979): 3-19.

Moran, Theodore. "Modeling OPEC Behavior: Economic and Political Alternatives," In *OPEC Behavior and World Oil Prices*, edited by James M. Griffen and David J. Teece, 94-130. London: George Allen & Unwin, 1982.

Morgenstern, Oscar. "Thirteen Critical Points in Contemporary Economic Theory: An Interpretation." *Journal of Economic Literature* 10 (1972): 1163-1189.

Morrison, Edward F. "Energy Tax Legislation: The Failure of the 93rd Congress." *Harvard Journal on Legislation* 12 (1979): 369-414.

Mueller, Dennis. *Public Choice*. Cambridge: Cambridge University Press, 1979.

Murphy, Frederic H., Michael A. Toman, and Howard J. Weiss. "An Integrated Analysis of US Oil Security Policies." *Energy Journal* 7 (1986): 67-82.

Musgrave, Richard A. "Another Look at Depletion," *National Tax Journal* 15 (June 1962): 205-208.

Nadel, Mark V. *The Politics of Consumer Protection*. Indianapolis: Bobbs-Merrill, 1971.

Nash, Gerald. *United States Oil Policy 1890-1964*. Pittsburgh: University of Pittsburgh Press, 1968.

National Journal, various issues.

Nelson, James R. "Prices, Costs, and Conservation in Petroleum." *American Economic Review* (May 1958): 502-515.

Nelson, Robert H. "The Economics Profession and the Making of Public Policy." *Journal of Economic Literature* (March 1987): 49-98.

Netschert, Bruce C. *The Future Supply of Oil and Gas*. Baltimore: John Hopkins Press, 1958.

Neuman, W. Russell. *The Paradox of Mass Politics*. Cambridge: Harvard University Press, 1986.

Newberry, D.M.G. "Oil Prices, Cartels, and the Problem of Dynamic Inconsistency." *Economic Journal* 91 (September 1981): 617-646.

Newsweek, various issues

New York Times

Nie, Norman, Sidney Verba and John Petrocik. *The Changing American Voter*. Cambridge: Harvard University Press, 1979.

Niskanen, William A. "Economists and Politicians." *Journal of Policy Analysis and Management* 5 (1986): 234-244.

Nivola, Pietro S. "Energy Policy and the Congress: The Politics of the Natural Gas Policy Act of 1978." *Public Policy* 28 (Fall 1980): 491-543.

Noll, Roger G., and Bruce M. Owen. *The Political Economy of Deregulation*. Washington D.C.: American Enterprise Institute, 1983.

Nordhaus, William D. "Oil and Economic Performance in Industrial Countries." *Brookings Papers on Economic Activity* 2 (1980): 341-399.

Nordhauser, Norman E. *The Quest For Stability: Domestic Oil Regulation 1917-1935*. New York: Garland Publishing, 1979.

North, Douglas. *Structure and Change in Economic History*. New York: W.W. Norton, 1981.

________."Three Approaches to the Study of Institutions." in *Neoclassical Political Economy*. Edited by David C. Colander, 33-40. Cambridge, Mass.: Ballinger Publishing Co., 1984.

Odell, Peter, and Kenneth E. Rosing. *The Future of Oil*. London: Hogan Page, 1983.

Oil and Gas Journal, various issues.

Oil Daily, various issues.

Oleszek, Walter J. *Congressional Procedures and the Policy Process*. Washington D.C.: Congressional Quarterly Press, 1978.

Olsen, Marvin, Eugene Rosa, Riley Dunlap, Robert Howell and Don Dillman. "Public Opinion Versus Governmental Policy on National Energy Issues." *Research in Political Sociology*, Vol. 1, p. 189-210.

Oppenheimer, Bruce Ian. *Oil And The Congressional Process*. Lexington, Mass.: D.C. Heath and Co., 1974.

________. "Policy Effects of US House Reform: Decentralization and the Capacity to Resolve Energy Issues." *Legislative Studies Quarterly* 5 (February 1980): 5-30.

Ordeshook, Peter. *Game Theory and Political Theory*. Cambridge: Cambridge University Press, 1986.

Ornstein, Norman J. "Causes and Consequences of Congressional Change: Subcommittee Reforms in the House of Representatives," in *Congress in Change*, edited by Norman J. Ornstein, 88-114. New York: Praeger Publishers, 1975.

Owen, Edgar Wesley. *Trek of the Oil Finders: A History of Exploration For Oil*. Tulsa: American Association of Petroleum Geologists, 1975.

Pabst, William R. "Unstable Conditions of Competition and Monopoly in Exhaustible Resource Industries." *Journal of Political Economy* 50 (October 1942): 739-749.

Page, Benjamin, and Robert Shapiro. "Effects of Public Opinion on Policy." *American Political Science Review* 72 (March 1983): 175-190.

Page, Benjamin I., Robert Y. Shapiro and Glenn R. Dempsey. "What Moves Public Opinion?" *American Political Science Review* 81 (March 1987): 23-43.

Paish, F.W. "Causes of Changes in Gold Supply." *Economica* 20 (November 1938): 379-407.

Pastor, Robert A. *Congress and the Politics of US Foreign Economic Policy: 1929-1976*. Berkeley: University of California Press, 1980.

Pastor, Robert A. "The Cry-and-Sigh Syndrome: Congress and Trade Policy." In *Making Economic Policy in Congress*, edited by Allen Schick, 158-195. Washington DC: American Enterprise Institute, 1983.

Pearson, Drew, and Jack Anderson. *The Case Against Congress*. New York: Simon and Schuster, 1968.

Peltzman, Sam. "Toward A More General Theory of Regulation." *Journal of Law and Economics* 19 (August 1976): 211-244.

________. "Constituent Interest and Congressional Voting." *Journal of Law and Economics* 27 (April 1984): 181-210.

Penrose, Edith. *The Large International Firm in the Developing Countries*. Cambridge: MIT Press, 1968.

Peters, B. Guy, and Brian W. Hogwood. "In Search of the Issue-Attention Cycle." *Journal of Politics* 47 (February 1985): 238-253.

Petroleum Economist, various issues.

Phillips, Kevin. "The Energy Battle: Why the White House Misfired." *Public Opinion* 1 (May/June 1978): 9-14.

Pierce, Richard J. "The Choice Between Adjudicating and Rulemaking for Formulating and Implementing Energy Policy," *Hastings Law Journal* 31 (September 1979): 1-101.

Pindyck, Robert S. "Gains to Producers From the Cartelization of Exhaustible Resources." *Review of Economics and Statistics* 60 (May 1978): 238-251.

________. "The American Energy Debate." *Public Interest* (Spring 1980): 100-105.

Pindyck, Robert, and Daniel Rubinfeld. *Econometric Models and Economic Forecasts*. New York: McGraw-Hill, 1981.

Poole, Keith T. "Dimensions of Interest Group Evaluation of the US Senate, 1969-1978." *American Journal of Political Science* 25 (February 1981): 49-67.

Porter, Richard, H. "The Federal Trade Commission v. the Oil Industry: An Autopsy on the Commission's Shared Monopoly Case Against the Nation's Eight Largest Oil Companies." *Antitrust Bulletin* 27 (Winter 1982): 753-820.

Posner, Richard A. "Theories of Economic Regulation." *Bell Journal of Economics* 5 (August 1974): 335-357.

———. "The Social Costs of Monopoly and Regulation." *Journal of Political Economy* 83 (1975): 807-827.

President's Materials Policy Commission, *Resources For Freedom, Vol. 1, Foundations for Growth and Security*, five vol., Report to the President, June 1952.

President's Materials Policy Commission, *Resources For Freedom, Vol. 3, The Outlook For Energy Sources*, five vol., Report to the President, June 1952.

Price, David E. "Professionals and 'Entrepreneurs': Staff Orientations and Policy Making on Three Senate Committees." *Journal of Politics* 33 (1971): 316-336.

Prindle, David F. *Petroleum Politics and the Texas Railroad Commission*. Austin: University of Texas Press, 1981.

"Public Interest in Petroleum Products--Discussion." *American Economic Review* (May 1958): 520-524.

Quirk, Paul J. "In Defense of the Politics of Ideas." *Journal of Politics* 50 (February 1981): 31-41.

Randall, Donna, and Robert De Fillippi. "Media Coverge of Corporate Malfeasance in the Oil Industry." *Social Science Journal* 24 (1987): 31-42.

Ravazi, Hossein. "Effects of Uncertainty on Oil Extraction Decisions." *Journal of Economic Dynamics and Control* 5 (1983): 359-370.

Renshaw, Edward F. "A Note On Petroleum Taxes To Increase Domestic Welfare And Reduce The Monopoly Profits Of The International Oil Cartel." *Public Finance* 31 (1976): 132-142.

Resichauer, Robert D. "Getting, Using, and Misusing Economic Information." In *Making Economic Policy in Congress*, edited by Allen Schick, 38-68. Washington, D.C.: American Enterprise Institute, 1983.

Richman, Al. "The Polls: Public Attitudes Toward the Energy Crisis." *Public Opinion Quarterly* 43 (Winter 1979): 577.

Riddlesperger, James W., and James D. King, "Energy Votes in the US Senate, 1973-1980," *Journal of Politics* 44 (August 1982): 838-847.

Ripley, Randall B., *Party Leaders in the House of Representatives.* Washington D.C.: Brookings, 1967.

Rivers, Douglas, and Nancy L. Rose, "Passing the President's Program: Public Opinion and Presidential Influence in Congress." *American Journal of Political Science* 29 (May 1985): 183-196.

Robinson, Michael J. "Three Faces of Congressional Media." In *The New Congress*, edited by Thomas E. Mann and Norman J. Ornstein, 73-90. Washington DC: American Enterprise Institute, 1981.

Romer, Thomas, and Howard Rosenthal. "Modern Political Economy and the Study of Regulation." In *Public Regulation: New Perspectives on Institutions and Policies*, edited by Elizabeth Bailey, 73-116. Cambridge: MIT Press, 1987.

Rostow, Eugene V. *A National Policy For the Oil Industry.* New Haven: Yale University Press, 1948.

________. "A Reply." *Journal of Political Economy* 57 (February 1949): 60-68.

Roth, William to Lee White, June 1, 1965, Office of the Special Representative for Trade Negotiations, declassified 1/19/78.

Rouhani, Fuad. *A History of OPEC.* New York: Praeger Publishers, 1971.

Rudder, Catherine E. "Committee Reform and the Revenue Process." In *Congress Reconsidered*, 3rd Ed., edited by Lawrence C. Dodd and Bruce I. Oppenheimer, 118-132. New York: Praeger Publishers, 1977.

Sabato, Larry J. *The Rise of Political Consultants.* New York: Basic Books, 1981.

________. *PAC Power.* New York: W.W. Norton, 1984.

Salant, Stephen W. "Exhaustible Resources and Industrial Structure: A Nash-Cournot Approach to the World Oil Market." *Journal of Political Economy* 84 (1976): 1079-1093.

Salisbury, Robert H. "An Exchange Theory of Interest Groups." *Midwest Journal of Political Science* 13 (February 1969): 1-32.

Salisbury, Robert H., John P. Heinz, Edward O. Laumann and Robert L. Nelson. "Who Works With Whom? Interest Group Alliances and Opposition." *American Political Science Review* 81 (December 1987): 1217-1234.

Sampson, Anthony. *The Seven Sisters*, 3rd ed. New York: Bantam Books, 1983.

Sanders, Elizabeth. "The Regulatory Surge of the 1970s in Histroical Perspective." In *Public Regulation: New Perspectives on Institutions and Policies*, edited by Elizabeth Bailey, 117-150. Cambridge: MIT Press, 1987.

Schattschneider, *E.E. Politics, Pressures and the Tariff.* Hamden Conn.: Archon Books, 1963.

________. *The Semisovereign People.* Rev. ed., Hinsdale, Illinois: Dryden Press, 1975.

Schlesinger, James R. *The Political Economy of National Security.* New York: Frederick A. Praeger, 1960.

Schlozman, Kay L., and John T. Tierney, "More of the Same: Washington Pressure Group Activity in a Decade of Change." *Journal of Politics* 45 (May 1983): 351-377.

Schneider, Jerrold E. *Ideological Coalitions In Congress.* Westport Conn.: Greenwood Press, 1979.

Schneider, Steven A. *The Oil Price Revolution.* Baltimore: John Hopkins University Press, 1983.

Schumpeter, Joseph A. *Capitalism, Socialism and Democracy*, 3rd ed. New York: Harper and Row, 1950.

Schurr, Sam H., and Bruce C. Netschert. *Energy in the American Economy 1850-1975.* Baltimore: Johns Hopkins Press, 1960.

Scott, Anthony. *Natural Resources: Economics of Conservation.* Toronto: University of Toronto Press, 1955.

Sears, David, Tom Tyler, Jack Citrin, and Donald Kinder. "Political System Support and Public Response to the Energy Crisis." *American Journal of Political Science* 22 (February 1978): 56-83.

Sears, David, Richard Lau, Tom Tyler and Harris Allen. "Self-Interest vs. Symbolic Politics in Policy Attitudes and Presidential Voting." *American Political Science Review* 74 (1980): 670-684.

Sen, Amartya K. "Rational Fools: A Critique of the Behavioral Foundations of Economic Theory." *Philosophy & Public Affairs* (Summer 1977): 317-344.

Sharkansky, Ira. "The Utility of Elazar's Political Culture: A Research Note." In *The Ecology of American Political Culture: Readings.* Edited by Daniel Elazar and Joseph Zikmund, 245-263. New York: Thomas Crowell, 1975.

Simon, Herbert. *Models of Bounded Rationality, Vol. 2,* Cambridge: MIT Press, 1982.

________."Testibility and Approximation." in *The Philosophy of Economics.* Edited by Daniel M. Hausman, 245-248. Cambridge: Cambridge University Press, 1984.

________. "Human Nature in Politics: The Dialogue of Psychology with Political Science." *American Political Science Review* 79: 293-303.

Sinclair, Barbara. *Congressional Realignment 1925-1978.* Austin: University of Texas Press, 1982.

Singer, S. Fred. "The Price of Oil." *Annual Review of Energy* 8 (1983): 451-508.

Smith, Rodney T., "In Search of the 'Just' U.S. Oil Policy: A Review of Arrow and Kalt and More." *Journal of Business* 54 (1981): 87-116.

Smith, Rodney T., and Charles E. Phelps, "The Subtle Impact of Price Controls on Domestic Oil Production." *American Economic Review* 68 (1978): 428-433.

Smith, Steve and Christopher Deering. *Committees in Congress.* Washington D.C.: Congressional Quarterly Press, 1984.

Steele, Henry. "Public Policy Problems of the Domestic Crude Oil Industry: Comment." *American Economic Review* 54 (March 1964): 119-124.

Steiner, Peter O. "Percentage Depletion and Resource Allocation," *Tax Revision Compendium*, submitted to the Committee on Ways and Means, Vol. 2, 949-966. Washington: GPO, 1959.

_______ . "The Non-Neutrality of Corporate Income Taxation--With and Without Depletion." *National Tax Journal* 16 (September 1963): 238-251.

_______ . "Rejoiner to McDonald." *National Tax Journal* 17 (March 1964): 105-107.

Stigler, George J. "The Theory of Economic Regulation." *Bell Journal of Economics* 2 (Spring 1971): 3-19.

_______ . "Economic Competition and Political Competition." *Public Choice* 13 (Fall 1972): 91-106.

_______ . "Free Riders and Collective Action: An Appendix to Theories of Economic Regulation," *Bell Journal of Economics* 5 (Autumn 1974): 364-365.

_______ . "Do Economists Matter?" *Southern Economic Journal* 42 (January 1976): 347-354.

Stobaugh, Robert, and Daniel Yergin, Editors. *Energy Future.* New York: Random House, 1979.

Stocking, George Ward. *The Oil Industry and the Competitive System.* Boston: Houghton Mifflin, 1925.

Stocking, George Ward, "Stabilization of the Oil Industry: Its Economic and Legal Aspects." *American Economic Review* 23 (March 1933): 55-70.

________. "Oil Proration: An Analysis and Evaluation." *Land Economics* 9 (May 1933): 135-144.

Stokes, Donald, and Warren E. Miller, "Party Government and the Saliency of Congress." *Public Opinion Quarterly* 26 (1962): 531-546.

Stockman, David. *The Triumph of Politics.* New York: Harper and Row, 1986.

Stone, Alan. "Capitalism and Case Studies." In *The Political Economy*, edited by Thomas Ferguson and Joel Rogers, 300-313. Armonk, New York: M. E. Sharpe, Inc., 1984.

Teece, David J. "OPEC Behavior: An Alternative View." In *OPEC Behavior and World Oil Prices*, edited by James M. Griffen and David J. Teece, 64-93. London: George Allen & Unwin, 1982.

Theberge, Leonard J., Editor. *TV Coverage Of The Oil Crises.* Washington, DC: The Media Institute, 1982.

Thorp, William L. "Trade Barriers and National Security," *American Economic Review* 50 (May 1960): 433-442.

Tribe, Lawrence. "Policy Science: Analysis or Ideology?" *Philosophy and Public Affairs* 2 (Fall 1972): 66–110.

Truman, David B. *The Governmental Process*, 2nd ed., New York: Alfred A. Knopf, 1971.

Tugwell, Franklin. *The Politics of Oil in Venezuela*. Stanford: Stanford University Press.

Tullock, Gordon. "Efficient Rent Seeking." reprinted in *Toward A Theory of the Rent-Seeking Society*. edited by James Buchanan, James, Robert Tollison, and Gordon Tullock, 97-112. College Station: Texas A&M University Press, 1980.

———. "More on the Welfare Cost of Transfers," reprinted in *Toward A Theory of the Rent-Seeking Society*, edited by James Buchanan, James, Robert Tollison, and Gordon Tullock, 288-292. College Station: Texas A&M University Press, 1980.

———. "The Welfare Costs of Tariffs, Monopolies, and Theft," reprinted in *Toward A Theory of the Rent-Seeking Society*, edited by James Buchanan, James, Robert Tollison, and Gordon Tullock, 39-50. College Station: Texas A&M University Press, 1980.

Turner, Louis. *Oil Companies in the International System*, 3rd ed., London: George Allen & Unwin, 1983.

Tussing, Arlon R. "Oil Prices Are Still Too High." *Energy Journal* 6 (1985): 9-12.

Tversky, Amos, and Daniel Kahneman. "Judgment Under Uncertainty: Heuristics and Biases." In *Judgment Under Uncertainty: Heuristics and Biases*, edited by Tversky, Kahneman and Paul Slovic, 1-20. Cambridge: Cambridge University Press, 1982.

Ulph, A.M., and G.M. Folie. "Exhaustible Resources And Cartels: An Intertemporal Nash-Cournot Model." *Canadian Journal of Economics* 13 (November 1980): 645-658.

US Congress, House, Select Committee on Small Business, *Hearings: Effects of Foreign Oil Imports on Independent Domestic Producers*. 81st Cong., 1st sess., May, November, 1949.

US Congress, House, Select Committee on Small Business, *Small Business Problems Created by Petroleum Imports, Hearings before Subcommittee No. 4*, 87th Cong., 1st sess., Part I, November 20-24, 1961.

US Congress, House, Committee on Ways and Means , *President's 1963 Tax Message, Hearings before the Committee on Ways and Means*, 88th Cong., 1st sess., 1963.

US Congress, Senate, Select Committee on Small Business, Oil *Import Allocations, Hearings before the Select Committee on Small Business*, 88th Cong., 2nd sess., August 10, 11, 1964.

US Congress. Senate, Committee on Finance, *Import Quotas Legislation, Hearings before the Committee on Finance*, 90th Cong., 1st sess., Part 1, October 18, 19, 1967.

US Congress, Senate, Committee on the Judiciary, *Governmental Intervention in the Market Mechanism, Hearings before the Subcommittee on Antitrust and Monopoly*, 91st Cong., 1st sess., The Petroleum Industry, Part 1, Economists' Views, March, April, 1969.

US Congress, House, Committee on Interior and Insular Affairs, *Oil Import Controls, Hearings before the Subcommittee on Mines and Mining*, 91st Cong., 2nd sess., March, 1970.

US Congress, Senate, Committee on Interior and Insular Affairs, *Oil And Gas Imports Issues, Hearings before the Committee on Interior and Insular Affairs*, 93rd Cong., 1st sess., January 11-22, 1973.

US Congress, Senate, Committee on Interior and Insular Affairs, *The President's Energy Message And S. 1570, Hearing before the Committee on Interior and Insular Affairs*, 93rd Cong., 1st sess., May 1, 1973.

US Congress, Senate, Committee on Interior and Insular Affairs, *Strategic Petroleum Reserves, Hearings before the Committee on Interior and Insular Affairs*, 93rd Cong., 1st sess., May, July, 1973.

US Congress, House, Committee on Interstate and Foreign Commerce, *Mandatory Fuels Allocation, Hearings before the Committee on Interstate and Foreign Commerce*, 93rd Cong,, 1st sess,, July, 1973.

US Congress, House, Committee on Ways and Means, *The Energy Crisis And Proposed Solutions, Panel Discussions before the Committee on Ways and Means*, 94th Cong., 1st sess., March, 1975.

US Congress, House, Committee on Interstate and Foreign Commerce, *Energy Conservation And Oil Policy, Hearings before the Subcommittee on Energy and Power*, 94th Cong., 1st sess., March 1975.

US Congress, Senate, Committee on Finance, *Energy Tax Act of 1977, Hearings before the Committee on Finance*, 95th Cong., 1st sess., August, September 1977.

US Congress, Senate, Committee on Finance, *Crude Oil Severance Tax, Hearings before the Subcommittee on Energy and Foundations*, 96th Cong., 1st sess., May 1979.

US Congress, House, Committee on Science and Technology, *Energy Demand Forecasting, Hearings before the Subcommittee on Investigations and Oversight*, 97th Cong., 1st sess., 1981.

US Congress, House, Committee on Energy and Commerce, *Strategic Petroleum Reserve, Hearing before the Subcommittee on Fossil and Synthetic Fuels*, 99th Cong., 2nd sess., March 4, 1986.

US Department of Energy, Office of International Affairs, *The Role of Foreign Governments in the Energy Industries*, Washington: GPO, 1977.

Uslaner, Eric M. "Policy Entrepreneurs and Amateur Democrats in the House of Representatives: Toward a More Party-Oriented Congress?" In *Legislative Reform*, edited by Leroy N. Rieselbach, 105-116. Lexington: Lexington Books, 1978.

Verleger, Philip. "The Determinants of Official OPEC Crude Prices." *Review of Economics and Statistics* 64 (May 1982): 177-183.

________. "The Evolution of Oil as a Commodity." In *Energy: Markets and Regulation*, edited by Richard Gordon, Henry Jacoby, and Martin Zimmerman, 161-186. Cambridge: MIT Press, 1987.

Vietor, Richard. *Energy Policy In America Since 1945*. Cambridge: Cambridge University Press, 1984.

Vogel, David. "The Public-Interest Movement and the American Reform Tradition." *Political Science Quarterly* 95 (Winter 1980-81): 607-627.

Waddams, Frank C. *The Libyan Oil Industry*. Baltimore: Johns Hopkins Press, 1980.

Wagner, Richard E. "Pressure Groups and Political Entrepreneurs: A Review Article." In *Papers on Non-Market Decision-Making*, edited by Gordon Tullock, 161-170. Charlottesville: University of Virginia, 1966.

Walker, Jack L., "Performance Gaps, Policy Research, and Political Entrepreneurs: Toward a Theory of Agenda Setting," *Policy Studies Journal* (Autumn 1974): 112-116.

———. "The Diffusion of Knowledge, Policy Communities and Agenda Setting: The Relationship of Knowledge and Power." In *New Strategic Perspectives on Social Policy*, edited by John E. Tropman, Milan J. Dluhy and Robert Lind, 75-96. New York: Pergamon Press, 1981.

———. "The Origins and Maintenance of Interest Groups in America." *American Political Science Review* 77 (June 1983): 390-405.

Walker, Thomas G. *Political Parties, Interest Groups, and Public Policy*. Englewood Cliffs, NJ: Prentice-Hall, 1980.

Watkins, Myron W. *Oil: Stabilization or Conservation?* New York: Harper & Brothers, 1937.

Weatherford, M. Stephen. "The Interplay of Ideology and Advice in Economic Policy-Making: The Case of Political Business Cycles." *Journal of Politics* (November 1987): 925-952.

Weisberg, Herbert F. "Evaluating Theories of Congressional Roll-Call Voting." *American Journal of Political Science* 22 (August 1978): p. 554-577.

Weisberg, Richard C. *The Politics of Crude Oil Pricing in the Middle East, 1970-1975*. Institute of International Studies, University of California at Berkeley, 1977.

Weiss, Carol H. "What America's Leaders Read," *Public Opinion Quarterly* 38 (Spring 1974): 1-22.

________. "Research for Policy's Sake: The Enlightenment Function of Social Research." *Policy Analysis* 3 (Fall 1977): 531-545.

Weyant, John P. "Quantitative Models in Energy Policy." *Policy Analysis* 6 (Spring 1980): 211-234.

Wheaton, William C. "The Long-run Structure of Transportation and Gasoline Demand." *Bell Journal of Economics* 13 (Autumn 1982): p. 439-454.

Wiggins, Steven N., and Gary D. Libecap. "Oil Field Unitization: Contractual Failure in the Presence of Imperfect Information." *American Economic Review* 75 (1985): 368-385.

———. "Firm Heterogeneities and Cartelization Efforts in Domestic Crude Oil." *Journal of Law, Economics, and Organization* 3 (Spring 1987): 1-25.

Wildavsky, Aaron. "Choosing Preferences by Constructing Institutions: A Cultural Theory of Preference Formation." *American Political Science Review* 81 (March 1987): 3-21.

Wildavsky, Aaron, and Ellen Tenenbaum. *The Politics of Mistrust*. Beverly Hills: Sage Publications, 1981.

Williamson, H.F., Andreano, R.L., Daum, A.R. and G.C. Klose. *The American Petroleum Industry: The Age of Energy 1899-1959*. Evanston, Illinois: Northwestern University Press, 1963.

Williamson, Harold F., and Ralph L. Andreano. "Integration and Competition in the Oil Industry: A Review Article." *Journal of Political Economy* 69 (August 1961): 381-385.

Wilson, James Q., editor. *The Politics of Regulation*. New York: Basic Books, 1980.

———. "Policy Intellectuals and Public Policy." *Public Interest* 64 (Summer 1981): 31-46.

Wright, John. "PACs, Contributions and Roll Calls: An Organizational Perspective." *American Political Science Review* 79 (1985): 400-414.

Yager, Joseph A. "The Energy Battles of 1979." In *Energy Policy in Perspective: Today's Problems, Yesterday's Solutions*, edited by Craufurd D. Goodwin, 601-636. Washington D.C.: Brookings Institution, 1981.

Zerbe, Richard O., Jr., and Nicole Urban. "Including the Public Interest in Theories of Regulation." *Research in Law and Economics* 11 (1988): 1-23.

Zimmerman, Erich W. *World Resources and Industries*. New York: Harper & Brothers, 1933.

———. *Conservation in the Production of Petroleum*. New Haven: Yale University Press, 1957.

Zweir, Robert, "The Search for Information: Specialists and Nonspecialists in the U.S. House of Representatives." *Legislative Studies Quarterly* 4 (February 1979): 31-42.

Index

Abourezk, James, 145 295, 321, 322, 325, 412
Achnacarry agreement, 8
Adelman, M.A., 77-78, 82, 115-116, 209, 211, 213-214, 218, 221, 223, 406
Ad Hoc Committee on Energy, 147-148, 324
Agenda Setting, 383, 391-400, 406
Akins, James, 214
Albert, Carl, 138-140, 300, 317
American Farm Bureau, 83, 89, 107
Allred, James
  Attorney General, 32, 40-41, 43
  Governor, 46, 48-52, 55
Amazon Petroleum, 41, 43
Anderson, John, 274, 295, 325
Anderson, Robert B., 85, 98
American Petroleum Institute (API), 13, 30, 34-36, 39, 84, 106, 109, 112-113, 115-116, 127, 145, 153, 163-164, 170-171, 254, 377, 380, 408, 411
  conservation, 13-20
  federal control, 47
Aramco (Arabian-American Oil Co.), 7, 76-77, 85, 104, 129
Archer, Bill, 350, 352, 393, 394
Ash, Roy, 133, 134, 136, 138
Atlantic Refining, 38-39, 78, 90, 99
Atlantic Richfield (ARCO), 259

Bain, Joe, 199-200, 203
Bartlett, Dewey, 143, 291, 301, 307, 316, 321
Becker, Gary, 372, 403
Bentsen, Lloyd, 142, 149, 157-158, 162, 171-172, 178-179, 301, 316-319, 325, 328, 335-336, 352
Beaty, Amos 13-14
Benedum, Michael, 35, 40
Boatwright, John, 200-202
Boggs, Hale, 288, 386
Bolling, Richard, 148, 324, 350
Boren, David, 335, 338, 340, 346
Bradley, Bill, 160, 162, 165, 335-336, 339, 345-346, 351-352
British Petroleum (BP), 103-104, 130
Brown, Clarence, 312, 317, 321, 324-325, 349
Brown, Russell, 19, 50
Bryant, Randolph, 27, 29, 31, 33, 42, 47
Buchanan, James, 364, 368-369
Bumpers, Dale, 335, 338, 342
Bureau of Mines, 10, 13, 20, 40, 42, 45, 51
Bush, George, 162, 173-175, 287
Bush Administration, 173-174, 177-178
Byrd, Harry, 92, 385, 402
Byrd, Robert, 149, 162, 325, 336, 340, 350

Campaign Financing, 376-377
  Money and elections, 375-379
  Oil Money, 375-379, 404-407

Campaign Financing (cont'd)
Political Action Committees (PACs), 242, 377-379
Canada, 95, 98, 102, 108-110, 402
Capper, Arthur, 37, 39
Carlson, Frank, 84, 95
Carson, Matthew, 98, 100, 102
Carter, Jimmy, 144-147, 149-154, 157, 159-161, 164-165, 242, 245-246, 249-250, 260-261, 323, 325, 327, 330-331, 334, 341, 349-350, 352, 410-412
Carter Administration, 149-152, 154, 157, 161, 323-324, 330, 336, 349-350, 398, 410
Cassady, Ralph, 200, 202-203
Chamber of Commerce (US), 91, 107, 172, 331, 381, 409
Champlin Refining, 27, 29
Charles Rivers Associates (CRA), 117, 209, 216, 406
Chemical Manufacturers Association, 170-171
Chevron, 201; *see also* Socal, California Standard
Cities Service, 78, 99, 112
Coal industry, 19, 75, 86, 88-90, 106, 110, 112, 178, 403
Commission on Foreign Economic Policy, 75, 92
Common Cause, 388, 396, 401, 409
Conable, Barber, 335, 337
Congressional Budget Office (CBO), 156, 163, 398
Congressional Research Service, 397, 427
Congressional staff, 383, 397, 391, 400
Congressional voting, 375, 378-379, 381, 387, 401
Connally, Tom, 19, 37, 44, 51, 279
Connally Hot Oil Act, 37, 51
Consumer Federation of America (CFA), 133, 142, 150, 171, 331, 395-396, 407-408
Consumer Movement, 389, 391, 393, 396
Coolidge, Calvin, 13-14, 16, 35
Corporate average fuel efficiency standards (CAFE), 173, 175-179
Council of Economic Advisors (CEA), 94, 101, 109, 127-128, 136, 138-139, 146, 163, 170
Cranfill, Tom, 22, 25, 27
Crude Oil Equalization Tax, 147-152, 324-329, 331, 334

Davidson, Paul, 206-208
De Chazeau, Melvin, 202-203, 205
Democratic Caucus, 138, 142, 143, 153, 288, 300, 332, 382, 389
Democratic Study Group, 128, 387
Department of Energy (DOE), 127, 153, 160-161, 165-167, 171, 174-176
Depletion allowance, 83-85, 107, 109, 113, 116, 172, 285-291, 300-302, 306, 377, 380, 385-386, 397, 401, 404, 406-407, 413
economists, 116, 189, 201-203, 213
independents, 113, 158, 162
repeal, 138, 140, 142, 158, 162
see also Finance Committee, Ways and Means
DiBona, Charles, 129, 132-134
Dingell, John 153, 160, 163, 165, 327, 331, 335-336, 338, 390, 427, 434
Disney, Wesley, 44, 48, 49, 51
Divestiture, 144, 145, 149
Doherty, Henry, 13-16

Dole, Robert, 150-151, 157, 161, 163, 165, 321, 331, 335-336, 338, 340, 349-350, 412-413
Douglas, Paul, 85, 285-286
Downs, Anthony, 364-366
Eagleton, Thomas 132, 290
East Texas Oil Field, 13
  discovery, 23-24
  martial law, 28-29
  proration, 25-34, 40-52, 83
Eckhardt, Bob, 144, 148, 295, 317, 320, 324, 332-333
Economic regulation, 359, 360, 369-374, 376, 403
Edwards, James, 160, 161, 163
Ehrlichman, John, 127, 129
Eisenhower, 75, 85, 91-92, 94, 96, 97, 99, 105
Eisenhower Administration, 85-86, 102-103
Eizenstat, Stuart, 127, 159
Eldridge, Douglas, 203, 205
Ellender, Allen, 289-290
Elliott, John, 36-37
Emergency Natural Gas Bill, 316-317
Emergency Petroleum Allocation Act (EPAA), 136-137, 143, 291, 307, 309, 312
Energy and Natural Resources Committee, 149, 171, 175-176
Energy and Power Subcommittee, 153, 156, 175-176, 309, 317, 332, 388, 390
Energy Conservation, 131, 132, 134, 136-138, 140-141, 143-147, 149-151, 155, 158-161, 172, 175-179, 294-295, 304, 310, 324-325
Energy Crisis, 130, 133, 146-147, 243, 245, 248-249, 253, 255, 261, 407, 434
Energy Modeling Forum, 221, 226
Energy Policy and Conservation Act (EPCA), 144-145, 151, 173, 309, 313, 331
Energy Prices, 237, 245-246, 248-250, 252-255, 260-261
Entitlements, 403, 426, 430-432
Erickson, Edward, 209, 215-216, 427
Escape clause, 15, 91, 92, 95, 108
Estes, Carl, 25, 27, 42
Exhaustible resource theory, 18, 190, 192-194, 218
Exxon, 76, 77, 86, 87, 104, 129, 171, 173, 259

Fannin, Paul, 137, 294-295, 316
Farish, William, 12, 16, 20, 22, 35, 39
Federal Energy Administration (FEA), 138-139, 142, 144, 146, 160, 294-295, 321, 409, 425, 430
Federal Energy Office (FEO), 126, 255
Federal Oil Conservation Board (FOCB), 13-18, 20, 34, 36, 40
Federal Power Commission (FPC), 130, 160, 281-282, 284, 316-317, 320, 395, 404-405, 411, 425
Federal Trade Commission (FTC), 133-134, 251, 254, 291, 407, 432
Finance Committee, 75, 82, 115, 142-143, 157, 161-162, 171, 279, 385-387, 402, 411
  Depletion allowance, 82-84, 142, 289-290
  Gasoline taxes, 327
  Oil import fee, 170, 349-351
  Oil import quotas, 92-95, 100
  Windfall profit tax, 152, 156-158, 294, 301, 335, 338, 340, 345
Fisher, Franklin, 115, 209, 215, 216

Fiorina, Morris, 271-272, 375
Flemming, Arthur S., 94-97
Ford, Gerard, 139-140, 143-147, 165, 274, 300, 304-308, 312, 314, 316, 324, 326, 330, 349
Foreign Oil Policy Committee, 108
Frankel, Paul, 76, 198, 207
Franklin, Wirt, 18-19, 21, 36
Freeman, S. David, 127, 146

Gallup Poll, 245-246, 249
Gas guzzler tax, 147-149, 151, 324
Gasoline rationing, 136-137, 140-142, 153, 294-296, 332
Gasoline shortages, 150, 153, 161
Gasoline taxes, 134, 136, 140-143, 146-151, 154, 163, 172-173, 175-176, 178-179, 248-249, 256, 304-307, 323-24, 410, 413-414
GATT, 74, 98, 102, 110
General Accounting Office (GAO), 145, 153, 161, 258, 291, 397-398
Geological Survey, 10, 14, 17
Gibbons, Sam, 142, 288, 300, 305, 306, 350
Glenn, John, 307, 310, 316, 318, 411
Gramm, Phil, 162, 165, 332-333, 345, 379
Gravel, Mike, 307, 309
Green, William, 138, 142, 285, 300-301
Gulf Oil, 3, 5, 7-9, 10, 12, 13, 19, 22, 24, 25, 39, 76, 78, 94, 100, 377
Gulf War, 174-175, 179, 414

Halleck, Charles A., 92, 282
Harberger, Arnold, 203, 204, 206
Harris, Oren, 89, 405
Harris Poll, 245, 249, 264, 266,
Hart, Gary, 145, 170-171, 319, 352
Hart, Philip, 115, 138, 144-145, 304, 321-322, 407
Hartke, Vance 142-143, 301, 303
Heating oil, 128-130, 134, 161, 170, 172, 179
Herrington, John, 167, 171
Hodel, Donald, 166, 171
Hollings, Ernest, 142, 301, 316-319
Hoover, Herbert, 12, 17-18, 20
Hotelling, Harold, 190-192
Hot Oil, 30-32, 35-37, 40-50, 52-55, 57
Hughes, Charles, 15, 48
Humble Oil, 12, 22, 24-27, 31-33, 56
Humphrey, Hubert, 84, 271, 285, 291
Hutcheson, Joseph, 27, 29, 31, 42

Ickes, Harold, 35-37, 40-41, 43, 45, 47, 49, 51, 57-58
Ideology, 360, 366, 370, 373-374, 378, 402
Ikard, Frank, 96, 99-100, 102, 254, 377, 386
Independent Petroleum Association of America (IPAA), 18, 19, 36, 39, 52, 87, 89-91, 93-96, 100, 106, 109, 112-116, 119, 129, 138, 147, 153, 164, 166, 170-172, 178
Independent Petroleum Association of Texas (IPAT), 25, 31-33, 36, 41
Independent Refiners Coalition, 170, 172
Indiana Standard, 22, 25, 27, 39, 78, 85, 98-99, 106, 112
see also Amoco
Interest Groups, 359-363, 366-368, 372, 374, 378, 380, 384, 386-391, 393-394, 396-398, 402, 406, 408, 410, 413
Interior and Insular Affairs Committee, 294, 307, 322, 385, 388, 389, 407

Interior Department, 18, 37, 40, 45-47, 389, 403
Interstate and Foreign Commerce Committee, 294, 307, 307, 309, 317, 332, 348, 385, 388, 390
Interstate Commerce Commission (ICC), 15, 37, 59
Interstate Oil Compact, 18, 25, 34, 48, 56
Iraq, 130, 155, 156, 159, 173-176
Iran, 7, 9, 130, 135, 152, 155, 156, 159, 164
Ise, John, 195-196
Issue-Attention Cycle, 243-245

Jackson, Henry, 131, 133, 137, 140, 149, 151, 153, 157, 254-255, 291-292, 294, 307-309, 325, 327, 329, 331, 349, 389, 427
Jersey Standard, 6, 7-10, 12, 18, 25, 35, 39
see also Exxon
Johnson, Hugh, 38-40
Johnson, Lyndon, 84, 85, 87, 93, 109-110, 112, 114, 377, 383, 394, 402, 405, 406
Johnston, Bennett, 171, 172, 175-178, 295, 307-309, 311, 322, 350
Joiner, C.M. "Dad," 24
Jones, James, 155, 324, 334-335, 337
Justice Department, 16, 18, 45, 47 59, 432

Kahn, Alfred, 82, 202-203, 205, 208, 406
Kalt, Joseph, 272-274, 431
Kansas, 27, 36, 46, 48, 56, 80, 219, 281, 308, 330
Kassebaum, Nancy, 166, 347-349
Kau, James, 272, 274
Kelliher, Tom, 46-47
Kennedy, Edward, 115, 140, 142, 145, 149, 159, 301, 325, 327-328, 411-413
Kennedy, John, 84-85, 107-108, 286, 394
Kennedy Administration, 106, 109
Kerr, Robert, 84, 107, 377, 402, 405
Krueger, Robert, 144, 271, 310-311, 317, 320-321
Kuwait, 7-8, 155, 172-175

Laird, Melvin, 134, 136
Lamont, Robert, 25-26
Landon, Alf, 36, 48
Levy, Walter, 77, 119, 145, 155, 214
Libya, 129, 130, 135, 155
Lincoln, George, 119-120
Logrolling, 362, 369, 381, 386
Long, Russell, 84, 100, 108, 113-144, 137, 142, 149-150, 154, 156-157, 271, 289, 294-295, 301, 325, 327, 330, 334-336, 349, 385, 389, 412
Louisiana, 80-81, 86, 114
Love, John, 133, 134, 136, 145
Lovejoy, Wallace, 209, 213, 215

MacDougall, Kent, 257, 259
McDonald, Stephen, 82, 204-207, 213-214, 406
Mckie, James, 82, 116-117, 207, 406
Magnolia, 22, 39
Mancke, Richard, 210, 216
Manes, Renee, 205, 208
Manley, John, 74, 122, 271
Mansfield, Mike, 140, 142, 286, 301, 319
Marshall, Howard, 42, 47
Matsunaga, Spark M., 383, 385
McClure, James, 160, 166, 348, 352
McIntyre, Thomas, 291, 327, 329
Mead, Walter, 127, 164, 213, 221

Media, 238-240, 243-244, 361, 367, 375, 380
elections, 241-243, 390
energy, 250-260, 400
political entrepreneurship, 391, 393-397, 399-400
Metzenbaum, Howard, 158, 161-162, 327, 331, 336, 342, 413
Mexico, 6-7, 9-10, 170
Mid-Continent Region, 15, 19, 25, 29, 35, 39-40, 46, 56-58
Middle East, 5, 7, 10, 76-78, 85-87, 90, 95-96, 98, 102, 104-105, 119
Millikin, Eugene, 75, 92, 271, 280
Mills, Wilbur, 93, 109, 114, 119, 288, 300, 385-386, 389, 407
Mobil, 3, 153, 259
see also Socony
Moffett, Toby, 153, 232
Moffett, James, 35, 36, 40
Moore, W. Henson, 156, 173, 334-335
Morton, Rogers, 131, 134, 139, 288
Moss, Frank, 291, 316, 319, 377
Moynihan, Daniel Patrick, 158, 335, 336

Nader, Ralph, 249, 260, 394-396
National Energy Act, 147
National Energy Emergency Act (NEEA), 137, 294-297, 314
National Energy Plan, 161, 324
National Industrial Recovery Act (NIRA), 39, 48, 57
National Petroleum Council (NPC), 77, 85, 95, 113, 127, 136, 198
National Recovery Adminstration (NRA), 39-40, 43
Oil Code, 39-43, 45, 47, 51
National security, 162, 171-173, 189, 201, 203, 212-214,
Natural Gas, 90, 94, 108, 114, 130, 131, 133, 139, 142, 145-146, 157, 160-161, 164, 173, 177-178, 278, 301-303, 305, 406, 411, 413
Deregulation, 132, 139, 147-151, 161, 171, 281-285, 304, 316-328, 300-331, 346-349, 376, 390, 400-401, 404, 411, 413-414, 425, 427
Natural Gas Act of 1938, 287, 404
Natural Gas Policy Act of 1978 (NGPA), 347, 411-412
Neely, Matthew, 77-78, 83, 89-90, 281
Nelson, James, 201-202, 227
Neopopulist, 433-434
New England, 91, 95, 98, 106, 109-110, 112, 402-403, 411
Newsweek, 238, 250-252
New York Times, 238, 240, 246-247
Nixon, Richard, 128-129, 132, 134, 136-138, 231, 290, 295, 300, 377, 397, 410
Nixon Administration, 127, 132-133, 291

Office of Civil and Defense Mobilization (OCDM), 101, 106-107
Office of Defense Mobilization (ODM), 94, 96, 100, 101
Office of Emergency Planning (OEP), 107-108, 115, 118, 119, 129, 131
Office of Management and Budget (OMB), 133, 136-139, 146, 171, 174-175
Oil Companies
international Majors, 3, 6, 9, 19, 20, 82, 85-87, 89, n93, 104-105, 111, 113, 129-131, 402
see also Seven Sisters

Oil Companies (cont'd)
large, integrated Companies, 12-14, 16, 26, 39, 41, 54-55, 58-59, 84, 129, 133, 136, 138, 142, 145, 150, 153, 156, 161, 379-381, 407, 409, 411
media coverage, 251, 253-261, 263
public opinion, 246, 248-250, 264-265
second-Tier, 76, 86, 89, 98, 102-104
small producers, 79, 82, 86, 89, 129, 134, 140, 142-143, 157-158, 162, 171, 177-178, 277, 280, 283, 286, 288, 290, 295-296, 301, 304, 317, 320, 323, 404, 405, 408, 411, 413
Oil Embargo, 130, 135, 145-146
Oil Import Administration, 111, 119
Oil Import Fee, 139-140, 143 151, 159, 162-164, 170-173, 175, 178, 413-414
Oil Import Quotas, 82, 189, 199, 201, 202, 206
Allocation 99, 102, 106, 108, 110-111, 113-114, 127, 129, 372, 385, 401-403, 406
Carter, 141-143, 151, 157-159, 163,
Inland Refiner, 98, 100, 102, 106-107
Mandatory, 96, 101, 103-110, 207-210
Politics, 89-92, 94-96, 99-100, 112, 119
Presidential Proclamation 3279, 101-102
Small refiner, 100, 102,-103, 110, 115
Value of rights, 102, 103, 105, 109, 112, 115
Voluntary, 95-100, 201, 213
Oil Import Tariffs, 99, 108, 112, 118, 120, 199, 201, 210, 226-227
Oil Imports,
1930s, 3, 20, 25, 34, 36-37, 47
1950s, 73, 75-79, 82, 84, 86-90, 96, 99
1970s, 127, 129, 131-133, 139-141, 143, 145-146, 151-153, 157
1980s, 166-168, 170-173
residual, 88-95, 98, 100-102, 106-110, 112, 114, 116, 119, 143
Oil Lobbying, 376, 377, 379, 381, 386, 408, 411
Oil Price Decontrol, 134, 136-138, 140-141, 143, 146-147, 151-154, 156, 159-161, 292, 299, 304-309, 311-312, 314, 316, 318, 320-324, 331-332, 334, 342, 346, 390, 409, 412-413
Oil States Advisory Committee (OSAC) 25-26, 31, 34
Oklahoma, 10, 13, 15, 18-19, 22-24, 27-28, 35-36, 41
Oklahoma City field, 3, 21, 23, 28, 42
Oklahoma Corporate Commission, 21, 23-24, 28
Olson, Mancur, 360, 364. 366-367, 73, 391
O'Neill, Tip, 147-148, 153-154, 324, 331-332, 349, 412
OPEC, 105, 130, 135, 151, 154-156, 163-165, 168-170, 172-173, 175
economic theories, 192, 212, 214, 218-224
media coverage, 248, 253, 256-257
public opinion, 237
Overproduction, 4, 10, 14, 16, 19, 29, 34, 45, 57, 50, 56

Packwood, Bob, 200
Paley Report, 75, 194, 212, 217
Pastore, John, 105, 271, 304
Peril point, 75, 91
Persian Gulf, 76, 77
Planning and Coordinating (P&C) Committee, 40
Panama Refining, 42, 48
Parten, J.R., 31, 36
Pearson, James, 295, 316-319, 325, 328
Peltzman, Sam,272, 352-353, 372, 403
Penn, Robert, 22, 25
Petroleum Administration Board (PAB), 40, 42-45, 51, 57
Petroleum Administration for War (PAW), 60, 82
Petroleum Study Committee (PSC), 107-109, 406.
Pew, J. Edgar, 13-14, 39
Phillips Petroleum, 19, 77, 95, 99, 111, 377, 405
Pickle, Jake, 291, 305
Pindyck, Robert, 219-220
Platt's Oilgram, 76, 167
Pluralism, 351, 361, 364, 367
Political Entrepreneurs, 369, 391, 393-396, 399-400
Polls, 140, 150, 153, 159, 175-176, 238, 245, 247, 260-261, 263-264, 266
Price Controls, 128-132, 134
Production Allowables, 79, 81, 89, 96, 114
Prorationing, 78-79, 81-82, 87, 89, 103, 108-109, 116-117
  economic theory, 196, 197, 199-202, 207-210, 212
  economic waste, 18-22, 27-28, 31
Proxmire, William, 84, 105, 412
Puerto Rico, 109, 111
Public Opinion, 141, 153, 174, 237-240, 243-245
  business, 238, 251-252, 256-258, 261-264
  profits, 257, 259, 262-264

Railroads, 90, 93, 101
Randall, Clarence, 92, 100
Randall Commission, 92, 93
Randolph, Jennings, 108, 389
Rayburn, Sam, 37, 44, 48, 51, 92-93, 281-282, 288, 382, 386, 402
Reciprocal Trade Acts, 74, 75, 87, 90-91, 93-94, 99, 107, 109
Reagan, Ronald, 145, 160-161, 163-164, 166-167, 170-172, 249, 341-342, 344, 346-347, 352-353
Reagan Administration, 160-162, 165, 171, 266, 341
Regan, Donald, 163, 170
Red Line Agreement, 7
Reed, Dan, 75, 83, 92, 100
Rent Seeking, 12, 359, 369, 372-373, 401, 403, 410
Refiners,
  east texas and hot oil, 30, 45, 47, 56-58
  small, independent, 20, 36, 42-43, 57, 100, 115, 132-134, 136, 138, 150, 161, 170,172, 290-291, 323, 341-342, 353, 403, 407-410, 413
Requa, Mark, 12-14, 16, 18, 20, 35
Ribicoff, Abraham, 291, 336, 340
Rochester, E.S., 17-18
Roosevelt, Franklin D., 34-41, 44, 49, 51, 58, 74, 83, 86
Rostenkowski, Dan, 148, 156, 162, 178, 334, 350
Rostow, Eugene, 198-200
Roth, William, 271, 319, 327
Rule of Capture, 4, 13
Rule 37, 43, 52-53, 55

Rules Committee, 382, 384, 387-389, 412

Sadler, Jerry, 43, 53, 56
Saudi Arabia, 7, 86, 105, 130, 135, 155, 163, 165, 168, 170, 175
Sawhill, John, 136, 138-139, 147
Schattschneider, E.E., 73, 74,
Schlesinger, James, 146, 149, 150, 213, 349
Schultz, George, 115-116, 118-119, 129, 134, 136
Schurr, Sam, 215-217
Section 9(c), 48-49
Seminole Field, 3, 10, 15, 21, 24
Seven Sisters, 3, 68
Sharp, Philip, 166, 175-178
Shell, 3, 6, 8-10, 19, 22, 25, 39, 76-78, 104, 259
Sherman Act, 18, 58-59
Simon, William, 129, 134, 136, 141, 255
Simpson, Richard, 75, 91-93, 100
Sinclair, Harry, 12, 14, 25, 28, 39-40
Sinclair Oil, 56, 78, 99
Smoot-Hawley Act, 73-75
Socal 3, 8, 77, 86, 377 *see also* Chevron
Socony-Vacuum, 92, *see also* Mobil
Special Committee to Investigate Oil Imports, 97-98, 100
Spot market, 135, 155, 163, 167-168
Staggers, Harley, 294, 296, 311, 313, 317, 332, 390
Standard Oil Trust, 3, 5
Stanley, E.N., 43, 53
Standby Energy Power Act, 307-310
Steed, Tom, 105, 113
Steele, Henry, 208, 222
Stein, Herbert, 136, 163
Sterling, Ross, 25-28, 31, 33
Stevenson, Adlai III, 316, 318
Storage, 4, 10,14, 21, 25-27, 195, 198, 201, 212
    Earthen, 22, 23, 30, 54
Strategic Petroleum Reserve (SPR), 117-118, 131, 133, 139, 140-141, 146, 148-149, 164-167, 174-176, 210, 212, 214, 226-227, 304, 404
Stripper wells, 19, 22-23, 37, 49, 81-82, 135-136, 138, 142, 144, 157-158, 162, 171, 205, 209, 213, 277, 280, 291, 295, 297, 299, 322-323, 330
Stocking, George, 195-197, 203
Suez Canal, 96, 104
Sun Oil, 98-99, 111, 377

Talmadge, Herman, 155, 349
Tariff Commission, 75, 91-92, 107
Tariffs, 73-74, 93, 107
Task Force on Oil Import Control, 115-119, 209, 214, 406-407
Tax breaks, 82, 89, 91, 137, 172, 401, 403-404, 406, 409, 413-414
    see also Depletion Allowance
    expensing, 83, 89, 137, 141, 148, 172, 201, 204, 206, 210, 213, 380, 401
    foreign tax credit, 84-85, 113, 138, 140-143, 290, 300-304
Teagle, Walter, 8, 12-16, 35, 39, 43
Teapot Dome, 14
Television, 238-240, 243, 257-260
Tender Board
    Federal, 46-47, 49, 55
    State, 44, 46, 48
Terrell, C.V., 26, 41, 56
Texas, 5, 10, 18, 21-23, 26-28, 33, 35-36, 39-40, 42, 44, 46-49, 52, 56, 60, 76-77, 79-81, 86, 90, 93, 101, 110, 114, 252, 279, 281, 305, 330
Texas Company, 8, 9 , 33, 46
Texaco, 3, 5, 77, 86, 95, 100, 103, 112

Texas Independent Petroleum and Royalty Owners Association (TIPRO), 93, 94, 97, 106, 109, 112-116, 119, 129, 170
Texas Legislature, 42, 49, 50, 55-56
  House, 36, 50, 52
  Senate, 33, 45, 50
Texas Petroleum Council (TPC) 48, 51-52
Texas Railroad Commisision (TRC), 3, 22, 25-27, 29-31, 33, 40-46, 50, 52, 55, 60, 77-80, 89, 96, 132
Thomas, Elmer, 219, 34, 44, 50-51, 88, 278-280
Thomas-Disney Bill, 44, 47, 50
Thompson, E.O., 31-32, 39-44, 46, 51-56
Tidewater Oil, 98-99, 112
Time, 238, 250
Tower, John, 301, 304, 412
Truman, David, 360-362, 391
Truman, Harry, 73-75, 83, 87, 281, 285, 405
Tullock, Gordon, 364, 368, 369, 372-373

Udall, Stewart L., 106, 109-112, 114
Ullman, Al, 138, 140, 142, 156, 389-390
Union Oil, 78, 93, 378
United Mine Workers, 90, 101
Unitization, 79, 81

Vanik, Charles, 106, 157, 288, 350
Veasey, James, 20, 22
Venezuela, 7, 9, 20, 76-78, 85, 88, 91, 95-96, 98, 102-105, 109-111, 130
Verleger, Philip, 220, 223

Waggonner, Joseph,305, 307, 330
Wallop, Malcolm, 170, 175-176, 379
Ways and Means Committee, 19, 37, 73-75, 82, 84, 91-94, 128, 141, 143,147, 177, 304-305, 382, 385-386, 388-390
  Depletion allowance, 83, 109, 138, 142, 285, 287-288, 300, 407, 413
  Gasoline taxes, 148, 163, 172
  Oil quotas and tariffs, 89, 92, 94, 100, 119, 274, 278, 282, 305, 350, 407
  Oil import fees, 140, 163, 172, 304, 350, 352
  Windfall profit taxes, 152, 154, 156, 162, 334-335, 337, 340, 390
Well Spacing, 29, 53, 60
  see also Rule 37
West Texas, 10, 22, 26, 49, 53, 96
White, Lee, 133, 142, 395
Wilbur, Ray Lyman, 17-21
Williams, John, 84, 285-287, 290
Wilson, Charles, 142, 148, 301, 310, 313, 324
Windfall Profits Tax, 136, 139, 141, 144, 151-154, 156-158, 161-162, 164, 171, 274, 294-295, 297, 308, 313, 334-335, 337-338, 340, 342, 345
Wirth, Tim, 173, 176-177
Work, Hubert, 13, 14, 16
Wright, Jim, 140, 148, 171

Yamani, 130, 155-156
Young, Stephen, 289-290

Zimmerman, Erich, 193-194, 200-203